The Neutron Star – Black Hole Connection

NATO Science Series

A Series presenting the results of activities sponsored by the NATO Science Committee. The Series is published by IOS Press and Kluwer Academic Publishers, in conjunction with the NATO Scientific Affairs Division.

A. Life Sciences	IOS Press
B. Physics	Kluwer Academic Publishers
C. Mathematical and Physical Sciences	Kluwer Academic Publishers
D. Behavioural and Social Sciences	Kluwer Academic Publishers
E. Applied Sciences	Kluwer Academic Publishers
F. Computer and Systems Sciences	IOS Press

1. Disarmament Technologies	Kluwer Academic Publishers
2. Environmental Security	Kluwer Academic Publishers
3. High Technology	Kluwer Academic Publishers
4. Science and Technology Policy	IOS Press
5. Computer Networking	IOS Press

As a consequence of the restructuring of the NATO Science Programme in 1999, the NATO Science Series has been re-organized and new volumes will be incorporated into the following revised sub-series structure:

I. Life and Behavioural Sciences	IOS Press
II. Mathematics, Physics and Chemistry	Kluwer Academic Publishers
III. Computer and Systems Sciences	IOS Press
IV. Earth and Environmental Sciences	Kluwer Academic Publishers
V. Science and Technology Policy	IOS Press

NATO-PCO-DATA BASE

The NATO Science Series continues the series of books published formerly in the NATO ASI Series. An electronic index to the NATO ASI Series provides full bibliographical references (with keywords and/or abstracts) to more than 50000 contributions from international scientists published in all sections of the NATO ASI Series.
Access to the NATO-PCO-DATA BASE is possible via CD-ROM "NATO-PCO-DATA BASE" with user-friendly retrieval software in English, French and German (WTV GmbH and DATAWARE Technologies Inc. 1989).

The CD-ROM of the NATO ASI Series can be ordered from: PCO, Overijse, Belgium

Series C: Mathematical and Physical Sciences – Vol. 567

The Neutron Star –
Black Hole Connection

edited by

Chryssa Kouveliotou

NASA/MSFC,
Huntsville, Alabama, U.S.A.

Joseph Ventura

Department of Physics,
University of Crete, Heraklion, Crete, Greece

and

Ed van den Heuvel

Astronomical Institute "Anton Pannekoek",
University of Amsterdam, Amsterdam, The Netherlands

Kluwer Academic Publishers

Dordrecht / Boston / London

Published in cooperation with NATO Scientific Affairs Division

Proceedings of the NATO Advanced Study Institute on
The Neutron Star – Black Hole Connection
Elounda, Crete, Greece
7 – 18 June 1999

A C.I.P. Catalogue record for this book is available from the Library of Congress.

ISBN 1-4020-0204-1

Published by Kluwer Academic Publishers,
P.O. Box 17, 3300 AA Dordrecht, The Netherlands.

Sold and distributed in North, Central and South America
by Kluwer Academic Publishers,
101 Philip Drive, Norwell, MA 02061, U.S.A.

In all other countries, sold and distributed
by Kluwer Academic Publishers,
P.O. Box 322, 3300 AH Dordrecht, The Netherlands.

Printed on acid-free paper

Στον Γιαν

Io non morii ma non rimasi vivo

Ariosto

Table of Contents

PREFACE

Set against the background of beautiful Mirabello Bay, astronomers from fourteen countries met at Elounda, Crete in the period 7-18 June, 1999 to debate some of the most compelling issues of present day astrophysics.

Neutron stars and black holes have been at the forefront of astrophysics for over thirty years. As recently as ten years ago it was still being debated whether galactic stellar-mass black holes existed or not. It is now generally accepted that many (possibly a thousand) stellar-mass black holes – most of them still undetected – lie in low mass X-ray binary (LMXB) systems; a few of them are detected every year as X-ray or gamma-ray transients. These objects are more massive than 3 $M_\odot$, the maximum possible mass for a neutron star, and show none of the tell-tale signs of neutron stars, such as X-ray bursts and X-ray pulsations.

It is quite remarkable that all LMXBs display a similar temporal and spectral behaviour, independently of whether the accreting compact object is a neutron star or a black hole. A broad debate on these similarities and differences naturally constituted one of the main focal points during the Elounda meeting. Evidence on these aspects has been forthcoming from the Compton Gamma-ray Observatory (CGRO), the ROSAT and ASCA satellites, the Rossi X-Ray Timing Explorer (RXTE), and from the Beppo-SAX Observatory.

Several reports zeroed in onto the very rich phenomenology of the transient X-ray source GRS 1915+105, a black hole, also found to be a microquasar expelling superluminal plasma jets at regular intervals. This source also displays an interesting pattern of fast spectral and time variations, and has been singled out as a unique prototype for the study of accretion-disk instabilities, possibly at work in other accreting black holes. An observing run of this source with RXTE was actually taking place while the meeting was in progress, and a direct internet connection to the experiment enabled observer Tomaso Belloni to obtain the latest light curves and variability patterns. There are indications that we are seeing emission from very close to the black hole event horizon, possibly at the location of the last stable orbit, and excitement over the possibility of observing direct manifestations of general relativity in this and related objects is quite strong.

Predictably, the mysterious gamma ray bursts (GRB), and the recently discovered magnetars added two more important focal points to the Elounda meeting. GRBs are thought to be catastrophic, one-time-only events, resulting in the total disruption of the initial system. Recent success in following up GRB afterglows has led to identifications relating these most powerful explosions to faint galaxies at cosmological distances. The two prevailing models for these events describe them either as the result of catastrophic mergers of neutron star binaries/neutron star - black hole binaries, or due

ix

to the collapse of supermassive stars ('collapsars'). Both mechanisms were reviewed at the meeting.

Contrary to GRBs, the distinct class of the so called soft gamma repeaters (SGRs), numbering only four sources (three located in our Galaxy and one in the nearby Large Magelanic Cloud), do exhibit repeated sporadic eruptions displaying very soft gamma ray spectra. These objects have been linked to magnetic neutron stars with rather long, 5 to 10 second spin periods. These long periods along with the measured rates of spin decay point to ultra-strong magnetic fields of the order of 10^{15} Gauss. Are these objects related to another class of low luminosity - long period X-ray pulsars, known as anomalous X-ray pulsars? This is still a point of detailed investigation and debate involving theory and observation.

These, and a multitude of related issues were reviewed, analysed, and debated in Elounda:

– Can magnetospheric beat frequency models explain some of the neutron star QPOs?

– QPOs were recently detected during thermonuclear X-ray burst events in accreting neutron stars. Do these quasiperiodicities relate to the propagation velocity of burning fronts as they move across the neutron star's surface?

– How is one to interpret the evident similarities of accreting neutron stars and black holes in low mass X-ray binaries?

– Do NS magnetic fields evolve?

– Do we see the surface thermal emission of isolated neutron stars?

– Do we observe all the neutron stars predicted by the current counts of supernova events in our Galaxy?

– Which evolutionary scenarios give rise to NS and BH binary systems?

– Could a sub-class of GRBs be due to the catastrophic release of the rotational energy of the neutron star in some odd, accreting low mass X-ray binaries in distant galaxies?

All these debates certainly served in re-focussing the observing strategies to be followed with the new and powerful Chandra and XMM observatories launched within the months following the Elounda meeting. The meeting has thus offered an excellent opportunity for reviewing the capabilities of these and other future space-borne missions.

One characteristic that made this event memorable to the participants was the special effort lecturers and speakers placed in preserving a broad tutorial character in their presentations. This aspect is specially important in meetings bringing together researchers with very disparate backgrounds, as it serves in unifying the audience, and furthering the cross-fertilization of ideas. The authors contributing to this volume were asked explicitly to preserve the tutorial character in their manuscripts. We therefore expect that this volume can also serve the function of a graduate level text on the astrophysics and phenomenology of stellar mass compact objects.

The Scientific Organizing Committee consisted of Chryssa Kouveliotou (cochair), Jan van Paradijs (cochair), Joseph Ventura (Director), Jerry Fishman, Peter Mészáros, Joachim Trümper, and Ed van den Heuvel. Following the debilitating illness of Jan during the spring of 1999, Ed van den Heuvel readily agreed to assume a leading role in managing the last phases of the ASI. This volume is dedicated to Jan van Paradijs.

We are grateful to the NATO Scientific Affairs Division for its generous funding of this Advanced Study Institute. We are also grateful to NASA, the University of Amsterdam, and the University of Crete for substantial financial support which improved the level of available student support, and helped cover the cost of the overall organization of the meeting. The company MITOS S.A. secured for us low cost accommodation and help with travel, while the local authorities provided excellent excursions and a memorable conference banquet and barbeque nights. The National Science Foundation provided travel grants which enabled the participation of five students from the U.S. to attend the ASI. Special thanks are also due to the members of the Local Organizing Committee, and especially to Nick Kylafis, and Sifis Papadakis who took it upon themselves to ensure a smooth and uniquely successful meeting. Finally we thank Valerie Connaughton, and Fotis Mavromatakis for their skillful editorial assistance.

Chryssa Kouveliotou, Joseph Ventura

List of Participants

Marek ABRAMOWICZ, Chalmers Institute of Technology/Astrophysics, Sweden, marek@fy.chalmers.se

Ali ALPAR, Middle East Technical University, Turkey, alpar@sabanciuniv.edu*

Elena AMATO, University of Florence, Italy, amato@arcetri.astro.it

Askin ANKAY, Middle East Technical University, Turkey, askin@astroa.physics.metu.edu.tr

Tomaso BELLONI, Brera Observatory, Italy, belloni@merate.mi.astro.it

Dipankar BHATTACHARYA, Raman Research Institute, India, dipankar@rri.ernet.in

Lars BILDSTEN, UC-Berkeley, USA, bildsden@fire.berkeley.edu

Olga BITZARAKI, University of Athens, Greece, obitzar@atlas.uoa.gr

Steffen BLUM, Universitaet Tuebingen, Germany, blum@tat.physik.uni-tuebingen.de

S.V. BOGOVALOV, Moscow Engineering Physics Institute, Russia, bogoval@mpi-hd.mpg.de

Edward BROWN, University of California Berkeley, USA, brown@flash.uchicago.edu*

S. Cagdas INAM, Middle East Technical University, Turkey, inam@newton.physics.metu.edu.tr

Alessandro COLETTA, Telespazio s.p.a., Italy, coletta@saxnet.sdc.asi.it

Remon CORNELISSE, SRON, The Netherlands, cornelis@sron.nl

Vladimir Nikolov DAMGOV, Space Research Institute at the Bulgarian Academy of Sciences, Bulgaria, vdamgov@bas.bg

Tiziana DI SALVO, Universita di Palermo, Italy, disalvo@gifco.fisica.unipa.it

Stefan DIETERS, University of Alabama Huntsville, USA, sdieters@awc.cc.az.us*

Anton DORODNITSYN, Space Research Institute, Russia, dora@mx.iki.rssi.ru

Ene ERGMA, Tartu University, Estonia, ene@physic.ut.ee

Robert FENDER, University of Amsterdam, The Netherlands, rpf@astro.uva.nl

Gerald FISHMAN, NASA/Marshall Space Flight Center, USA, jerry.fishman@msfc.nasa.gov

Warren FOCKE, University of Maryland/ NASA GSFC, USA,
warren@lheamail.gsfc.nasa.gov

Eric C. FORD, University of Amsterdam, The Netherlands,
ecford@astro.uva.nl

Lucia M. FRANCO, University of Chicago, USA,
Lucia_Franco@Mckinsey.com*

Chris FRYER, UC Santa Cruz, USA, cfryer@ucolick.org

Yael FUCHS, CEA Saclay, France, yfuchs@discovery.saclay.cea.fr

Giangiacomo GANDOLFI, Istituto Astrofisica Spaziale, CNR, Italy,
gandolfi@hal9000.ias.rm.cnr.it

Ioannis GEORGANTOPOULOS, National Observatory of Athens, Greece,
ig@astro.noa.gr

Marek GIERLINSKI, Jagiellonian University Observatory, Poland,
Marek.Gierlinski@durham.ac.uk*

Aaron GOLDEN, NUI Galway, Ireland, agolden@itc.nuigalway.ie

Jonathan GRANOT, Hebrew University, Israel,
jgranot@merger.fiz.huji.ac.il

Emilios HARLAFTIS, National Obs. of Athens, Greece,
ehh@titan.astro.noa.gr

Jeroen HOMAN, University of Amsterdam, The Netherlands,
homan@astro.uva.nl

Zach IOANNOU, Keele University, UK, zac@astro.as.utexas.edu*

P.G. JONKER, University of Amsterdam, The Netherlands,
peterj@astro.uva.nl

Menas KAFATOS, George Mason University, USA, mkafatos@gmu.edu

Vicky KALOGERA, Harvard-Smithsonian Centre for Astrophysics, USA,
vkalogera@cfa.harvard.edu

Demosthenes KAZANAS, NASA/Goddard Space Flight Center, USA,
kazanas@milkyway.gsfc.nasa.gov

Wlodzimierz KLUZNIAK, Copernicus Astronomical Center, Poland,
wlodek@cow.physics.wisc.edu

Lydie KOCH MIRAMOND, CEA Saclay, France,
lkoch@discovery.saclay.cea.fr

Sushan KONAR, Inter University Centre for Astronomy and Astrophysics,
India, sushan@iucaa.ernet.in

Chrysa KOUVELIOTOU, NASA/MSFC, USA,
chryssa.kouveliotou@iss.msfc.nasa.gov

Nick KYLAFIS, University of Crete, Greece, kylafis@physics.uch.gr

Frederick K. LAMB, University of Illinois at Urbana-Champaign, USA, f-lamb@uiuc.edu

Chang-Hwan LEE, SUNY at Stony Brook, USA, chlee@silver.physics.sunysb.edu

William LEE, Instituto de Astronomia - UNAM, Mexico, wlee@astroscu.unam.mx

Kseniya P. LEVENFISH, Ioffe Physical Technical Institute, Russia, ksen@astro.ioffe.rssi.ru

Nicole M. LLOYD, Stanford University, USA, nicole@urania.stanford.edu

Duncan R. LORIMER, Arecibo Observatory, USA, drl@jb.man.ac.uk*

Vanessa MANGANO, SISSA-ISAS, Italy, mangano@sissa.it

A. MASTICHIADIS, University of Athens, Greece, amastich@atlas.uoa.gr

Fotis MAVROMATAKIS, University of Crete, Greece, fotis@physics.uch.gr

Diana MAXWELL, The Open University, UK, D.H.Maxwell@open.ac.uk

Mariano MENDEZ, University of Amsterdam, The Netherlands, M.Mendez@sron.nl*

Sandro MEREGHETTI, Instituto di Fisica Cosmica G. Occhialini, CNR, Italy, sandro@ifctr.mi.cnr.it

Andrea MERLONI, Institute of Astronomy, UK, am@ast.cam.ac.uk

Bronson MESSER, U. Tenn. Oak Ridge National Laboratory, USA, bmesser@utk.edu

Peter MESZAROS, Pennsylvania State University, USA, nnp@astro.psu.edu

Roberto MIGNANI, STECF-ESO, Germany, rmignani@eso.org

Felix MIRABEL, CE Saclay, France, mirabel@discovery.saclay.cea.fr

Raquel MORALES, Institute of Astronomy, UK, rm@ast.cam.ac.uk

Herman J. MOSQUERA CUESTA, International Centre for Theoretical Physics, Italy, herman@ictp.trieste.it

Ignacio NEGUERUELA, SAX SDC, Italia Space Agency, Italy, ignacio@isaac.u-strasbg.fr*

Gijs NELEMANS, University of Amsterdam, The Netherlands, gijsn@astro.uva.nl

Kieran O' BRIEN, University of St. Andrews, UK, kso@astro.uva.nl*

Gordon OGILVIE, Max-Planck Institut fur Astrophysik, Germany, gogilvie@ast.cam.ac.uk*

Feryal OZEL, Harvard University - CfA, USA, fozel@cfa.harvard.edu

Iossif E. PAPADAKIS, University of Crete, Greece, jhep@physics.uch.gr

Hara PAPATHANASSIOU, SISSA, Italy, hara@sun1.sms.port.ac.uk*

Santina PIRAINO, IFCAI/CNR and Harvard-Smithsonian CfA, Italy and USA, piraino@ifcai.pa.cnr.it*

Luigi PIRO, Instituto Astrofisica Spaziale, CNR, Italy, pito@ias.rm.cnr.it

Juri POUTANEN, Stockholm Observatory, Sweden, juri@astro.su.se

V. RADHAKRISHNAN, University of Amsterdam, The Netherlands, rad@astro.uva.nl

R. RAMACHANDRAN, University of Amsterdam, The Netherlands, ramach@astro.uva.nl

Saul A. RAPPAPORT, M.I.T., USA, sar@mit.edu

Fred RASIO, Massachusetts Institute of Technology, USA, rasio@ensor.mito.edu*

Pablo REIG, University of Crete, Greece, pablo@physics.uch.gr

Marc RIBO, Universitat de Barcelona, Spain, marc@fajmpp0.am.ub.es

Evert ROL, University of Amsterdam, The Netherlands, evert@astro.uva.nl

Norbert S. SCHULZ, MIT, USA, nss@space.mit.edu

Nikolai SHAKURA, Sternberg Astronomical Institute, Russia, shakura@sai.msu.ru

Stephen L. SKINNER, University of Colorado, USA, skinner@jila.colorado.edu

Andrea L. SOMER, UC Berkeley, USA, andrea@astro.berkeley.edu

Henk C. SPRUIT, MPI f. Astrophysics, Germany, henk@mpa-garching.mpg.de

Ben STAPPERS, University of Amsterdam, The Netherlands, bws@astro.uva.nl

Adamantios STAVRIDIS, Aristotle University of Thessaloniki, Greece, astavrid@astro.auth.gr

Danny STEEGHS, University of St. Andrews, UK, ds@astro.soton.ac.uk*

Rashid SUNYAEV, Max-Planck Institute fuer Astrophysik, Germany, and Space Research Institute of Russia, Russia, rs@star.iki.rssi.ru, sunyaev@mpa-garching.mpg.de

Firoza K. SUTARIA, Inter University Center for Astronomy and Astrophysics (IUCAA), India, F.K.Sutaria@open.ac.uk*

M. Ozgur TASKIN, Middle East Technical University, Turkey, ozgur@astroa.physics.metu.edu.tr

Thomas M. TAURIS, University of Amsterdam, The Netherlands,

tauris@nordita.dk*

Christopher THOMPSON, University of North Carolina at Chapel Hill, USA, act@physics.unc.edu

Ulf TORKELSSON, Chalmers University of Technology, Sweden, torkel@fy.chalmers.se

Joachim TRUEMPER, MPI f. Extraterrestrische Physik, Germany, jtrumper@mpe-garching.mpg.de

Roberto TUROLLA, University of Padova, Italy, turolla@pd.infn.it

Edward VAN DEN HEUVEL, University of Amsterdam, The Netherlands, edvdh@astro.uva.nl

Michiel VAN DER KLIS, University of Amsterdam, The Netherlands, michiel@astro.uva.nl

Jan VAN PARADIJS, University of Amsterdam, The Netherlands, jvp@astro.uva.nl

Joseph VENTURA, University of Crete, Greece, ventura@physics.uch.gr

Lodie VOUTE, Universiteit van Amsterdam, The Netherlands, lodie@astro.uva.nl

Paul VREESWIJK, University of Amsterdam, The Netherlands, pmv@astro.uva.nl

Martin C. WEISSKOPF, NASA/Marshall Space Flight Center, USA, martin@smoker.msfc.nasa.gov

Norbert WEX, Max-Planck-Institut fuer Radioastronomie, Germany, norbert.wex@eplus-online.de*

Rudy WIJNANDS, University of Amsterdam, The Netherlands, rudy@space.mit.edu*

Peter WOODS, Univ. of Alabama in Huntsville, USA, Peter.Woods@msfc.nasa.gov

Silvia ZANE, University of Oxford, UK, sz@mssl.ucl.ac.uk*

* New e-mail owing to recent change of address

1. Radio Pulsars and Neutron Stars

RADIO PULSARS — AN OBSERVER'S PERSPECTIVE

D.R. LORIMER

National Astronomy and Ionospheric Center
Arecibo Observatory
HC3 Box 53995,
Arecibo PR 00612, USA

Abstract. Pulsar astronomy is currently enjoying one of the most productive phases in its history. In this review, I outline some of the basic observational aspects and summarise some of the latest results of searches for pulsars in the disk of our Galaxy and its globular cluster system.

1. Preamble

Pulsar astronomy began serendipitously in 1967 when Jocelyn Bell and Antony Hewish discovered periodic signals originating from distinct parts of the sky via pen-chart recordings taken during an interplanetary scintillation survey of the radio sky at 81.5 MHz (Hewish et al. 1968). This remarkable phenomenon has since been unequivocally linked with the radiation produced by a rotating neutron star (Gold 1968; Pacini 1968). Baade & Zwicky (1934) were the first to hypothesise the existence of neutron stars as a stable configuration of degenerate neutrons formed from the collapsed remains of a massive star after it has exploded as a supernova.

Although the theory of pulsar emission is complex, the basic idea can be simply stated as follows: as a neutron star rotates, charged particles are accelerated out along its magnetic poles and emit electromagnetic radiation. The combination of rotation and the beaming of particles along the magnetic field lines means that a distant observer records a pulse of emission each time the magnetic axis crosses his/her line of sight, i.e. one pulse per stellar rotation. Like many things in life, the emission does not come for free. It takes place at the expense of the neutron star's rotational kinetic energy — one of the key predictions of the Gold/Pacini theory.

3

C. Kouveliotou et al. (eds.), The Neutron Star – Black Hole Connection, 3–19.
© 2001 *Kluwer Academic Publishers. Printed in the Netherlands.*

Measurements of the secular increase in pulse period through pulsar timing techniques (§4) are in excellent agreement with this idea.

Pulsar astronomy has come a long way in a remarkably short time. Systematic surveys with the world's largest radio telescopes over the last 30 years have revealed more than 1200 pulsars in a rich variety of astrophysical settings. The present "zoo" of objects is summarised in Figure 1. Some

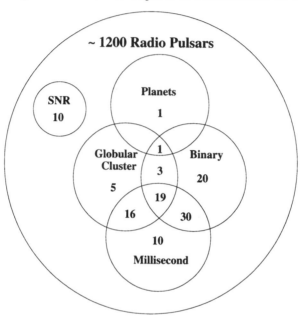

Figure 1. An adaptation of Dick Manchester's Venn diagram showing the various types of radio pulsars. SNR denotes pulsars likely to be associated with supernova remnants.

of the highlights so far include: (1) The original *binary pulsar* B1913+16 (Hulse & Taylor 1975) — a pair of neutron stars in a 7.75-hr eccentric orbit. The measurement of the orbital decay due to gravitational radiation from the binary system resulted in the 1993 Physics Nobel Prize. (2) The original *millisecond pulsar*, B1937+21, discovered by Backer et al. (1982) has a period of only 1.5578 ms. The implied spin frequency of 642 Hz means that this neutron star is close to being torn apart by centrifugal forces. (3) Pulsars with planetary companions. Well before the discoveries of Jupiter-mass planets by optical astronomers, Wolszczan & Frail (1992) discovered the millisecond pulsar B1257+12 accompanied by three Earth-mass planets — the first planets discovered outside our Solar system. (4) Pulsars in globular clusters. Dense globular clusters are unique breeding grounds for exotic binary systems and millisecond pulsars. Discoveries of 40 or so "cluster pulsars" have permitted detailed studies of pulsar dynamics in the cluster potential, and of the cluster mass distribution.

The plan for the rest of this review is as follows: §2 covers pulse dispersion and what it tells us about pulsars and the interstellar medium; §3 discusses pulse profiles and their implications. In §4 the essential aspects of pulsar timing observations are reviewed. §5 reviews the techniques employed by pulsar searchers. Finally, in §6, we summarise some of the recent results from searches at Parkes. More complete discussions on the observational aspects covered here can be found in Lyne & Smith (1998).

2. Pulse Dispersion and the Interstellar Medium

Newcomers to pulsar astronomy would do well to begin their studies by reading the discovery paper (Hewish et al. 1968), a classic article packed with observational facts and their implications.

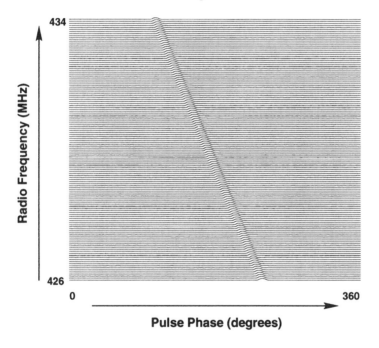

Figure 2.   Pulse dispersion shown in this recent 30-s Arecibo observation of PSR B1933+16 across an 8-MHz passband centred at 430 MHz. The period of this pulsar is 358.7 ms. The dispersive time delay between 434 MHz and 426 MHz is 133 ms.

One of the phenomena clearly noted in the discovery paper was pulse dispersion — pulses at higher radio frequencies arrive earlier at the telescope than their lower frequency counterparts. An example of this is shown in Figure 2. Hewish et al. (1968) correctly interpreted the effect as the frequency dependence of the group velocity of radio waves as they propagate through the interstellar medium — a cold ionised plasma.

Applying standard plasma physics formulae, it can be shown (see e.g. Lyne & Smith 1998) that the difference in arrival times Δt between a high frequency $\nu_{\rm hi}$ (MHz) and a low frequency $\nu_{\rm lo}$ (MHz) is given by

$$\Delta t = 4.15 \times 10^6 \ {\rm ms} \ \times (\nu_{\rm lo}^{-2} - \nu_{\rm hi}^{-2}) \times {\rm DM}, \tag{1}$$

where the dispersion measure DM ($\rm cm^{-3}$ pc) is the integrated column density of free electrons along the line of sight:

$$\mathrm{DM} = \int_0^d n_{\rm e} \, dl. \tag{2}$$

Here, d is the distance to the pulsar (pc) and $n_{\rm e}$ is the free electron density ($\rm cm^{-3}$). Pulsars at large distances have higher column densities, and therefore larger DMs, than pulsars closer to Earth so that, from Eq. 1, the dispersive delay across the bandwidth is greater. In the original discovery paper, Hewish et al. (1968) measured a delay for the first pulsar (B1919+21) of $\simeq 0.2$ s between 81.5 and 80.5 MHz. From Eq. 1, we infer a DM of about 13 $\rm cm^{-3}$ pc. Assuming, as a first-order approximation, that the mean Galactic electron density is 0.03 $\rm cm^{-3}$ (Ables & Manchester 1976), this implies a distance of about 0.4 kpc[1].

The most straightforward method to compensate for pulse dispersion is to use a filterbank to sample the passband as a number of contiguous channels and apply successively larger time delays (calculated from Eq. 1)

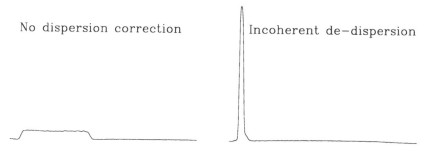

No dispersion correction   Incoherent de-dispersion

Figure 3.   Left: the "raw profile" of B1933+16 formed by the direct addition of individual filterbank channels in Figure 2. Right: incoherently de-dispersed profile formed by delaying the filterbank channels before addition. Both displays have the same scale.

to higher frequency channels *before* summing over all channels to produce a sharp "de-dispersed" profile. This can be carried out either in hardware or in software. Figure 3 shows the clear gain in signal-to-noise ratio and time resolution achieved when the data from Figure 2 are properly de-dispersed, as opposed to a simple detection over the whole bandwidth.

---

[1]Hewish et al. (1968) assumed 0.2 $\rm cm^{-3}$ which resulted in an underestimated distance, but still clearly demonstrated that the source is located well beyond the solar system.

The fact that the free electrons in the Galaxy are finite in extent is well demonstrated by Figure 4 which shows the dispersion measures of 700 pulsars plotted against the absolute value of their respective Galactic latitudes. It is straightforward to show that, for a simple slab model of free electrons with a mean density n and half-height H, the maximum DM for a given line of sight along a latitude b is $Hn/\sin b$. The solid curve in Figure 4 shows fairly convincingly that this simple model accounts for the trend rather well implying that $Hn \sim 30$ pc cm^{-3}. Taking $n = 0.03$ cm^{-3} as before gives us a first-order estimate of the thickness of the electron layer — about 1 kpc.

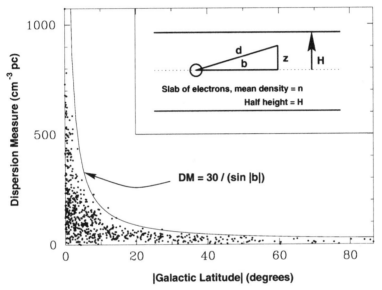

Figure 4. Dispersion measures plotted against Galactic latitudes. Inset: a simple slab model to explain the envelope of points in terms of a finite electron layer (solid curve).

Independent measurements of pulsar distances can, for a large enough sample, be fed back into Eq. 2 to calibrate the Galactic distribution of free electrons. There are three basic distance measurement techniques: neutral hydrogen absorption, trigonometric parallax (measured either with an interferometer or through pulse time-of-arrival techniques) and from associations with objects of known distance (i.e. supernova remnants, globular clusters and the Magellanic Clouds). Together, these provide measurements of (or limits on) the distances to over 100 pulsars. Taylor & Cordes (1993) have used these distances, together with measurements of interstellar scattering for various Galactic and extragalactic sources, to calibrate an electron density model. In a statistical sense, the model can be used to provide distance estimates with an uncertainty of $\sim 30\%$. Although the model is free of large systematic trends, its use to estimate distances to individual pulsars may result in systematic errors by as much as a factor of two.

3. Erratic Individual Pulses and Stable Integrated Profiles

Pulsars are weak radio sources. Mean flux densities, usually quoted in the literature at a radio frequency of 400 MHz, vary between 1 and 100 mJy (1 $Jy = 10^{-26}$ W m^{-2} Hz^{-1}). This means that the addition of many thousands of pulses is required in order to produce a discernible profile. Only a handful of sources presently known are strong enough to allow studies of individual pulses. A remarkable fact from these studies is that, although the individual pulses vary quite dramatically, at any particular observing frequency the integrated profile is very stable. This is illustrated in Figure 5.

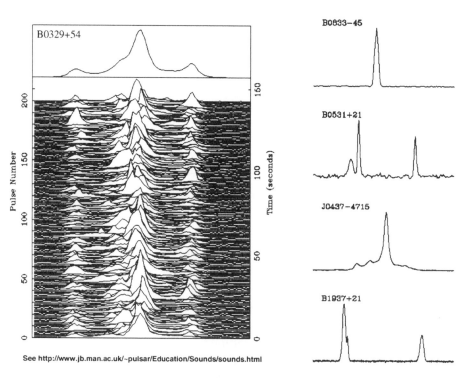

See http://www.jb.man.ac.uk/~pulsar/Education/Sounds/sounds.html

Figure 5.   Left: a sequence of 200 individual pulses received from the strong pulsar B0329+54 showing the rich diversity of emission from one pulse to the next. The sum of all the pulses forms a characteristic "integrated profile" shown for this pulsar in the box at the top. Right: some other examples of these stable waveforms.

In the above examples of stable pulse profiles, which have been normalised to represent 360 degrees of rotational phase, the astute reader will notice two examples of so-called interpulses — a secondary pulse separated by about 180 degrees from the main pulse. The most natural interpretation for this phenomenon is that the two pulses originate from opposite magnetic poles of the neutron star (see however Manchester & Lyne 1977). Geometrically speaking, this is a rather unlikely situation. As a result, the fraction of the known pulsars with interpulses is only a few percent.

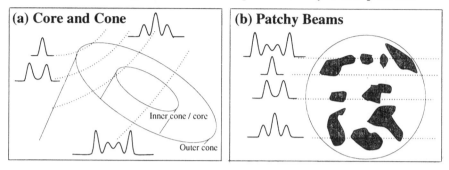

Figure 6. Phenomenological models for pulse shape morphology produced by different line-of-sight cuts of the beam. (Figure designed by M. Kramer and A. von Hoensbroech).

The integrated pulse profile should really be thought of as a unique "fingerprint" of the radio emission beam of each neutron star. The rich variety of pulse shapes can be attributed to different line-of-sight cuts through the radio beam of the neutron star as it sweeps past the Earth. Two contrasting phenomenological models which account for this are shown in Figure 6. The "core and cone" model, proposed by Rankin (1983), depicts the beam as a core surrounded by a series of nested cones. Alternatively, the "patchy beam" model, championed by Lyne & Manchester (1988), has the beam populated by a series of emission regions.

4. Pulsar Timing Basics

Soon after their discovery, it became clear that pulsars are excellent celestial clocks. Hewish et al. (1968) demonstrated that the period of the first pulsar, B1919+21, was stable to one part in 10^7 over a time-scale of a few months. Following the discovery of the millisecond pulsar, B1937+21, in 1982 (Backer et al. 1982) it was demonstrated that its period could be measured to one part in 10^{13} or better (Davis et al. 1985). This unrivaled stability leads to a host of applications including time keeping, probes of relativistic gravity and natural gravitational wave detectors. Subsequently, a whole science has developed to accurately measure the pulse time-of-arrival in order to extract as much information about each pulsar as possible.

Figure 7 summarises the essential steps involved in a pulse "time-of-arrival" (TOA) measurement. Incoming pulses emitted by the rotating neutron star traverse the interstellar medium before being received by the radio telescope. After amplification by high sensitivity receivers, the pulses are de-dispersed (§2) and added to form a mean pulse profile. During the observation, the data regularly receive a time stamp, usually based on a maser at the observatory, plus a signal from the GPS (Global Positioning System of satellites) time system. The TOA of this mean pulse is then defined as the arrival time of some fiducial point on the profile. Since the mean profile has a stable form at any given observing frequency (§3) the TOA can be accurately determined by a simple cross-correlation of the observed profile with a high signal-to-noise "template" profile — obtained from the addition of many observations of the pulsar at a particular observing frequency.

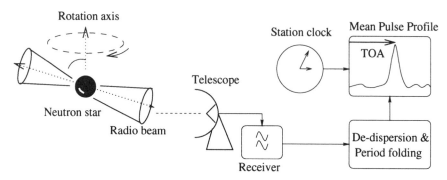

Figure 7.   Schematic showing the main stages involved in pulsar timing observations.

In order to properly model the rotational behaviour of the neutron star, we require TOAs as measured by an inertial observer. Due to the Earth's orbit around the Sun, an observatory located on Earth experiences accelerations with respect to the neutron star. The observatory is therefore not in an inertial frame. To a very good approximation, the centre-of-mass of the solar system, the solar system barycentre, can be regarded as an inertial frame. It is standard practice to transform the observed TOAs to this frame using a planetary ephemeris.

Following the accumulation of about ten to twenty barycentric TOAs from observations spaced over at least several months, a surprisingly simple model can be applied to the TOAs and optimised so that it is sufficient to account for the arrival time of any pulse emitted during the time span of the observations, and predict the arrival times of subsequent pulses. The model is based on a Taylor expansion of the angular rotational frequency about a model value at some reference epoch to calculate a model pulse phase as a function of time. Based upon this simple model, and using initial estimates of the position, dispersion measure and pulse period, a "timing residual" is calculated for each TOA as the difference between the observed and predicted pulse phases (see e.g. Lyne & Smith 1998).

Ideally, the residuals should have a zero mean and be free from any systematic trends (Figure 8a). Inevitably, however, due to our *a-priori* ignorance of the rotational parameters, the model needs to be refined in a bootstrap fashion. Early sets of residuals will exhibit a number of trends indicating a systematic error in one or more of the model parameters, or a parameter not initially incorporated into the model. For example, a parabolic trend results from an error in the period time derivative (Figure 8b). Additional effects will arise if the assumed position of the pulsar used in the barycentric time calculation is incorrect. A position error of just one arcsecond results in an annual sinusoid (Figure 8c) with a peak-to-peak amplitude of about 5 ms for a pulsar on the ecliptic; this is easily measurable for typical TOA uncertainties of order one milliperiod or better. Similarly, the effect

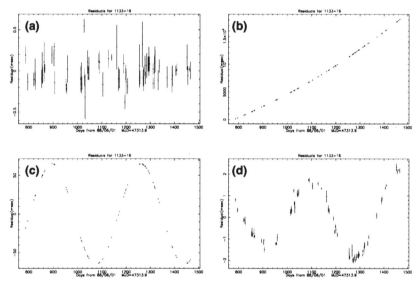

Figure 8. Timing model residuals versus date for PSR B1133+16. Case (a) shows the residuals obtained from the best fitting model which includes period, period derivative, position and proper motion. Case (b) is the result of setting the period derivative term to zero in this model. Case (c) shows the effect of a 1 arcmin error in the assumed declination. Case (d) shows the residuals obtained assuming zero proper motion.

of a proper motion produces an annual sinusoid of linearly increasing magnitude (Figure 8d). Proper motions are, for many long-period pulsars, more difficult to measure due to the confusing effects of timing noise (a random walk process seen in the timing residuals; see e.g. Cordes & Helfand 1980). For these pulsars, interferometric techniques can be used to obtain proper motions (see Ramachandran's contribution to this volume and references therein).

In summary, a phase-connected timing solution obtained over an interval of a year or more will, for an isolated pulsar, provide an accurate measurement of the period, the rate at which the neutron star is slowing down, and the position of the neutron star. Presently, measurements of these essential parameters are available for about 600 pulsars. The implications of these measurements for the ages and magnetic fields of neutron stars will be discussed in my other contribution to this volume.

For binary pulsars, the model needs to be extended to incorporate the additional radial accelerations of the pulsar as it orbits the common centre-of-mass of the binary system. Treating the binary orbit using just Kepler's laws to refer the TOAs to the binary barycentre requires five additional model parameters: the orbital period, projected semi-major orbital axis, orbital eccentricity, longitude of periastron and the epoch of periastron passage. The Keplerian description of the orbit is identical to that used for spectroscopic binary stars where a characteristic orbital "velocity curve" shows the radial component of the star's velocity as a function of time. The analogous plot for pulsars is the apparent pulse period against time. For circular orbits the behaviour is sinusoidal whilst for eccentric orbits the curve has a "saw-tooth" appearance. Two examples are shown in Figure 9.

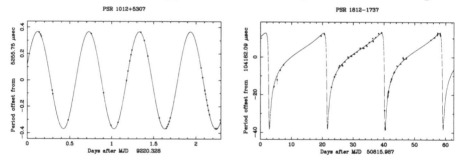

Figure 9. Two different binary pulsars. Left: PSR J1012+5307 — a 5.25-ms pulsar in a 14.5-hour circular orbit around a low-mass white dwarf companion (Nicastro et al. 1995). Right: PSR J1811−1736 (Lyne et al. 2000). A 104-ms pulsar in a highly eccentric 18.8-day orbit around a massive companion (probably another neutron star).

Also by analogy with spectroscopic binaries, constraints on the mass of the orbiting companion can be placed by combining the projected semi-major axis $a_p \sin i$ and the orbital period P_o to obtain the mass function:

$$f(m_p, m_c) = \frac{4\pi^2}{G} \frac{(a_p \sin i)^3}{P_o^2} = \frac{(m_c \sin i)^3}{(m_p + m_c)^2}, \tag{3}$$

where G is the universal gravitational constant. Assuming a pulsar mass m_p of 1.35 $M_\odot$ (see below), the mass of the orbiting companion m_c can be estimated as a function of the (initially unknown) angle i between the orbital plane and the plane of the sky. The minimum companion mass m_{min} occurs when the orbit is assumed edge-on ($i = 90°$).

Further information on the orbital inclination and component masses may be obtained by studying binary systems which exhibit a number of relativistic effects not described by Kepler's laws. Up to five "post-Keplerian" parameters exist within the framework of general relativity. Three of these parameters (the rate of periastron advance, a gravitational redshift parameter, and the orbital period derivative) have been measured for the original binary pulsar, B1913+16 (Taylor & Weisberg 1989), allowing high-precision measurements of the masses of both components, as well as stringent tests of general relativity. A further two post-Keplerian parameters related to the Shapiro delay in the double neutron star system PSR B1534+12 have now also been measured (Stairs et al. 1998). Based on these results, and other radio pulsar binary systems, Thorsett & Chakrabarty (1999) have recently demonstrated that the range of neutron star masses has a remarkably narrow underlying Gaussian distribution with a mean of 1.35 $M_\odot$ and a standard deviation of only 0.04 $M_\odot$.

5. Pulsar Searching

Pulsar searching is, conceptually at least, a rather simple process — the detection of a dispersed, periodic signal hidden in a noisy time series data taken with a radio telescope. In what follows we give only a brief description of the basic search techniques. Further discussions can be found in Lyne (1988), Nice (1992) and Lorimer (1998) and references therein.

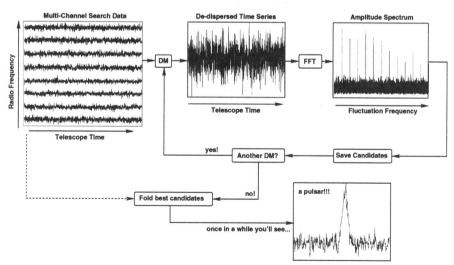

Figure 10.   Schematic summarising the essential steps in a "standard" pulsar search.

Most pulsar searches can be pictured as shown in Figure 10. The multi-channel search data is typically collected using a filterbank or a correlator (see e.g. Backer et al. 1990), either of which usually provides a much finer channelisation than the eight channels shown for illustrative purposes in Figure 10. The channels are then incoherently de-dispersed (see §2) to form a single noisy time series. An efficient way to find a periodic signal in these data is to take the Fast Fourier Transform (FFT) and plot the resulting amplitude spectrum. For a narrow duty cycle the spectrum will show a family of harmonics which show clearly above the noise. To detect weaker signals still, a harmonic summing technique is usually implemented at this stage (see e.g. Lyne 1988). The best candidates are then saved and the whole process is repeated for another trial DM.

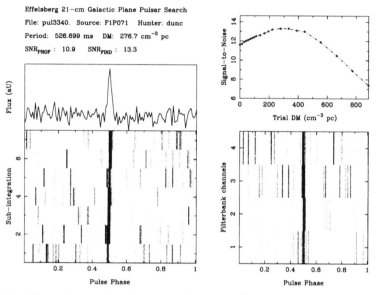

Figure 11. Example search code output showing PSR J1842–0415 — the first pulsar ever discovered with the 100-m Effelsberg radio telescope (Lorimer et al. 2000).

After a sufficiently large number of DMs have been processed, a list of pulsar candidates is compiled and it is then a matter of folding the raw time series data at the candidate period. Figure 11 is an excellent example of the characteristics of a strong pulsar candidate. The high signal-to-noise integrated profile (top left panel) can be seen as a function of time and radio frequency in the grey scales (lower left and right panels). In addition, the dispersed nature of the signal is immediately evident in the upper right hand panel which shows the signal-to-noise ratio as a function of trial DM. This combination of diagnostics proves extremely useful in differentiating between pulsar candidates and spurious interference.

In the discussion hitherto we have implicitly assumed that the apparent pulse period remains constant throughout the observation. For searches with long integration times (Figure 11 represents a 35-min observation), this assumption is only valid for solitary pulsars, or those in binary systems where the orbital periods are longer than about a day. For shorter-period binary systems, as noted by Johnston & Kulkarni (1992), the Doppler shifting of the period results in a spreading of the total signal power over a number of frequency bins in the Fourier domain. Thus, a narrow harmonic becomes smeared over several spectral bins.

To quantify this effect, consider the resolution in fluctuation frequency (the width of a Fourier bin) $\Delta f = 1/T$, where T is the length of the integration. It is straightforward to show that the drift in frequency of a signal due to a constant acceleration[2] a during this time is $aT/(Pc)$, where P is the true period of the pulsar and c is the speed of light. Comparing these two quantities, we note that the signal will drift into more than one spectral bin if $aT^2/(Pc) > 1$. Thus, without due care, long integration times potentially kill off all sensitivity to short-period pulsars in exciting tight orbits where the line-of-sight accelerations are high!

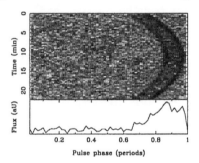 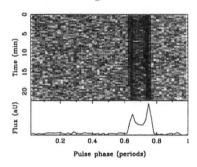

Figure 12. Left: a 22.5-min Arecibo observation of the binary pulsar B1913+16. The assumption that the pulsar has a constant period during this time is clearly inappropriate given the drifting in phase of the pulse during the integration (grey scale plot). Right: the same observation after applying an acceleration search. This shows the effective recovery of the pulse shape and a significant improvement (factor of 7) in the signal-to-noise ratio.

As an example of this effect, as seen in the time domain, Figure 12 shows a 22.5-min search mode observation of Hulse & Taylor's binary pulsar B1913+16. Although this observation covers only about 5% of the orbit (7.75 hr), the effects of the Doppler smearing on the pulse signal are very apparent. Whilst the search code nominally detects the pulsar with a signal-to-noise ratio of 9.5 for this observation, it is clear that the Doppler shifting of the pulse period seen in the individual sub-integrations results in a significant reduction in the signal-to-noise ratio.

Pulsar searches of distant globular clusters are most prone to this effect since long integration times are required to reach a reasonable level of

---

[2]The smearing is even more severe if a varies — i.e. extremely short orbital periods.

sensitivity. Since one of the motivations for searching clusters is their high specific incidence of low-mass X-ray binaries, it is likely that short orbital period pulsars will also be present. Anderson et al. (1990) were the first to really address this problem during their survey of a number of globular clusters using the Arecibo radio telescope. In the so-called acceleration search, the pulsar is assumed to have a constant acceleration (a) during the integration. Each de-dispersed time series can then be re-sampled to refer it to the frame of an observer with an identical acceleration. This transformation is readily achieved by applying the Doppler formula to relate a time interval in the pulsar frame, τ, to that in the observed frame at time t, as $\tau(t) = \tau_0(1 + at/c)$, where a is the observed radial acceleration of the pulsar along the line-of-sight, c is the speed of light, and τ_0 is a normalising constant (for further details, see Camilo et al. 2000a). If the correct acceleration is chosen, then the net effect is a time series containing a signal with a constant period which can be found using the standard pulsar search outlined above. An example of this is shown in the right panel of Figure 12 where the time series has been re-sampled assuming a constant acceleration of –17 m s^{-2}. The signal-to-noise ratio is increased to 67!

The true acceleration is, of course, *a-priori* unknown, meaning that a large number of acceleration values must be tried in order to "peak up" on the correct value. Although this necessarily adds an extra dimension to the parameter space searched it can pay handsome dividends, particularly in globular clusters where the dispersion measure is well constrained by pulsars already discovered in the same cluster. Anderson et al. (1990) used this technique to find PSR B2127+11C — a double neutron star binary in M15 which has parameters similar to B1913+16. Camilo et al. (2000a) have recently applied the same technique to 47 Tucanae, a globular cluster previously known to contain 11 millisecond pulsars, to aid the discovery of a further 9 binary millisecond pulsars in the cluster.

The new discoveries in 47 Tucanae include a 3.48-ms pulsar in 96-min orbit around a low-mass companion (Camilo et al. 2000a). Whilst this is presently the shortest orbital period for a radio pulsar binary, the mere existence of this pulsar, as well as the 11-min X-ray binary X1820−303 in the globular cluster NGC6624 (Stella et al. 1987), strongly suggests that there is a population of extremely short-period radio pulsar binaries residing in globular clusters just waiting to be found. As Camilo et al. (2000a) demonstrate, the assumption of a constant acceleration during the observation clearly breaks down for such short orbital periods, requiring alternative techniques.

One obvious extension is to include a search over the time derivative of the acceleration. This is currently being tried on some of the 47 Tucanae data. Although this does improve the sensitivity to short-period binaries, it is computationally rather costly. An alternative technique developed by Ransom, Cordes & Eikenberry (2000) seems to be particularly efficient at finding binaries whose orbits are so short that many orbits can take place during an integration. This *phase modulation technique* exploits the fact that the periodic signals from such a binary are modulated by the orbit to create a family of periodic sidebands around the nominal spin period of the pulsar. This technique appears to be extremely promising and is currently being applied to radio and X-ray search data.

6. Recent Survey Highlights — The Parkes Multibeam Survey

No current review on pulsar searching would be complete without summarising the revolution in the field that is presently taking place at the 64-m Parkes radio telescope in New South Wales, Australia. With 13×20-cm 25-K receivers on the sky, along with $13 \times 2 \times 288$-MHz filterbanks, the telescope is presently making major contributions in a number of different pulsar search projects. In its main use for a Galactic plane survey (Camilo et al. 2000b), the system achieves a sensitivity of 0.15 mJy in 35 min and covers about one square degree of sky per hour of observing — a standard that is far beyond the present capabilities of any other observatory.

The staggering total of over 500 new pulsars has come from an analysis of about half the total data. Such a large haul is resulting in significant numbers of interesting individual objects: several of the new pulsars are observed to be spinning down at high rates, suggesting that they are young objects with large magnetic fields. The inferred age for the 400-ms pulsar J1119−6127, for example, is only 1.6 kyr. Another member of this group is the 4-s pulsar J1814−1744, an object that may fuel the ever-present "injection" controversy surrounding the initial spin periods of neutron stars (see however my other contribution to these proceedings).

A number of the new discoveries from the survey have orbiting companions. Several low-eccentricity systems are known where the likely companions are white dwarf stars. Two possible double neutron star systems are presently known: J1811−1736 (see Figure 9), while J1141−65 has a lower eccentricity but an orbital period of only 4.75 hr. The fact that J1141−65 may have a characteristic age of just over 1 Myr implies that the likely birth-rate of such objects may be large. Although tempting, it is premature to extrapolate the properties of one object. It is, however, clear that these binary systems, and the many which will undoubtably come from this survey, will teach us much about the still poorly-understood population of double neutron star systems (see Kalogera's contribution in this volume).

The Parkes multibeam system has not only been finding young, distant pulsars along the Galactic plane. Edwards et al. (2000) have been using the same system to search intermediate Galactic latitudes ($5° \leq |b| \leq 15°$). The discovery of 8 short-period pulsars during this search, not to mention 50 long-period objects, strongly supports a recent suggestion by Toscano et al. (1998) that L-band (λ 20 cm) searches are an excellent means of finding relatively distant millisecond pulsars.

The most massive binary system yet from either of these two multibeam surveys is J1740−3052, whose orbiting companion must be at least 11 M$_\odot$! Recent optical observations (Manchester et al. 2000) reveal a K-supergiant as being the likely companion star in this system. With such high-mass systems in the Galaxy, surely it is only a matter of time before a radio pulsar will be found orbiting a stellar-mass black hole.

Having finally made the connection between neutron stars and black holes, I will finish by reiterating that pulsar astronomy is currently enjoying one of the most productive phases in its history. The new discoveries are sparking off a variety of follow-up studies of all the exciting new objects. There will surely be plenty of surprises in the coming years and new students are encouraged to join this hive of activity.

Acknowledgements

Many thanks to Chris Salter and Fernando Camilo for their comments on an earlier version of this manuscript. The Arecibo Observatory is operated by Cornell University under a cooperative agreement with the NSF.

References

Ables, J.G. and Manchester, R.N. (1976), *A&A* **50**, 177.
Anderson, S.B. et al. (1990), *Nature* **346**, 42.
Baade, W. and Zwicky, F. (1934), *Proc. Nat. Acad. Sci.* **20**, 254.
Backer, D.C. et al. (1982), *Nature* **300**, 615.
Backer, D.C., Clifton, T.R., Kulkarni, S.R., and Wertheimer, D.J. (1990), *A&A* **232**, 292.
Camilo, F. et al. (2000a), *ApJ* **535**, 975.
Camilo, F. et al. (2000b), to appear in *Pulsar Astronomy — 2000 and Beyond*, (astro-ph/9911185).
Cordes, J.M. and Helfand, D.J. (1980), *ApJ* **239**, 640.
Davis, M.M., Taylor, J.H., Weisberg, J.M., and Backer, D.C. (1985), *Nature* **315**, 547.
Edwards, R.T. (2000), in preparation (astro-ph/9911221).
Gold, T. (1968), *Nature* **218**, 731.
Hewish, A., Bell, S.J., Pilkington, J.D.H., Scott, P.F., and Collins, R.A. (1968), *Nature* **217**, 709.
Hulse, R.A. and Taylor, J.H. (1975), *ApJ* **195**, L51.
Johnston, H.M. and Kulkarni, S.R. (1992), *ApJ* **393**, L17.
Lorimer, D.R. (1998), in M. Davier, P. Hello (eds.), *Second Workshop on Gravitational Wave Data Analysis*, Editions Frontiers, p121 (astro-ph/9801091).
Lorimer, D.R. et al. (2000), *A&A* **358**, 169.
Lyne, A.G. and Manchester, R.N. (1988), *MNRAS* **234**, 477.

Lyne, A.G. and Smith, F.G. (1998) *Pulsar Astronomy*, Cambridge University Press.

Lyne, A.G. (1988), in B. Schutz (ed.), *Gravitational Wave Data Analysis*, Dordrecht, p95.

Lyne, A.G. et al. (2000), *MNRAS* **312**, 698.

Manchester, R.N. and Lyne, A.G. (1977), *MNRAS* **181**, 761.

Manchester, R.N. et al. (2000), to appear in *Pulsar Astronomy — 2000 and Beyond*, (astro-ph/9911319).

Nicastro, L. et al. (1995), *MNRAS* **273**, L68.

Nice, D.J. (1992), *PhD thesis*, Princeton University.

Pacini, F. (1968), *Nature* **219**, 145.

Rankin, J.M. (1983), *ApJ* **274**, 333.

Ransom, S.M., Cordes, J.M., and Eikenberry S. (2000), in preparation (astro-ph/9911073).

Stairs, I.H. et al. (1998), *ApJ* **505**, 352.

Stella, L., Priedhorsky, W., and White, N.E. (1987), *ApJ* **312**, L17.

Taylor, J.H. and Cordes, J.M. (1993), *ApJ* **411**, 674.

Taylor, J.H. and Weisberg, J.M. (1989), *ApJ* **345**, 434.

Thorsett, S.E. and Chakrabarty, D. (1999), *ApJ* **512**, 288.

Toscano, M., Bailes, M., Manchester, R.N., and Sandhu, J. (1998), *ApJ* **506**, 863.

Wolszczan, A. and Frail, D.A. (1992), *Nature* **355**, 145.

NEUTRON STAR BIRTH RATES

D.R. LORIMER

National Astronomy and Ionospheric Center
Arecibo Observatory
HC3 Box 53995,
Arecibo PR 00612, USA

Abstract. A crucial test any proposed evolutionary scenario must pass is *can the birth rate of the sources we see be sustained by the proposed progenitor population?* In this review, I investigate the methods used to determine the birth rates of normal and millisecond radio pulsars and summarise recent results for these two distinct neutron star populations.

1. Introduction

In principle, estimating the birth rate of a population of sources is trivial: divide the total number of sources by the mean lifetime of the population. In practice, however, for the neutron star population, precise estimates of both the number and lifetime of the sources are hard to obtain. The sample of Galactic neutron stars is heavily biased by a number of observational selection effects which must be properly accounted for in a birth rate analysis.

In this review I shall be mainly concerned with the birth rates of two distinct sub-populations of the neutron star zoo: the normal pulsars and the millisecond pulsars. §2 defines what is meant by these two classes. §3 reviews the "standard" evolutionary scenarios which connect normal pulsars to supernova remnants and millisecond pulsars to low-mass X-ray binaries. §4 discusses the various selection effects known to significantly bias these populations. §5 summarises techniques to correct for these effects. This leads naturally to an estimate of the birth rates (§6) which are discussed in context with the proposed progenitor populations in §7. Some suggestions for future work are also made in this section.

C. Kouveliotou et al. (eds.), The Neutron Star – Black Hole Connection, 21–40.
© 2001 *Kluwer Academic Publishers. Printed in the Netherlands.*

2. Normal and Millisecond Pulsars

As I alluded to in my other contribution to this volume, the period P and period derivative $\dot{P}$ for each pulsar can be obtained to very high levels of precision through a series of timing measurements. Perhaps the most oft-plotted figure from these data is the "P—$\dot{P}$ diagram" shown in Figure 1.

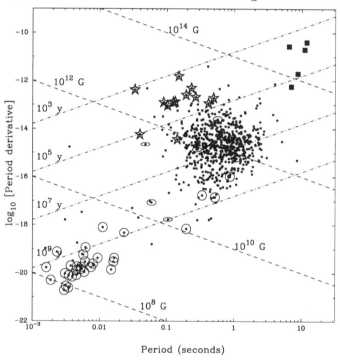

Period (seconds)

Figure 1. The ubiquitous $P - \dot{P}$ diagram shown for a sample consisting of radio pulsars (the black dots) and anomalous X-ray pulsars (the black squares — see review by Mereghetti in this volume). Pulsars known to be members of binary systems are highlighted by a circle (for low-eccentricity orbits) or an ellipse (for elliptical orbits). Pulsars thought to be associated with supernova remnants are highlighted by the starred symbols.

Amongst other things, the diagram demonstrates clearly the distinction between the "normal pulsars" ($P \sim 0.5$ s and $\dot{P} \sim 10^{-15}$ s/s and populating the "island" of points) and the "millisecond pulsars" ($P \sim 3$ ms and $\dot{P} \sim 10^{-20}$ s/s and occupying the lower left part of the diagram).

The differences in P and $\dot{P}$ imply fundamentally different typical ages and magnetic field strengths for the two populations. For the "canonical neutron star" of mass 1.4 $M_\odot$, radius 10 km and moment of inertia 10^{45} g cm^2 (see for example Shapiro & Teukolsky 1983), it can be shown that the surface dipole magnetic field strength in units of 10^{12} Gauss, $B_{12} \simeq (P_s \dot{P}_{-15})^{1/2}$, where P_s is the period in seconds and $\dot{P}_{-15} = 10^{15}\dot{P}$. Integrating this first-order differential equation over time, assuming a constant magnetic field, results in an expression for the age of the pulsar $t = \tau \left[1 - (P_0/P)^2\right]$, where $\tau = P/2\dot{P}$ is the so-called "characteristic age" and P_0 is the period of the pulsar at $t = 0$. Under the assumption that the neutron star has slowed down significantly since birth $(P_0 \ll P)$, the characteristic age τ is a good approximation to the true age t. Lines of constant B and τ are drawn on Figure 1. From these, we infer typical magnetic fields and ages of 10^{12} G and 10^7 yr for the normal pulsars and 10^8 G and 10^9 yr for the millisecond pulsars respectively.

Whilst the consensus of evidence generally supports the assumptions in the characteristic age estimates, viz: dipolar spin down from a small initial spin period and a constant magnetic field (see e.g. Bhattacharya et al. 1992; Lorimer et al. 1993), we should caution that, rather like dispersion-measure-derived distances, individual characteristic ages should not be taken as precise values. It is therefore of interest to look for independent age constraints. Strong support of the characteristic age estimates for old pulsars comes from the observations of asymmetric drift seen in the proper motions of normal pulsars with large characteristic ages (Hansen & Phinney 1997), and overwhelmingly in the millisecond pulsars (Toscano et al. 1999). See Ramachandran's review in this volume for further discussion.

A very important additional difference between normal and millisecond pulsars is binarity. Orbital companions are much more commonly observed around millisecond pulsars ($\sim 80\%$ of the observed sample) than the normal pulsars ($\lesssim 1\%$). As discussed in my other contribution to these proceedings, timing measurements constrain the masses of orbiting companions which, when supplemented by observations at other wavelengths, tell us a great deal about their nature. The present sample of orbiting companions are either white dwarfs, main sequence stars, or other neutron stars. Binary pulsars with low-mass companions ($\lesssim 0.5$ M$_\odot$ — predominantly white dwarfs) usually have millisecond spin periods and essentially circular orbits: $10^{-5} \lesssim e \lesssim 10^{-1}$. Measurements of the "cooling ages" of the white dwarfs (see e.g. van Kerkwijk 1996) provide further evidence that millisecond pulsars have typical ages of a few Gyr. The binary pulsars with high-mass companions ($\gtrsim 1$ M$_\odot$ — neutron stars or main sequence stars) have larger spin periods ($\gtrsim 30$ ms) and are in much more eccentric orbits: $0.2 \lesssim e \lesssim 0.9$.

3. Evolutionary Scenarios

We now briefly review the various end-products that are implied by standard models for the formation and evolution of single and binary radio pulsars which basically assume that every neutron star in the disk of our Galaxy was formed during the core-collapse phase of a supernova explosion of a massive star (see also van den Heuvel's contribution to this volume).

The simplest scenario is shown in Figure 2 and begins in the final moments of the life of a single massive star that is about to explode as a supernova. The neutron star formed during the core collapse will receive

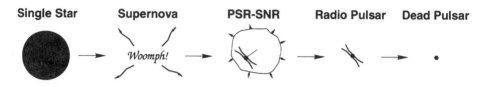

Single Star   **Supernova**   **PSR-SNR**   **Radio Pulsar**   **Dead Pulsar**

Figure 2.   Cartoon showing the evolutionary sequence of a single neutron star.

an impulsive kick (see e.g. Spruit & Phinney 1998) if the explosion is not symmetric and, as a result, begin to move away from the centre of the explosion. In the meantime, the outer layers of the star are expanding into the surrounding space at velocities of up to 10,000 km s^{-1}. The result is a pulsar–supernova remnant association (PSR-SNR) which may be visible as a pair for up to 10^5 yr after the explosion. Eventually, the expanding shell becomes so diffuse that it is no longer visible as a supernova remnant. The pulsar, on the other hand, may produce radio emission for a further 10^7 yr or more as it gradually spins down to longer periods. Presently, the longest period for a radio pulsar is 8.5 s (Young, Manchester & Johnston 1999). At some point the rotational energy of the neutron star is insufficient to induce pair production in its magnetosphere and the radio emission ceases.

The presently favoured model to explain the formation of the various types of binary systems has been developed over the years by a number of authors (Bisnovatyi-Kogan & Komberg 1974; Flannery & van den Heuvel 1975; Smarr & Blandford 1976; Alpar et al. 1982). The model is sketched in Figure 3 and can be qualitatively summarised as follows: starting with a binary star system, the neutron star is formed during the supernova explosion of the initially more massive star (the primary) which has an inherently shorter main sequence lifetime. From the virial theorem, it follows that the binary system gets disrupted if more than half the total pre-supernova mass is ejected from the system during the explosion. In addition, the fraction of surviving binaries is affected by the magnitude and direction of the impulsive kick velocity the neutron star receives at birth (Hills 1983; Bailes 1989). Those binary systems that disrupt produce a high-velocity isolated neutron star and an OB runaway star (Blaauw 1961). Rather like the fortunate survivor of a crash, the isolated pulsar has no recollection of the binary system it once belonged to and behaves from the moment of release as a single pulsar discussed above. The high binary disruption probability explains why so few normal pulsars are observed with orbiting companions.

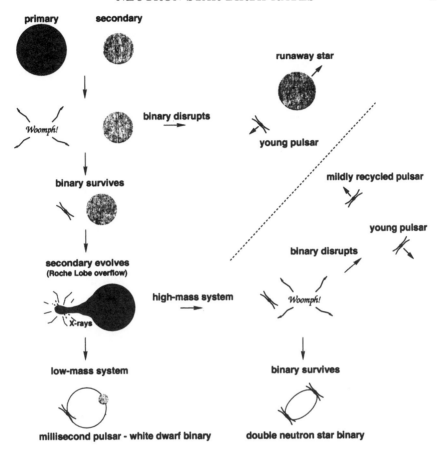

Figure 3. Cartoon showing various evolutionary scenarios involving binary pulsars.

For those few binaries that remain bound, and where the companion star is sufficiently massive to evolve into a giant and overflow its Roche lobe, the old spun-down neutron star can gain a new lease of life as a pulsar by accreting matter and therefore angular momentum from its companion (Alpar et al. 1982). The term "recycled pulsar" is often used to describe such objects. During this accretion phase, the X-rays liberated by heating the material falling onto the neutron star mean that such a system is expected to be visible as an X-ray binary system.

Two classes of X-ray binaries relevant to binary and millisecond pulsars exist: neutron stars with high-mass or low-mass companions. The high-mass companions are massive enough to explode themselves as a supernova, producing a second neutron star. If the binary system is lucky enough to survive this explosion, it ends up as a double neutron star binary. The classic example is PSR B1913+16, a 59-ms radio pulsar with a characteristic age of $\sim 10^8$ yr which orbits its companion every 7.75 hr (Hulse &

Taylor 1975). In this formation scenario, PSR B1913+16 is an example of the older, first-born, neutron star that has subsequently accreted matter from its companion. Presently, there are no clear observable examples of the second-born neutron star in these systems. This is probably reasonable when one realises that the observable lifetimes of recycled pulsars are much larger than normal pulsars. As discussed by Kalogera in this volume, double neutron star binary systems are very rare in the Galaxy — another indication that the majority of binary systems get disrupted when one of the components explodes as a supernova. Systems disrupted after the supernova of the secondary form a midly-recycled isolated pulsar and a young pulsar (formed during the explosion of the secondary).

By definition, the companions in the low-mass X-ray binaries evolve and transfer matter onto the neutron star on a much longer time-scale, spinning the star up to periods as short as a few ms (Alpar et al. 1982). This model has recently gained strong support from the detection of Doppler-shifted 2.49-ms X-ray pulsations from the transient X-ray burster SAX J1808.4–3658 (Wijnands & van der Klis 1998; Chakrabarty & Morgan 1998). At the end of the spin-up phase, the secondary sheds its outer layers to become a white dwarf in orbit around a rapidly spinning millisecond pulsar. A number of binary millisecond pulsars now have compelling optical identifications of the white dwarf companion. The existence of solitary millisecond pulsars in the Galactic disk (which comprise just under 20% of all Galactic millisecond pulsars) cannot easily be explained in the context of this model and alternative formation scenarios need to be developed.

4. Selection Effects

Having gotten a flavour for the likely evolutionary scenarios, we now turn our attention to the first piece in the birth rate puzzle — how many active radio pulsars exist in the Galaxy? Although over 1200 sources are presently known, the total Galactic population is hidden from us by a number of selection effects. As a result, the true number of pulsars is likely to be substantially larger than the observed number.

The most prominent selection effect at play in the observed pulsar sample is the inverse square law, i.e. for a given luminosity the observed flux density falls off as the inverse square of the distance. This results in the observed sample being dominated by nearby and/or bright objects. This effect is well demonstrated by the clustering of known pulsars around our location when projected onto the Galactic plane shown in Figure 4. Although this would be consistent with Ptolemy's geocentric picture of the heavens, it is clearly at variance with what we now know about the Galaxy, where the massive stars show a radial distribution about the Galactic centre.

The extent to which the sample is incomplete is shown in the right panel of Figure 4 where the cumulative number of pulsars is plotted as a function of the projected distance from the Sun. The observed distribution (solid line) is compared to the expected distribution (dashed line) for a uniform disk population in which there are errors in the distance scale, but no such selection effects. We see that the observed sample becomes strongly deficient in terms of the number of sources for distances beyond a few kpc.

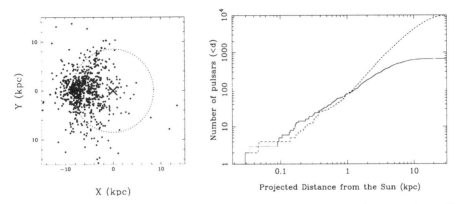

Figure 4. Left: the observed sample of pulsars projected onto the Galactic plane. The Galactic centre is at (0,0) and the Sun is at (−8.5,0). Right: Cumulative number of observed pulsars (solid line) as a function of projected distance from the Sun. The dashed line shows the expected distribution for a hypothetical uniform disk Galaxy (see text).

Beyond distances of a few kpc from the Sun, the apparent flux density falls below the flux thresholds S_{min} of most surveys. Following Dewey et al. (1984), we may parameterise the survey threshold by:

$$S_{min} = \frac{\beta \, \sigma_{min} \, (T_{rec} + T_{sky})}{A_e \, \sqrt{N_p \, \Delta\nu \, t_{int}}} \sqrt{\frac{W}{P - W}}. \tag{1}$$

In this expression β is a correction factor ~ 1.3 which reflects losses to hardware limitations, σ_{min} is the threshold signal-to-noise ratio (typically 7–10), T_{rec} and T_{sky} are the receiver and sky noise temperatures (K), A_e is the effective area of the antenna in K/Jy (1 K/Jy = 2760 m^2), N_p is the number of polarisations observed, $\Delta\nu$ is the observing bandwidth, t_{int} is the integration time, W is the detected pulse width and P is the pulse period. This expression is valid as long as $W < P$.

If the *detected pulse width* W equals the pulse period, the pulsar is of course no longer detectable as a periodic radio source. The detected pulse width is larger than the intrinsic value for a number of reasons: finite sampling effects, pulse dispersion, as well as scattering due to the presence of free electrons in the interstellar medium. From Eq. 1 of my other article in this volume, it is easy to show that the dispersive smearing scales as $\Delta\nu/\nu^3$, where the bandwidth $\Delta\nu$ is assumed to be much smaller than ν, the observing frequency. As discussed in the other article, this can largely be removed via the de-dispersion process. The smearing across the individual channels, however, still remains and becomes significant when searching for short-period ($P \lesssim 200$ ms) pulsars located at large distances.

As well as the dispersion broadening effect, free electrons in the interstellar medium can scatter the pulses causing an additional broadening due to the different arrival times of scattered pulses. A simple scattering model is shown in Figure5 in which the scattering electrons are assumed to lie in a thin screen between the pulsar and the observer (Scheuer 1968). At

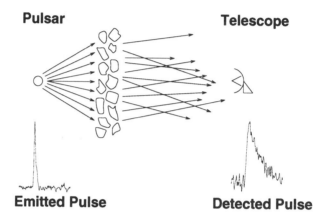

Figure 5. Pulse scattering caused by irregularities in the interstellar medium. The difference in path lengths and therefore in arrival times of the scattered rays result in a "scattering tail" in the observed pulse profile which lowers its signal-to-noise ratio.

observing frequencies $\lesssim 400$ MHz, scattering becomes particularly important for pulsars with DMs $\gtrsim 200$ cm^{-3} pc. The increased column density of free electrons can cause a significant tail in the observed pulse profile as shown in Figure 5, reducing the effective signal-to-noise ratio and overall search sensitivity. Multi-path scattering results in a one-sided broadening due to the delay in arrival times which scales roughly as ν^{-4}. This cannot be removed by instrumental means.

Dispersion and scattering become most severe for distant pulsars in the inner Galaxy as the number of free electrons along the line of sight becomes large. The strong inverse frequency dependence of both effects means that they are considerably less of a problem for surveys at observing frequencies $\gtrsim 1400$ MHz compared to the usual 400 MHz search frequency. An added bonus for such observations is the reduction in T_{sky}, since the spectral index of the non-thermal emission is about -2.8 (Lawson et al. 1987). Pulsars themselves have steep radio spectra. Typical spectral indices are

−1.6 (Lorimer et al. 1995), so that flux densities are an order of magnitude lower at 1400 MHz compared to 400 MHz. This can usually be compensated somewhat by the use of larger receiver bandwidths at higher radio frequencies. The tremendous success of the multibeam surveys at Parkes is due to a combination of all these factors which allow the survey to probe deeper into the Galaxy than has been possible hitherto (Camilo et al. 2000).

A selection effect that we simply have to live with is beaming: the fact that the emission beams of radio pulsars are narrow means that only a fraction of 4π steradians is swept out by the radio beam during one rotation. A first-order estimate of the so-called "beaming factor" or "beaming fraction" (f) is 20%; this assumes a beam width of 10 degrees and a randomly distributed inclination angle between the spin and magnetic axes.

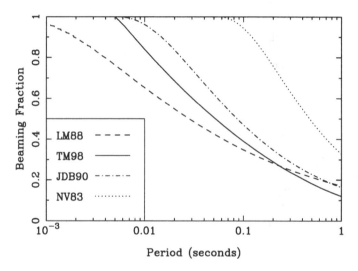

Period (seconds)

Figure 6. Beaming fraction plotted against pulse period for four different beaming models: Tauris & Manchester (1998; TM88), Lyne & Manchester (1988; LM88), Biggs (1990; JDB90) and Narayan & Vivekanand (1983; NV83).

More detailed studies show that short-period pulsars have wider beams and therefore larger beaming fractions than their long-period counterparts (Narayan & Vivekanand 1983; Lyne & Manchester 1988; Biggs 1990; Tauris & Manchester 1998). It must be said, however, that a consensus on the beaming fraction-period relation has yet to be reached. This is shown in Figure 6 where we compare the period dependence of f as given by a number of models. Adopting the Lyne & Manchester model, pulsars with periods ~ 0.1 s beam to about 30% of the sky compared to the Narayan & Vivekanand model in which pulsars with periods below 0.1 s beam to the entire sky. When many of these models were proposed, the sample of millisecond pulsars was $\lesssim 5$ and hence their predictions about the beaming fractions of short-period pulsars relied largely on extrapolations from the normal pulsars. A recent analysis of a large sample of millisecond pulsar profiles (Kramer et al. 1998) suggests that the beaming fraction of millisecond pulsars lies between 50 and 100%.

Before concluding this section, it is appropriate to mention two other selection effects which we will not discuss in detail: smearing of the pulse due to motion in a binary system, and nulling of long-period pulsars. The former effect is discussed in part in my other contribution to these proceedings and is most relevant to the double neutron star population (see Kalogera, this volume). The latter effect, nulling, refers to the apparent spasmodic switch off in the pulsar emission process first observed by Backer (1970) in a study of single pulses at Arecibo. The obvious connotation here is that pulsars which spend significant periods in the "null state" will be missed by any systematic searches, particularly those with short integration times. For example, the existence of PSR B0826-34, was extremely difficult to confirm since its null state often lasts many hours (Durdin et al. 1979).

In the context of neutron star birth rates, the selection effect caused by pulse nulling is of little significance, however, since it is fairly certain from the work of e.g. Ritchings (1976) that nulling occurs predominantly in older, long-period pulsars. In the context of pulsar surveys in general, however, this effect is potentially important. Nice (1999) has recently reanalysed archival Arecibo search data to look for dispersed single pulses. By comparison to the Fourier-transform-based searches outlined in my other article, the search analysis employed by Nice was trivial. One new 2-s pulsar, J1918+08, that emitted only one detectable pulse (!) during the original survey integration (67 s) was discovered. Nice's results suggest that further long-period pulsars which only occasionally emit detectable pulses may be found by future searches with only a modest amount of additional effort.

5. Accounting for Selection Effects

Now that we know how severe the selection effects are, how do we correct for them? There are three basic techniques: (1) source counting; (2) Monte Carlo simulations; (3) scale factor calculations.

Counting sources (see e.g. Kundt 1992) is by far the crudest method, but nevertheless easy and instructive. In essence the trick is to look at the cumulative distribution shown in Figure 4 and count sources out to a distance where you think the sample is reasonably complete and then, by assuming some underlying distribution function, extrapolate this number to get the total number of sources in the Galaxy. Based on Figure 4 we count 100 objects out to 1 kpc, i.e. a mean surface density of $100/(\pi \times 1\text{kpc}^2) \simeq 30 \text{ kpc}^{-2}$. If pulsars have a radial distribution similar to that of other stellar populations, the corresponding local-to-Galactic scale factor is then 1000 ± 250 kpc^2 (Ratnatunga & van den Bergh 1989). With this factor, we estimate there to be of order 30,000 potentially observable pulsars in the Galaxy. Assuming a beaming fraction of 20% scales this to a total of 150,000.

Whilst this simple technique gives a rough answer, and can be done on the back of an envelope, it is clearly not making good use of all the available information contained in the observed sample. A considerably more rigorous approach, shown in Figure 7, is to carry out a Monte Carlo

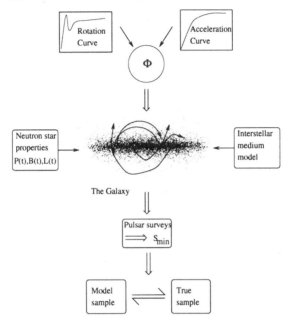

Figure 7.   The Full Monty — an idealised neutron star population synthesis scheme.

simulation of the Galaxy, the insterstellar medium and the neutron star population. Numerous accounts of this technique can be found in the literature, see e.g. Stollman (1987); Narayan (1987); Emmering & Chevalier (1989); Narayan & Ostriker (1990); Bhattacharya et al. (1992).

In brief, the simulation seeds a model galaxy with pulsars having a given set of initial parameters (period, magnetic field strength, space velocity). The model pulsars are then allowed to evolve kinematically in a model of the Galactic gravitational potential, and rotationally (based on a model for pulsar spin-down and luminosity) for a given time. The detailed models of the pulsar surveys then produce a "model observed sample" which can then be directly compared to the sample of 1200 objects that we actually observe. By varying the input model parameters and repeating the simulation to maximise the agreement between the model and actually observed samples, it is possible to directly constrain the physical parameters of the Galactic neutron star population.

Although these Monte Carlo simulations assume a pulsar birth rate, and in this sense can be used to obtain a "best-fitting value", problems are often encountered in dealing with the large number of assumptions required about pulsar spin-down and the radio luminosity evolution. What these simulations do best is to teach us about selection effects and allow useful tests of ideas hypothesised from the observable population e.g. magnetic field decay (Bhattacharya et al. 1992) and the velocity-magnetic moment correlation (Lorimer, Bailes & Harrison 1997).

The most model-independent way of constraining the number of pulsars in the Galaxy and ultimately the birth rate is, following Phinney & Blandford (1981) and Vivekanand & Narayan (1981), to define a scaling function ξ as the ratio of the total Galactic volume weighted by pulsar density to the volume in which a pulsar could be detected by various systematic surveys:

$$\xi(P, L) = \frac{\int\int_{\text{Galaxy}} \Sigma(R, z)\, R\, dR\, dz}{\int\int_{P,L} \Sigma(R, z)\, R\, dR\, dz}. \tag{2}$$

In this expression, $\Sigma(R, z)$ is the assumed pulsar density in terms of galactocentric radius R and height above the Galactic plane z. Note that ξ is primarily a function of period P and luminosity L such that short period/low-luminosity pulsars have smaller detectable volumes and therefore higher ξ values than their long-period/high-luminosity counterparts.

The scaling function is calculated in practice for each pulsar individually using a Monte Carlo simulation to model the volume of the Galaxy probed by the major surveys (Narayan 1987). For a sample of N_{obs} observed pulsars above a minimum luminosity L_{min}, the total number of pulsars in the Galaxy with luminosities above this value is

$$N_{\text{gal}} \approx \sum_{i=1}^{N_{\text{obs}}} \frac{\xi_i}{f_i}, \tag{3}$$

where f is the model-dependent beaming fraction discussed above. The beaming fraction is, effectively, the only big unknown in this calculation and it is often useful to quote potentially observable pulsar number estimates (i.e. those obtained before any beaming model has been applied).

The most recent analysis to use the scaling function approach to derive the characteristics of the true normal and millisecond pulsar populations is based on the sample of pulsars within a cylinder of radius 1.5 kpc centred on the Sun (Lyne et al. 1998). The rationale for this cut-off is that, within this region, the selection effects are well understood and easier to quantify by comparison with the rest of the Galaxy. These calculations should, at the very least, give reliable estimates for the *local pulsar population*.

The luminosity distributions obtained from this analysis are shown in Figure 8. For the normal pulsars, integrating the corrected distribution above 1 mJy kpc^2 and dividing by $\pi(1.5)^2$ kpc^2 yields a local surface density, assuming Biggs' (1990) beaming model, of 156 ± 31 pulsars kpc^{-2}. The same analysis for the millisecond pulsars, assuming a mean beaming fraction of 75% (Kramer et al. 1998), leads to a local surface density of 38 ± 16 pulsars kpc^{-2} for luminosities above 1 mJy kpc^2.

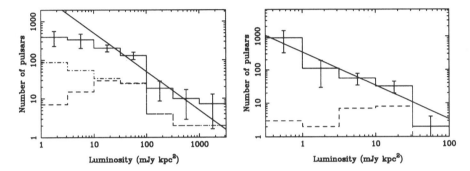

Figure 8.   Left: The corrected luminosity distribution (solid histogram with error bars) for normal pulsars. The corrected distribution *before* the beaming model has been applied is shown by the dot-dashed line. Right: The corresponding distribution for millisecond pulsars. In both cases, the observed distribution is shown by the dashed line and the thick solid line is a power law with a slope of -1. The difference between the observed and corrected distributions highlights the severe under-sampling of low-luminosity pulsars.

Scaling these local surface densities of pulsars over the whole Galaxy using the factor 1000 ± 250 kpc^2 discussed above leads us to an estimate of $(160\pm50)\times10^3$ active normal pulsars and $(40\pm20)\times10^3$ millisecond pulsars in the Galaxy. These numbers are to be compared with our back-of-the-envelope estimate of 150,000 pulsars (of all types) made at the beginning of this section. The robust analysis using scale factors and more realistic beaming models suggests a larger total population of objects.

6. Birth Rates

Simple birth rate estimates based on the above numbers and assuming some mean lifetimes may now be made. For the millisecond pulsars, we recall from the discussion in §2 that their characteristic ages are a few Gyr. This is corroborated by the observations of white dwarf cooling ages and the effect of asymmetric drift in the proper motions. Since millisecond pulsars cannot, by definition, be older than the age of the Galactic disk itself (which we take to be 10 Gyr; see e.g. Jimenez, Flynn & Kotoneva

1998 and references therein) we set a lower limit on the millisecond pulsar birth rate of $40,000/10^{10} = 4 \times 10^{-6}$ yr^{-1}. We discuss this in detail in §7.

For the normal pulsars, which are clearly much younger than the age of the Galactic disk, we make use of a relatively model-free approach to estimate the birth rate of normal pulsars that was first suggested by Phinney & Blandford (1981) and applied by Vivekanand & Narayan (1981). The technique used the scale factor determinations described above to calculate the flow of pulsars from short to long periods — that is the pulsar "current".

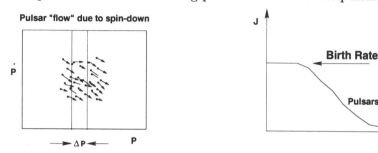

Figure 9. Schematic representation of pulsar current. Left: pulsar flow across the $P - \dot{P}$ plane. Right: pulsar current (number of objects per unit time) as a function of period.

The relevance of pulsar current theory to birth rate calculations is shown in Figure 9 — a sketch of the expected situation for a pulsar population born with short initial spin periods. Following the initial "plateau", there is a steady decline (reflecting luminosity and/or beaming evolution) until pulsars no longer radiate sufficient energy to be detected. The birth rate of this population is then simply proportional to the height of the plateau. The major advantage of this approach over previous methods is that it assumes nothing about the nature of pulsar spin–down and in this sense is model-free. The only two assumptions underlying the pulsar current theory are: (1) pulsars are spinning down to longer periods; (2) the population has reached steady state i.e. the mean age of the population is much younger than the lifetime of the Galaxy. It should be clear by now that both these assumptions are justified for normal pulsars in the disk of the Galaxy. As a result, the pulsar current analysis does not require any *a-priori* knowledge of the mean lifetime of the population to estimate the birth rate.

When Vivekanand & Narayan (1981) first calculated the current, they found a discontinuity in the distribution which showed a significant increase in the current at a period of ~ 0.5 s. They concluded that, in order to account for this anomaly, a large number of pulsars must be "injected" into the population with periods of about 0.5 s, clearly challenging the conventional view that radio pulsars begin their lives with short periods.

A number of authors have investigated the injection issue further, with contrasting results. Narayan (1987) and Narayan & Ostriker (1990) undertook detailed simulations and presented evidence in favour of injection. Lorimer et al. (1993), however, found little evidence to support these claims.

Regardless of whether there is injection into the radio pulsar sample, the discovery of long-period (5-12 s) anomalous X-ray pulsars in supernova remnants (see e.g. Mereghetti's review), which are apparently radio-quiet, has important implications for the spin periods and magnetic fields of neutron stars. It is likely that the anomalous X-ray pulsars will radically change our picture of neutron star birth properties (Gotthelf & Vasisht 2000)

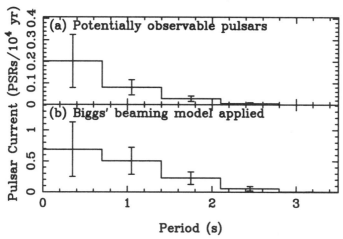

Figure 10. The results of the pulsar current analysis carried out by Lyne et al. (1998). The top panel (a) shows the current derived with no corrections for beaming. The lower panel (b) shows the data after applying Bigg's (1990) beaming model.

The results of the pulsar current analysis by Lyne et al. (1998) are shown in Figure 10. To derive the birth rate from this figure, we take the pulsar current from the first bin (0.007 pulsars per century assuming Biggs' beaming model) and divide this by $\pi(1.5)^2$ kpc^2 to scale it to unit surface area. Multiplying this number by the local-to-Galactic scale factor of 1000 ± 250 kpc^2 discussed above, the overall birth rate of normal pulsars is 1.0 ± 0.7 pulsars per century, about 2500 times larger than the lower limit obtained for the millisecond pulsars. This highly significant difference in the birth rates of the two species is further indication that the vast majority of binary systems disrupt at the moment of the first supernova.

The relatively large error bars in the birth rate estimate for normal pulsars reflects the model-free nature of the pulsar current analysis adopted. Other authors (see e.g. Narayan 1987) have attempted to reduce the uncertainties by making additional assumptions about pulsar luminosity. Whilst this is a commendable approach, there is presently no convincing evidence to support these assumptions. The interested reader is referred to §4.2 of Lorimer et al. (1993) for further discussion which is not repeated here.

7. Summary, Implications and Suggestions for Future Work

Having made a lengthy tour, with many diversions and stops on the way, we have finally achieved our goal of obtaining (some) birth rate estimates for the neutron star population. In this closing section, we discuss the implications of these results in the context of the proposed progenitor populations.

7.1. NORMAL PULSARS AND CORE-COLLAPSE SUPERNOVAE

From the results of Lyne et al. (1998), a normal pulsar is born on average once every 60–330 yr in the Galaxy. More recently, Brazier & Johnston (1999) have derived a rate of one birth every 90 yr from a local sample of X-ray detected neutron stars. This latter value is independent of the radio pulsar beaming model. Based on an analysis of historical supernovae by Tammann (see §8.3.2 of van den Bergh and Tammann 1991) the rate of Galactic supernovae is one event every 10–30 yr. Thus, *in the sense that the radio pulsar production rate does not significantly exceed the supernova rate*, this is consistent with our expectation from the standard model that every normal radio pulsar we see began its life in a supernova. Two important caveats are intimately entwined with this statement:

(1) The radio pulsar birth rate applies to objects with luminosities above 1 mJy kpc^2, and needs to be boosted by some factor if a significant part of the population is born with lower luminosities. The study of Lyne et al. (1998) shows that the radio luminosity function is flattening between 1 and 10 mJy kpc^2 (Figure 8). Lorimer et al. (1993) suggested that there may be no need to have pulsars *born* with luminosities below 10 mJy kpc^2. The implication of this is that the faint pulsars have undergone a significant luminosity evolution since birth. Further studies of this issue are warranted.

(2) The supernova rate is based on a sample which is a mixture of all supernovae types. It should be scaled by the fraction of core-collapse supernovae. Based on the death rate of stars more massive than 6.5 $M_\odot$, from the discussion in §8.5.1 of van den Bergh and Tammann (1991), one can deduce the Galactic rate of core-collapse supernova to be one event every 10–50 yr.

Assuming that few radio pulsars are born with exceedingly low luminosities, it seems that the birth rate of radio pulsars is significantly below the rate of core-collapse supernovae. Presently, there seems to be little evidence that the birth rates of soft gamma-ray repeaters (Kouveliotou et al. 1994) and anomalous X-ray pulsars (Gotthelf & Vasisht 1998) will add significantly to the rates discussed here, although further studies of the selection effects imposed on these neutron star populations are warranted. Future attempts to place tighter constraints on the pulsar birth rate (for example based on a statistical analysis of the Parkes multibeam survey results) should also clarify this situation.

7.2. MILLISECOND PULSARS AND LOW-MASS X-RAY BINARIES

As derived in §6, based on the analysis of Lyne et al. (1998), and assuming a beaming fraction of 75%, a lower limit to Galactic birth rate of millisecond pulsars is 4×10^{-6} yr^{-1}. This number is based on a sample of 18 sources within 1.5 kpc of the Sun with 400 MHz luminosities $\gtrsim 1$ mJy kpc^2. The entire sample from this paper is summarised in Table 1. This includes the aforementioned objects, plus a further three solitary millisecond pulsars with luminosities below 1 mJy kpc^2 that also lie within 1.5 kpc.

TABLE 1. The sample of 21 millisecond pulsars used by Lyne et al. (1998). For each pulsar, we list the type of system, 400 MHz luminosity and derived scale factor. Type is either: sB (short-period binary), lB (long-period binary), Si (single pulsar) or Pl (planetary system). The boundary between orbital periods of sB and lB systems is 25 days. Note that this sample is confined to a cylinder of radius 1.5 kpc centred on the Sun and does therefore not contain the few more distant millisecond pulsars (e.g. PSR B1937+21).

PSR	Type	L_{400} mJy kpc^2	Scale Fact	PSR	Type	L_{400} mJy kpc^2	Scale Fact
J0034−0534	sB	16	6	J1744−1134	Si	0.4	210
J0437−4715	sB	19	3	J1804−2717	sB	14	2
J0711−6830	Si	11	7	B1855+09	sB	12	3
J1012+5307	sB	5	9	J2019+2425	lB	2	40
J1022+1001	sB	5	10	J2033+1734	lB	10	6
J1024−0719	Si	0.6	180	J2051−0827	sB	9	7
B1257+12	Pl	8	5	J2124−3358	Si	0.4	520
J1455−3330	lB	5	9	J2145−0750	sB	18	2
J1713+0747	lB	15	4	J2229+2643	lB	13	5
J1730−2304	Si	4	10	J2317+1439	sB	68	2
				J2322+2057	Si	1	70

In the context of the standard formation scenario discussed in §2, the binary millisecond pulsars descend from low-mass X-ray binaries. The sum of the scale factors of *just* the binary systems (sB and lB types) listed in Table 1, is 108. From the sample definition of Lyne et al., this translates to a local surface density of $108/(\pi(1.5)^2) \simeq 15$ binary millisecond pulsars kpc^{-2}. Assuming, as before, that these pulsars beam to 75% of the sky, and using the Galactic scale factor of 1000 kpc^2 (Ratnatunga & van den Bergh 1989; see §5), we estimate the Galactic population of binary millisecond pulsars to be 20,000. Taking the maximum lifetime of this population to be 10 Gyr results in a lower limit to the birth rate for this population of

2×10^{-6} yr^{-1}. This number is comfortably below the birth rate for low-mass X-ray binaries ($\gtrsim 7 \times 10^{-6}$ yr^{-1}) derived by Coté & Pylyser (1989).

In the original millisecond pulsar birth rate analysis, Kulkarni & Narayan (1988) found a significant discrepancy between the birth rate of short-orbital period ($\lesssim 25$ day) millisecond pulsar binaries and their progenitor low-mass X-ray binary systems, i.e. the pulsar birth rate was two orders of magnitude too large. Based on the larger sample here, we can easily re-investigate this issue. Repeating the above analysis for the 9 short-period (sB) binaries in Table 1, we find the local surface density to be $44/(\pi(1.5)^2) \simeq 6$ short-period binaries kpc^2. Making the same beaming and Galactic scaling correction as in the previous calculation, we find a Galactic birth rate for these sources of $\gtrsim 8 \times 10^{-7}$ yr^{-1}. Whilst this is nominally a factor of at least four larger than Coté & Pylyser's estimate for short-period low-mass X-ray binary systems (2×10^{-7} yr^{-1}), the discrepancy does not seem to be as pronounced as previously thought. Based on the uncertainties in both birth rates (factors of a few), one could conclude that the populations are consistent with each other.

Although all seems to be well with the connection between low-mass X-ray binaries and millisecond pulsars, we should re-iterate the statement made in §2 that it is presently still something of a mystery how the isolated millisecond pulsars are formed. Based on the 6 single millisecond pulsars (Si) in Table 1, we deduce a Galactic birth rate of 2×10^{-5} yr^{-1}. This is significantly larger than the birth rates for the binary pulsars and is primarily a reflection of the low luminosities observed for PSRs J1024–0719, J1744–1134 and J2124–3358 which results in relatively large scale factor estimates. Indeed, these pulsars dominate the lowest bin in the luminosity function (Figure 8). It is presently marginally significant that solitary millisecond pulsars have, on average, lower luminosities than binary millisecond pulsars (Bailes et al. 1997; Kramer et al. 1998). If confirmed by future discoveries, this may be a clue to the origin of these mysterious objects.

As a final remark, we estimate from Table 1 that the Galactic population of millisecond pulsars with planetary systems is of order 900. This estimate is based upon only one object and therefore has a 100% uncertainty! It does appear, from the lack of discoveries of similar systems, that the birth rate of pulsar planetary systems in the Galactic disk is rather small.

Acknowledgements

I wish to thank the organisers of this meeting for putting together an exciting programme, and providing a splendid venue in which to hold it. Many thanks to Mike Davis and Fernando Camilo for extremely useful comments on an earlier version of this manuscript.

References

Alpar, M.A., Cheng, A.F., Ruderman, M.A., and Shaham, J. (1982), *Nature* **300**, 728.

Backer, D.C. (1970), *Nature* **228**, 42.

Bailes, M. (1989), *ApJ* **342**, 917.

Bailes, M. et al. (1997), *ApJ* **481**, 386.

Bhattacharya, D., Wijers, R.A.M.J., Hartman, J.W., and Verbunt, F. (1992), *A&A* **254**, 198.

Biggs, J.D. (1990), *MNRAS* **245**, 514.

Bisnovatyi-Kogan, G.S. and Komberg, B.V. (1974), *Soviet Ast.* **18**, 217.

Blaauw, A. (1961), *Bull. Astr. Inst. Netherlands* **15**, 265.

Brazier, K.T.S. and Johnston, S. (1999), *MNRAS* **305**, 671.

Camilo, F. et al. (2000), to appear in *Pulsar Astronomy — 2000 and Beyond*, (astro-ph/9911185).

Chakrabarty, D. and Morgan, E.H. (1998), *Nature* **394**, 346.

Coté, J., Pylyser, E.P.H. (1989), *A&A* **218**, 131.

Dewey, R.J. et al. (1984), in S. Reynolds and D. Stinebring (eds.), *Millisecond Pulsars*, p234.

Emmering, R.T. and Chevalier, R.A. (1989), *ApJ* **345**, 931.

Flannery, B.P. and van den Heuvel, E.P.J. (1975), *A&A* **39**, 61.

Gotthelf, E.V. and Vasisht, G. (1998), *New Astronomy* **3**, 293.

Gotthelf, E.V. and Vasisht, G. (2000) to appear in *Pulsar Astronomy — 2000 and Beyond*, (astro-ph/9911344).

Hansen, B. and Phinney, E.S. (1997), *MNRAS* **291**, 569.

Hills, J.G. (1983), *ApJ* **267**, 322.

Hulse, R.A. and Taylor, J.H. (1975), *ApJ* **195**, L51.

Jimenez, R., Flynn, C., and Kotoneva, E. (1998), *MNRAS* **299**, 515.

Kouveliotou, C. et al. (1994), *Nature* **368**, 125.

Kramer, M. et al. (1998), *ApJ* **501**, 270.

Kulkarni, S.R. and Narayan, R. (1988), *ApJ* **335**, 755.

Kundt, W. (1992), in T.H. Hankins, J.M. Rankin, and J. Gil (eds.), *IAU Colloquium 128*, Pedagogical University Press, Zielona Góra, Poland, p86.

Lawson, K.D., Mayer, C.J., Osborne, J.L., and Parkinson, M.L. (1987), *MNRAS* **225**, 307.

Lorimer, D.R., Bailes, M., and Harrison, P.A. (1997), *MNRAS* **289**, 592.

Lorimer, D.R., Bailes, M., Dewey, R.J., and Harrison, P.A. (1993), *MNRAS* **263**, 403.

Lorimer, D.R., Yates, J.A., Lyne, A.G., and Gould, D.M. (1995), *MNRAS* **273**, 411.

Lyne, A.G. and Manchester, R.N. (1988), *MNRAS* **234**, 477.

Lyne, A.G. et al. (1998), *MNRAS* **295**, 743.

Narayan, R. and Ostriker, J.P. (1990), *ApJ* **352**, 222.

Narayan, R. and Vivekanand, M. (1983), *A&A* **122**, 45.

Narayan, R. (1987), *ApJ* **319**, 162.

Nice, D.J. (1999), *ApJ* **513**, 927.

Phinney, E.S. and Blandford, R.D. (1981), *MNRAS* **194**, 137.

Ratnatunga, K.U. and van den Bergh, S. (1989), *ApJ* **343**, 713.

Ritchings, R.T. (1976), *MNRAS* **176**, 249.

Scheuer, P.A.G. (1968), *Nature* **218**, 920.

Shapiro, S.L. and Teukolsky, S.A. (1983), *The physics of Compact Objects*, Wiley (New York).

Smarr, L.L. and Blandford, R. (1976), *ApJ* **207**, 574.

Spruit, H. and Phinney, E.S. (1998), *Nature* **393**, 139.

Stollman, G.M. (1987), *A&A* **178**, 143.

Tauris, T.M. and Manchester, R.N. (1998), *MNRAS* **298**, 625.

Toscano, M., Sandhu, J.S., Bailes, M., Manchester, R.N., Britton, M.C., Kulkarni, S.R., Anderson, S.B., and Stappers, B.W. (1999), *MNRAS* **307**, 925.

Van den Bergh, S. and Tammann, G.A. (1991), *Annual Review of Astronomy and Astro-physics* **29**, 363.

Van Kerkwijk, M.H. (1996), in S. Johnston, M.A. Walker, and M. Bailes (eds.), *Pulsars: Problems and Progress, IAU Coll 160*, Astronomical Society of the Pacific (San Francisco), p489.

Vivekanand, M. and Narayan, R. (1981), *Journal of Astrophysics and Astronomy Supp.* **2**, 315.

Wijnands, R. and van der Klis, M. (1998), *Nature* **394**, 344.

Young, M.D., Manchester, R.N., and Johnston, S. (1999), *Nature* **400**, 848.

KINEMATICS OF RADIO PULSARS

R. RAMACHANDRAN
Netherlands Foundation for Research in Astronomy
Postbus 2, 7990 AA Dwingeloo, The Netherlands.

1. Introduction

A very good fraction of stars in the sky are believed to be in binary (or multiple) systems. Therefore, we expect a good fraction of pulsars to have originated in binary systems. However, observations of pulsars have shown that most of the pulsars are not in a binary system now. This would imply that these pulsars must have become solitary after the disruption of their progenitor binary system during (or shortly after) the supernova explosion.

Well before the discovery of pulsars, the work of Blaauw (1961) showed that when a star explodes in a binary system, if the mass lost in that process is greater than half the initial total mass, then the final system will be unbound. This would imply that the resultant pulsar will have a spatial velocity equal to the orbital velocity of the exploded star at the time of the explosion.

After the discovery of pulsars, from statistical analysis of the pulsar population, Trimble and Rees (1970) showed that in order to explain the distribution of pulsars, one needs larger spatial velocities for pulsars compared to what would have been imparted to them from the orbital velocities of their progenitor binary systems.

Following this, two main mechanisms were suggested to explain the origin of pulsar velocities (other than from the binary orbital velocities).

1. Shklowskii (1970) conjectured that supernova explosions are asymmetric in nature, and that helps pulsars to acquire substantial velocities during their birth.
2. In 1975, Harrison & Tademaru came up with another mechanism, where pulsars acquire their velocities by the "Rocket-effect", by having oblique-offcentred magnetic dipole.

C. Kouveliotou et al. (eds.), The Neutron Star – Black Hole Connection, 41–56.

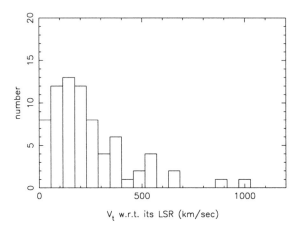

Figure 1. Distribution of pulsar observed transverse velocities after correcting for differential Galactic rotation.

The latter showed that if we have such a magnetic field arrangement the radiation pattern of that dipole is asymmetric with respect to the two poles defined by the rotation axis. The reaction force on the neutron star in such a case is proportional to the fifth power of the angular velocity (Ω^5). This implies two important results: (a) Young new-born fast-rotating pulsars can be accelerated to very high velocities, (b) This effect is strong enough to disrupt the binary system where the pulsars were born! Therefore, with this mechanism, the authors could explain both the problems (that pulsars have large spatial velocities, and they are solitary).

Although this method, in principle could give qualitative explanation to the nature of pulsar population, it failed a very important observational test. As Morris, Radhakrishnan & Shukre (1976) showed, the acceleration of a pulsar in the direction of its rotation axis would necessarily mean that in the plane of the sky the projected direction of rotation axis is the same as that of the measured proper motion direction. From the measured proper motion and polarisation data, they proved that this correlation is not seen (see also Anderson & Lyne 1983; Lorimer et al. 1995; Deshpande, Ramachandran, Radhakrishnan 1999). Therefore, we can assume that this elegant mechanism is not relevant for understanding the origin of pulsar velocities.

In this paper, I summarise the main developments in this subject so far.

2. Observational facts

There are two "direct" ways of measuring pulsar proper motions. The first is by an interferometer, where observations at two well separated epochs can give an accurate measurement of positional displacement of the pulsar in

TABLE 1. Pulsars with possible transverse velocities greater than 500 km/sec. Columns 2 & 3 give the galactic longitude and latitude, col. 4 gives the distance calculated using the Taylor & Cordes (1993) model, 5–8 give the measured proper motions along RA and Dec and their errors, and the last column gives the transverse velocity.

| Jname | l | b | d | μ_α | err | μ_δ | err | V_t |
	(deg)	(deg)	(kpc)	(mas/yr)		(mas/yr)		km/s
0152–1637	179.3	–72.5	.79	–10	50	–150	50	563
0452–1759	217.1	–34.1	>3.14	19	8	35	18	>593
0525+1115	192.7	–13.2	>7.68	30	7	–4	5	>1102
0601–0527	212.2	–13.5	>7.54	18	8	–16	7	>861
0738–4042	254.2	–9.2	>11.03	–56	9	46	8	>3789
0826+2637	197.0	31.7	.38	61	3	–90	2	644
0922+0638	225.4	36.4	>2.97	13	29	64	37	>919
1509+5531	91.3	52.3	1.93	–73	4	–68	3	913
1604–4909	332.2	2.4	3.59	–30	7	–1	3	511
1645–0317	14.1	26.1	2.90	41	17	–25	11	660
1709–1640	5.8	13.7	1.27	75	20	147	50	993
1720–0212	20.1	18.9	>5.43	26	9	–13	6	>748
1935+1616	52.4	–2.1	7.94	2	3	–25	5	944
2149+6329	104.3	7.4	>13.64	14	3	10	4	>1112
2225+6535	108.6	6.8	1.95	144	3	112	3	1686
2305+3100	97.7	–26.7	>3.93	13	8	–33	6	>661

the sky. The second is by timing measurements, where the positional errors (due to the proper motion) lead to annual oscillations in timing residuals, and this gives an estimate of their proper motion. So far, we have proper motion measurements for about 100 pulsars by the above two methods.

Figure 1 shows the observed transverse velocity of pulsars after correcting for the differential galactic rotation. As we can see, the maximum velocities go all the way to about 1000 km/sec, but the majority of pulsars have velocities of about 150 to 200 km/sec. While understanding this distribution, apart from the measurement errors, it is also important to appreciate the uncertainties in the estimated distances to these pulsars ($V_t = \mu \times D$, where μ and D are the measured proper motion and distance to the pulsar). Table 1 gives a list of pulsars for which transverse velocities are suspected to be greater than 500 km/sec. Distances to these objects are estimated with the help of the Galactic free-electron density distribution model of Taylor & Cordes (1993). It is clear that for half of them, we have only a lower limit on their distance. Even for these pulsars, a "lower limit" doesn't necessarily mean that the actual distance is greater than the lower limit! As Deshpande & Ramachandran (1998) showed from their scattering measurements, the distance to PSR J0738–4042 is as small as about 4 kpc, whereas the lower limit from the Taylor & Cordes model is > 11 kpc. After accounting for all these uncertainties, what we can conclude from this table is that there are a few pulsars, like 0826+2637, 1509+5531, and 2225+6535, which have sufficiently well measured proper motions and well determined distances, which are moving with transverse velocities well in excess of 500 km/sec. Therefore, we do see some fraction of pulsars having such high velocities.

Many observational arguments have emerged over the years to justify the idea that one needs an impulse ("kick") due to asymmetric supernova explosion. These are:

- The rotation axis of the Be-star companion of PSR J0045–7319 is misaligned with the orbital angular momentum axis, strongly suggesting that the supernova explosion must have been asymmetric (Kaspi et al. 1996; Lai 1996).
- Large orbital eccentricities of Be X-ray binaries
- Low incidence of double neutron star binary systems in the Galaxy. Observational estimates show a birthrate of $\leq 10^{-5}$ yr^{-1} (Phinney 1991; Narayan et al. 1991; van den Heuvel & Lorimer 1996). To reproduce this, we need a kick speed of a few hundred km/sec (Portegies Zwart & Spreew 1996; Ramachandran 1996; Lipunov et al. 1996; Bagot 1996; Fryer & Kalogera 1997).
- Low incidence of Low Mass X-ray Binaries, and their kinematics in the Galaxy (van Paradijs & White 1995; Brandt & Podsiadlowski 1995; Ramachandran & Bhattacharya 1997; Cordes & Chernov 1997; Kalogera 1996; 1998; Tauris & Bailes 1996).
- It is necessary to have kick velocities to produce systems like PSR B1913+16 (Wex et al. 2000)

Considering all the above observational constraints, it seems that an asymmetric supernova explosion leading to a "kick" seems to be most probable. However, it is also important to understand the fractional contribution of binary orbital velocities to the observed pulsar velocities.

3. Distribution of pulsar speeds

To understand the intrinsic distribution of pulsar speeds in the Galaxy, we need to do a detailed statistical analysis of the pulsar population. There are many observational selection effects biasing our sample of pulsars, and they need to be understood and corrected for, before attempting to quantify any statistical property of the pulsar population. A brief description of the

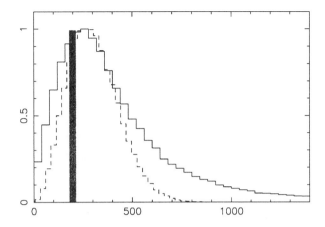

Figure 2. Derived distribution of three dimensional velocities of pulsars. Solid line is by Lyne & Lorimer (1994), Dashed line, by Hansen & Phinney (1997), and the shaded distribution by Blaauw & Ramachandran (1998).

selection effects is as follows (see Lorimer's talk in this volume for further details).

- Pulsar radio luminosities have a wide ranging distribution. Faint objects are not detected, biasing our sample against the 'fainter' side of the luminosity distribution.
- The probability of detecting farther pulsars is reduced by smearing of the pulse due to dispersion and scattering of radio signals in the interstellar medium.
- Slow pulsars do not migrate to large heights above the plane, whereas fast pulsars do. This gives an over-estimation of the scale factors for slow pulsars (Hansen & Phinney 1997).
- Farther slow pulsars do not show significant (measurable) proper motion, biasing the sample against slow pulsars.
- Distances are over-estimated due to the presence of (unaccounted) HII regions (or Strömgren spheres of OB stars) which introduce significant extra dispersion measure. This results in over-estimation of the transverse velocities (Deshpande & Ramachandran 1998).

Any statistical study of pulsar population must take into account all these effects to get an unbiased idea about its properties.

Figure 2 shows the distribution of three dimensional velocities of the pulsar population estimated by three different works. Lyne & Lorimer (1994) considered a sample of 28 pulsars whose ages are less than 3 Myr for their analysis. With the assumption that there are no selection effects on such a sample of young pulsars, they estimated the speed distribution, which has a long 'tail' extending to well above 1000 km/sec. This predicts an average speed of about 450 km/sec, with an r.m.s. speed of 535 km/sec and the peak of the distribution around 250 km/sec (solid line in the figure). Hansen & Phinney (1997), from their detailed analysis of the Galactic population of pulsars came to the conclusion that the pulsar population is quite consistent with a Maxwellian distribution with an one-dimensional r.m.s. velocity of about 190 km/sec (dashed curve). With their detailed analysis of the local population of pulsars, Blaauw & Ramachandran (1998) concluded that the speed distribution of pulsars in the solar neighbourhood is consistent with even a 'delta-function' distribution at about 200 km/sec! The idea that a predominant fraction of pulsars may be moving with substantially lower velocities ($\sim 200-250$ km/sec) was also supported by the analysis by Hartman (1997). Therefore from all these analyses, we can conclude that the bulk of the population is moving with 3−D speeds of about 200 − 250 km/sec.

4. Kinematics of pulsars in the Galaxy

From the above analysis, it is clear that pulsars do have large peculiar speeds of the order of about 200 − 250 km/sec. With the addition of the Galactic rotation, the resultant velocities of these objects could be very much comparable to the escape velocity of the Galaxy. Therefore, a study of the evolution of these objects in the Galactic potential becomes important. I will present here a 'quick' Monte Carlo simulation to understand some important aspects of their evolution. We will address some of the questions related to their kinematical properties like 'asymmetric drift' in the Galactic plane, their migration along the Galactocentric radius, etc. Let us generate a large number of pulsars in the Galaxy with the following properties:

- The surface density distribution of these objects along the Galactocentric radius, R, is a Gaussian, with an r.m.s. of $\sigma_R = 4.5$ kpc.
- Along the height from the galactic plane, z, let us assume a scale-height of 75 pc.
- Evolve each pulsar after including the Galactic rotation contribution corresponding to its place of birth, for a total length of time T, where T is distributed uniformly between zero and a maximum of 2×10^9 yrs. After evolving each object, store the values of its initial and final coordinates, and its final velocity components in the Galaxy.
- Study the spatial and velocity distributions, after correcting for the observational selection effects as indicated in section 3.

TABLE 2. Parameters of the Galactic potential function by Kuijken & Gilmore (1989).

Parameter	Disc-Halo	Nucleus	Bulge
Mass ($M_\odot$)	1.45×10^{11}	9.3×10^{11}	1×10^{10}
β_1	0.4		
β_2	0.5		
β_3	0.1		
h_1	0.325		
h_2	0.090		
h_3	0.125		
a	2.4		
b	5.5	0.25	1.5

The Galactic potential function is assumed to be the one given by Kuijken & Gilmore (1989):

$$\Phi(R, z) = \frac{-M}{\left[\left(a + \sqrt{z^2 + h^2}\right)^2 + b^2 + R^2\right]^{1/2}} \tag{1}$$

This potential function has three components, namely the Disc-Halo, Nucleus, and the Central Bulge. For the Disc-Halo component, $\sqrt{z^2 + h^2} = \sum_{i=1}^{3} \beta_i \sqrt{z^2 + h_i^2}$. The values of all the constants are tabulated in Table 2.

4.1. ASYMMETRIC DRIFT

When a pulsar evolves in the galaxy, due to its peculiar velocity it migrates along both z and R. When a pulsar is born in the Galaxy, its initial angular momentum is determined by the Galactocentric radius at which the pulsar is born, and its velocity around the galactic centre. Given this, and the flat rotation curve of the Galaxy, when the pulsar migrates to a different galactocentric radius, it either leads ahead, or lags behind the local flow, depending on whether the present galactocentric radius is less or more than its initial radius. Since, on the average, objects flow to the outer portions of the Galaxy from the inner portions, we expect more objects laging behind. In other words, any virialised population with a significant radial velocity component will revolve more slowly around the Galactic centre. This effect is known as 'asymmetric drift', and is seen in many stellar populations in the Galaxy (Mihalas & Binney 1981).

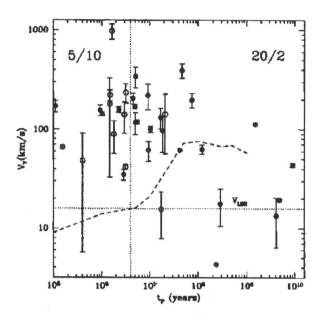

Figure 3.   Plot of characteristic age *vs* azimuthal velocity (V_y) in the Galaxy (Hansen
& Phinney 1997). Solid circles indicate positive V_y. About 90% of the pulsars with ages
greater than 4 Myr seem to have positive V_y, strongly indicating asymmetric drift. The
dashed line indicates the evolution of the mean asymmetric drift velocity from their
calculations.

Pulsars, although relatively young objects, have far greater velocity than
almost all stellar populations. Therefore, even within a relatively short time
(~ 10 Myr) they exhibit this important property. Hansen & Phinney (1997)
showed through their detailed analysis that 90% of those pulsars whose ages
are greater than about 4 Myr show this effect.

Figure 3 shows the plot of Hansen & Phinney of the characteristic age
and the V component (velocity in the azimuthal direction with respect to
the Galactic centre) derived from their proper motion measurements.
Though they have calculated only the azimuthal velocity with respect to
the position of the Sun (not with respect to the position of each of the
pulsars), it gives a clear idea about the significance of the asymmetric drift.

To get an idea about how severe is the migration of pulsars along the
radial direction, in Figure 4 I have plotted the distribution of the initial
galactocentric radius of those pulsars which end up in the Solar neighbourhood ($R \sim 6 - 10$ kpc) after their evolution. We can see that the migration
becomes very significant even in 10 Myr, for velocities of 250 km/sec. At
100 Myr, practically the objects come from everywhere to the solar neighbourhood. This suggests that a complete understanding of the kinematical

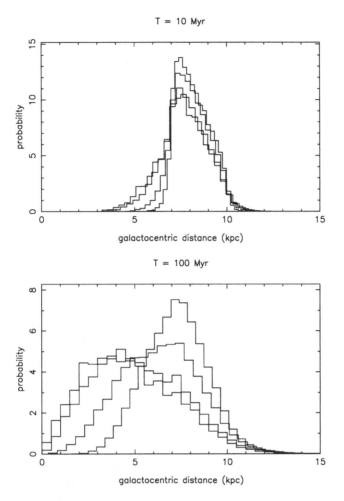

Figure 4. Distribution of the initial galactocentric radius for those objects which end up in the Solar neighbourhood after their evolution. The two panels correspond to a maximum evolution time of 10 Myr and 100 Myr, respectively. The distributions (from the narrowest to the broadest) correspond to the assumed initial Maxwellian speed distributions with 1-D r.m.s. of 50, 100, 250, and 500 km/sec, respectively.

properties of pulsars is possible only if we study their evolution in the full galactic potential, and not in the approximated 'local potential'.

5. Velocities of millisecond pulsars

The population of millisecond pulsars (MSPs) differs from the ordinary pulsars in many properties. First of all, they are much older (more than 10^9 years old). Moreover, a good fraction of them are in binary systems. These make the kinematic properties of millisecond pulsars quite distinctly

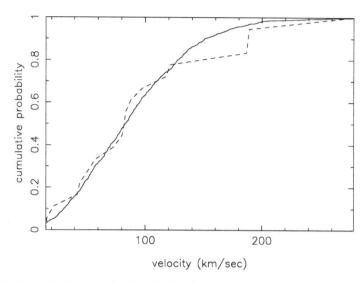

Figure 5.   Cumulative distribution of the observed transverse velocities of millisecond pulsars (dashed line). Solid line is the result of the simulation, with a Maxwellian speed distribution of 1–D r.m.s. of 100 km/sec, and maximum age of about 10^9 years. The K–S probability of this fit is 61%. See text for details.

different from the ordinary population.

In the standard evolutionary scenario of MSPs (see Bhattacharya & van den Heuvel 1991 for a detailed review), they originate from Low Mass X-ray Binaries (LMXBs). These LMXBs acquire their space velocities during the primary (and the only) Supernova explosion in the system. Even if the supernova explosion is a symmetric one, the surviving binaries will acquire some space velocity, since the explosion is symmetric only with respect to the exploding star, and not with respect to the centre of mass of the binary system. During this process, a significant fraction of the binary systems may get disrupted. The survival probability gets reduced further if the explosion is an asymmetric one (Flannery & van den Heuvel 1975; Hills 1983; Ramachandran & Bhattacharya 1997; Tauris & Bailes 1996).

Over the past few years, many detailed analyses have been done to understand the kinematic properties of these two populations. Through their analysis, van Paradijs and White (1995) argued that to explain the distribution of LMXBs we need high asymmetric kick velocities, and the distribution is consistent with the kick speed distribution given by Lyne & Lorimer (1994). Through their Monte Carlo simulation, Brandt & Podsiad-lowski (1995) supported this conclusion. Tauris & Bailes (1996), from their evolutionary calculations, showed that it is almost impossible to produce

TABLE 3. List of measured proper motions of millisecond pulsars. Columns 2 & 3 give the galactic longitude and latitude, 4–7 give the proper motions along RA and Dec and their errors, columns 8 & 9 give the distance calculated using the Taylor & Cordes (1993) model, and transverse velocity, respectively. For those pulsars for which we know only a lower limit to distance, velocity has been specified as "Undef" (undefined).

Jname	l (deg)	b (deg)	μ_α (mas/yr)	err	μ_δ (mas/yr)	err	d (km/s)	V_t
J0437-4715	253.394	-41.964	114	2	-72	4	0.14	88
J0613-0200	210.41	-9.30	2.0	0.4	-7	1	2.19	58
J0711-6830	279.53	-23.28	-15.7	0.5	15.3	0.6	1.03	118
J1024-0719	251.70	40.52	-41	2	-70	3	0.35	121
J1045-4509	280.85	12.25	-5	2.0	6	1	3.24	188
J1300+1240	311.301	75.414	46.4	0	-82.9	0	0.62	279
J1455-3330	330.72	22.56	5	6.0	24	12	0.74	97
J1603-7202	316.63	-14.50	-3.5	0.3	-7.8	0.5	1.64	73
J1643-1224	5.67	21.22	3.0	1.0	-8.0	5	>4.86	Undef
J1713+0747	28.751	25.223	4.9	0.3	-4.1	1.0	0.89	22
J1730-2304	3.14	6.02	20.5	0.4	0.0	0	0.51	51
J1744-1134	14.79	9.18	18.64	0.08	-10.3	0.5	0.17	15
J1857+0943	42.290	3.061	-2.94	0.04	-5.41	0.06	1.00	16
J1911-1114	25.14	-9.58	-6	4	23	13	1.59	187
B1937+21	57.509	-0.290	-0.130	0.008	-0.464	0.009	3.58	80
B1957+20	59.197	-4.697	-16.0	0.5	-25.8	0.6	1.53	189
J2019+2425	64.746	-6.624	-9.9	0.7	-21.3	1.4	0.91	83
J2124-3358	10.93	-45.44	-14	1	-47	1	0.24	45
J2129-5721	338.01	-43.57	7	2	-4	3	>2.55	Undef
J2145-0750	47.78	-42.08	-9.1	0.7	-15	2	0.50	43
J2322+2057	96.515	-37.31	-17	2	-18	3	0.78	82.24

any MSP system with speeds $\gtrsim 250$ km/sec. Later, Ramachandran & Bhattacharya (1997) showed through their detailed Monte Carlo analysis of the evolution of LMXBs and MSPs in the Galactic potential, that the spatial distribution of LMXBs and the speed distribution of MSPs are consistent with the Lyne & Lorimer distribution, but it is even more consistent with even 'zero' kick speed! This uncertainty (mainly by low number statistics) was again shown by the analysis by Kalogera (1998) where she showed that the distribution is consistent with speeds in the range 100 - 500 km/sec.

Given all these, let us see what we can infer from our simple simulation with new proper motion measurements of many MSPs. Table 3 gives a list of MSPs with their measured proper motions and transverse velocities. It is understandable that these objects, on the average, are moving slower than the ordinary pulsars, since high velocities acquired during the supernova explosion would have disrupted the binary system.

In order to get an idea about their intrinsic birth speeds, I have evolved the sample pulsars for about 2×10^9 years. Then, I have compensated for all the observational selection effects. For this process, I have assumed that the Parkes 70 cm survey (Manchester et al. 1996) can observe all directions in the Galaxy. This is just to improve the statistics in the final 'observable' distribution of pulsars in the simulation. Then, from the velocities of those 'observable' pulsars, I have calculated their transverse velocity as measured from the position of the Sun, so that they can be directly compared with the measured transverse velocities of MSPs. Figure 5 shows the cumulative distribution of the 'true' distribution (dashed line) and the simulated distribution for an assumed initial Maxwellian speed distribution with 1-D r.m.s. of 100 km/sec (solid line). The Kolmogorov-Smirnov probability (K-S probability) of this fit comes to about 61%. The K-S probability of many other distributions seems to be less than this value. For instance, for $\sigma_v = 50$ km/sec, it is 20%. And for σ_v of 250 and 500 km/sec, the probabilities were 12% and 0%, respectively.

6. Observational evidence pertinent to kick mechanisms

From various reasons presented above, it is clear that the formation process of neutron stars must be asymmetric, so that pulsars gain significant velocities. The reasons, as we saw, are purely empirical, with different kinds of observations pointing to the existence of an impulsive transfer of momentum to the protoneutron star at birth (Shklowskii 1970; Gunn & Ostriker 1970; van den Heuvel & van Paradijs 1997). The mechanisms suggested range from hydrodynamical instabilities to those in which asymmetric neutrino emission is postulated (Burrows 1987; Keil et al. 1996; Horowitz & Li 1997; Lai & Qian 1998; Spruit & Phinney 1998).

Whatever is the mechanism to create such an asymmetry, it is important to understand if the direction of asymmetry is random, or it is associated with some physical property of the system such as the rotation and the magnetic dipole axes of the neutron star. Many mechanisms suggested to produce the asymmetry have invoked both these axes for the direction of the asymmetry (Harrison & Tademaru 1975a,b; Burrows & Hayes 1996; Kusenko & Segre 1996).

The recent analysis by Deshpande, Ramachandran & Radhakrishnan (1999) explores the observational tests to prove (or disprove) the mechanisms predicting any relation between the direction and magnitude of the observed velocities and the magnetic and the rotation axis of the star. As they show, observations do not support any relation between (i) the magnitudes of velocities and magnetic field, and (*ii*) the direction of velocities and the magnetic axis and rotation axis.

The recent work by Spruit & Phinney (1998) suggests that the cores of the progenitors of neutron stars cannot have the angular momentum to explain the rotation of pulsars. They propose that the rotation of pulsars and their spatial velocities must have a common origin. This suggestion was first made by Burrows et al. (1995). As Spruit & Phinney state, if the asymmetry is not directed radially during the formation, then the star acquires both linear and angular momentum. Cowsik (1998) has also advanced such a possibility.

In order to test this hypothesis, we need to understand a number of possibilities. In this case, the variables are, (*i*) number of impulses which the (proto-) neutron star receives, (*ii*) duration of each impulse, (*iii*) direction of each impulse. Let us assume that the direction of the impulses are random. From the numerical simulations of Deshpande et al. (1999), it is clear that if we have only one impulse (of any duration), the resultant direction of the velocity is perpendicular to the direction of the spin axis. After taking account of the projection effects in the plane of the sky, they show that the angle between the rotation and the proper motion axes should be biased towards 90°. However, this is *not* what is seen observationally. Figure 6a shows that the distribution is consistent with a uniform distribution from zero to 90 degrees. Therefore, single impulses being responsible for both the velocity and the rotation of the star can be ruled out.

If the number of impulses are more, then the problem becomes complicated, and is a sensitive function of the duration of each of the kicks. First of all, we expect that the strength of the impulses required to produce the observed range of velocities goes down by $\sqrt{N}$, where N is the number of impulses. The velocity dispersion as a function of the duration of a momentum impulse of a given magnitude remains constant up to a certain critical duration (τ_c), above which the azimuthal averaging reduces the resultant velocities. For impulse durations much smaller than τ_c, both the linear and angular momenta grow as $\sqrt{N}$, but the angle between them becomes random. However, for relatively longer duration impulses a significant preference of the direction of the linear momentum develops towards the spin axis, which itself is evolving. This is shown in Figure 6b, where the angle has a preferential bias towards zero degrees. Since observations do not support this either, we can conclude that long duration multiple impulses can also be ruled out.

To summarise, observationally we can conclude the following: (1) velocity magnitudes are *not* correlated to the magnetic field strength, (2) pulsar velocities are *not* in the direction of their magnetic axis or rotation axis, (3) if non-radial kicks produce both pulsar velocities and rotation, then the

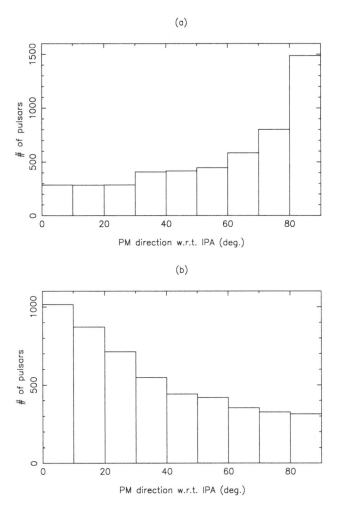

Figure 6. Distribution of angles between the proper motion and the spin axis projected in the plane of the sky (Deshpande, Ramachandran & Radhakrishnan 1999). (a) For a single non-radial impulse of any duration. The distribution is biased towards 90 degrees, (b) for a number of long-duration impulses. The bias is towards zero degrees.

single kicks of *any duration* are ruled out, and (4) multiple kicks of long duration are ruled out.

7. Concluding remarks

I have given a brief summary of the kinematic properties of pulsars in this article. Though there are some pulsars which have significantly large spatial velocities (of the order of 1000 km/sec), on the average, there is strong evidence that the population has a three dimensional speed of about 200 to 250 km/sec. All the derived speed distributions show that the predominant number of pulsars move with such speeds. There is much empirical evidence that pulsars acquire a significant fraction of their velocities through an asymmetric supernova explosion.

Pulsar population shows a great deal of migration along both galacto-centric radius and height from the plane. Because of this, they show significant asymmetric drift. This is one of the remarkable properties observed on pulsar population.

Millisecond pulsars, due to their evolutionary history, have significantly lower space velocities. Their average velocity is around 100 km/sec, and their kinematic properties in the Galaxy match well with those of the Low Mass X-Ray Binaries, supporting the idea that MSPs are born from LMXBs.

There are many models which have advanced many different ways of producing asymmetries during supernova explosions. Given the latest high-quality proper motion measurements and polarisation information with which we can determine the direction of rotation axis in the plane of the sky, one can test these models. These tests reveal that:

- Mechanisms predicting a correlation between the rotation axis and the pulsar velocities are ruled out.
- There is no correlation between the direction of magnetic axis and the direction of velocities.
- There is no correlation between the strength of the magnetic fields and the magnitude of velocities.
- If non-radial asymmetric impulses is responsible for both velocity and rotation of pulsars, then the scenario where the star receives a single impulse (of any duration) can be ruled out.
- Even multiple impulses of long durations can be ruled out.

In principle, multiple short-duration impulses are not ruled out, as they would produce randomly oriented rotation and velocity axis. Similarly, we cannot rule out the possibility of having completely radial kicks as the origin of pulsar velocities. In this case, the rotation of the pulsar will have nothing to do with the asymmetric explosion.

Acknowledgements

I would like to thank D. Bhattacharya, A. Blaauw, A. A. Deshpande, V. Radhakrishnan and E. P. J. van den Heuvel, for many fruitful discussions.

References

Anderson, B. and Lyne, A.G. (1983), *Nature* **303**, 597.
Bagot, P. (1996), *A&A* **314**, 576.
Bhattacharya, D. and van den Heuvel, E.P.J. (1991), *Phys. Rep.* **203**, 1.
Blaauw, A. (1961), *Bull. Astr. Inst. Netherlands* **15**, 265.
Blaauw, A. and Ramachandran, R. (1998), *JApA* **19**, 19.
Brandt, W.N. and Podsiadlowski, P. (1995), *MNRAS* **277**, 35.
Burrows, A. (1987), *ApJ* **318**, L57.
Burrows, A. and Hayes, J. (1996), *PRL* **76**, 352.
Burrows, A., Hayes, J., and Fryxell, B.A. (1995), *ApJ* **450**, 830.
Cordes, J.M. and Chernoff, D.F. (1997), *ApJ* **482**, 971.
Cowsik, R. (1998), *A&A* **340**, L65.
Deshpande, A.A. and Ramachandran, R. (1998), *MNRAS* **300**, 577.
Deshpande, A.A., Ramachandran, R., and Radhakrishnan, V. (1999), *A&A* **351**, 195.
Flannery, B.P. and van den Heuvel, E.P.J. (1975), *A&A* **39**, 61.
Fryer, C. and Kalogera, V. (1997), *ApJ* **489**, 244.
Gunn, J.E. and Ostriker, J.P. (1970), *ApJ* **160**, 979.
Hansen, B.M.S. and Phinney, E.S. (1997), *MNRAS* **291**, 569.
Harrison, E.R. and Tademaru, E. (1975a), *ApJ* **201**, 447.
Harrison, E.R. and Tademaru, E. (1975b), *Nature* **254**, 676.
Hills, J.G. (1983), *ApJ* **267**, 322.
Horowitz, C.J. and Li, G. (1998), *PRL* **80**, 3694.
Kalogera, V. (1996), *ApJ* **471**, 352.
Kalogera, V. (1998), in N. Shibazaki et al. (eds.), *Neutron stars and Pulsars*, held in Rikkoyo University, Japan, p27.
Kaspi, V.M. Bailes, M., Manchester, R.N. et al. (1996), *Nature* **381**, 584.
Keil, W., Janka, H.-Th., and Muller, E. (1996), *ApJ* **473**, L111.
Kuijken, K. and Gilmore, G. (1989), *MNRAS* **239**, 571.
Kusenko A. and Segre, G. (1996), *PRL* **77**, 24.
Lai, D. (1996), *ApJ* **466**, 35.
Lai, D. and Qian, Y. (1998), *ApJ* **495**, 103.
Lipunov, V.M., Postnov, K.A., and Prokhorov, M.E. (1996), *A&A* **310**, 489.
Lorimer, D.R., Lyne, A.G., and Anderson, B. (1995), *MNRAS* **275**, L16.
Lyne, A.G. and Lorimer, D.R. (1994), *Nature* **369**, 127.
Manchester, R.N., Lyne, A.G., D'Amico, N. et al. (1996), *MNRAS* **279**, 1235.
Mihalas, D. and Binney, J. (1981) *Galactic Astronomy*, W. H. Freeman (New York).
Morris, D., Radhakrishnan, V.B., and Shukre, C.S. (1976), *Nature* **260**, 124.
Narayan, R., Piran, T., and Shemi, A. (1991), *ApJ* **379**, 17.
Phinney, E.S. (1991), *ApJ* **380**, 17.
Portegies Zwart, S.F. and Spreeuw, H.N. (1996), *A&A* **312**, 670.
Ramachandran, R. (1996), *Ph.D. Thesis*, Osmania University.
Ramachandran, R. and Bhattacharya, D. (1997), *MNRAS* **288**, 565.
Shklowskii, I.S. (1970), *Astr. Zu.* **46**, 715.
Spruit, H.C. and Phinney, E.S. (1998), *Nature* **393**, 139.
Tauris, T. and Bailes, M. (1996), *A&A* **315**, 432.
Taylor, J.H. and Cordes, J.M. (1993), *ApJ* **411**, 674.
Trimble, V. and Rees, M. (1971), in R.D. Davies and F. Graham-Smith (eds.), *IAU Symposium no. 46*, Dordrecht, Reidel, p273.
Van den Heuvel, E.P.J. and Lorimer, D.R. (1996), *MNRAS* **283**, 37.
Van den Heuvel, E.P.J. and van Paradijs, J. (1997), *ApJ* **483**, 399.
Van Paradijs, J. and White, N. (1995), *ApJ* **447**, 33.
Wex, N., Kalogera, V., and Kramer, M. (2000), *ApJ* **528**, 401.

SUPERFLUID DYNAMICS AND ENERGY DISSIPATION IN NEUTRON STARS

M. ALI ALPAR

Sabancı University, Tuzla 81474, Istanbul, Turkey

Abstract. We review the basic properties of the neutron superfluid and the proton superconductor in the neutron star, and of the quantized neutron vortices and proton flux lines. Energy must be dissipated in the rotational dynamics of neutron stars, at rates that yield observable luminosities of dim thermal neutron stars.

1. Introduction

A superfluid is a system in which a macroscopic fraction of the total number of particles occupy a single quantum mechanical state. Single particle excitations out of this ground state face an energy gap. For fermions the free particle ground state, the Fermi sphere, is unstable to a ground state consisting of Cooper pairs in the presence of interactions. At low enough temperatures systems of interacting fermions are in the superfluid phase characterized by this ground state. At finite temperatures below the superfluid transition temperature a fraction $e^{-\Delta/kT}$ of the particles are in excited states. A superfluid is described as a two fluid system, the particles in the ground state constituting the superfluid component and the particles in the excited states, with closely spaced energy levels and therefore allowed collisions and dissipative processes, constituting the normal fluid component. In neutron stars the normal fluid components of the neutron and proton superfluids are negligible because the temperatures are far below the energy gap Δ .

In the neutron star the neutrons are superfluid at all densities above the neutron drip density, $\rho \geq 4 \times 10^{11}$ g cm^{-3}, the density in the crust of the neutron star above which neutrons start to populate extended states. The superfluidity is caused by attractive nuclear interactions at the av-

C. Kouveliotou et al. (eds.), The Neutron Star – Black Hole Connection, 57–70.

erage interparticle spacing. Proton superfluidity (superconductivity) arises also from the attractive strong interactions between protons. These become effective at higher total densities, in the neutron star's core where the protons are also in the extended states but they make up a small fraction of the baryons, of the order of 5% .

Superconductivity and superfluidity do not effect the equation of state, the energy difference per particle between the superfluid and normal states being of order Δ^2/E_F , which is much less than the Fermi energy E_F. Superfluidity and superconductivity do have characteristic and dominant effects on the dynamical, magnetic, thermal and transport properties of the neutron star. This lecture concentrates on some of the basic dynamical properties. Excellent general introductions to superconductivity and superfluidity are to be found in Richard Feynman's books, "Lectures on Physics, Vol. III" and "Statistical Mechanics" (Feynman 1964, 1972). An (also excellent) introduction to neutron star superconductivity and superfluidity is provided in the lecture by Jim Sauls published in the proceedings of the first of this ASI series, "Timing Neutron Stars", (Sauls 1989).

2. Constraints and Conservation Laws

Minimizing the free energy F = E - TS gives the equilibrium properties of a system in thermodynamic equilibrium. This is minimizing the energy while taking into account the likeliest ways of distributing the energy into microscopic states of the system. The density of states is an intrinsic constraint here, representing the physics of the particular system. Hence the form of the free energy, where the energy E is supplemented by the term -TS, with the temperature T acting as the Lagrange multiplier and S, the entropy, representing the intrinsic constraint.

The neutron star superfluids are, to a very good approximation, at T = 0, in the sense that the physical temperatures extending to 10^9 - 10^{10} K in the cores of young neutron stars, are still very low compared to all physical energy scales, in particular the neutron Fermi energy. In the approximation of effectively zero temperature, the fraction of particles in excited states is negligible (though for thermal properties of the superfluids, for ordinary viscosity, transport and dissipative processes this normal fluid component must be taken into account). Taken as a zero temperature system, the neutron star contains superfluid neutrons, superconducting protons and normal matter, which includes electrons throughout the star, all matter in the outer crust, and a crystal of nuclei in the inner crust, where some of the neutrons, together with all protons , are bound in these nuclei. Excess neutrons in the inner crust are in extended states, making up part of the neutron superfluid. The neutron superfluid carries most of the moment of inertia of the star and plays the determining role in its rotational dynamics. The proton superconductor in the core of the star, coupling to the crust through the normal electrons in the core, plays a dominant role in determining the magnetic properties of the star as well as its rotational dynamics. To understand the rotational and magnetic properties of the neutron star superfluids, let us start with the general thermodynamic arguments on minimizing the free energy of a zero temperature system under the constraints of constant angular momentum and of constant magnetic moment.

3. Rotation of the Neutron Superfluid

Under the constraint of constant total angular momentum $\mathbf{J}$, the free energy to be minimized is:

$$E' = E - \mathbf{\Omega}.\mathbf{J}. \tag{1}$$

The velocity distribution that minimizes this free energy is rigid body rotation

$$\mathbf{v}(\mathbf{r}) = \mathbf{\Omega} \times \mathbf{r} \tag{2}$$

where the uniform rotation rate $\mathbf{\Omega}$, which appears in the free energy as a Lagrange multiplier accompanying the angular momentum constraint, turns out to be

$$\mathbf{\Omega} = \mathbf{J}/I_{\text{total}}. \tag{3}$$

In rigid body rotation the velocity field has a uniform curl $\nabla \times \mathbf{v} = 2\mathbf{\Omega}$.

On the basis of this general thermodynamic argument, a superfluid, like any other system, would have minimum energy in a state of rigid body rotation. But can a superfluid reach a state of rigid body rotation? Since in the superfluid ground state a macroscopic number of particles (at T = 0, all particles) are described by a single quantum mechanical wave function

$$\psi(\mathbf{r}) = \psi_0(\mathbf{r}) \, e^{[i\theta(\mathbf{r})]} \tag{4}$$

the gradient of the phase of this function determines the particle current and macroscopic fluid velocity field $\mathbf{v}(\mathbf{r})$:

$$\mathbf{v}(\mathbf{r}) = (\hbar/M) \, \nabla\theta(\mathbf{r}). \tag{5}$$

(For the neutron superfluid, M = 2m, the mass of a Cooper pair of neutrons.) Thus superfluid flow is curl-free potential flow, $\nabla \times \mathbf{v} = 0$, unless there are singularities in θ. Strictly speaking, a superfluid cannot reach the exact rigid body rotation state. A superfluid does manage to reach a minimum free energy state with a macroscopic velocity field very close to rigid body rotation almost everywhere by forming an array of singular structures called quantized vortex lines. Consider a line singularity on a line parallel to the rotation axis; around the corresponding singular point on a perpendicular plane, the phase θ should change by $2\pi n$, where n is an integer, as the polar angle ϕ around the singularity changes by 2π in order that the wave function is single valued. Thus

$$\theta = n\phi \tag{6}$$

$$\mathbf{v}(\mathbf{r}) = (\hbar/2m)\,\nabla\theta = (n\hbar/2m)\,\mathbf{e}_\phi/\,r. \tag{7}$$

$$\oint \mathbf{v}\cdot\mathbf{dl} = nh/2m \equiv n\kappa, \tag{8}$$

$$\nabla\times\mathbf{v} = n\kappa\,\delta(\mathbf{r}). \tag{9}$$

The curl of the velocity (the vorticity) vanishes everywhere except on these vortex singularities. Each vortex can carry n units of the vorticity quantum $\kappa = 2\times10^{-3}$ cm^2 s^{-1}. The vortex is actually a cylindrical structure with a core radius ξ, called the superfluid coherence length. Within the vortex core many neutrons are in the normal state even at T = 0. In order to estimate ξ one notes that at radii smaller than ξ the kinetic energy per neutron of the superfluid circulation around the vortex line is larger than the condensation energy per neutron of the superfluid phase, E_c, so that it is energetically favorable to have particles in the normal phase, not taking part in the circulation associated with the quantum mechanical phase of the superfluid state. Thus, at the core radius ξ

$$E_c = 3/8\,\Delta^2/E_F = 1/2 \quad mv(\xi)^2 = \kappa^2/8\pi^2\xi^2. \tag{10}$$

An exact calculation in BCS theory gives

$$\xi = (2/\pi)(E_F/k_F\Delta) \tag{11}$$

where E_F, k_F and Δ are the Fermi energy, Fermi wavenumber and the superfluid's energy gap respectively.

A distribution of many vortices, with an area density n(r) at distance r from the rotation axis will give a macroscopic velocity field $\mathbf{V}(\mathbf{r})$. The velocity field is axially symmetric with circulation:

$$\oint \mathbf{V}(\mathbf{r})\cdot\mathbf{dl} = \int \nabla\times\mathbf{V}\cdot\mathbf{dS} = \int \sum_i \delta(\mathbf{r}-\mathbf{r}_i)\kappa\cdot\mathbf{dS} = \kappa\int n(r)2\pi rdr. \tag{12}$$

Defining the rotation rate, $\Omega(r)\,r = V(r)$, we obtain the general relation between the rotation rate (or the vorticity, which is the left hand side in the equation below) and the axially symmetric vortex density n(r)

$$2\Omega + r\partial\Omega/\partial r = \kappa n(r). \tag{13}$$

For rigid body rotation this reduces to

$$2\Omega = \kappa n_o. \tag{14}$$

A superfluid can thus reach its state of minimum free energy, the state of macroscopic rigid body rotation, simply by setting up an array of quantized vortex lines at uniform density n_o. Microscopically the velocity field deviates from rigid body rotation only very close to a particular vortex line, where the local circulation set up by that vortex line (Equation (7)) dominates. The energy of this state is only slightly higher than the energy of the classical rigidly rotating fluid, by the energy cost of the vortex lines and the local velocity fields around each vortex line. As the student can verify easily, this difference is truly negligible compared to the gain in energy afforded by this closest possible approach to rigid rotation. The rigid rotation state was observed and the array of vortex lines was imaged and photographed in rotating superfluid helium 4 in some beautiful experiments performed in the 1950s and 1960s. The average spacing $l_v \equiv (n_o)^{-1/2}$ between vortex lines is 3×10^{-3} cm in a neutron star rotating at $\Omega = 100$ rad/s.

4. Rotation and Magnetic Field of the Proton Superconductor

For protons in the superconducting state, the relation between the macroscopic velocity and the phase χ of the wave function is the canonical relation for charged particles:

$$\mathbf{v} = (\hbar/2m_p)\nabla\chi - (e/m_p c)\mathbf{A} \tag{15}$$

where e and m_p are the charge and mass of the proton, c is the velocity of light and $\mathbf{A}$ is the vector potential. A charged superfluid can rotate without vortex singularities. Take $\nabla\chi = 0$ everywhere. Then

$$\mathbf{v} = -(e/m_p c)\mathbf{A}. \tag{16}$$

The rigid rotation state $\mathbf{v} = \Omega \times \mathbf{r}$, needed to minimize the free energy under the constraint of angular momentum conservation, is obtained simply by admitting a uniform vector potential

$$\mathbf{A} = -(m_p c/e)\Omega \times \mathbf{r} \tag{17}$$

and the associated magnetic field

$$\mathbf{B}_L = -2m_p c/e \; \Omega \tag{18}$$

throughout the superconductor (the core regions of the neutron star where protons are superconducting). This magnetic field, called the London field, is set up by currents in the boundary of the superconducting core. Its energy cost is negligible in comparison to the gain in rotational energy achieved by rigid body rotation. The cyclotron frequency corresponding to the London field is twice the rotation frequency, so that for a neutron star rotating at 100 rad/s the London field is only 2×10^{-4} G. One might ask "What about the Meissner effect?" . A superconductor at rest expels magnetic fields (the Meissner effect) because, as Equation (15) shows, the presence of a magnetic field in the superconductor requires a nonzero kinetic energy. In the case of a rotating system, the constraint of conserved angular momentum actually requires a specific amount of kinetic energy, that of the rigid body rotation, in order to minimize the free energy. So, the Meissner effect pertains to superconductors at rest, and not to rotating superconductors.

The intrinsic dipole magnetic fields of neutron stars are of the order of 10^9 - 10^{12} G. These fields have to be accommodated in the superconducting proton core for reasons of minimizing the free energy, analogous to the arguments on minimizing the rotational free energy under constant angular momentum. The magnetic moment $\mathbf{M} = \int \mathbf{B}(\mathbf{r}) \, d^3r$ is a conserved quantity, approximately a constant changing only on evolutionary timescales of the neutron star (or its progenitor) to the extent allowed by the high conductivity of the stellar material. The free energy to be minimized under the constraint of constant $\mathbf{M}$ is:

$$E' = E - \mathbf{B} \cdot \mathbf{M}. \tag{19}$$

When the free energy is minimized, the vector $\mathbf{B}$, introduced here as a Lagrange multiplier, of course turns out to be a uniform magnetic field. In the state of minimum free energy the magnetic field is uniform throughout the system such that $\mathbf{M} = \mathbf{B} \times$ (volume). For a given constant magnetic moment the minimum free energy state is that in which the magnetic moment is distributed as a uniform field $\mathbf{B}$. But to introduce this uniform field throughout the superconductor without any singularities would give an induced velocity to the superconductor, according to Eq.(16), corresponding to rigid rotation at half the cyclotron frequency! This brings in a kinetic energy, and the total free energy is far from being minimum. What is needed is a magnetic field configuration that minimizes the magnetic free energy without introducing a velocity field (beyond the rigid rotation of the star required for minimum rotational free energy and set up by the independent London field). To have a solution with $\mathbf{v} = 0$ corresponding to the magnetic field configuration that meets the constraint of total magnetic moment $\mathbf{M}$, we must have

$$\mathbf{A} = (\hbar c/2e)\nabla\chi \tag{20}$$

according to Eq.(15). The rest of the argument is parallel to the argument for quantized neutron vortices required for rotating the neutron superfluid.

Since $\nabla \times \mathbf{A} = \mathbf{B}$, the phase gradient $\nabla \chi$ must also have a nonzero curl, which means that $\nabla \chi$ must be singular. In terms of polar coordinates ϕ and r around the singularity

$$\chi = n\phi \tag{21}$$

$$\oint \mathbf{A} \cdot \mathbf{dl} = nhc/2e \equiv n\Phi_o, \tag{22}$$

$$\mathbf{B} = \nabla \times \mathbf{A} = n\Phi_o \delta(\mathbf{r}). \tag{23}$$

This structure is a quantized flux line. $\Phi_o = hc /2e = 2 \times 10^{-7}$ Gauss-cm^2 is the fundamental quantum of flux. A uniform density array of these quantized flux lines, with n = 1 to minimize the energy contribution of the line structure itself, sets up a uniform macroscopic $\mathbf{B}$ field and the corresponding macroscopic vector potential $\mathbf{A} = 1/2 \ \mathbf{B} \times \mathbf{r}$. This state differs from the exact minimum of magnetic free energy, the uniform $\mathbf{B}$ state microscopically only very near one of the quantized flux lines, within the length scales flux line core, which are the proton superfluid coherence length ξ and the commensurate but independent electromagnetic length scale, the London length $\lambda = c / \omega_p$, where ω_p is the proton plasma frequency. The energy is only slightly higher than the free energy of the exact uniform $\mathbf{B}$ state, due to the energy cost of the local structures of the flux lines, which are formed at an energy cost that is only a small fraction of the energy gain achieved by mimicking a uniform $\mathbf{B}$. At a magnetic field of 10^{12} Gauss in the neutron star the average spacing between the flux lines is about 5×10^{-10} cm. The flux line array is much denser than the vortex line array. As the flux lines are parallel to the magnetic axis of the star while neutron vortex lines are parallel to the rotation axis, these two arrays should intersect each other during spindown or spinup, which require motion of the vortex lines, or for magnetic field decay which requires migration of the flux lines. How the flux line–vortex line interactions effect the neutron star's evolution is a fundamental issue that has attracted much attention since the initial suggestion of Srinivasan et al. (1990).

5. Spindown

The neutron superfluid spins down through radially outward motion of the quantized vortices. In cylindrical symmetry, for a circle of radius R from the rotation axis, we find

$$\partial/\partial t \oint \mathbf{v} \cdot \mathbf{dl} = \partial/\partial t \int n\kappa dS = - \int (\nabla \cdot \mathbf{j}_{vor}) dS = - \oint j_{vor,r} dl \tag{24}$$

employing vortex number conservation. Here $\mathbf{j}_{vor}$ is the vortex current, $j_{vor,r}$ $= \kappa \ n \ V_r$ is the vortex current in the radial direction away from the rotation axis, V_r being the mean vortex speed in the radially outward direction. Thus the equation of motion governing the spindown of a rotating superfluid with macroscopic rotation rate Ω is

$$2\pi R^2 \partial\Omega/\partial t = -\kappa n V_r 2\pi R \tag{25}$$

$$\dot{\Omega} = -\kappa n V_r/R. \tag{26}$$

The motion of the vortex lines depends on their interaction with normal matter, in particular with the electrons in the neutron star core, and the lattice of nuclei with which the neutron superfluid coexists in the inner crust of the neutron star. The equation of motion for individual vortex lines, the Magnus equation, shows that the vortex lines move in the radial direction in response to azimuthal drag forces. The resulting mean radial velocity V_r of the vortex lines depends on the angular velocity lag $\omega = \Omega - \Omega_c$ between the superfluid's rotation rate Ω and the crust and normal matter rotation rate Ω_c. Note the difference from the spindown of a normal viscous fluid under the Navier-Stokes equation, where the viscous term is the contribution of the divergence of the vorticity current to the spindown. For the normal viscous fluid the vorticity is a continuous (distributed) quantity and not lumped in quantized carriers like the vortex lines of the superfluid. The vorticity current in the normal fluid is simply proportional to the viscosity times the local gradient of the rotation rate, $j_{vor,r} \sim -\nu\nabla\Omega_c$. In the superfluid the vortex current deoends on the difference of rotation rates between the superfluid and the normal fluid.

The crust and normal matter (electrons) in the star's core approximately corotate on observed timescales of neutron star spindown. The physics of the normal matter–vortex line coupling, reflected through the dependence of $V_r(\omega)$ on $\omega = \Omega - \Omega_c$ defines the nature of the spindown effected through coupled equations for the crust and normal matter and for the superfluid:

$$I_c \dot{\Omega}_c = N_{ext} - I_s \dot{\Omega} = N_{ext} + I_s \,\kappa n\, V_r(\Omega - \Omega_c)/r \tag{27}$$

$$\dot{\Omega} = -\,\kappa n\, V_r(\Omega - \Omega_c)/r. \tag{28}$$

The total external torque on the neutron star is denoted by N_{ext}. I_c and I_s are the moments of inertia of the crust-normal matter and superfluid components of the star respectively. Observations indicate that the crust and the superfluid core of the neutron star corotate on the shortest resolvable timescales of glitches and postglitch relaxation, as the moment of inertia I_c is seen to contain effectively the entire moment of inertia of the star, the bulk of which resides in the superfluid core neutrons. This tight coupling is understood in terms of a superfluid effect of spontaneous magnetization of the neutron vortex lines by superconducting proton drag currents (Alpar, Langer and Sauls 1985). $I_s / I_c \sim 10^{-2}$ according to evaluations of postglitch timing observations with model equations (27) and (28), which is taken to imply that the superfluid responsible for glitches and postglitch dynamics is the superfluid in the inner crust of the neutron star. For simplicity we illustrate here a neutron star with a single superfluid component. Actually there are several different superfluid parts with moments of inertia I_i and rotation rates Ω_i. Each of these interacts with the crust-normal matter component through Eq. (28). The character of the spindown is determined by the functional dependence of V_r on $\Omega - \Omega_c$, which reflects the physics of the vortex normal matter interactions.

Equations (27) and (28) lead to a steady state in which both components spin down at the same rate, determined by the external torque:

$$\dot{\Omega} = \dot{\Omega}_c = N_{ext}/(I_s + I_c) \equiv \dot{\Omega}_\infty. \tag{29}$$

The value of $V_r(\Omega - \Omega_c)$ to achieve steady state spindown is determined through Eq.(28). This rate of vortex flow is set up by the steady state lag ω_∞ between the rotation rates Ω and Ω_c :

$$\dot{\Omega}_\infty = -\kappa n V_r(\omega_\infty)/r \cong -2\Omega_o V_r(\omega_\infty)/r. \tag{30}$$

The steady state lag is found by inverting Eq(30); usually the approximation of a uniform vortex density corresponding to a reference value Ω_o of the superfluid rotation rate is sufficient. From any initial conditions the system will evolve asymptotically towards steady state, and will remain in the steady state unless perturbed, since both components are spinning down at the same rate sustaining the required lag. Models of postglitch relaxation are solutions for the response of this dynamical model to perturbations introduced by the glitch. The location of the crust superfluid justifies replacing the radius coordinate r with the star's radius R in Eqs. (27),(28) and (30).

6. Energy Dissipation

The exchange of angular momentum between the superfluid and the crust-normal matter components of the star entails energy dissipation. The energy dissipation rates are essentially determined by the steady state lag. Multiplying Eq.(27) by Ω_c we find that:

$$N_{ext}\Omega_c = I_c\,\Omega_c\,\dot{\Omega}_c \,+\, I_s\,\Omega\,\dot{\Omega} \,+\, \dot{E}_{diss} \tag{31}$$

where the rate of energy dissipation is

$$\dot{E}_{diss} = I_s\,\omega\,|\dot{\Omega}| \cong I_s\,\omega_\infty\,|\dot{\Omega}|_\infty \tag{32}$$

where the last expression gives the steady state energy dissipation rate. As variations of $\dot{\Omega}_c$ are of order $I_s / I_c \sim 10^{-2}$, average observed crust spindown rates can be used for $\dot{\Omega}$. The energy dissipation rate depends on the nature of the dynamical coupling between the superfluid and normal components through the steady state value of the lag ω. There are two opposite regimes. If $V_r(\omega)$ is linear in the lag ω, as is expected in the neutron star core superfluids (Alpar, Langer and Sauls 1985), in the weakest vortex pinning situations in the crust (Alpar, Cheng and Pines 1989) and for vortex-phonon coupling in the crust superfluid (Jones 1990), then Eq(28) has the form:

$$\dot{\Omega} = -(\Omega - \Omega_c)/\tau \qquad (33)$$

where τ is a relaxation time reflecting the linear dependence of $V_r(\omega)$. The perturbations in Ω and Ω_c then relax exponentially. The steady state lag and energy dissipation rate are:

$$\omega_\infty = |\dot{\Omega}|_\infty \, \tau \qquad (34)$$

$$\dot{E}_{diss} \cong I_s(|\dot{\Omega}|_\infty)^2\tau. \qquad (35)$$

With linear coupling the observed postglitch relaxation times and inferred moments of inertia give negligible energy dissipation rates: $\dot{E}_{diss} \sim 10^{29}$ erg/s for the Vela pulsar, which is below the observed thermal luminosity of this pulsar due to cooling. $\dot{E}_{diss}$ is much lower for older pulsars due to the dependence on the square of the spindown rate. Limits on the coupling time of the neutron star core superfluid give even smaller $\dot{E}_{diss}$ for the neutron star core superfluids.

Higher energy dissipation rates which can be observable in neutron stars after their initial cooling will result from nonlinear dynamical coupling. For all glitching pulsars with measured second derivatives of the rotation rate these second derivatives scale with the behaviour of the Vela pulsar in relation to glitch magnitudes in a simple way that points at nonlinear response to the glitch. Let us first examine what nonlinear coupling entails. The superfluid in the inner crust of the neutron star coexists with a lattice of normal matter nuclei which provides pinning centers, local centers of attraction for the vortex line cores. Vortex lines move through the lattice in the presence of pinning centers and energy barriers through thermal activation or quantum tunnelling. The vortex current has an essentially exponential dependence on the lag ω for both of these modes of vortex motion; indeed a linear dependence on the lag obtains only for weak pinning constants and very small values of the steady state lag.

Thermally activated vortex motion has an average expected radial vortex velocity

$$V_r(\omega) \cong V_0 \, e^{-E_p/kT} \, e^{E_p\omega/kT\omega_{cr}} \qquad (36)$$

where V_0 is a microscopic vortex velocity, E_p a typical pinning energy, and ω_{cr} is a critical lag at which the vortex lines will become unpinned. While there are questions to be resolved about the possibility of vortex pinning, unpinning and creep in the neutron star crust (Jones 1997, 1998), we will use it here as a simple generic model of nonlinear coupling. The steady state lag ω_∞ is related to the steady state spindown rate $\dot{\Omega}$ through

$$\dot{\Omega}_\infty \cong - (2\Omega V_0/r) \exp(-E_p/kT) \exp(E_p\omega_\infty/kT\omega_{cr}) \qquad (37)$$

$$\omega_\infty = \omega_{cr}[1 - (kT/E_p) \ln(2\Omega V_0/r|\dot{\Omega}_\infty|)\,] \cong \omega_{cr}. \qquad (38)$$

In a glitch the rotation rate of the crust and normal matter component increases by $\Delta\Omega_c$ while the rotation rate of the superfluid decreases by $\delta\Omega$. The lag decreases from its steady state value ω_∞ by an amount

$$\delta\omega = \Delta\Omega_c + \delta\Omega. \qquad (39)$$

After a glitch the vortex flow current and superfluid spindown is cut down by the exponential factor $\exp(-E_p/kT\,\delta\omega/\omega_{cr})$. The superfluid spindown practically stops, and the superfluid is temporarily uncoupled from the effect of the external torque. The crust therefore spins down at an increased rate. The crust's spindown eventually restores the lag to its steady state value, after a waiting time

$$t_g = \delta\omega/|\dot{\Omega}|_c. \qquad (40)$$

The signature of this behaviour would be a step-like increase in the observed spindown rate of the neutron star,

$$\dot{\Omega}_c = \dot{\Omega}_\infty + \Delta\dot{\Omega}_c = [\,1 + (I_s/I_c)\,]\,\dot{\Omega}_\infty \qquad (41)$$

with the spindown rate returning to the steady state value after the waiting time t_g. What is observed in the Vela pulsar between glitches (Alpar et al 1993) is not a step-like offset that persists for an interval t_g, but a triangle shaped return to steady state:

$$\Delta\dot{\Omega}_c(t) = (I_s/I_c)\dot{\Omega}_\infty\,[\,1 - t/t_g\,]. \qquad (42)$$

The completion of this response at $t = t_g$, the return to the pre-glitch spindown trend, signals the occurrence of the next glitch with about 20% accuracy in the Vela pulsar, repeatedly for all interglitch intervals. This behaviour seems to be universal. In all other glitching pulsars with measured second derivatives of the rotation rate (Shemar and Lyne 1996) the second derivatives scale with glitch parameters in accordance with the model developed for the Vela pulsar (Alpar 1998). There is a simple and natural explanation, as the nonlinear response of superfluid regions throughout which vortices have unpinned catastrophically during the glitch. All such regions, with total moment of inertia I_s, suffer a cutback in the continuous vortex current V_r and superfluid spindown $\dot{\Omega}$ as nonlinear response to the sudden change in superfluid rotation rate brought about by the vortex un- pinning at the glitch. This results in a steplike change in the observed crust spindown rate, as in Eq.(41). However, the glitch induced change $\delta\Omega(r)$ varies with position within the superfluid regions that suffered the unpin- ning event at the glitch. Thus different pieces of moment of inertia have different waiting times before rejoining the steady state vortex flow and spindown conditions. This leads to a stacked response of many step-like offsets. The triangle response observed corresponds to a uniform average density of unpinned vortices, the lowest order and dominant part of their distribution in r. Local deviations from the average are likely to be re- sponsible for the observed small deviations from the triangle response in time.

The triangle response means a constant second derivative $\ddot{\Omega}_c$ that scales with glitch parameters as predicted with the model. This behaviour seems to be universal among glitching pulsars. This can be taken as evidence for the presence of nonlinear coupling. For nonlinear response, Eq (32) and (38) lead to an energy dissipation rate

$$\dot{E}_{diss} = I_s\omega_\infty|\dot{\Omega}|_\infty \cong I_s\omega_{cr}|\dot{\Omega}|_c. \tag{43}$$

From glitch observations I_s is inferred to be of the order of 10^{43} g cm^2. The decrease in the superfluid rotation rate $\delta\Omega$ is inferred to be of the order of 10^{-2} rad/s . This must be less than the steady state or critical values of the lag ω between superfluid and normal matter rotation rates. Thus we obtain a lower bound on the energy dissipation rate in a neutron star under spindown by an external torque. An upper bound can be obtained from the detected thermal luminosity of radio pulsars that are old enough to have cooled away their initial heat content so that their current thermal energy source is dynamical energy dissipation inside the star as we consider here. Such analysis of EXOSAT data from PSR 1929+10 yields an upper limit of 1 rad/s for ω_{cr} (Alpar et al 1987). This is in agreement with ROSAT and ASCA observations of PSR 1929+10 and PSR 0950+08 (Yancopou- los, Hamilton and Helfand 1994; Manning and Willmore 1994; Wang and Halpern 1997). The energy dissipation rate in a neutron star with spindown rate $|\dot{\Omega}|$ (rad s^{-2}) is bounded between the limits

$$10^{41}\,|\dot{\Omega}|\ \text{erg/s} \leq \dot{E}_{diss} \leq 10^{43}\,|\dot{\Omega}|\ \text{erg/s}. \tag{44}$$

The energy dissipation due to nonlinear coupling will determine the thermal luminosity of pulsars beyond ages of a few times 10^5 yrs, after the initial cooling luminosity starts to decrease rapidly with the start of the photon cooling era. The energy dissipation will continue after the end of the pulsar activity. There is no evidence that the magnetic dipole moments of isolated neutron stars decay. The star will continue to spin down because of the dipole radiation torque. The energy dissipation may be the sole means of detecting isolated neutron stars which are not active as pulsars. Scaling from the Vela pulsar, the dipole spindown rate of a neutron star with a magnetic field in the 10^{12} Gauss range is

$$|\dot{\Omega}|_{-13} \sim (t_6)^{-3/2} \tag{45}$$

where the age of the star is t_6 million years, and the spindown rate is given in units of 10^{-13} rad s^{-2}. The lower limits to the energy dissipation rate-thermal luminosity and surface blackbody temperature of a neutron star of age 10^6 yrs are 10^{28} erg/s and 6×10^4 degrees, a very dim luminosity in the soft x-ray range. Taking the number of neutron stars of age t years to be 10^{-2} t, there are about ten thousand such stars in the Galaxy. There are 10^5 neutron stars of age 10^7 years, with dissipation luminosities greater than 3×10^{26} erg/s and surface blackbody temperatures exceeding 2.4×10^4 ; 10^6 neutron stars of age 10^8 years, with dissipation luminosities greater than 10^{25} erg/s and surface blackbody temperatures exceeding 10^4 and 10^7 neutron stars of age 10^9 years, with dissipation luminosities greater than 3×10^{23} erg/s and surface blackbody temperatures exceeding 4×10^3, in the optical. With increasing numbers, there may be chance nearby candidates for detection, but these objects are excessively dim and soft radiators.

Energy dissipation is the likely agent to power the thermal luminosities of the dim neutron star candidates RXJ 185635–3754 (Walter et al. 1996) and RXJ 0720.4–3125 (Haberl et al. 1997) discovered by ROSAT. These are very nearby sources, and are therefore likely to be members of an abundant old population of neutron stars. This argues against initial cooling of young neutron stars as the source of the luminosity. The spindown rates inferred from the energy dissipation rates are between a few times 10^{-12} and 10^{-10} rad s^{-2}. This is the first instance of estimating neutron star spindown rates from observations of thermal luminosity. If these neutron stars are spinning down through dipole radiation torques, they would be rather young radio pulsars, which is not very likely in view of their proximity. A young age, together with the realtively long 8.4 s pulse period of RXJ 0720.4–3125 would imply that this source is a magnetar. Alternatively, an old neutron star with an ordinary 10^{12} Gauss magnetic field can spin down at these rates through the propeller mechanism if it is surrounded by material that cannot accrete because of the centrifugal barrier. Dynamical energy dissipation in neutron stars may turn out to be a useful first indicator of this fascinating astrophysical process.

Acknowledgements

This research has been supported by the Turkish Academy of Sciences, and by the Scientific & Technical Research Council of Turkey, TÜBİTAK, under the grant TBAG-Ü18. I thank Joseph Ventura and the local organizers for support and hospitality at this inspiring ASI, and my previous institution, Middle East Technical University, for travel support.

References

Alpar, M.A. (1998), in R. Buccheri, J. van Paradijs, and M.A. Alpar (eds.), *The Many Faces of Neutron Stars*, Proc. NATO-ASI, Kluwer, p59.

Alpar, M.A., Brinkmann, W., Kızıloğlu, Ü., Ögelman, H. and Pines, D. (1987), *A&A* **177**, 101.

Alpar, M.A., Chau, H.F., Cheng, K.S., and Pines, D. (1993), *ApJ* **409**, 345.

Alpar, M.A., Cheng, K.S., and Pines, D. (1989), *ApJ* **346**, 823.

Alpar, M.A., Langer, S.A., and Sauls, J.A. (1985), *ApJ* **282**, 533.

Feynman, R.P. (1964), *Lectures on Physics*, Vol.III, Addison-Wesley.

Feynman, R.P. (1972), *Statistical Mechanics*, W.A. Benjamin.

Haberl, F., Motch, C., Buckley, D.A.H., Zickgraf, F.-J., and Pietsch, W. (1997), *A&A* **326**, 662.

Jones, P.B. (1990), *MNRAS* **243**, 257.

Jones, P.B. (1997), *Phys. Rev. Lett.* **79**, 792.

Jones, P.B. (1998), *Phys. Rev. Lett.* **81**, 4560.

Manning, R.A. and Willmore, A.P. (1994), *MNRAS* **266**, 635.

Sauls, J.A. (1989), in H. Ögelman and E.P.J. van den Heuvel (eds.), *Timing Neutron Stars*, Proc. NATO-ASI, Kluwer, p457.

Shemar, S.L. and Lyne, A.G. (1996), *MNRAS* **282**, 677.

Srinivasan, G., Bhattacharya, D., Muslimov, A.G., and Tsygan, A.I. (1990), *Curr. Sci.* **59**, 31.

Walter, F.M., Wolk, S.J., and Neuhäuser, R. (1996), *Nature* **379**, 233.

Wang, F. and Halpern, J.P. (1997), *ApJ* **482**, L159.

Yancopoulos, S., Hamilton, T.T., and Helfand, D.J. (1994), *ApJ* **429**, 832.

MAGNETIC FIELDS OF NEUTRON STARS

S. KONAR

Inter-University Centre for Astronomy & Astrophysics, Pune

AND

D. BHATTACHARYA

Raman Research Institute, Bangalore

Abstract. The evolution of the magnetic field is investigated for isolated as well as binary neutron stars. The overall nature of the field evolution is seen to be similar for an initial crustal field and an expelled flux. The major uncertainties of the present models of field evolution and the directions in which further investigation are required are also discussed in detail.

1. Introduction

There is no consensus regarding the generation of the magnetic field in neutron stars. The field could either be a fossil remnant from the progenitor star in the form of Abrikosov fluxoids of the core proton super-conductor (Baym, Pethick & Pines 1969; Ruderman 1972; Bhattacharya & Srinivasan 1995), or it could be generated after the formation of the neutron star in which case the currents would be entirely confined to the solid crust (Blandford, Applegate & Hernquist 1983). Evidently, the nature of the evolution would depend very much on the internal field configuration. Observations and statistical analyses of existing pulsar data, nevertheless, indicate that significant decay of magnetic field is achieved only if the neutron star is a member of an interacting binary (Bailes 1989; Bhattacharya 1991; Hartman et al. 1997).

The processes that are responsible for the field evolution in neutron stars in binaries are: a) expulsion of the magnetic flux from the super-conducting core during the phase of propeller spin-down, b) screening of the field by accreted matter and c) rapid ohmic decay of the crustal field in an accretion-

C. Kouveliotou et al. (eds.), The Neutron Star – Black Hole Connection, 71–76.
© 2001 *Kluwer Academic Publishers. Printed in the Netherlands.*

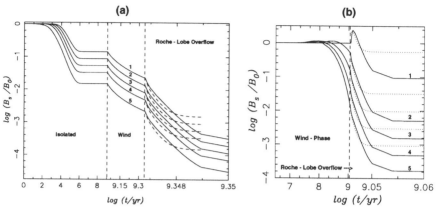

Figure 1.   Evolution of the surface magnetic field in LMXBs. **a)** For initial crustal currents with an wind accretion rate of $\dot{M} = 10^{-14}$ M$_\odot$ yr^{-1}. Curves 1 to 5 correspond to initial current configurations centred at $\rho = 10^{13}, 10^{12.5}, 10^{12}, 10^{11.5}, 10^{11}$ g cm^{-3}. All curves correspond to an impurity content, $Q = 0.0$. A standard cooling has been assumed for the isolated phase. The wind and Roche-contact phases are plotted in expanded scales. The dashed and the solid curves correspond to accretion rates of $\dot{M} = 10^{-9}, 10^{-10}$ M$_\odot$ yr^{-1} in the Roche-contact phase. **b)** For an expelled flux. The dotted and the solid curves correspond to accretion rates of $\dot{M} = 10^{-9}, 10^{-10}$ M$_\odot$ yr^{-1} in the Roche contact phase. The curves 1 to 5 correspond to $Q = 0.0, 0.01, 0.02, 0.03$ and 0.04, respectively. All curves correspond to a wind accretion rate of $\dot{M} = 10^{-16}$ M$_\odot$ yr^{-1}.

heated crust. Diamagnetic screening of the field by accreted matter does not seem likely to have any long-term effect (Konar 1997) and we shall exclude it from the present discussion. The other models invoke ohmic decay of the current loops for a permanent decrease in the field strength. In either case, the effect of accretion is two-fold. The heating reduces the electrical conductivity and consequently the ohmic decay time-scale inducing a faster decay. At the same time the material movement, caused by the deposition of matter on top of the crust, pushes the original current carrying layers into deeper and denser regions where the higher conductivity slows the decay down. The mass of the crust of a neutron star changes very little with a change in the total mass; accretion therefore implies assimilation of the original crust into the super-conducting core. When the original current carrying regions undergo such assimilation, further decay is stopped altogether. Both the purely crustal model as well as the model assuming an expelled flux have been investigated by many authors. The important difference between our work and that of the other investigators lies in our assumption of a *flux freezing* upon the assimilation of the original current carrying layers into the super-conducting core.

TABLE 1. Purely Crustal Field

system	final field and period	comment
isolated radio pulsars	high field, long period	no significant field decay in 10^9 years
HMXB	high field, long period	high-mass binary pulsars and solitary counterparts
	low field, long period	not active as pulsars
LMXB	high field, long period	high field low-mass binary pulsars and solitary counterparts
	low field, short period	low field low-mass binary pulsars and solitary counterparts, millisecond pulsars

2. Nature of Field Evolution

2.1. GENERIC FEATURES

The qualitative features of field evolution, as outlined below, are similar for a) an initial crustal field and b) an expelled flux.

Pure Ohmic Decay in Isolated Neutron Stars (Konar 1997)

1. A slow/fast cooling of neutron star implies a fast/slow decay; hence a low/high final field.

2. Initial crustal currents concentrated at lower/higher densities gives rise to low/high final surface fields.

3. Large impurity content makes the decay rapid and gives rise to smaller final fields.

Accretion-Induced Field Decay in Accreting Neutron Stars (Konar & Bhattacharya 1997 - KBI)

1. In an accreting neutron star the field undergoes an initial rapid decay, followed by slow down and an eventual *freezing*.

2. A positive correlation between the rate of accretion and the final field strength is observed, giving rise to higher final saturation field strengths for higher rates of accretion.

Magnetic Field and Spin Period

We have investigated the nature of the final 'magnetic field-spin period' combination. Our results agree well with the observations and are summarised in Table 1 (KBI; Konar & Bhattacharya 1999a - KBII). The nature of field evolution is similar for the model of spin-down induced flux

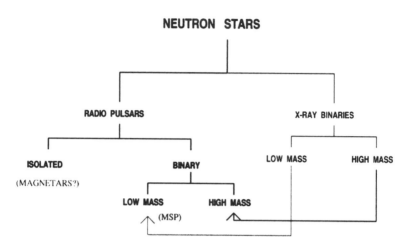

Figure 2. The arrows indicate the expected evolutionary link between X-ray binaries and binary radio pulsars.

expulsion (Konar & Bhattacharya 1999b - KBIII), though there is one major difference as can be seen from Figure 1. To produce millisecond pulsars in LMXBs for an expelled flux large values of impurities, in the prior-to-accretion original crust, are required, but this would result in extremely small surface fields in old isolated pulsars. This is in complete contrast to a purely crustal model and expectation from statistical analyses of pulsar data.

2.2. RANGES OF PHYSICAL PARAMETERS

The paradigm of field evolution that has emerged out of observations, statistical analyses and theoretical expectations has been summarized in Figure 2, where the connection between the radio pulsars and their binary counterparts, namely the X-ray binaries is indicated. In Table 2 we indicate the constraints on various physical parameters in the field evolution models placed by the requirement to match observed properties in a variety of systems (KBII, KBIII). The parameters discussed here are the density at which the initial crustal current distribution is located (ρ_c), the impurity strength in the crust (Q), the duration of wind-accretion phase in different binary systems and the rate of accretion in the Roche-contact phase for LMXBs.

TABLE 2. Constraints on Physical Parameters

parameter	model	system	requirement	parameter range
ρ_0	crustal	HMXB	high field	high ρ_0
Q	crustal	isolated radio pulsar	no decay over active pulsar life-time	$Q \lesssim 0.01$ for standard cooling, $Q \lesssim 0.05$ for accelerated cooling
	expelled flux	LMXB	millisecond pulsar generation	$Q \gtrsim 0.05$ with wind accretion, $Q \gg 1$ without wind accretion
duration of wind accretion	crustal	HMXB	high field	short
$\dot{M}$ in Roche-phase	crustal	LMXB	high field	Eddington rate

3. Uncertainties and Future Directions

The results and conclusions stated above suffer from a number of uncertainties regarding the micro-physics of the neutron star as have been listed below. Moreover, a lot of the new theoretical results as well as the observational facts have recently become available. In this section we mention some of the more important aspects that need to be incorporated in any future work on the evolution of the magnetic fields of neutron stars.

Thermal Behaviour

1. Isolated Phase - The present data can be made to fit scenarios with both a *slow* or an *accelerated* cooling. Therefore, it is not clear which is the correct cooling behaviour of an isolated neutron star.
2. Accreting Phase - The crustal temperature corresponding to a given rate of accretion has not been determined with any degree of certainty. Also, the existing results are limited in their scope.
3. Post-Accretion Phase - No calculation exists for the thermal behaviour of this phase at all.

Transport Properties

Several factors affect the transport properties and hence both thermal and magnetic field evolution. Prominent among them is the change in the **chemical composition** due to a) accretion and b) spin-down. Recently it has been shown that the impurity content of the accreted crust for near-Eddington accretion rates could be extremely large (Schatz et al. 1999).

This, along with a temperature inversion near or beyond the neutron drip (Brown 2000) might modify the transport properties significantly. Moreover the presence of dislocations, defects, non-spherical nuclei have so far not been taken into account in the calculation of transport properties. These are also expected to have an impact on the field evolution in isolated as well as in binary pulsars.

Multi-polar Structure

All of our and similar investigations have been based on an assumption of a pure dipolar model for the magnetic field. Though calculations for isolated neutron stars do not show any appreciable change in multi-polar structures (Mitra, Konar & Bhattacharya 1999) the situation would change in the presence of accretion or a very strong magnetic field (Geppert et al. 1999) due to the importance of the Hall term requiring further investigation.

The Magnetar Question

Amongst some of the more recent developments the magnetars pose a great challenge for the existing theories of field evolution since they require a very rapid field evolution in isolated neutron stars. Though some work has already been done in this area (Heyl & Kulkarni 1998; Geppert et al. 1999) more detailed investigation is needed.

References

1. Baiko, D. and Haensel, P. (1999), *Acta Phys. Polon.* **B30**, 1097.
2. Bailes, M. (1989), *ApJ* **342**, 917.
3. Baym, G., Pethick, C., and Pines, D. (1969), *Nature* **223**, 673.
4. Bhattacharya, D. (1991), in E. Ventura and D. Pines (eds.), *Neutron Stars : Theory and Observations*, Kluwer Academic Publishers, p219.
5. Bhattacharya, D. and Srinivasan, G. (1995), in W.H.G. Lewin, J. van Paradijs, and E.P.J. van den Heuvel (eds.), *X-Ray Binaries*, Cambridge University Press, p495.
6. Blandford, R.D., Applegate, J.H., and Hernquist, L. (1983), *MNRAS* **204**, 1025.
7. Brown, E.F. (2000), *ApJ* **531**, 988.
8. Geppert, U., Page, D., Colpi, M., and Zannis T. (1999), in N. Kramer, N. Wex, and R. Wielebinski (eds.), *Pulsar Astronomy - 2000 and Beyond*, ASP Conference Series, in press.
9. Hartman, J.W., Verbunt, F., Bhattacharya, D., and Wijers, R.A.M.J. (1997), *A&A* **322**, 477.
10. Heyl., J.S. and Kulkarni, S.R. (1998), *MNRAS* **300**, 599.
11. Konar S. (1997), *Ph.D. thesis*, Indian Institute of Science, Bangalore.
12. Konar, S. and Bhattacharya D. (1997), *MNRAS* **284**, 311.
13. Konar, S. and Bhattacharya D. (1999), *MNRAS* **303**, 588.
14. Konar, S. and Bhattacharya D. (1999), *MNRAS* **308**, 795.
15. Mitra, D., Konar, S., and Bhattacharya D. (1999), *MNRAS* **307**, 459.
16. Ruderman., M.A., (1972), *ARA&A* **10**, 427.
17. Schatz, H., Bildsten, L., Cumming, A., and Wiescher, M. (1999), *ApJ* **524**, 1014.

2. The Neutron Star – Black Hole Connection

NEUTRON STAR AND BLACK HOLE FORMATION

C.L. FRYER

Lick Observatory, University of California Observatories
Santa Cruz, CA 95064

Abstract. Although the primary formation scenario of black holes and neutron stars (the core-collapse of massive stars) has been known for 65 years, understanding the details of this process has been a long, uphill battle. Although a complete understanding of core-collapse has not been achieved, the past 30 years have seen much progress in this field. Here we present a review of the mechanism behind core-collapse and apply our knowledge to gain better insight into the formation of neutron stars and black holes.

1. Introduction

In 1934, Baade & Zwicky proposed a formation scenario for compact objects by arguing that supernovae were caused by the collapse of massive stars into neutron stars. This scenario is now believed to be the dominant formation mechanism of both black holes and neutron stars. Determining which stars actually *do* collapse to form these compact remnants has occupied both theorists and observers for the past 65 years. Theoretical models of stellar evolution have produced a general picture of compact object formation in which stars with masses above $\sim 8 \, M_\odot$ end their lives by collapsing into neutron stars or black holes (see Woosley & Weaver 1986 for a review). Observations of compact binaries and nucleosynthetic yields suggest that stars with mass greater than $\sim 25 \, M_\odot$ collapse to form black holes, whereas the lower mass stars ($8 - 25 M_\odot$) form neutron stars (see Fryer 1999 for a review). However, in all cases, tying the observational constraints to the theory requires a series of uncertain assumptions, and none of the evidence which determines the black hole mass limit is conclusive.

A more direct method of determining the black hole mass limit is to study the core-collapse mechanism itself. Unfortunately, understanding the

C. Kouveliotou et al. (eds.), The Neutron Star – Black Hole Connection, 79–88.
© 2001 *Kluwer Academic Publishers. Printed in the Netherlands.*

mechanism which extracts the gravitational potential energy released as a massive star's core collapses down to a 10 km neutron star has proven to be extremely difficult, requiring the implementation of a broad range of physics from neutrino interactions and transport to general relativity and to convection. Currently, no one has a model of the core-collapse of a massive star which contains all of this physics completely. However, the sophistication of the models has progressed significantly over the past 30 years, and core-collapse models can now dramatically increase our understanding of neutron star and black hole formation. Most reviews of this subject are limited to the glimpses obtained through stellar evolution models. Instead, this paper discusses the formation of black holes and neutron stars through the eyes of a core-collapse theorist. Current models of core-collapse not only allow us to estimate the mass limits of neutron star and black hole formation, but also to predict the mass distribution of these compact remnants.

2. Core-Collapse Supernovae

The understanding of neutron star and black hole formation has remained intimately linked to that of core-collapse (Type Ib, Ic, II) supernovae since the proposal of Baade & Zwicky (1934). Most neutron stars and black holes are the remnants of supernova explosions, and understanding supernovae is necessary to understanding the formation of these compact remnants. To the non-practitioner, the nature of core-collapse supernovae may seem to contain numerous contradictions and unsolved mysteries. This is primarily the result of poor communication between core-collapse theorists and the astronomy community and not because of major disagreements between core-collapse theorists themselves. Core-collapse theorists tend to be more interested in basic physics rather than the equally important comparisons to observation, and they have generally let others translate their results and make the observational comparisons. This attitude has allowed these "translators" to introduce many misconceptions about the nature of supernova explosions such as: a) more massive stars have larger supernova explosion energies (due to an erroneous assumption that the supernova explosion energy is proportional to the potential energy released in the core collapse), b) no neutrino-driven mechanism of core-collapse supernova works, and hence some piece of crucial physics must certainly be missing.

Theorists are more interested in the physics of core-collapse primarily because simulations have shown that the success or failure of neutrino-driven explosions depends upon many aspects of physics. At this time, no simulation has included all of the necessary physical details and no final answer is known, but the basic neutrino-driven mechanism seems to work.

Indeed, core-collapse simulations have progressed to an extent that we can make predictions about the formation of neutron stars and black holes. To make these predictions, however, we must first understand the core-collapse paradigm. The collapse of massive stars occurs when the densities and temperatures at the core are sufficiently high to cause iron to dissociate and electrons to capture onto protons, both of which reduce the pressure in the core. This sudden decrease in pressure causes the core to collapse nearly at free-fall, and the collapse stops only when nuclear densities are reached and nucleon degeneracy pressure once again stabilizes the core and drives a bounce shock back out through the core. The bounce shock stalls at roughly 100-400 km as neutrino emission and iron dissociation sap its energy. It leaves behind an entropy profile which is unstable to convection, and sets up a convective region from ~50 km out to ~400 km (Figure 1) capped by the accretion shock of infalling material. The convective region must overcome this accretion cap to launch a supernova explosion. The ram pressure of the infalling material plays a major role on the strength of supernova explosion, and hence, the compact remnant mass. The magnitude of the ram pressure, in turn, depends upon the structure of the initial massive star progenitor (Section 3).

Neutrino heating deposits considerable energy into the bottom layers of the convective region. If this material were not allowed to convect (which is the case for most of the 1-dimensional simulations), it would then re-emit this energy via neutrinos producing a steady state with no net energy gain. In the meantime, the increase in pressure as more material piles up at the accretion shock and the decrease in neutrino luminosity as the proto-neutron star cools make it increasingly difficult to launch an explosion. In the "delayed-neutrino" supernova mechanism (Wilson & Mayle 1988; Miller, Wilson, & Mayle 1993; Herant et al. 1994; Burrows, Hayes, & Fryxell 1995; Janka & Müller 1996, Mezzacappa et al. 1998, Messer et al. 1998), convection aids the explosion in two ways: a) as the lower layers of the convective region are heated, that material rises and cools adiabatically and converts the energy from neutrino deposition into kinetic and potential energy rather than re-radiating it as neutrinos, and b) the material does not simply pile up at the shock but instead convects down to the surface of the proto-neutron star where it either accretes onto the proto-neutron star providing additional neutrino emission or is heated and rises back up. Thus, convection both increases the efficiency at which neutrino energy is deposited into the convective region and reduces the energy required to launch an explosion by reducing the pressure at the accretion shock.

The fact that convection plays a role at all is an indication of how sensitive the core-collapse explosion mechanism is to the details. In astronomy, theorists often assume that the broad brush approach is sufficient to model

Spherically Symmetric Collapse

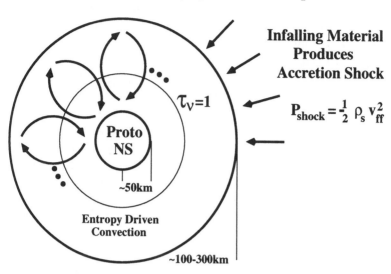

Rotating Collapse

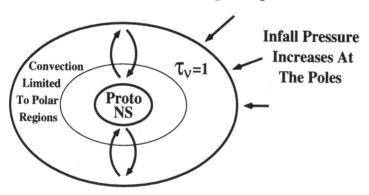

Figure 1. Entropy-driven convection has been found to enhance the energy deposition within the accretion shock. As the density of the infalling matter decreases, the convective region is able to push the shock outward, launching a supernova explosion. The angular momentum profile in rotating core-collapses stabilizes the convection along the equator.

nature. Nature, however, is amazingly good at proving theorists wrong. In the case of core-collapse models, there remains quite a bit of evidence that the pressure in the convective region is balanced very closely by that of the infalling ram pressure. If the shock moves inward, the ram pressure increases and the explosion tends to fizzle. Once the accretion shock begins to move outward, the ram pressure decreases dramatically and an explosion is immediate. But if the shock moves out too soon, the core masses will be too low to match the masses of the observed neutron stars ($\sim 1.4 M_\odot$). So the convective and ram pressures must balance for $\sim 0.2 - 0.8$s before driving an explosion. It is no surprise that this balance requires accurate calculations of the details.

2.1. UNCERTAINTIES

Core-collapse supernovae challenge many facets of physics, making the calculation of the details difficult (see Burrows 1999, Mezzacappa 1999, for a review). As the star collapses, it reaches densities and temperatures that exceed anything that can be studied on the earth. The behavior of this matter, and the production, scattering, and absorption of neutrinos in these conditions are some of the major uncertainties in the simulations. The densities are sufficiently high that neutrinos are trapped in the core, and much of the current work on core-collapse supernovae is concentrated on correctly modelling the transport of neutrinos out of the core. Since neutrinos are the medium through which gravitational energy is converted to explosion energy, the accurate transport of the neutrinos is essential. And, of course, convection has forced theorists to move to multi-dimensional simulations. Even the effects of general relativity must be included! Despite all of the uncertainties and effects, the basic picture of core-collapse supernovae can be coupled to massive progenitors to make some quantitative predictions about the formation of compact objects.

3. Massive Stars (aka Supernova Progenitors)

Let's now take a look at supernova progenitors through the eyes of a core-collapse theorist. The supernova explosion occurs within ~ 1s of the collapse of a massive star, before most of the star has even realized that the core collapsed. In fact, the structure of the inner $2-3\,M_\odot$ of a massive star alone decides the fate of the star, its explosion, and its resultant remnant. Winds, metallicity, binary companions, etc. do not effect the supernova explosion (and the formation of a neutron star or black hole) unless they change the structure of the core.

One of the current myths that has somehow become standard lore is that the mass of the iron core determines whether you form a neutron star or a black hole. Like many such old-wives' tales, the statement holds true, but the physical reason often gets lost in the generations of retelling. The physical reason can be understood through models of core-collapse. Recall from Figure 1 that, to launch an explosion, the convective region must overcome the pressure of the infalling matter. The pressure of this infalling matter depends upon the position of the shock and the density of the infalling material which is determined by the density structure of the

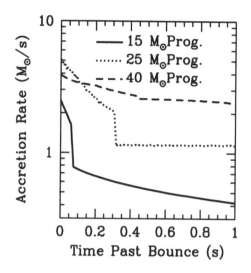

Figure 2.   Mass infall rates for 3 separate progenitor masses: 15,25,40 M$_\odot$. The mass infall rate for the 15 M$_\odot$ progenitor drops to 1/5th that of the 25 and 40 M$_\odot$ models in 100 ms. This allows it to explode sooner, leaving behind a smaller core. The infall rates of the 25 and 40 M$_\odot$ progenitors stay roughly the same for 300 ms past bounce, and hence their explosion energies are similar.

core just prior to collapse. More massive progenitor stars (with their higher iron core masses) have higher mass accretion rates at the shock (Figure 2), and it takes longer for the convective region to overcome the pressure at this accretion shock (Burrows & Goshy 1993). This delay affects both the explosion energy and the mass of the compact remnant.

3.1. REMNANT VS. PROGENITOR MASS

The cores of massive stars vary dramatically from 8 M$_\odot$ up to the most massive stars, which leads to a range of supernova explosion energies and compact remnant masses. Stars in the mass range of 8–11 M$_\odot$ mark the boundary between white dwarf formation and core-collapse. These stars tend to form low-mass cores (~ 1.39M$_\odot$) with very little mass to produce significant ram pressure (Timmes, Woosley, & Weaver 1996). In the collapse models of these stars (which closely resemble the core-collapse of accreting white dwarfs), neutrino heating revives the shock shortly after the shock stalls. The explosion occurs early, but it is weaker because it does not have time to build up energy in a convective regime (Figure 3). More massive stars (~ 15M$_\odot$) are delayed significantly and have strong explosions. Be-

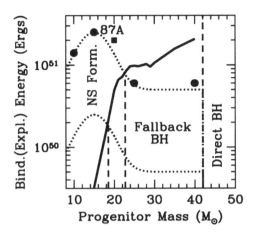

Figure 3. Binding energy (solid line) and explosion energy (dots) vs. mass of progenitor. This binding energy includes all but the inner 3 $M_\odot$ core of the star. If the energy available to unbind the core (dotted lines) is less than the binding energy, the compact remnant will exceed 3 $M_\odot$ and collapse to form a black hole. The explosion energy drops and the binding energy rises with increasing progenitor mass; their net effect is to create a fairly narrow range of uncertainty in the transition mass from neutron star formation and black hole formation from fallback. For reference, supernovae 1987A is placed on the figure (square).

yond $\sim 20 M_\odot$, the iron core increases dramatically as carbon is burned radiatively rather than convectively (Timmes et al. 1996). The entropy of these more massive cores is higher and the mass infall increases dramatically (Figure 2). The higher accretion ram pressure delays the explosion even further. Neutrino cooling begins to overcome heating in the convective region, and the resultant explosion is much weaker. Beyond $\sim 40 M_\odot$, it is likely that no explosion is produced at all.

Figure 3 summarizes the results of core-collapse simulations for a range of progenitor masses. The fit to the simulations (dotted line) shows the basic trends in the explosion energy. The solid line shows the binding energy of the outer layers of the star down to $3 M_\odot$ for comparison. If the explosion energy is below the binding energy (roughly at $22 M_\odot$), the remnant will exceed $3 M_\odot$ and form a black hole.

3.2. UNCERTAINTIES

There are a number of outstanding uncertainties in stellar evolution: e.g. the algorithm used to model mixing in the star, nuclear reaction rates, mass

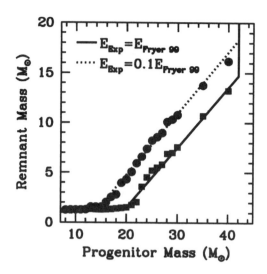

Figure 4.   Remnant mass vs. progenitor mass for unbinding energies set to 10% and 100% that of the explosion energy.

loss from winds, rotation, binary effects, and opacities. Particularly for the massive stars in the mass range from $8-11\,M_\odot$, the uncertainties in stellar evolution make it difficult to determine whether these stars form neutron stars or white dwarfs (see Timmes et al. 1996 for a review). Note that this narrow range in progenitor masses may produce nearly half the neutron stars in the galaxy. Rotation does not change the structure of the core significantly, but it weakens the bounce and constrains the convection. Thus, for a given progenitor mass, as the rotation speed increases, the supernova explosion energy decreases and the remnant mass increases (Fryer & Heger 2000). Whether or not mass loss from winds affects the core of massive stars is still a matter of debate. In fact, black hole masses are likely to be the best constraint on winds (Section 4).

4.   Compact Remnant Masses

Comparing the supernova explosion energies with their progenitor's binding energy allows us to estimate the resultant compact remnant mass for each progenitor (Figure 4). As the supernova shock progresses out of the star, its energy is used to unbind the star. The very energetic explosions of low mass progenitors ($< 20 M_\odot$) will eject the entire star, leaving behind only the proto-neutron star that powered the supernova shock. But with

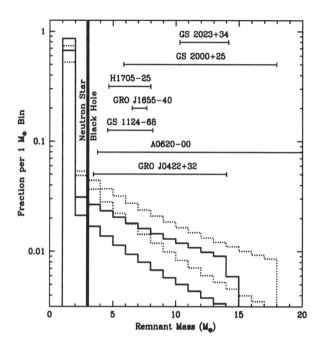

Figure 5. Mass distribution of compact remnants for "unbinding" energies set to 100% (solid lines) and 10% (dotted lines) of the explosion energy and two initial mass functions ($\alpha_{IMF} = 2.35, 2.7$). Note that most of the neutron stars have masses between $1-2M_\odot$. Indeed, most of the neutron stars have masses very close to $1.4M_\odot$. The black holes, on the other hand, have a large range of masses. Both the explosion energy and the initial mass function vary the formation rates of black holes, but the ratio of black holes in the mass range of $5-10M_\odot$ and those black holes more massive than $10M_\odot$ is ~ 2 for all our calculations. The observed black hole masses are from Bailyn et al. (1998).

increasing progenitor mass, the star's explosion energy decreases and its binding energy increases, both causing more and more mass to fall back onto the proto-neutron star. All of the explosion energy can not be used to unbind the stars, because some of it is converted into kinetic energy of the ejecta. As a probe of the sensitivity of the remnant masses on the energy available to unbind the star, we vary this "unbinding" energy by a factor of 10 (10-100% that of the explosion energy). Although this variation does vary the remnant mass by up to $4\,M_\odot$, in calculating the black hole mass distribution, this difference is relatively minor.

Combining our fits from Figure 4 with an initial mass function, we produce a mass distribution of compact remnants in the Galaxy (Figure 5). Assuming (somewhat arbitrarily) that the maximum neutron star mass is $\sim 3M_\odot$, we find that the number of black holes in the Galaxy is $\sim 10-40\%$ that of neutron stars (this number increases slightly if one includes black holes formed from accreting neutron stars). Over 80% of the neutron stars all huddle around a mass of $\sim 1.4 M_\odot$, with the other 20% in a continuous distribution up to the maximum neutron star mass. Black hole masses, on the other hand, are much more spread out, but over half fill the range between $5 M_\odot$ and $10 M_\odot$.

The biggest uncertainty in estimates of black hole masses comes from the uncertainties in stellar winds. If Wolf-Rayet winds are as strong as Wellstein & Langer (1999) believe, then the number of black holes in binaries goes down by a factor of $\sim 10 - 100$ (this number is extremeley sensitive to our values of stellar radii) and the total number of black holes goes down by a factor of 2. In addition, stellar winds prevent the production of black holes with masses beyond $\sim 10 M_\odot$. If the mass of GS 2023+34 exceeds $10\,M_\odot$, we begin to constrain the mass loss from winds better than the constraints set by observations of Wolf-Rayet stars. In this paper, we assumed the effects of winds were minimal.

Acknowledgements

This research has been supported by NASA (NAG5-8128), and the US DOE ASCI Program (W-7405-ENG-48). It is a pleasure to thank Alex Heger, Stan Woosley, and Andrew MacFadyen for encouragement and advice.

References

Baade, W. and Zwicky, F. (1934), *Phys. Rev.* **45**, 138.
Bailyn, C.D., Jain, R.K., Coppi, P., and Orosz, J.A. (1998), *ApJ* **499**, 367.
Burrows, A. (1998) to appear in E. Müller and W. Hillebrandt (eds.), *On the Systematics of Core-Collapse Explosions, the proceedings of the 9'th Workshop on Nuclear Astrophysics*, held at the Ringberg Castle, Germany, March 23-29.
Burrows, A. and Goshy, J. (1993), *ApJ* **416**, L75.
Burrows, A., Hayes, J., and Fryxell, B.A. (1995), *ApJ* **450**, 830.
Fryer, C.L. (1999), *ApJ* **522**, 413.
Fryer, C.L. and Heger, A. (2000), *ApJ* **541**, 1033.
Herant, M., Benz, W., Hix, W.R., Fryer, C.L., and Colgate, S.A. (1994), *ApJ* **435**, 339.
Janka, H.-T. and Müller, E. (1996), *A&A* **306**, 167.
Messer, O.E.B., Mezzacappa, A., Bruenn, S.W., and Guidry, M.W. (1998), *ApJ* **507**, 353.
Mezzacappa, A., Calder, A.C., Bruenn, S.W., Blondin, J.M., Guidry, M.W., Strayer, M.R., and Umar, A.S. (1998), *ApJ* **495**, 911.
Mezzacappa, A. (1998), to appear in *Memoirs of the Italian Astronomical Society as the Proceedings of Future Directions of Supernova Research: progenitors to remnants*, held September 29-October 2, 1998, Gran Sasso.
Miller, D.S., Wilson, J.R., and Mayle, R.W. (1993), *ApJ* **415**, 278.
Timmes, F.X., Woosley, S.E., and Weaver, T.A. (1996), *ApJ* **457**, 834.
Wellstein, S. and Langer, N. (1999), *A&A* **350**, 148.
Woosley, S.E. and Weaver, T.A. (1986), *Ann. Rev. Astron.& Astrophys.* **301**, 601.
Wilson, J.R. and Mayle, R.W. (1988), *Phys. Rep.* **163**, 63.

COALESCENCE RATES OF COMPACT OBJECTS

V. KALOGERA

Harvard-Smithsonian Center for Astrophysics
60 Garden St., Cambridge, MA 02138, USA

Abstract. Current estimates of the Galactic coalescence rates of close binaries with two compact objects (neutron stars or black holes) are reviewed, in the context of gravitational wave detection. The uncertainties involved in obtaining both theoretical and empirical estimates are discussed as well as empirical ways of obtaining upper limits on the coalescence rate of double neutron star systems.

1. Introduction

The spiralling in and final coalescence of close binaries with two compact objects, neutron stars (NS) or black holes (BH), are considered to be major sources of gravitational waves for the laser interferometers currently under construction (e.g., LIGO, VIRGO). At present only systems containing two neutron stars have been discovered, PSR B1913+16 being the prototypical NS–NS binary (Hulse & Taylor 1975). This binary radio pulsar has provided a remarkable empirical confirmation of general relativity with the measurement of orbital decay due to gravitational radiation (Taylor & Weisberg 1982). As spiralling in proceeds for similar systems in other galaxies, the frequency of the gravitational wave signal will enter the operational window of ground-based observatories and could be detected. Similar BH–NS and BH–BH spirallings in, which have been predicted theoretically, are considered as possible sources.

The expected detection rate of inspiralling events depends on (i) the strength of the gravitational wave signal, or else the maximum distance out to which such an event could be detected, and (ii) the rate of mergers in the Galaxy and by extrapolation out to that maximum distance. In what follows I review the current estimates (both theoretical and empirical) of Galactic coalescence rates and discuss the associated uncertainties. For the

C. Kouveliotou et al. (eds.), The Neutron Star – Black Hole Connection, 89–94.
© 2001 *Kluwer Academic Publishers. Printed in the Netherlands.*

case of NS–NS binaries, I also discuss ways of obtaining empirical upper
bounds on their coalescence rate. The scale for all these rates is set by
requiring that the detection rate for the "enhanced" LIGO is $\sim 2-3$
events per year. Given the amplitude of the signals (e.g., Thorne 1996), the
requirement for the NS–NS Galactic rate is $\sim 10^{-5}\,\mathrm{yr}^{-1}$ (detected out to
$\sim 200\,\mathrm{kpc}$) and for the BH–BH Galactic rate $\sim 2 \times 10^{-7}\,\mathrm{yr}^{-1}$ (detected out
to $\sim 700\,\mathrm{kpc}$).

2. Theoretical Estimates

The formation rate of coalescing binary compact objects (tight enough binaries that coalesce within a Hubble time) can be calculated, given a sequence
of evolutionary stages leading to binary compact object formation. Over the
years a relatively standard picture has been formed describing the birth of
such systems based on considerations of NS–NS binaries (van den Heuvel
1976), although more recently variations have also been discussed (Brown
1995). In all versions, however, the main picture remains the same: the initial binary progenitor consists of two binary members massive enough to
eventually collapse into a NS or a BH. The evolutionary path involves multiple phases of stable or unstable mass transfer, common-envelope phases,
and accretion onto compact objects, as well as two core collapse events.

Such theoretical modeling has been undertaken by various authors by
means of population synthesis. This provides us with *ab initio* predictions of
coalescence rates. The evolution of an ensemble of primordial binaries with
assumed initial properties is followed through specific evolutionary stages
until a coalescing binary is formed. The changes in the properties of the
binaries at the end of each stage are calculated based on our current understanding of the various processes involved: wind mass loss from massive
hydrogen- and helium-rich stars, mass and angular-momentum losses during mass transfer phases, dynamically unstable mass transfer and common-
envelope evolution, effects of highly super-Eddington accretion onto NS,
and supernova explosions with kicks imparted to newborn NS or even BH.
Given our limited understanding of several of these phases, the results of
population synthesis are expected to depend on the assumptions made in
the treatment of the various processes. Therefore, exhaustive parameter
studies are required by the nature of the problem.

Recent studies of the formation of compact objects and calculations
of their Galactic coalescence rates (Lipunov et al. 1997; Fryer et al. 1998;
Portegies-Zwart & Yungel'son 1998; Brown & Bethe 1998; Fryer et al. 1999)
have explored the input parameter space and the robustness of the results
at different levels of (in)completeness. Almost all have studied the sensitivity of the coalescence rate to the average magnitude of the kicks imparted

to compact objects at birth. The range of predicted NS–NS Galactic rates obtained by varying the kick magnitude is $< 10^{-7} - 5 \times 10^{-4} \, \text{yr}^{-1}$. This large range indicates the importance of supernovae (two in this case) in binaries. Variations in the assumed mass-ratio distribution for the primordial binaries can *further* change the predicted rate by about a factor of 10, while assumptions of the common-envelope phase add another factor of about $10 - 100$. Variation in other parameters typically affects the results by factors of two or less. Predicted rates for BH–NS and BH–BH binaries lie in the ranges $< 10^{-7} - 10^{-4} \, \text{yr}^{-1}$ and $< 10^{-7} - 10^{-5} \, \text{yr}^{-1}$, respectively, when the kick magnitude to both NS and BH is varied. Other uncertain factors such as the critical progenitor mass for NS and BH formation lead to variations of the rates by factors of $10 - 50$.

It is evident that recent theoretical predictions for coalescence rates cover a wide range of values (typically 3–4 orders of magnitude). Note that binary properties other than the coalescence rate, such as orbital sizes, eccentricities, center-of-mass velocities, are much less sensitive to the various input parameters and assumptions; the latter affect more severely the absolute normalization (birth rate) of the population. Given these results it seems fair to say that population synthesis calculations have a rather limited predictive power and provide fairly loose constraints on coalescence rates.

3. Empirical Estimates

In the case of NS–NS coalescence, there is another way to estimate the rate, using the properties of the observed coalescing NS–NS (only two systems: PSR B1913+16 and PSR B1534+12) combined with models of selection effects in radio pulsar surveys. For each observed object, a scale factor is calculated based on the fraction of the Galactic volume within which pulsars with properties identical to those of the observed pulsar could be detected, in principle, by any of the radio pulsar surveys, given their detection thresholds. This scale factor is a measure of how many more pulsars like those detected in the coalescing NS–NS systems exist in our galaxy. The coalescence rate can then be calculated based on the scale factors and estimates of detection lifetimes summed up for all the observed systems. This basic method was first used by Phinney (1991) and Narayan et al. (1991) who estimated the Galactic rate to be $\sim 10^{-6} \, \text{yr}^{-1}$.

Since then, estimates of the coalescence rate have decreased significantly primarily because of (i) the increase of the Galactic volume covered by radio pulsar surveys with no additional coalescing NS–NS being discovered (Curran & Lorimer 1995), (ii) the increase of the distance estimate for PSR B1534+12 based on measurements of post-Newtonian parameters (Stairs et al. 1998) (iii) changes in the lifetime estimates for the observed systems (van den Heuvel & Lorimer 1996; Arzoumanian et al. 1999). In addition, a significant upward correction factor ($\sim 7 - 10$) has been used recently to account for the faint (undetected) end of the radio pulsar luminosity function (see also Kalogera et al. 2000 for such a correction and the effect of a small-number observed sample). The most recently published study (Arzoumanian et al. citeyearA99) gives a lower limit of $2 \times 10^{-7}\,\mathrm{yr}^{-1}$ and a "best" estimate of $\sim 6 - 10 \times 10^{-7}\,\mathrm{yr}^{-1}$, which agrees with other recent estimates of $2 - 3 \times 10^{-6}\,\mathrm{yr}^{-1}$ (Stairs et al. 1998; Evans et al. 1999). Additional uncertainties arise from estimates of pulsar ages and distances, the pulsar beaming fraction, the spatial distribution of DNS in the Galaxy.

Despite all these uncertainties the empirical estimates of the NS–NS coalescence rate appear to span a range of $\sim 1 - 2$ orders of magnitude, which is narrow compared to the range covered by the theoretical estimates.

4. Limits on the NS–NS Coalescence Rate

Observations of NS-NS systems and (related to their formation) isolated pulsars allow us to obtain upper limits to their coalescence rate. Depending on how their value compares to the "enhanced" LIGO requirements such limits can provide us with valuable information about the prospects of gravitational wave detection.

Bailes (1996) used the absence of any young pulsars detected in NS–NS systems and obtained a rough upper limit to the rate of $\sim 10^{-5}\,\mathrm{yr}^{-1}$, while recently Arzoumanian et al. (1999) reexamined this in more detail and claimed a more robust upper limit of $\sim 10^{-4}\,\mathrm{yr}^{-1}$.

An upper bound to the rate can also be obtained by combining our theoretical understanding of orbital dynamics (for supernovae with NS kicks in binaries) with empirical estimates of the birth rates of *other* types of pulsars related to NS–NS formation (Kalogera & Lorimer 2000). Progenitors of NS–NS systems experience two supernova explosions when the NS are formed. The second supernova explosion (forming the NS that is *not* observed as a pulsar) provides a unique tool for the study of NS–NS formation, since the post-supernova evolution of the system is simple, driven only by gravitational wave radiation. There are three possible outcomes after the second supernova: (i) a coalescing NS–NS is formed (CB), (ii) a wide NS–NS (with a coalescence time longer than the Hubble time) is formed (WB), or (iii) the binary is disrupted (D) and a single pulsar similar to the ones seen in NS–NS systems is ejected. Based on supernova orbital dynamics we can accurately calculate the probability branching ratios for these three outcomes, P_{CB}, P_{WB}, and P_{D}. For a given kick magnitude, we can calculate the maximum ratio $(P_{\mathrm{CB}}/P_{\mathrm{D}})^{\mathrm{max}}$ for the complete range of pre-

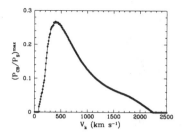

Figure 1. Maximum probability ratio for the formation of coalescing DNS and the disruption of binaries as a function of the kick magnitude at the second supernova.

supernova parameters defined by the necessary constraint $P_{CB} \neq 0$ (Figure 1). Given that the two types of systems have a common parent progenitor population, the ratio of probabilities is equal to the ratio of the birth rates (BR_{CB}/BR_D).

We can then use (i) the absolute maximum of the probability ratio (≈ 0.26 from Figure 1) and (ii) an empirical estimate of the birth rate of single pulsars similar to those in NS–NS based on the current observed sample to obtain an upper limit to the coalescence rate. The selection of this sample involves some subtleties (Kalogera & Lorimer 2000), and the analysis results in $BR_{CB} < 1.5 \times 10^{-5}\,\mathrm{yr}^{-1}$ (Kalogera & Lorimer 2000). Note that this number could be increased because of the small-number sample and luminosity bias affecting this time the empirical estimate of BR_D by a factor of $2 - 6$ (Kalogera et al. 2000).

This is an example of how we can use observed systems other than NS–NS to improve our understanding of their coalescence rate. A similar calculation can be done using the wide NS–NS systems instead of the single pulsars (Kalogera & Lorimer 2000).

5. Conclusions

A comparison of the NS–NS coalescence rate studies indicates that theoretical estimates based on modeling of their formation have a rather limited predictive power. The range of rates exceeds 3 orders of magnitude and most importantly includes the value of $10^{-5}\,\mathrm{yr}^{-1}$ required for an "enhanced" LIGO detection rate of 2–3 events per year. This means that at the two edges of the range the conclusion swings from no detection to many per month. In other words the detection prospects of NS–NS coalescence cannot be assessed firmly. On the other hand empirical estimates derived based on the observed sample appear to be more robust (estimates are all within a factor smaller than 100). Given these we would expect a detection of one event every few years up to even ten events per year.

For coalescence rates of BH–NS and BH–BH systems we have to rely solely on our theoretical understanding of their formation. As in the case of NS–NS binaries, the model uncertainties are significant and the ranges extend to more than 2 orders of magnitude. However, the requirement on the Galactic rate is less stringent for $10\,M_\odot$ BH–BH binaries, only $\sim 2 \times 10^{-7}\,\mathrm{yr}^{-1}$. Therefore, even with the pessimistic estimates for BH–BH coalescence rates ($\sim 10^{-7}\,\mathrm{yr}^{-1}$), we would expect at least a few detections per year, which is quite encouraging. We note that a very recent examination of dynamical BH–BH formation (Portegies-Zwart & McMillan 2000) in globular clusters leads to detection rates as high as a few per day.

Acknowledgments

I am grateful to the organizers of the meeting for inviting me and supporting my participation financially. I also acknowledge full support by the Smithsonian Institute in the form of a CfA Post-doctoral Fellowship.

References

Arzoumanian, Z. et al. (1999), *ApJ* **520**, 696.

Bailes, M. (1996), in J. van Paradijs, E.P.J. van den Heuvel, and E. Kuulkers (eds.), *Compact Stars in Binaries*, IAU Symp. No. 165, Kluwer Academic Publishers, Dordrecht, p213.

Brown, G.E. (1995), *ApJ* **440**, 270.

Brown, G.E. and Bethe, H. (1998), *ApJ* **506**, 780.

Curran, S.J. and Lorimer, D.R. (1995), *MNRAS* **276**, 347.

Evans, T. et al. (1999) to appear in the proceedings of the XXXIVth Rencontres de Moriond on "Gravitational Waves and Experimental Gravity", Les Arcs, France.

Fryer, C.L. et al. (1998), *ApJ* **496**, 333.

Fryer, C.L. et al. (1999), *ApJ* **526**, 152.

Hulse, R.A. and Taylor, J.H. (1975), *ApJ* **195**, L51.

Kalogera, V. and Lorimer, D.R. (2000), *ApJ* **530**, 890.

Kalogera, V. et al. (2000), *ApJ*, to be submitted.

Lipunov, V.M. et al. (1997), *MNRAS* **288**, 245.

Narayan, R. et al. (1991), *ApJ* **379**, L17.

Phinney, E.S. (1991), *ApJ* **380**, L17.

Portegies-Zwart, S.F. and McMillan, S.L.W. (2000), *ApJ* **528**, L17.

Portegies-Zwart, S.F. and Yungel'son, L.R. (1998), *A&A* **332**, 173.

Stairs, I.H. et al. (1998), *ApJ* **505**, 352.

Taylor, J.H. and Weisberg, J.M. (1992), *ApJ* **253**, 908.

Thorne, K.S. (1996), in J. van Paradijs, E.P.J. van den Heuvel, and E. Kuulkers (eds.), *Compact Stars in Binaries*, IAU Symp. No. 165, Kluwer Academic Publishers, Dordrecht, p153.

Van den Heuvel, E.P.J. (1976), in P. Eggleton, S. Mitton, and J. Whelan (eds.), *Structure and Evolution of Close Binary Systems*, IAU Symp. No. 73, Kluwer Academic Publishers, Dordrecht.

Van den Heuvel, E.P.J. and Lorimer, D.R. (1996), *MNRAS* **283**, 37.

THE FINAL FATE OF COALESCING COMPACT BINARIES: FROM BLACK HOLE TO PLANET FORMATION

F.A. RASIO
Department of Physics,
Massachusetts Institute of Technology,
Cambridge, MA 02139, USA

1. Introduction

The coalescence and merging of two compact stars into a single object is a very common end-point of close binary evolution. Dissipation mechanisms such as friction in common envelopes, tidal dissipation, or the emission of gravitational radiation, are always present and cause the orbits of close binary systems to decay. This review will concentrate on the coalescence of compact binaries containing either two neutron stars (hereafter NS) or two white dwarfs (WD).

1.1. DOUBLE NEUTRON STARS

Many theoretical models of gamma-ray bursts (GRBs) rely on coalescing NS binaries to provide the energy of GRBs at cosmological distances (e.g., Eichler et al. 1989; Narayan, Paczyński, & Piran 1992; Mészáros & Rees 1992; for recent reviews see Mészáros 1999 and Piran 1999). The close spatial association of some GRB afterglows with faint galaxies at high redshifts may not be inconsistent with a NS binary merger origin, in spite of the large recoil velocities acquired by NS binaries at birth (Bloom, Sigurdsson, & Pols 1999; but see also Bulik & Belczynski 2000). Currently the most popular models all assume that the coalescence of two NS leads to the formation of a rapidly rotating black hole (BH) surrounded by a torus of debris. Energy can then be extracted either from the rotation of the Kerr BH or from the material in the torus so that, with sufficient beaming, the gamma-ray fluxes observed from even the most distant GRBs can be explained (Mészáros, Rees, & Wijers 1999). However, it is important to understand the hydrodynamic processes taking place during the final

C. Kouveliotou et al. (eds.), The Neutron Star – Black Hole Connection, 95–108.

coalescence before making assumptions about its outcome. In particular, as will be argued below (§2.2), it is not clear that the coalescence of two $1.4\,M_\odot$ NS forms an object that will collapse to a BH on a dynamical timescale, and it is not certain either that a significant amount of matter will be ejected during the merger to form an outer torus around the central object (Faber & Rasio 2000).

Coalescing NS binaries are also important sources of gravitational waves that may be directly detectable by the large laser interferometers currently under construction, such as LIGO (Abramovici et al. 1992; see Barish & Weiss 1999 for a recent pedagogical introduction) and VIRGO (Bradaschia et al. 1990). In addition to providing a major new confirmation of Einstein's theory of general relativity (GR), including the first direct proof of the existence of black holes (see, e.g., Flanagan & Hughes 1998; Lipunov, Postnov, & Prokhorov 1997), the detection of gravitational waves from coalescing binaries at cosmological distances could provide accurate independent measurements of the Hubble constant and mean density of the Universe (Schutz 1986; Chernoff & Finn 1993; Marković 1993). Expected rates of NS binary coalescence in the Universe, as well as expected event rates in laser interferometers, have now been calculated by many groups. Although there is some disparity between various published results, the estimated rates are generally encouraging (see Kalogera 2000 for a recent review).

Many calculations of gravitational wave emission from coalescing binaries have focused on the waveforms emitted during the last few thousand orbits, as the frequency sweeps upward from $\sim 10\,\mathrm{Hz}$ to $\sim 300\,\mathrm{Hz}$. The waveforms in this frequency range, where the sensitivity of ground-based interferometers is highest, can be calculated very accurately by performing high-order post-Newtonian (PN) expansions of the equations of motion for two *point masses* (see, e.g., Owen & Sathyaprakash 1999 and references therein). However, at the end of the inspiral, when the binary separation becomes comparable to the stellar radii (and the frequency is $\gtrsim 1\,\mathrm{kHz}$), hydrodynamics become important and the character of the waveforms must change. Special purpose narrow-band detectors that can sweep up frequency in real time will be used to try to catch the last ~ 10 cycles of the gravitational waves during the final coalescence (Meers 1988; Strain & Meers 1991). These "dual recycling" techniques are being tested right now on the German-British interferometer GEO 600 (Danzmann 1998). In this terminal phase of the coalescence, when the two stars merge together into a single object, the waveforms contain information not just about the effects of GR, but also about the interior structure of a NS and the nuclear equation of state (EOS) at high density. Extracting this information from observed waveforms, however, requires detailed theoretical knowledge about all relevant hydrodynamic processes. If the NS merger is followed by the formation of a BH, the corresponding gravitational radiation waveforms will also provide direct information on the dynamics of rotating core collapse and the BH "ringdown" (see, e.g., Flanagan & Hughes 1998).

1.2. DOUBLE WHITE DWARFS

Coalescing WD binaries have long been discussed as possible progenitors of Type Ia supernovae (Iben & Tutukov 1984; Webbink 1984; Paczyński 1985; see Branch et al. 1995 for a recent review). To produce a supernova, the total mass of the system must be above the Chandrasekhar mass. Given evolutionary considerations, this requires two C-O or O-Ne-Mg WD. Yungelson et al. (1994) showed that the expected merger rate for close WD pairs with total mass exceeding the Chandrasekhar mass is consistent with the rate of Type Ia supernovae deduced from observations. Alternatively, a massive enough merger may collapse to form a rapidly rotating NS (Nomoto & Iben 1985; Colgate 1990). Chen & Leonard (1993) speculated that most millisecond pulsars in globular clusters might have formed in this way. In some cases planets may also form in the disk of material ejected during the coalescence and left in orbit around the central pulsar (Podsiadlowski, Pringle, & Rees 1991). Indeed the very first extrasolar planets were discovered in orbit around a millisecond pulsar, PSR B1257+12 (Wolszczan & Frail 1992). A merger of two magnetized WD might lead to the formation of a NS with extremely high magnetic field, and this scenario has been proposed as a source of GRBs (Usov 1992).

Close WD binaries are expected to be extremely abundant in our Galaxy, even though their direct detection remains very challenging (Han 1998; Saffer, Livio, & Yungelson 1999). Iben & Tutukov (1984, 1986) predicted that $\sim 20\%$ of all binary stars produce close WD pairs at the end of their stellar evolution. More recently, theoretical estimates of the double WD formation rate in the Galaxy have converged to a value $\simeq 0.1 \, \mathrm{yr}^{-1}$, with an uncertainty that may be only a factor of two (Han 1998; Kalogera 2000). The most common systems should be those containing two low-mass helium WD. Their final coalescence can produce an object massive enough to start helium burning. Bailyn (1993 and references therein) and others have suggested that some "extreme horizontal branch" stars in globular clusters may be such helium-burning stars formed by the coalescence of two WD. Planets in orbit around a massive WD may also form following the binary coalescence (Livio, Pringle, & Saffer 1992).

Coalescing WD binaries are also important sources of low-frequency gravitational waves that should be easily detectable by future space-based laser interferometers. The currently planned LISA (Laser Interferometer

Space Antenna; see Folkner 1998) should have an extremely high sensitivity (down to a characteristic strain $h \sim 10^{-23}$) to sources with frequencies in the range $\sim 10^{-4} - 1\,\text{Hz}$. Han (1998) estimated a WD merger rate $\sim 0.03\,\text{yr}^{-1}$ in our own Galaxy. Individual coalescing systems and mergers may be detectable in the frequency range ~ 10–$100\,\text{mHz}$. In addition, the total number ($\sim 10^4$) of close WD binaries in our Galaxy emitting at lower frequencies ~ 0.1–$10\,\text{mHz}$ (the emission lasting for $\sim 10^2$–$10^4\,\text{yr}$ before final merging) should provide a continuum background signal of amplitude $h \sim 10^{-20}$–10^{-21} (Hils et al. 1990). The detection of the final burst of gravitational waves emitted during an actual merger would provide a unique opportunity to observe in "real time" the hydrodynamic interaction between the two degenerate stars, possibly followed immediately by a supernova explosion, nuclear outburst, or some other type of electromagnetic signal.

2. Coalescing Binary Neutron Stars

2.1. HYDRODYNAMICS OF NEUTRON STAR MERGERS

The final hydrodynamic merger of two NS is driven by a combination of relativistic and fluid effects. Even in Newtonian gravity, an innermost stable circular orbit (ISCO) is imposed by *global hydrodynamic instabilities*, which can drive a close binary system to rapid coalescence once the tidal interaction between the two stars becomes sufficiently strong. The existence of these global instabilities for close binary equilibrium configurations containing a compressible fluid, and their particular importance for binary NS systems, were demonstrated for the first time by Rasio & Shapiro (1992, 1994, 1995; hereafter RS1-3) using numerical hydrodynamic calculations. These instabilities can also be studied using analytic methods. The classical analytic work for close binaries containing an incompressible fluid (e.g., Chandrasekhar 1969) was extended to compressible fluids in the work of Lai, Rasio, & Shapiro (1993a,b, 1994a,b,c, hereafter LRS1-5). This analytic study confirmed the existence of dynamical instabilities for sufficiently close binaries. Although these simplified analytic studies can give much physical insight into difficult questions of global fluid instabilities, fully numerical calculations remain essential for establishing the stability limits of close binaries accurately and for following the nonlinear evolution of unstable systems all the way to complete coalescence.

A number of different groups have now performed such calculations, using a variety of numerical methods and focusing on different aspects of the problem. Nakamura and collaborators (see Nakamura & Oohara 1998 and references therein) were the first to perform 3D hydrodynamic calculations of binary NS coalescence, using a traditional Eulerian finite-difference code.

Instead, RS used the Lagrangian method SPH (Smoothed Particle Hydro-dynamics). They focused on determining the ISCO for initial binary models in strict hydrostatic equilibrium and calculating the emission of gravitational waves from the coalescence of unstable binaries. Many of the results of RS were later independently confirmed by New & Tohline (1997) and Swesty, Wang, & Calder (2000), who used completely different numerical methods but also focused on stability questions, and by Zhuge, Centrella, & McMillan (1994, 1996), who also used SPH. Zhuge et al. (1996) also explored in detail the dependence of the gravitational wave signals on the initial NS spins. Davies et al. (1994) and Ruffert et al. (1996, 1997) have incorporated a treatment of the nuclear physics in their hydrodynamic calculations (done using SPH and PPM codes, respectively), motivated by models of GRBs at cosmological distances. All these calculations were performed in *Newtonian gravity*, with some of the more recent studies adding an approximate treatment of energy and angular momentum dissipation through the gravitational radiation reaction (e.g., Janka et al. 1999; Rosswog et al. 1999), or even a full treatment of PN gravity to lowest order (Ayal et al. 2000; Faber & Rasio 2000).

All recent hydrodynamic calculations agree on the basic qualitative picture that emerges for the final coalescence (see Figure 1). As the ISCO is approached, the secular orbital decay driven by gravitational wave emission is dramatically accelerated (see also LRS2, LRS3). The two stars then plunge rapidly toward each other, and merge together into a single object in just a few rotation periods. In the corotating frame of the binary, the relative radial velocity of the two stars always remains very subsonic, so that the evolution is nearly adiabatic. This is in sharp contrast to the case of a head-on collision between two stars on a free-fall, radial orbit, where shock heating is very important for the dynamics (RS1; Shapiro 1998). Here the stars are constantly being held back by a (slowly receding) centrifugal barrier, and the merging, although dynamical, is much more gentle. After typically $1-2$ orbital periods following first contact, the innermost cores of the two stars have merged and a secondary instability occurs: *mass shedding* sets in rather abruptly. Material (typically $\sim 10\%$ of the total mass) is ejected through the outer Lagrange points of the effective potential and spirals out rapidly. In the final stage, the spiral arms widen and merge together, forming a nearly axisymmetric thick disk or torus around the inner, maximally rotating dense core.

In GR, strong-field gravity between the masses in a binary system is alone sufficient to drive a close circular orbit unstable. In close NS binaries, GR effects combine nonlinearly with Newtonian tidal effects so that the ISCO is encountered at larger binary separations and lower orbital frequency than predicted by Newtonian hydrodynamics alone, or GR alone

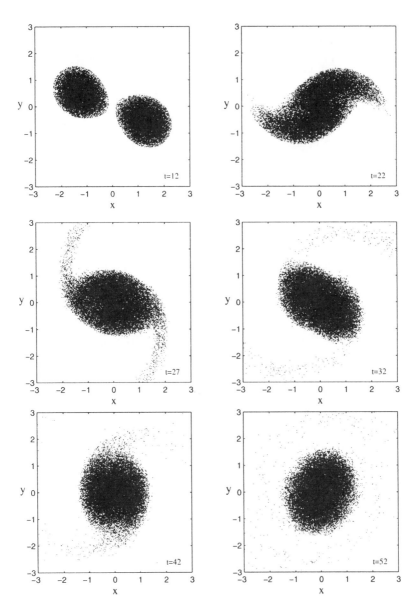

Figure 1. Post-Newtonian SPH calculation of the coalescence of two identical neutron stars modeled as simple $\Gamma = 3$ polytropes. Projections of a subset of all SPH particles onto the orbital $(x - y)$ plane are shown at various times. Units are such that $G = M = R = 1$ where M and R are the mass and radius of each star initially. The orbital rotation is counter-clockwise. From Faber & Rasio (2000).

for two point masses. The combined effects of relativity and hydrodynamics on the stability of close compact binaries have only very recently begun to be studied, using both analytic approximations (basically, PN generalizations of LRS; see, e.g., Lai & Wiseman 1997; Lombardi, Rasio, & Shapiro 1997; Shibata & Taniguchi 1997), as well as numerical calculations in 3D incorporating simplified treatments of relativistic effects (e.g., Baumgarte et al. 1998; Marronetti, Mathews & Wilson 1998; Wang, Swesty, & Calder 1998; Faber & Rasio 2000).

Several groups have been working on a fully general relativistic calculation of the final coalescence, combining the techniques of numerical relativity and numerical hydrodynamics in 3D (Baumgarte, Hughes, & Shapiro 1999; Landry & Teukolsky 2000; Seidel 1998; Shibata & Uryu 2000). However this work is still in its infancy, and only very preliminary results of test calculations have been reported so far.

2.2. BLACK HOLE FORMATION

The final fate of a NS–NS merger depends crucially on the NS EOS, and on the extraction of angular momentum from the system during the final merger. For a stiff NS EOS, it is by no means certain that the core of the final merged configuration will collapse on a dynamical timescale to form a BH. One reason is that the Kerr parameter J/M^2 of the core may exceed unity for extremely stiff EOS (Baumgarte et al. 1998), although Newtonian and PN hydrodynamic calculations suggest that this is never the case (see, e.g., Faber & Rasio 2000). More importantly, the rapidly rotating core may in fact be dynamically stable.

Take the obvious example of a system containing two identical $1.35\,M_\odot$ NS. The total baryonic mass of the system for a stiff NS EOS is then about $3\,M_\odot$. Almost independent of the spins of the NS, all hydrodynamic calculations suggest that about 10% of this mass will be ejected into the outer torus, leaving at the center a *maximally rotating* object with baryonic mass $\simeq 2.7\,M_\odot$ (Any hydrodynamic merger process that leads to mass shedding will produce a maximally rotating object since the system will have ejected just enough mass and angular momentum to reach its new, stable quasi-equilibrium state). Most stiff NS EOS (including the well-known "AU" and "UU" EOS of Wiringa et al. 1988; see Akmal et al. 1998 for a recent update) allow stable, maximally rotating NS with baryonic masses exceeding $3\,M_\odot$ (Cook, Shapiro, & Teukolsky 1994), i.e., well above the mass of the final merger core. Differential rotation (not taken into account in the calculations of Cook et al. 1994) can further increase this maximum stable mass very significantly (see Baumgarte, Shapiro, & Shibata 2000). Thus the hydrodynamic merger of two NS with stiff EOS and realistic masses is not expected to produce a BH. This expectation is confirmed by the preliminary full-GR calculations of Shibata & Uryu (2000), for polytropes with $\Gamma = 2$, which indicate collapse to a BH only when the two NS are initially very close to the maximum stable mass.

For *slowly rotating* stars, the same stiff NS EOS give maximum stable baryonic masses in the range $2.5 - 3\,M_\odot$, which may or may not exceed the total merger core mass. Therefore, collapse to a BH could still occur on a timescale longer than the dynamical timescale, following a significant loss of angular momentum. Indeed, processes such as electromagnetic radiation, neutrino emission, and the development of various secular instabilities (e.g., r-modes), which may lead to angular momentum losses, take place on timescales much longer than the dynamical timescale (see, e.g., Baumgarte & Shapiro 1998, who show that neutrino emission is probably negligible). These processes are therefore decoupled from the hydrodynamics of the coalescence. Unfortunately their study is plagued by many fundamental uncertainties in the microphysics.

2.3. THE IMPORTANCE OF THE NEUTRON STAR SPINS

The question of the final fate of the merger could also depend crucially on the NS spins and on the evolution of the fluid vorticity during the final coalescence. Close NS binaries are likely to be *nonsynchronized*. Indeed, the tidal synchronization time is almost certainly much longer than the orbital decay time (Kochanek 1992; Bildsten & Cutler 1992). For NS binaries that are far from synchronized, the final coalescence involves some new, complex hydrodynamic processes (Rasio & Shapiro 1999).

Consider for example the case of an irrotational system (containing two nonspinning stars at large separation; see LRS3). Because the two stars appear to be counter-spinning in the corotating frame of the binary, a *vortex sheet* (where the tangential velocity jumps discontinuously by $\Delta v \sim 0.1\,c$) appears when the stellar surfaces come into contact. Such a vortex sheet is Kelvin-Helmholtz unstable on all wavelengths and the hydrodynamics are therefore extremely difficult to model accurately given the limited spatial resolution of 3D calculations. The breaking of the vortex sheet generates some turbulent viscosity so that the final configuration may no longer be irrotational. In numerical simulations, however, vorticity is quickly generated through spurious shear viscosity, and the merger remnant is observed to evolve rapidly (in just a few rotation periods) toward uniform rotation.

The final fate of the merger could be affected drastically by these processes. In particular, the shear flow inside the merging stars (which supports a highly triaxial shape; see Rasio & Shapiro 1999) may in reality persist long enough to allow a large fraction of the total angular momentum in the system to be radiated away in gravitational waves during the hydrodynamic phase of the coalescence. In this case the final merged core may resemble a Dedekind ellipsoid, i.e., it will have a triaxial shape supported entirely by internal fluid motions, but with a stationary shape in the inertial frame (so that it no longer radiates gravitational waves). This state will be reached on the gravitational radiation reaction timescale, which is no more than a few tens of rotation periods. On the (much longer) *viscous timescale*, the core will then evolve to a uniform, slowly rotating state and will probably collapse to a BH. In contrast, in all 3D numerical simulations performed to date, the shear is quickly dissipated, so that gravitational radiation never gets a chance to extract more than a small fraction ($\lesssim 10\%$) of the angular momentum, and the final core appears to be a uniform, maximally rotating object (stable to collapse) exactly as in calculations starting from synchronized binaries. However this behavior is most likely an artefact of the large spurious shear viscosity present in the 3D simulations.

In addition to their obvious significance for gravitational wave emission, these issues are also of great importance for models of GRBs that depend on energy extraction from a torus of material around the central BH. Indeed, if a large fraction of the total angular momentum is removed by the gravitational waves, rotationally-induced mass shedding may not occur at all during the merger, eventually leaving a BH with no surrounding matter, and no way of extracting energy from the system. Note also that, even without any additional loss of angular momentum through gravitational radiation, PN effects tend to reduce drastically the amount of matter ejected during the merger (Faber & Rasio 2000).

3. Coalescing White Dwarf Binaries

3.1. HYDRODYNAMICS OF WHITE DWARF MERGERS

The results of RS3 for polytropes with $\Gamma = 5/3$ show that hydrodynamics also play an important role in the coalescence of two WD, either because of dynamical instabilities of the equilibrium configuration, or following the onset of dynamically unstable mass transfer. Systems with mass ratios $q \approx 1$ must evolve into deep contact before they become dynamically unstable and merge. Instead, equilibrium configurations for binaries with q sufficiently far from unity never become dynamically unstable. However, once these binaries reach their Roche limit, dynamically unstable mass transfer occurs and the less massive star is completely disrupted after a small number (< 10) of orbital periods (see also Benz et al. 1990). In both cases, the final merged configuration is an axisymmetric, rapidly rotating object with a core – thick disk structure similar to that obtained for coalescing NS (RS2, RS3; see also Mochkovitch & Livio 1989).

3.2. THE FINAL FATE: COLLAPSE TO A NEUTRON STAR? PLANETS?

For two massive enough WD, the merger product may be well above the Chandrasekhar mass M_{Ch}. The object may therefore explode as a (Type Ia) supernova, or perhaps collapse to a NS. The rapid rotation and possibly high mass (up to $\sim 2M_{Ch}$) of the object must be taken into account for determining its final fate. Unfortunately, rapid rotation and the possibility of starting from an object well above the Chandrasekhar limit have not been taken into account in most previous theoretical calculations of "accretion-induced collapse" (AIC), which consider a nonrotating WD just below the Chandrasekhar limit accreting matter slowly and quasi-spherically (e.g., Canal et al. 1990; Nomoto & Kondo 1991; see Fryer et al. 1999 for a recent 2-D SPH calculation including rotation). Under these assumptions it is found that collapse to a NS is possible only for a narrow range of initial conditions. In most cases, a supernova explosion follows the ignition of the nuclear fuel in the degenerate core. However, the fate of a much more massive object with substantial rotational support and large deviations from spherical symmetry (as would be formed by dynamical coalescence) may be very different.

If a NS does indeed form, and later accretes some of the material ejected during the coalescence, a millisecond radio pulsar may emerge. Planets around this millisecond pulsar may be formed at large distances $\sim 1\,\mathrm{AU}$ following the viscous evolution of the remaining material in the outer disk (Podsiadlowski, Pringle & Rees 1991; Phinney & Hansen 1993). This is one of the possible formation scenarios for the extraordinary planetary system discovered around the millisecond pulsar PSR B1257+12 (see Wolszczan 1999 for a recent update; Podsiadlowski 1993 for alternative planet formation scenarios). This system contains three confirmed Earth-mass planets in quasi-circular orbits (Wolszczan & Frail 1992; Wolszczan 1994). The planets have masses of $0.015/\sin i_1\,\mathrm{M_\oplus}$, $3.4/\sin i_2\,\mathrm{M_\oplus}$, and $2.8/\sin i_3\,\mathrm{M_\oplus}$, where i_1, i_2 and i_3 are the inclinations of the orbits with respect to the line of sight, and are at distances of 0.19 AU, 0.36 AU, and 0.47 AU, respectively, from the pulsar. In addition, the unusually large second and third frequency derivatives of the pulsar suggest the existence of a fourth, more distant and massive planet in the system (Wolszczan 1999). The simplest interpretation of the present best-fit values of the frequency derivatives implies a mass of about $100/\sin i_4\,\mathrm{M_\oplus}$ (i.e., comparable to Saturn's mass) for the fourth planet, at a distance of about 38 AU (i.e., comparable to Pluto's distance from the Sun), and with a period of about 170 yr in a circular, coplanar orbit (Wolszczan 1996; Joshi & Rasio 1997). However, if, as may well be the case, the first pulse frequency derivative is not entirely acceleration-induced, then the fourth planet can have a wide range of masses (Joshi &

Rasio 1997). In particular, it can have a mass comparable to that of Mars (at a distance of 9 AU), Uranus (at a distance of 25 AU) or Neptune (at a distance of 26 AU). The presence of this fourth planet, if confirmed, would place strong additional constraints on possible formation scenarios, as both the minimum mass and minimum angular momentum required in the protoplanetary disk would increase considerably (see Phinney & Hansen 1993 for a general discussion).

Acknowledgements

This work was supported by NSF Grant AST-9618116, NASA ATP Grant NAG5-8460, and by an Alfred P. Sloan Research Fellowship. Our computational work is supported by the National Computational Science Alliance.

References

Abramovici, M. et al. (1992), *Science* **256**, 325.
Akmal, A., Pandharipande, V.R., and Ravenhall, D.G. (1998), *Phys.Rev.C* **58**, 1804.
Ayal, S., Piran, T., Oechslin, R., Davies, M.B., and Rosswog, S. (2000), *ApJ* **550**, 846.
Bailyn, C.D. (1993), in S.G. Djorgovski and G. Meylan (eds.), *Structure and Dynamics of Globular Clusters*, ASP Conf. Series 50 (San Fransisco), p191.
Barish, B.C. and Weiss, R. (1999), *Physics Today* **52,10**, 44.
Baumgarte, T.W., Cook, G.B., Scheel, M.A., Shapiro, S.L., and Teukolsky, S.A. (1998), *Phys.Rev.D* **57**, 7299.
Baumgarte, T.W., Hughes, S.A., and Shapiro, S.L. (1999), *Phys.Rev.D* **60**, 087501.
Baumgarte, T.W. and Shapiro, S.L. (1998), *ApJ* **504**, 431.
Baumgarte, T.W., Shapiro, S.L., and Shibata, M. (2000), *ApJ* **528**, L29.
Benz, W., Cameron, A.G.W., Press, W.H., and Bowers, R.L. (1990), *ApJ* **348**, 647.
Bildsten, L. and Cutler, C. (1992), *ApJ* **400**, 175.
Bloom, J.S., Sigurdsson, S., and Pols, O.R. (1999), *MNRAS* **305**, 763.
Bradaschia, C. et al. (1990), *Nucl.Instr.Methods A* **289**, 518.
Branch, D., Livio, M., Yungelson, L.R., Boffi, F.R., and Baron, E. (1995), *PASP* **107**, 1019.
Bulik, T. and Belczynski, K. (2000), in R.M. Kippen, R. Mallozzi, G.J. Fishman (eds.), *Proc. 5th Huntsville Symposium*, AIP CP526, p648.
Canal, R., Garcia, D., Isern, J., and Labay, J. (1990), *ApJ* **356**, L51.
Chandrasekhar, S. (1969), *Ellipsoidal Figures of Equilibrium*, Yale University Press (New Haven); Revised Dover edition 1987.
Chen, K. and Leonard, P.J.T. (1993), *ApJ* **411**, L75.
Chernoff, D.F. and Finn, L.S. (1993), *ApJ* **411**, L5.
Colgate, S.A. (1990), in S.E. Woosley (ed.), *Supernovae*, Springer-Verlag (New York), p585.
Cook, G.B., Shapiro, S.L., and Teukolsky, S.L. (1994), *ApJ* **424**, 823.
Danzmann, K. (1998), in H. Riffert et al. (eds.), *Relativistic Astrophysics*, Proc. of 162nd W.E. Heraeus Seminar, Vieweg Verlag (Wiesbaden), p48.
Davies, M.B., Benz, W., Piran, T., and Thielemann, F.K. (1994), *ApJ* **431**, 742.
Eichler, D., Livio, M., Piran, T., and Schramm, D.N. (1989), *Nature* **340**, 126.
Faber, J. and Rasio, F.A. (2000), *Phys.Rev.D*, submitted (gr-qc/9912097).
Flanagan, E.E. and Hughes, S.A. (1998), *Phys.Rev.D* **57**, 4566.
Folkner, W.M. (ed.) (1998), *Laser Interferometer Space Antenna, Second International LISA Symposium on the Detection and Observation of Gravitational Waves in Space,*

AIP CP456.

Fryer, C., Benz, W., Herant, M., and Colgate, S.A. (1999), *ApJ* **516**, 892.

Hils, D., Bender, P.L., and Webbink, R.F. (1990), *ApJ* **360**, 75.

Iben, I., Jr. and Tutukov, A.V. (1984), *ApJ Suppl.* **54**, 335.

Iben, I., Jr. and Tutukov, A.V. (1986), *ApJ* **311**, 753.

Janka, H., Eberl, T., Ruffert, M., and Fryer, C.L. (1999), *ApJ* **527**, L39.

Joshi, K.J. and Rasio, F.A. (1997), *ApJ* **479**, 948.

Kalogera, V. (2000), to appear in S. Meshkov (ed.), *Proceedings of the 3rd Amaldi Conference on Gravitational Waves*, (astro-ph/9911532).

Kochanek, C.S. (1992), *ApJ* **398**, 234.

Lai, D., Rasio, F.A., and Shapiro, S.L. (1993a) [LRS1], *ApJ Suppl.* **88**, 205.

Lai, D., Rasio, F.A., and Shapiro, S.L. (1993b) [LRS2], *ApJ* **406**, L63.

Lai, D., Rasio, F.A., and Shapiro, S.L. (1994a) [LRS3], *ApJ* **420**, 811.

Lai, D., Rasio, F.A., and Shapiro, S.L. (1994b) [LRS4], *ApJ* **423**, 344.

Lai, D., Rasio, F.A., and Shapiro, S.L. (1994c) [LRS5], *ApJ* **437**, 742.

Lai, D. and Wiseman, A.G. (1997), *Phys.Rev.D* **54**, 3958.

Landry, W. and Teukolsky, S.A. (2000), *Phys.Rev.D*, submitted (gr-qc/9912004).

Lipunov, V.M., Postnov, K.A., and Prokhorov, M.E. (1997), *AstL* **23**, 492.

Livio, M., Pringle, J.E., and Saffer, R.A. (1992), *MNRAS* **257**, 15P.

Lombardi, J.C., Rasio, F.A., and Shapiro, S.L. (1997), *Phys.Rev.D* **56**, 3416.

Marković, D. (1993), *Phys.Rev.D* **48**, 4738.

Marronetti, P., Mathews, G.J., and Wilson, J.R. (1998), *Phys.Rev.D* **58**, 107503.

Meers, B.J. (1988), *Phys.Rev.D* **38**, 2317.

Mészáros, P. (1999), *Prog.Theor.Phys.Supp.* **136**, 78, (astro-ph/9912546).

Mészáros, P. and Rees, M.J. (1992), *ApJ* **397**, 570.

Mészáros, P., Rees, M.J., and Wijers, R.A.M.J. (1999), *NewA* **4**, 303.

Mochkovitch, R. and Livio, M. (1989), *A&A* **209**, 111.

Nakamura, T. and Oohara, K. (1998), preprint (gr-qc/9812054).

Narayan, R., Paczyński, B., and Piran, T. (1992), *ApJ* **395**, L83.

New, K.C.B. and Tohline, J.E. (1997), *ApJ* **490**, 311.

Nomoto, K. and Iben, I., Jr. (1985), *ApJ* **297**, 531.

Nomoto, K. and Kondo, Y. (1991), *ApJ* **367**, L19.

Owen, B.J. and Sathyaprakash, B.S. (1999), *Phys.Rev.D* **60**, 022002.

Paczyński, B. (1985), in D.Q. Lamb and J. Patterson (eds.), *Cataclysmic Variables and Low-mass X-ray Binaries*, Reidel (Dordrecht), p1.

Phinney, E.S. and Hansen, B.M.S. (1993), in J.A. Phillips, S.E. Thorsett, and S.R. Kulkarni (eds.), *Planets around Pulsars*, ASP Conf. Series 36, p371.

Piran, T. (1999), *Phys. Rep.* **314**, 575, (astro-ph/9907392).

Podsiadlowski, P., Pringle, J.E., and Rees, M.J. (1991), *Nature* **352**, 783.

Podsiadlowski, P. (1993), in J.A. Phillips, S.E. Thorsett, and S.R. Kulkarni (eds.), *Planets around Pulsars*, ASP Conf. Series 36, p149.

Rasio, F.A. and Shapiro, S.L. (1992) [RS1], *ApJ* **401**, 226.

Rasio, F.A. and Shapiro, S.L. (1994) [RS2], *ApJ* **432**, 242.

Rasio, F.A., and Shapiro, S.L. (1995) [RS3], *ApJ* **438**, 887.

Rasio, F.A. and Shapiro, S.L. (1999), *CQG* **16**, 1.

Rosswog, S., Liebendoerfer, M., Thielemann, F.-K., Davies, M.B., Benz, W., and Piran, T. (1999), *A&A* **341**, 499.

Ruffert, M., Janka, H.-T., and Schäfer, G. (1996), *A&A* **311**, 532.

Ruffert, M., Rampp, M., and Janka, H.-T. (1997), *A&A* **321**, 991.

Saffer, R.A., Livio, M., and Yungelson, L.R. (1999), in S.-E. Solheim and E.G. Meistas (eds.), *11th European Workshop on White Dwarfs*, ASP Conf. Series 169, p260.

Schutz, B.F. (1986), *Nature* **323**, 310.

Seidel, E. (1998), in H. Riffert et al. (eds.), *Relativistic Astrophysics*, Proc. of 162nd W.E. Heraeus Seminar, Vieweg Verlag (Wiesbaden), p229.

Shapiro, S.L. (1998), *Phys.Rev.D* **58**, 103002.

Shibata, M. and Taniguchi, K. (1997), *Phys.Rev.D* **56**, 811.

Shibata, M. and Uryu, K. (2000), *Phys.Rev.D* **61**, 064001.

Strain, K.A. and Meers, B.J. (1991), *Phys.Rev.Lett.* **66**, 1391.

Swesty, F.D., Wang, E.Y.M., and Calder, A.C. (2000), *ApJ* **541**, 937.

Usov, V.V. (1992), *Nature* **357**, 472.

Wang, E.Y.M., Swesty, F.D., and Calder, A.C. (1998), in *Proceedings of the Second Oak Ridge Symposium on Atomic and Nuclear Astrophysics*, (astro-ph/9806022).

Webbink, R.F. (1984), *ApJ* **277**, 355.

Wiringa, R.B., Fiks, V., and Fabrocini, A. (1988), *Phys.Rev.C* **38**, 1010.

Wolszczan, A. and Frail, D.A. (1992), *Nature* **355**, 145.

Wolszczan, A. (1994), *Science* **264**, 538.

Wolszczan, A. (1996), in S. Johnston et al. (eds.), *IAU Colloquium 160, Pulsars: Problems and Progress*, ASP Conf. Series 105, p91.

Wolszczan, A. (1999), in Z. Arzoumanian, F. Van der Hooft, and E.P.J. van den Heuvel (eds.), *Pulsar Timing, General Relativity and the Internal Structure of Neutron Stars*, Koninklijke Nederlandse Akademie van Wetenschappen (Amsterdam), p101.

Yungelson, L.R., Livio, M., Tutukov, A.V., and Saffer, R.A. (1994), *ApJ* **420**, 336.

Zhuge, X., Centrella, J.M., and McMillan, S.L.W. (1994), *Phys.Rev.D* **50**, 6247.

Zhuge, X., Centrella, J.M., and McMillan, S.L.W. (1996), *Phys.Rev.D* **54**, 7261.

3. Accretion Discs

ACCRETION DISKS

H.C. SPRUIT

Max-Planck-Institut für Astrophysik
Postfach 1523, D-85740 Garching, Germany

Abstract. In this lecture the basic theory of accretion disks is introduced, with emphasis on aspects relevant for X-ray binaries and Cataclysmic Variables.

1. Introduction

Accretion disks are inferred to exist in objects of very different scales: km to millions of km in low Mass X-ray Binaries (LMXB) and Cataclysmic Variables (CV), solar radius-to-AU scale in protostellar disks, and AU-to-parsec scales for the disks in Active Galactic Nuclei (AGN).

An interesting observational connection exists between accretion disks and jets (such as the spectacular jets from AGN and protostars), and outflows (the 'CO-outflows' from protostars and possibly the 'broad-line-regions' in AGN). Lacking direct (i.e. spatially resolved) observations of disks, theory has tried to provide models, with varying degrees of success. Uncertainty still exists with respect to some basic questions. In this situation, progress made by observations or modeling of a particular class of objects is likely to have direct impact for the understanding of other objects, including the enigmatic connection with jets.

In this lecture I concentrate on the more basic aspects of accretion disks, but an attempt is made to mention topics of current interest, such as magnetic viscosity, as well. Emphasis is on those aspects of accretion disk theory that connect to the observations of LMXB and CV's. For other reviews on the basics of accretion disks, see Pringle (1981) and Treves et al. (1988). For a more in-depth treatment, see the textbook by Frank et al. (1992).

C. Kouveliotou et al. (eds.), The Neutron Star – Black Hole Connection, 111–139.
© 2001 *Kluwer Academic Publishers. Printed in the Netherlands.*

2. Accretion: general

Gas falling into a point mass potential

$$\Phi = -\frac{GM}{r}$$

from a distance r_0 to a distance r converts gravitational into kinetic energy, by an amount $\Delta\Phi = GM(1/r - 1/r_0)$. For simplicity, assuming that the starting distance is large, $\Delta\Phi = GM/r$. If the gas is then brought to rest, for example at the surface of a star, the amount of energy e dissipated per unit mass is

$$e = \frac{GM}{r} \qquad \text{(rest)}$$

or, if it goes into a circular Kepler orbit at distance r:

$$e = \frac{1}{2}\frac{GM}{r} \qquad \text{(orbit)}.$$

The dissipated energy goes into internal energy of the gas, and into radiation which escapes to infinity (usually in the form of photons, but neutrino losses can also play a role).

2.1. ADIABATIC ACCRETION

Consider first the case when radiation losses are neglected. This is *adiabatic* accretion. For an ideal gas with constant ratio of specific heats γ, the internal energy per unit mass is

$$e = \frac{P}{(\gamma - 1)\rho}.$$

With the equation of state

$$P = \mathcal{R}\rho T/\mu \qquad (1)$$

where $\mathcal{R}$ is the gas constant, and μ the mean atomic weight per particle, we find the temperature of the gas after the dissipation has taken place (assuming that the gas goes into a circular orbit):

$$T = \frac{1}{2}(\gamma - 1)T_{\text{vir}}, \qquad (2)$$

where T_{vir}, the *virial temperature* is given by

$$T_{\text{vir}} = \frac{GM\mu}{\mathcal{R}r}.$$

In an atmosphere with temperature near T_{vir}, the sound speed is close to the escape speed from the system, the hydrostatic pressure scale height is of the order of r, and such an atmosphere may evaporate on a relatively short time scale in the form of a stellar wind.

A simple example is *spherical* adiabatic accretion (Bondi 1952). An important result is that such accretion is possible only if $\gamma \leq 5/3$. The larger γ, the larger the temperature in the accreted gas (eq. 2), and beyond a critical value the temperature is too high for the gas to stay bound in the potential. A classical situation where adiabatic and roughly spherical accretion takes place is a supernova implosion: when the central temperature becomes high enough for the radiation field to start desintegrating nuclei, γ drops and the envelope collapses onto the forming neutron star via a nearly static accretion shock. Another case are Thorne-Zytkow objects (e.g. Cannon et al. 1992), where γ can drop to low values due to pair creation, initiating an adiabatic accretion onto the black hole.

Adiabatic spherical accretion is fast, taking place on the dynamical or free fall time scale

$$\tau_d = r/v_K = (r^3/GM)^{1/2}, \qquad (3)$$

where v_K is the Kepler orbital velocity.

When radiative loss becomes important, the accreting gas can stay cool irrespective of the value of γ, and Bondi's critical value $\gamma = 5/3$ plays no role. With such losses, the temperatures of accretion disks are usually much lower than the virial temperature. The optical depth of the accreting flow increases with the accretion rate $\dot{M}$. When the optical depth becomes large enough so that the photons are 'trapped' in the flow, the accretion just carries them in, together with the gas (Rees 1978, Begelman 1979). Above a certain critical rate $\dot{M}_c$, accretion is therefore adiabatic.

2.2. THE EDDINGTON LIMIT

Objects of high luminosity have a tendency to blow their atmospheres away due to the radiative force exerted when the outward traveling photons are scattered or absorbed. Consider a volume of gas on which a flux of photons is incident from one side. Per gram of matter, the gas presents a scattering (or absorbing) surface area of κ cm^2. The force exerted by the radiative flux F on one gram is $F\kappa/c$. The force of gravity pulling back on this one gram of mass is GM/r^2. The critical flux at which the two forces balance is

$$F_E = \frac{c}{\kappa} \frac{GM}{r^2} \qquad (4)$$

Assuming that the flux is spherically symmetric, this can be converted into a critical luminosity

$$L_E = 4\pi GMc/\kappa,\tag{5}$$

the Eddington luminosity (e.g. Rybicki and Lightman 1979). If the gas is fully ionized, its opacity is dominated by electron scattering, and for solar composition κ is then of the order 0.3 cm^2/g (about a factor 2 lower for fully ionized helium, a factor up to 10^3 higher for partially ionized gases). With these assumptions,

$$L_E \approx 1.7 \times 10^{38} \frac{M}{M_\odot} \text{ erg/s} \approx 4 \times 10^4 \frac{M}{M_\odot} L_\odot$$

If this luminosity results from accretion, it corresponds to the Eddington accretion rate $\dot{M}_E$:

$$\frac{GM}{r}\dot{M}_E = L_E \quad \rightarrow \quad \dot{M}_E = 4\pi rc/\kappa.\tag{6}$$

Whereas L_E is a true limit that can not be exceeded by a static radiating object except by geometrical factors of order unity (see chapter 10 in Frank et al. 1992), no maximum exists on the accretion rate. For $\dot{M} > \dot{M}_E$ the plasma is just swallowed whole, including the radiation energy in it (cf. discussion in the preceding section). With $\kappa = 0.3$:

$$\dot{M}_E \approx 1.3\,10^{18} r_6 \text{ g/s} \approx 2\,10^{-8} r_6 \ M_\odot \text{yr}^{-1},$$

where r_6 is the radius of the accreting object in units of 10 km.

3. Accretion with Angular Momentum

When the accreting gas has a zonzero angular momentum with respect to the accreting object, it can not accrete directly. A new time scale appears, the time scale for outward transport of angular momentum. Since this is in general much longer than the dynamical time scale, much of what was said about spherical accretion needs modification for accretion with angular momentum.

Consider the accretion in a close binary consisting of a compact (white dwarf, neutron star or black hole) primary of mass M_1 and a main sequence companion of mass M_2. The mass ratio is defined as $q = M_2/M_1$ (note: q is just as often defined the other way around).

If M_1 and M_2 orbit each other in a circular orbit and their separation is a, the orbital frequency Ω is

$$\Omega^2 = G(M_1 + M_2)/a^3.$$

The accretion process is most easily described in a coordinate frame that corotates with this orbit, and with its origin in the center of mass. Matter that is stationary in this frame experiences an effective potential, the *Roche potential* (Ch. 4 in Frank, King, and Raine 1992), given by

$$\phi_R(\mathbf{r}) = -\frac{GM}{r_1} - \frac{GM}{r_2} - \frac{1}{2}\Omega^2 r^2 \tag{7}$$

where $r_{1,2}$ are the distances of point $\mathbf{r}$ to stars 1,2. Matter that does *not* corotate experiences a very different force (due to the Coriolis force). The Roche potential is therefore useful only in a rather limited sense. For non-corotating gas intuition based on the Roche geometry is usually confusing. Keeping in mind this limitation, consider the equipotential surfaces of (7). The surfaces of stars $M_{1,2}$, assumed to corotate with the orbit, are equipotential surfaces of (7). Near the centers of mass (at low values of ϕ_R) they are unaffected by the other star, at higher Φ they are distorted and at a critical value Φ_1 the two parts of the surface touch. This is the critical Roche surface S_1 whose two parts are called the Roche lobes. Binaries lose angular momentum through gravitational radiation and a magnetic wind from the secondary (if it has a convective envelope). Through this loss the separation between the components decreases and both Roche lobes decrease in size. Mass transfer starts when M_2 fills its Roche lobe, and continues as long as the angular momentum loss from the system lasts. A stream of gas then flows through the point of contact of the two parts of S_1, the inner Lagrange point L_1. If the force acting on it were derivable entirely from (7) the gas would just fall in radially onto M_1. As soon as it moves however, it does not corotate any more and its orbit under the influence of the Coriolis force is different (Figure 1).

Since the gas at L_1 is very cold compared with the virial temperature, its sound speed is small compared with the velocity it gets after only a small distance from L_1. The flow into the Roche lobe of M_1 is therefore highly *supersonic*. Such hypersonic flow is essentially ballistic, that is, the stream flows along the path taken by free particles.

Though the gas stream on the whole follows an orbit close to that of a free particle, a strong shock develops at the point where the orbit intersects itself. [In practice shocks already develop shortly after passing the pericenter at M_1, when the gas is decelerated again. Supersonic flows that are decelerated by whatever means in general develop shocks (e.g. Courant and Friedrichs 1948; Massey 1968). The effect can be seen in action in the movie published in Różyczka and Spruit 1993]. After this, the gas settles into a ring, into which the stream continues to feed mass. If the mass ratio q is not too small this ring forms fairly close to M_1. An approximate value for its radius is found by noting that near M_1 the tidal force due to the

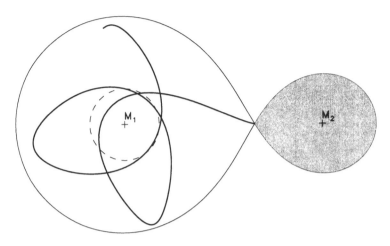

Figure 1.   Roche geometry for $q = 0.2$, with free particle orbit from L_1 (as seen in a frame corotating with the orbit). Dashed: circularization radius.

secondary is small, so that the angular momentum of the gas with respect to M_1 is approximately conserved. If the gas continues to conserve angular momentum while dissipating energy, it settles into the minimum energy orbit with the specific angular momentum j of the incoming stream. The radius of this orbit, the *circularization radius* r_c is determined from

$$(GM_1r_c)^{1/2} = j.$$

The value of j is found by a simple integration of the orbit starting at L_1 and measuring j at some point near pericenter. In units of the orbital separation a, the distances r_c and r_{L1} (measured from M_1 to L_1) are functions only of the mass ratio. As an example for $q = 0.2$, $r_{L1} \approx 0.66a$ and the circularization radius $r_c \approx 0.16a$. In practice the ring forms somewhat outside r_c, because there is some angular momentum redistribution in the shocks that form at the impact of the stream on the ring.

The evolution of the ring depends critically on nature and strength of the angular momentum transport processes. If sufficient 'viscosity' is present it spreads inward and outward to form a disk.

At the point of impact of the stream on the disk the energy dissipated is a significant fraction of the orbital kinetic energy, hence the gas heats up to a significant fraction of the virial temperature. For a typical system with $M_1 = 1M_\odot$, $M_2 = 0.2M_\odot$ having an orbital period of 2 hrs, the observed size of the disk (e.g. Wood et al. 1989b; Rutten et al. 1992) $r_d/a \approx 0.3$, the orbital velocity at r_d about 900 km/s, the virial temperature at r_d is 10^8K. The actual temperatures at the impact point are much lower, due to rapid cooling of the shocked gas. Nevertheless the impact gives rise to a prominent 'hot spot' in many systems, and an overall heating of the outermost part of the disk.

4. Thin disks: properties

4.1. FLOW IN A COOL DISK IS SUPERSONIC

Ignoring viscosity, the equation of motion in the potential of a point mass is

$$\frac{\partial \mathbf{v}}{\partial t} + \mathbf{v} \cdot \nabla \mathbf{v} = -\frac{1}{\rho}\nabla P - \frac{GM}{r^2}\hat{\mathbf{r}}, \tag{8}$$

where $\hat{\mathbf{r}}$ is a unit vector in the spherical radial direction r. To compare the order of magnitude of the terms, choose a position r_0 in the disk, and choose as typical time and velocity scales the orbital time scale $\Omega_0^{-1} = (r_0^3/GM)^{1/2}$ and velocity $\Omega_0 r_0$. The pressure gradient term is

$$\frac{1}{\rho}\nabla P = \frac{\mathcal{R}}{\mu}T\nabla \ln P.$$

In terms of the dimensionless quantities

$$\tilde{r} = r/r_0, \qquad \tilde{v} = v/(\Omega_0 r_0),$$

$$\tilde{t} = \Omega_0 t, \qquad \tilde{\nabla} = r_0 \nabla,$$

the equation of motion is then

$$\frac{\partial \tilde{\mathbf{v}}}{\partial \tilde{t}} + \tilde{\mathbf{v}} \cdot \tilde{\nabla}\tilde{\mathbf{v}} = -\frac{T}{T_{\rm vir}}\tilde{\nabla} \ln P - \frac{1}{\tilde{r}^2}\hat{\mathbf{r}}. \tag{9}$$

All terms and quantities in this equation are of order unity by the assumptions made, except the pressure gradient term which has the coefficient $T/T_{\rm vir}$. If cooling is important, so that $T/T_{\rm vir} \ll 1$, the pressure term is negligible to first approximation, and vice versa. Equivalent statements are also that the gas moves hypersonically on nearly Keplerian orbits, and that the disk is thin, as is shown next.

4.2. DISK THICKNESS

The thickness of the disk is found by considering its equilibrium in the direction perpendicular to the disk plane. In an axisymmetric disk, using cylindrical coordinates (ϖ, ϕ, z), measure the forces at a point $\mathbf{r}_0$ $(\varpi, \phi, 0)$ in the midplane, in a frame rotating with the Kepler rate Ω_0 at that point. The gravitational acceleration $-GM/r^2\,\hat{\mathbf{r}}$ balances the centrifugal acceleration $\Omega_0^2\varpi$ at this point, but not at some distance z above it because gravity and centrifugal acceleration work in different directions. Expanding both accelerations near r_0, one finds a residual acceleration toward the midplane of magnitude

$$g_z = -\Omega_0^2 z.$$

Assuming again an isothermal gas, the condition for equilibrium in the z direction under this acceleration yields a hydrostatic density distribution

$$\rho = \rho_0(\varpi) \exp\left(-\frac{z^2}{2H^2}\right).$$

H, the *scale height* of the disk, is given in terms of the isothermal sound speed $c_s = (\mathcal{R}T/\mu)^{1/2}$ by

$$H = c_s/\Omega_0.$$

We define $\delta \equiv H/r$, the *aspect ratio* of the disk, and find that it can be expressed in several equivalent ways:

$$\delta = \frac{H}{r} = \frac{c_s}{\Omega r} = M^{-1} = \left(\frac{T}{T_{\text{vir}}}\right)^{1/2},$$

where M is the Mach number of the orbital motion.

4.3. VISCOUS SPREADING

The shear flow between neighboring Kepler orbits in the disk causes friction due to viscosity. The frictional torque is equivalent to exchange of angular momentum between these orbits. But since the orbits are close to Keplerian, a change in angular momentum of a ring of gas also means it must change its disctance from the central mass. If the angular momentum is increased, the ring moves to a larger radius. In a thin disk angular momentum transport (more precisely a nonzero divergence of the angular momentum flux) therefore automatically implies redistribution of mass in the disk.

A simple example (Lüst 1952, see also Lynden-Bell and Pringle 1974) is a narrow ring of gas at some distance r_0. If at $t = 0$ this ring is released to evolve under the viscous torques, one finds that it first spreads into an asymmetric hump with a long tail to large distances. As $t \to \infty$ the hump flattens in such a way that almost all the *mass* of the ring is accreted onto the center, while a vanishingly small fraction of the gas carries almost all the *angular momentum* to infinity. As a result of this asymmetric behavior essentially all the mass of a disk can accrete, even if there is no external torque to remove the angular momentum.

4.4. OBSERVATIONS OF DISK VISCOSITY

Evidence for the strength of the angular momentum transport processes in disks comes from observations of variability time scales. This evidence is not good enough to determine whether the processes really have the same effect as a viscosity, but if this is assumed, estimates can be made of the magnitude of the viscosity.

Cataclysmic Variables give the most detailed information. These are binaries with white dwarf (WD) primaries and (usually) main sequence companions (for reviews see Meyer-Hofmeister and Ritter 1993; Cordova 1995; Warner 1995). A subclass of these systems, the Dwarf Novae, show semiregular outbursts. In the currently most developed theory, these outbursts are due to an instability in the disk (Smak 1971; Meyer and Meyer-Hofmeister 1981; for recent references see King 1995; Hameury et al. 1998). The outbursts are episodes of enhanced mass transfer of the disk onto the primary, involving a significant part of the whole disk. The decay time of the burst is thus a measure of the viscous time scale of the disk (the quantitative details depend on the model, see Cannizzo et al. 1988; Hameury et al. 1998):

$$t_{\mathrm{visc}} = r_{\mathrm{d}}^2 / \nu,$$

where r_{d} is the size of the disk. With decay times on the order of days, this yields viscosities of the order 10^{15} cm^2/s, about 14 orders of magnitude above the microscopic viscosity of the gas.

Other evidence comes from the inferred time scale on which disks around protostars disappear, which is of the order of 10^7 years (Strom et al. 1993).

4.5. α-PARAMETRIZATION

The process responsible for such a large viscosity has not been identified with certainty yet. Many processes have been proposed, some of which demonstrably work, though often not with an efficiency as high as the observations of CV outbursts seem to indicate. Other ideas, such as certain turbulence models, do not have much predictive power and are based on ad-hoc assumptions about hydrodynamic instabilities in disks. In order to compare the viscosities in disks under different conditions, one introduces a dimensionless viscosity α:

$$\nu = \alpha \frac{c_{\mathrm{s}}^2}{\Omega}, \tag{10}$$

where c_{s} is the isothermal sound speed as before. The quantity α was introduced by Shakura and Sunyaev (1973), as a way of parametrizing our ignorance of the angular momentum transport process (their definition is based on a different formula however, and differs by a constant of order unity).

How large can the value of α be, on theoretical grounds? As a simple model, let's assume that the shear flow between Kepler orbits is unstable to the same kind of shear instabilities found for flows in tubes, channels, near walls and in jets. These instabilities occur so ubiquitously that the fluid mechanics community considers them a natural and automatic consequence (e.g. DiPrima and Swinney 1981, p144 2nd paragraph) of a high Reynolds number:

$$\text{Re} = \frac{LV}{\nu}$$

where L and V are characteristic length and velocity scales of the flow. If this number exceeds about 1000 (for some forms of instability much less), instability and turbulence are generally observed. It has been argued (e.g. Zel'dovich 1981) that for this reason hydrodynamic turbulence is the cause of disk viscosity. Let's look at the consequences of this assumption. If an eddy of radial length scale l develops due to shear instability, it will rotate at a rate given by the rate of shear, σ, in the flow, here

$$\sigma = r\frac{\partial \Omega}{\partial r} \approx -\frac{3}{2}\Omega.$$

The velocity amplitude of the eddy is $V = \sigma l$, and a field of such eddies produces a turbulent viscosity of the order (leaving out numerical factors of order unity)

$$\nu_{\text{turb}} = l^2\Omega. \tag{11}$$

In compressible flows, there is a maximum to the size of the eddy set by causality considerations. The force that allows an instability to form an overturning eddy is the pressure, which transports information about the flow at the sound speed. The eddies formed by a shear instability can therefore not move faster than c_s, hence their size does not exceed $c_s/\sigma \approx H$. At the same time, the largest eddies formed also have the largest contribution to the turbulent viscosity. Thus we should expect that the turbulent viscosity is given by eddies with size of the order H:

$$\nu \sim H^2\Omega,$$

or

$$\alpha \sim 1.$$

Does hydrodynamical turbulence along these lines exist in disks? Unfortunately, this question is still open, but current opinion is leaning toward the view that the angular momentum transport in sufficiently ionized disks is due to a small scale magnetic field (Shakura and Sunyaev 1973). This is discussed briefly in section 8.

5. Thin Disks: equations

Consider a thin (= cool, nearly Keplerian, cf. section 4.2) disk, axisymmetric but not stationary. Using cylindrical coordinates (r, ϕ, z), (note that we
have changed notation from ϖ to r compared with section 4.2) we define
the *surface density* Σ of the disk as

$$\Sigma = \int_{-\infty}^{\infty} \rho \, dz \approx 2 H_0 \rho_0, \tag{12}$$

where ρ_0, H_0 are the density and scale height at the midplane. The approximate sign is used to indicate that the coefficient in front of H in the last
expression actually depends on details of the vertical structure of the disk.
Conservation of mass, in terms of Σ is given by

$$\frac{\partial}{\partial t}(r\Sigma) + \frac{\partial}{\partial r}(r\Sigma v_r) = 0. \tag{13}$$

(derived by integrating the continuity equation over z). Since the disk is
axisymmetric and nearly Keplerian, the radial equation of motion reduces
to

$$v_\phi^2 = GM/r. \tag{14}$$

The ϕ-equation of motion is

$$\frac{\partial v_\phi}{\partial t} + v_r \frac{\partial v_\phi}{\partial r} + \frac{v_r v_\phi}{r} = F_\phi, \tag{15}$$

where F_ϕ is the azimuthal component of the viscous force. By integrating
this over height z and using (13), one gets an equation for the angular
momentum balance:

$$\frac{\partial}{\partial t}(r\Sigma \Omega r^2) + \frac{\partial}{\partial r}(r\Sigma v_r \Omega r^2) = \frac{\partial}{\partial r}(Sr^3 \frac{\partial \Omega}{\partial r}), \tag{16}$$

where $\Omega = v_\phi/r$, and

$$S = \int_{-\infty}^{\infty} \rho \nu \, dz \approx \Sigma \nu. \tag{17}$$

The second approximate equality in (17) holds if ν can be considered independent of z. The right hand side of (16) is the divergence of the viscous
angular momentum flux, and is derived most easily with a physical argument, as described in, e.g. Pringle (1981) or Frank et al. (1992)[1].

Assume now that ν can be taken constant with height. For an isothermal disk (T independent of z), this is equivalent to taking the viscosity

---

[1]If you prefer a more formal derivation, the fastest way is to consult Landau and
Lifshitz (1959) chapter 15 (hereafter LL). Noting that the divergence of the flow vanishes

parameter α independent of z. As long as we are not sure what causes the viscosity this is a reasonable simplification. Note, however, that recent numerical simulations of magnetic turbulence suggest that the effective α, and the rate of viscous dissipation per unit mass, are higher near the disk surface than near the midplane (see also the discussion in section 8). While eq (16) is still valid for rotation rates Ω deviating from Keplerian (only the integration over disk thickness must be justifiable), we now use the fact that $\Omega \sim r^{-3/2}$. Then Eqs. (13-16) can be combined into a single equation for Σ:

$$r\frac{\partial \Sigma}{\partial t} = 3\frac{\partial}{\partial r}[r^{1/2}\frac{\partial}{\partial r}(\nu\Sigma r^{1/2})]. \tag{18}$$

Under the same assumptions, eq. (15) yields the mass flux $\dot{M}$ at any point in the disk:

$$\dot{M} = -2\pi r\Sigma v_r = 6\pi r^{1/2}\frac{\partial}{\partial r}(\nu\Sigma r^{1/2}). \tag{19}$$

Eq. (18) is the standard form of the *thin disk diffusion equation*. An important conclusion from this equation is: in the thin disk limit, all the physics which determine the time dependent behavior of the disk enter through one quantitity only, the viscosity ν. This is the main attraction of the thin disk approximation.

5.1. STEADY THIN DISKS

In a steady disk ($\partial/\partial t = 0$) the mass flux $\dot{M}$ is constant through the disk and equal to the accretion rate onto the central object. From (19) we get the surface density distribution:

$$\nu\Sigma = \frac{1}{3\pi}\dot{M}\left[1 - \beta\left(\frac{r_i}{r}\right)^{1/2}\right], \tag{20}$$

where r_i is the inner radius of the disk and β is a parameter appearing through the integration constant. It is related to the flux of angular mo-

for a thin axisymmetric disk, the viscous stress σ becomes (LL eq. 15.3)

$$\sigma_{ik} = \eta\left(\frac{\partial v_i}{\partial x_k} + \frac{\partial v_k}{\partial x_i}\right),$$

where $\eta = \rho\nu$. This can be written in cylindrical or spherical coordinates using LL eqs. (15.15-15.18). The viscous force is

$$F_i = \frac{\partial \sigma_{ik}}{\partial x_k} = \frac{1}{\eta}\frac{\partial \eta}{\partial x_k}\sigma_{ik} + \eta\nabla^2 v_i,$$

Writing the Laplacian in cylindrical coordinates, the viscous torque is then computed from the ϕ-component of the viscous force by multiplying by r, and is integrated over z.

mentum F_J through the disk:

$$F_J = -\dot{M}\beta\Omega_i r_i^2, \tag{21}$$

where Ω_i is the Kepler rotation rate at the inner edge of the disk. If the disk accretes onto an object with a rotation rate Ω_* *less* than Ω_i, one finds (Shakura and Sunyaev 1973; Lynden-Bell and Pringle 1974) that $\beta = 1$, independent of Ω_*. Thus the angular momentum flux (torque on the accreting star) is inward (spin-up) and equal to the accretion rate times the specific angular momentum at the inner edge of the disk. For stars rotating near their maximum rate ($\Omega_* \approx \Omega_i$) and for accretion onto magnetospheres, which can rotate faster than the disk, the situation is different (Sunyaev and Shakura 1977, Popham and Narayan 1991; Paczyński 1991; Bisnovatyi-Kogan 1993).

Accreting magnetospheres, for example, can *spin down* by interaction with the disk. This case has a surface density distribution (20) with $\beta < 1$ (see also Spruit and Taam 1993). The angular momentum flux is then outward, and the accreting star spins down. This is possible even when the interaction between the disk and the magnetosphere takes place *only* at the inner edge of the disk. Magnetic torques due to interaction with the magnetosphere may exist at larger distances in the disk as well, but are not necessary for creating an outward angular momentum flux. Recent numerical simulations of disk-magnetosphere interaction (Miller and Stone 1997) give an interesting new view of how such interaction may take place, and suggest it happens very differently from what is assumed in previous 'standard' models.

5.2. DISK TEMPERATURE

In this section I assume accretion onto not-too-rapidly rotating objects, so that $\beta = 1$. The surface temperature of the disk, which determines how much energy it loses by radiation, is governed primarily by the energy dissipation rate in the disk, which in turn is given by the accretion rate. From the first law of thermodynamics we have

$$\rho T \frac{dS}{dt} = -\text{div}\mathbf{F} + Q_v, \tag{22}$$

where S the entropy per unit mass, $\mathbf{F}$ the heat flux (including radiation and any form of 'turbulent' heat transport), and Q_v the viscous dissipation rate. For changes which happen on time scales longer than the dynamical time Ω^{-1}, the left hand side is small compared with the terms on the right hand side. Integrating over z, the divergence turns into a surface term and we get

$$2\sigma_r T_s^4 = \int_{-\infty}^{\infty} Q_v \mathrm{d}z, \tag{23}$$

where T_s is the surface temperature of the disk, σ_r is Stefan-Boltzmann's radiation constant $\sigma_r = a_r c/4$, and the factor 2 comes about because the disk has 2 radiating surfaces (assumed to radiate like black bodies). Thus the energy balance is *local* (for such slow changes): what is generated by viscous dissipation inside the disk at any radius r is also radiated away from the surface at that position. The viscous dissipation rate is equal to $Q_v = \sigma_{ij}\partial v_i/\partial x_j$, where σ_{ij} is the viscous stress tensor (see footnote in section 5), and this works out[2] to be

$$Q_v = 9/4\,\Omega^2 \nu\rho. \tag{24}$$

Eq. (23), using (20) then gives the surface temperature in terms of the accretion rate:

$$\sigma_r T_s^4 = \frac{9}{8}\Omega^2 \nu\Sigma = \frac{GM}{r^3}\frac{3\dot{M}}{8\pi}\left[1 - \left(\frac{r_i}{r}\right)^{1/2}\right]. \tag{25}$$

This shows that the surface temperature of the disk, at a given distance r from a steady accreter, depends *only* on the product $M\dot{M}$, and not on the highly uncertain value of the viscosity. For $r \gg r_i$ we have

$$T_s \sim r^{-3/4}. \tag{26}$$

These considerations only tell us about the surface temperature. The internal temperature in the disk is quite different, and depends on the mechanism transporting energy to the surface. Because it is the internal temperature that determines the disk thickness H (and probably also the viscosity), this transport needs to be considered in some detail for realistic disk models. This involves a calculation of the vertical structure of the disk. Because of the local (in r) nature of the balance between dissipation and energy loss, such calculations can be done as a grid of models in r, without having to deal with exchange of energy between neighboring models. Schemes borrowed from stellar structure computations are used (e.g. Meyer and Meyer-Hofmeister 1982; Pringle et al. 1986; Cannizzo et al. 1988).

An approximation to the temperature in the disk can be found when a number of additional assumptions is made. As in stellar interiors, the energy transport is radiative rather than convective at high temperatures.

[2]using, e.g. LL eq. 16.3

Assuming local thermodynamic equilibrium (LTE, e.g. Rybicki and Lightman 1979), the temperature structure of a radiative atmosphere is given, in the Eddington approximation by:

$$\frac{\mathrm{d}}{\mathrm{d}\tau}\sigma_r T^4 = \frac{3}{4}F. \tag{27}$$

The boundary condition that there is no incident flux from outside the atmosphere yields the approximate condition

$$\sigma_r T^4(\tau = 2/3) = F, \tag{28}$$

where $\tau = \int_z^\infty \kappa\rho\mathrm{d}z$ is the optical depth at geometrical depth z, and F the energy flux through the atmosphere. Assuming that most of heat is generated near the midplane (which is the case if ν is constant with height), F is approximately constant with height and equal to $\sigma_r T_s^4$, given by (25). Eq (27) then yields

$$\sigma_r T^4 = \frac{3}{4}(\tau + \frac{2}{3})F. \tag{29}$$

Approximating the opacity κ as remaining constant with z, the optical depth at the midplane is $\tau = \kappa\Sigma/2$. If $\tau \gg 1$, the temperature at the midplane is then:

$$T^4 = \frac{27}{64}\sigma_r^{-1}\Omega^2\nu\Sigma^2\kappa. \tag{30}$$

With the equation of state (1), valid when radiation pressure is small, we find for the disk thickness, using (20):

$$\begin{aligned}
\frac{H}{r} &= (\mathcal{R}/\mu)^{2/5}\left(\frac{3}{64\pi^2\sigma_r}\right)^{1/10}(\kappa/\alpha)^{1/10}(GM)^{-7/20}r^{1/20}(f\dot{M})^{1/5} \\
&= 5\,10^{-3}\alpha^{-1/10}r_6^{1/20}(M/M_\odot)^{-7/20}(f\dot{M}_{16})^{1/5}, \qquad (P_r \ll P) \quad (31)
\end{aligned}$$

where $r_6 = r/(10^6\,\mathrm{cm})$, $\dot{M}_{16} = \dot{M}/(10^{16}\mathrm{g/s})$, and

$$f = 1 - (r_i/r)^{1/2}. $$

From this we conclude that: i) the disk is thin in X-ray binaries, $H/r < 0.01$, ii) the disk thickness is relatively insensitive to the parameters, especially α, κ and r. It must be stressed, however, that this depends fairly strongly on the assumption that the energy is dissipated in the disk interior. If the dissipation takes place close to the surface [such as in some magnetic reconnection models (Haardt et al. 1994; Di Matteo et al. 1999 and references therein)], the internal disk temperature will be much closer to the surface temperature. The midplane temperature and H are even smaller in such disks than calculated from (31).

The viscous dissipation rate per unit area of the disk, $W_v = (9/4)\Omega^2 \nu \Sigma$ [cf. eq. 25)] can be compared with the local rate W_G at which gravitational energy is liberated in the accretion flow. Since half the gravitational energy stays in the flow as orbital motion, we have

$$W_G = \frac{1}{2\pi r} \frac{GM\dot{M}}{2r^2},$$ (32)

so that

$$W_v/W_G = 3f = 3[1 - (r_i/r)^{1/2}].$$ (33)

At large distances from the inner edge, the dissipation rate is *3 times larger than the rate of gravitational energy release*. This may seem odd, but becomes understandable when it is realized that there is a significant flux of energy through the disk associated with the viscous stress[3]. Integrating the viscous energy dissipation over the whole disk, one finds

$$\int_{r_i}^{\infty} 2\pi r W_v \mathrm{d}r = \frac{GM\dot{M}}{2r_i},$$ (34)

as expected. That is, globally, but not locally, half of the gravitational energy is radiated from the disk while the other half remains in the orbital kinetic energy of the accreted material.

What happens to this remaining orbital energy depends on the nature of the accreting object. If the object is a slowly rotating black hole, the orbital energy is just swallowed by the hole. If it has a solid surface, the orbiting gas slows down until it corotates with the surface, dissipating the orbital energy into heat in a boundary layer. Unless the surface rotates close to the orbital rate ('breakup'), the energy released in this way is of the same order as the total energy released in the accretion disk. The properties of this boundary layer are therefore crucial for accretion onto neutron stars and white dwarfs (see also section 9.1 and Inogamov and Sunyaev (1999)).

5.3. RADIATION PRESSURE DOMINATED DISKS

In the inner regions of disks in XRB, the radiation pressure can dominate over the gas pressure, which results in a different expression for the disk thickness. The total pressure P is

$$P = P_r + P_g = \frac{1}{3}aT^4 + P_g.$$ (35)

---

[3]See LL section 16

Defining a 'total sound speed' by $c_t^2 = P/\rho$ the relation $c_t = \Omega H$ still holds. For $P_r \gg P_g$ we get from (30), with (25) and $\tau \gg 1$:

$$cH = \frac{3}{8\pi}\kappa f \dot{M},$$

(where the rather approximate relation $\Sigma = 2H\rho_0$ has been used). Thus,

$$\frac{H}{R} \approx \frac{3}{8\pi}\frac{\kappa}{cR} f \dot{M} = \frac{3}{2} f \frac{\dot{M}}{\dot{M}_E}, \tag{36}$$

where R is the stellar radius and $\dot{M}_E$ the Eddington rate for this radius. It follows that the disk becomes thick near the star, if the accretion rate is near Eddington (though this is mitigated somewhat by the decrease of the factor f). Accretion near the Eddington limit is evidently not geometrically thin any more. In addition, other processes such as angular momentum loss by 'photon drag' have to be taken into account.

5.4. TIME SCALES IN A DISK

Three locally defined time scales play a role in thin disks. The dynamical time scale t_d is the orbital time scale:

$$t_d = \Omega^{-1} = (GM/r^3)^{-1/2}. \tag{37}$$

The time scale for radial drift through the disk over a distance of order r is the viscous time scale:

$$t_v = r/(-v_r) = \frac{2}{3}\frac{rf}{\nu} = \frac{2f}{3\alpha\Omega}(\frac{r}{H})^2, \tag{38}$$

(using (19 and (20), valid for steady accretion). Finally, there are *thermal* time scales. If E_t is the thermal energy content (enthalpy) of the disk per unit of surface area, and $W_v = (9/4)\Omega^2 \nu\Sigma$ the heating rate by viscous dissipation, we can define a heating time scale:

$$t_h = E_t/W_v. \tag{39}$$

In the same way, a cooling time scale is defined by the energy content and the radiative loss rate:

$$t_c = E_t/(2\sigma_r T_s^4). \tag{40}$$

For a thin disk, the two are equal since the viscous energy dissipation is locally balanced by radiation from the two disk surfaces. [In thick disks (ADAFs), this balance does not hold, since the advection of heat with the accretion flow is not negligible. In ADAFs, $t_c > t_h$ (see Spruit, this volume)].

Thus, we can replace both time scales by a single thermal time scale t_t, and find, with (24):

$$t_t = \frac{1}{W_v} \int_{-\infty}^{\infty} \frac{\gamma P}{\gamma - 1} dz, \qquad (41)$$

where the enthalpy of an ideal gas of constant ratio of specific heats γ has been used. Leaving out numerical factors of order unity, this yields

$$t_t \approx \frac{1}{\alpha \Omega}. \qquad (42)$$

That is, the thermal time scale of the disk is independent of most of the disk properties and of the order $1/\alpha$ times longer than the dynamical time scale. This independence is a consequence of the α-parametrization used. If α is not a constant, but dependent on disk temperature for example, the dependence of the thermal time scale on disk properties will become apparent again.

If, as seems likely from observations, α is generally < 1, we have in thin disks the ordering of time scales:

$$t_v \gg t_t > t_d. \qquad (43)$$

6. Comparison with CV observations

The number of meaningful quantitative tests between the theory of disks and observations is somewhat limited since in the absence of a theory for ν, it is a bit meagre on predictive power. The most detailed information perhaps comes from modeling of CV outbursts.

Two simple tests are possible (nearly) independently of ν. These are the prediction that the disk is geometrically quite thin (eq. 31) and the prediction that the surface temperature $T_s \sim r^{-3/4}$ in a steady disk. The latter can be tested in a subclass of the CV's that do not show outbursts, the nova-like systems, which are believed to be approximately steady accreters. If the system is also eclipsing, eclipse mapping techniques can be used to derive the brightness distribution with r in the disk (Horne 1985;1993). If this is done in a number of colors so that bolometric corrections can be made, the results (e.g. Rutten et al. 1992) show in general a *fair* agreement with the $r^{-3/4}$ prediction. Two deviations occur: i) a few systems show significantly flatter distributions than predicted, and ii) most systems show a 'hump' near the outer edge of the disk. The latter deviation is easily explained, since we have not taken into account that the impact of the stream heats the outer edge of the disk. Though not important for the total light from the disk, it is an important local contribution near the edge.

Eclipse mapping of Dwarf Novae in quiescence gives a quite different picture. Here, the inferred surface temperature profile is often nearly flat (e.g. Wood et al. 1989a;1992). This is understandable however since in quiescence the mass flux depends strongly on r. In the inner parts of the disk it is small, near the outer edge it is close to its average value. With eq. (25), this yields a flatter $T_s(r)$. The lack of light from the inner disk is compensated during the outburst, when the accretion rate in the inner disk is higher than average (see Mineshige and Wood 1989 for a more detailed comparison). The effect is also seen in the 2-dimensional hydrodynamic simulations of accretion in a binary by Różyczka and Spruit (1993). These simulations show an outburst during which the accretion in the inner disk is enhanced, between two episodes in which mass accumulates in the outer disk.

7. Comparison with LMXB observations: irradiated disks

In low mass X-ray binaries a complication arises because of the much higher luminosity of the accreting object. Since a neutron star is roughly 1000 times smaller than a white dwarf, it produces 1000 times more luminosity for a given accretion rate.

Irradiation of the disk by the central source leads to a different surface temperature than predicted by (25). The central source (star plus inner disk) radiates the total accretion luminosity $GM\dot{M}/R$ (assuming sub-Eddington accretion, see section 2). If the disk is *concave*, it will intercept some of this luminosity. If the central source is approximated as a point source the irradiating flux on the disk surface is

$$F_{\text{irr}} = \epsilon \frac{GM\dot{M}}{4\pi R r^2},\tag{44}$$

where ϵ is the angle between the disk surface and the direction from a point on the disk surface to the central source:

$$\epsilon = \mathrm{d}H/\mathrm{d}r - H/r.\tag{45}$$

The disk is concave if ϵ is positive. We have

$$\frac{F_{\text{irr}}}{F} = \frac{2}{3}\frac{\epsilon}{f}\frac{r}{R},$$

where F is the flux generated internally in the disk, given by (25). On average, the angle ϵ is of the order of the aspect ratio $\delta = H/r$. With $f \approx 1$, and our fiducial value $\delta \approx 5 \times 10^{-3}$, we find that irradiation in LMXB dominates for $r > 10^9$cm. This is compatible with observations (for reviews see van Paradijs and McClintock 1993), which show that the optical and UV are dominated by reprocessed radiation.

When irradiation by an external source is included in the thin disk model, the surface boundary condition of the radiative transfer problem, equation (28) becomes

$$\sigma_r T_s^4 = F + (1 - a) F_{irr},   \qquad (46)$$

where a is the X-ray albedo of the surface, i.e. $1 - a$ is the fraction of the incident flux that is absorbed in the *optically thick* layers of the disk (photons absorbed higher up only serve to heat up the corona of the disk). The surface temperature T_s increases in order to compensate for the additional incident heat flux. The magnitude of the incident flux is sensitive to the assumed disk shape $H(r)$, as well as on the assumed shape (plane or spherical, for example) of the central X-ray emitting region. The disk thickness depends on temperature, and thereby also on the irradiation. It turns out, however, that this dependence on the irradiating flux is small, if the disk is optically thick, and the energy transport is by radiation (Lyutyi and Sunyaev 1976). To see this, integrate (27) with the modified boundary condition (46). This yields

$$\sigma_r T^4 = \frac{3}{4} F(\tau + \frac{2}{3}) + \frac{(1 - a) F_{irr}}{F}.   \qquad (47)$$

The irradiation adds an additive constant to $T^4(z)$. At the midplane, this constant has much less effect than at the surface. For the midplane temperature and the disk thickness to be affected significantly, it is necessary that

$$F_{irr}/F \gtrsim \tau.   \qquad (48)$$

The reason for this weak dependence of the midplane conditions on irradiation is the same as in radiative envelopes of stars, which are also insensitive to the surface boundary condition. The situation is very different for convective disks. As in fully convective stars, the adiabatic stratification then causes the conditions at the midplane to depend much more directly on the surface temparture. The outer parts of the disks in LMXB with wide orbits may be convective, and their thickness affected by irradiation.

In the reprocessing region of the disks of LMXB, the conditions are such that $F \ll F_{irr} \approx \tau F$, hence we must use eq. (31) for H. This yields $\epsilon = (21/20) H/r \approx 5\,10^{-3}$, and $T_s \sim r^{0.5}$, and we still expect the disk to remain thin.

From the paucity of sources in which the central source is eclipsed by the companion one deduces that the companion is barely or not at all visible from the inner disk, presumably because the outer parts of the disk are much thicker than expected from the above arguments. This is consistent with the observation that the characteristic modulation of the optical light curve due to irradiation of the secondary's surface by the X-rays is not very strong in LMXB (with the exception of Her X-1, which has a large companion). The place of the eclipsing systems is taken by the so-called 'Accretion Disk Corona' (ADC) systems, where shallow eclipses of a rather extended X-ray source are seen instead of the expected sharp eclipses of the inner disk (for reviews of the observations, see Lewin et al. 1995). The conclusion is that there is an extended X-ray scattering 'corona' above the disk. It scatters a few per cent of the X-ray luminosity.

What causes this corona and the large inferred thickness of the disk ? The thickness expected from disk theory is a rather stable small number. To 'suspend' matter at the inferred height of the disk forces are needed that are much larger than the pressure forces available in an optically thick disk. A thermally driven wind, produced by X-ray heating of the disk surface, has been invoked (Begelman et al. 1983; Schandl and Meyer 1994). For other explanations, see van Paradijs and McClintock (1995). Perhaps a magnetically driven wind from the disk, such as seen in protostellar objects (e.g. Königl and Ruden 1993) can explain both the shielding of the companion and the scattering. Such a model would resemble magnetically driven wind models for the broad-line region in AGN (e.g. Emmering et al. 1992; Königl and Kartje 1994). A promising possibility is that the reprocessing region at the disk edge consists of matter 'kicked up' at the impact of the mass transfering stream (Meyer-Hofmeister et al. 1997; Armitage and Livio 1998; Spruit et al. 1998). This produces qualitatively the right dependence of X-ray absorption on orbital phase in ADC sources, and the light curves of the so-called supersoft sources.

7.1. TRANSIENTS

Soft X-ray transients (also called X-ray Novae) are believed to be binaries similar to the other LMXB, but somehow the accretion is episodic, with very large outbursts recurring on time scales of decades (sometimes years). There are many black hole candidates among these transients (see Lewin et al. 1995 for a review). As with the Dwarf Novae, the time dependence of the accretion in transients can in principle be exploited to derive information on the disk viscosity, assuming that the outburst is caused by an instability in the disk. The closest relatives of soft transients among the White Dwarf plus main sequence star systems are probably the WZ Sge stars (van Paradijs and Verbunt 1984; Kuulkers et al. 1996), which show (in the optical) similar outbursts with similar recurrence times (cf. Warner 1987; O'Donoghue et al. 1991). Like the soft transients, they have low mass ratios ($q < 0.1$).

For a given angular momentum loss, systems with low mass ratios have low mass transfer rates, so the speculation is that the peculiar behavior of these systems is somehow connected with a low mean accretion rate.

7.2. DISK INSTABILITY

The most developed model for outbursts is the disk instability model of Osaki (1974); Hōshi (1979); Smak (1971;1984); Meyer and Meyer-Hofmeister (1981); see also King (1995); Osaki (1993). In this model the instability that gives rise to cyclic accretion is due to a temperature dependence of the viscous stress. In any local process that causes an effective viscosity, the resulting α- parameter will be a function of the main dimensionless parameter of the disk, the aspect ratio H/r. If this is a sufficiently rapidly increasing function, such that α is large in hot disks and low in cool disks, an instability results by the following mechanism. Suppose we start the disk in a stationary state at the mean accretion rate. If this state is perturbed by a small temperature increase, α goes up, and by the increased viscous stress the mass flux $\dot{M}$ increases. By (25) this increases the disk temperature further, resulting in a runaway to a hot state. Since $\dot{M}$ is larger than the average, the disk empties partly, reducing the surface density and the central temperature (eq. 30). A cooling front then transforms the disk to a cool state with an accretion rate below the mean. The disk in this model switches back and forth between hot and cool states. By adjusting α in the hot and cool states, or by adjusting the functional dependence of α on H/r, outbursts are obtained that agree reasonably with the observations of soft transients (Lin and Taam 1984; Mineshige and Wheeler 1989). A rather strong increase of α with H/r is needed to get the observed long recurrence times.

Another possible mechanism for instability has been found in 2-D numerical simulations of accretion disks (Blaes and Hawley 1988; Różyczka and Spruit 1993). The outer edge of a disk is found, in these simulations, to become dynamically unstable to an oscillation which grows into a strong eccentric perturbation (a crescent shaped density enhancement which rotates at the local orbital period). Shock waves generated by this perturbation spread mass over most of the Roche lobe; at the same time the accretion rate onto the central object is strongly enhanced. This process is different from the Smak-Osaki-Hōshi mechanism, since it requires 2 dimensions, and does not depend on the viscosity (instead, the internal dynamics in this instability *generates* the effective viscosity that causes a burst of accretion).

7.3. OTHER INSTABILITIES

Instability to heating/cooling of the disk can be due to several effects. The cooling rate of the disk, if it depends on temperature in an appropriate way, can cause a thermal instability like that in the interstellar medium. Other instabilities may result from the dependence of viscosity on conditions in the disk. For a general treatment see Piran (1978), for a shorter discussion see Treves et al. (1988).

8. Sources of Viscosity

The high Reynolds number of the flow in accretion disks (of the order 10^{11} in the outer parts of a CV disk) would, to most fluid dynamicists, seem an amply sufficient condition for the occurrence of hydrodynamic turbulence. A theoretical argument against such turbulence often used in astrophysics (Kippenhahn and Thomas 1981; Pringle 1981) is that in cool disks the gas moves almost on Kepler orbits, which are quite stable (except for the orbits that get close to the companion or near a black hole). This stability is related to the known stabilizing effect that rotation has on hydrodynamical turbulence (Bradshaw 1969, for a discussion see Tritton 1992). Kippenhahn and Thomas also point out that the one laboratory experiment that comes close to the situation in accretion disks, namely the rotating Couette flow, does not become unstable for parameters like in disks (for the rather limited range in Reynolds numbers available). A (not very strong) observational argument is that hydrodynamical turbulence as described above would produce an α that does not depend on the nature of the disk, so that all objects should have the same value. This is unlikely to be the case. From the modeling of CV outbursts one knows, for example, that α probably increases with temperature (more accurately, with H/r, see previous section). Also, there are indications from the inferred life times and sizes of protostellar disks (Strom et al. 1993) that α may be rather small there, $\sim 10^{-3}$, whereas in outbursts of CV's one infers values of the order $0.1 - 1$.

The indeterminate status of the hydrodynamic turbulence issue is an annoying problem in disk theory. Direct 3-D numerical simulation of the hydrodynamics in accretion disks is possible, and so far has not shown the expected turbulence. In fact, Balbus and Hawley (1996), and Hawley et al (1999) argue, on the basis of such simulations and a physical argument, that disks are actually quite stable against hydrodynamic turbulence, as long as the specific angular momentum increases outward. [Such heresy would not pass a referee in a fluid mechanics journal.] If it is true that disks are stable to hydrodynamic turbulence it will be an uphill struggle to convince the fluid mechanics community, since it can always be argued that one should go to even higher Reynolds numbers to see the expected turbulence in the simulations or experiments.

The astrophysical approach has been to circumvent the problem by finding plausible alternative mechanisms that might work just as well. Among the processes that have been proposed repeatedly as sources of viscosity is convection due to a vertical entropy gradient (e.g. Kley et al. 1993), which may have some limited effect in convective parts of disks. Another class are *waves* of various kinds. Their effect can be global, that is, not reducible to a local viscous term because by traveling across the disk they can communicate torques over large distances. For example, waves set up at the outer edge of the disk by tidal forces can travel inward and by dissipating there can effectively transport angular momentum *outward* (e.g. Narayan et al. 1987; Spruit et al. 1987). A nonlinear version of this idea are self-similar spiral shocks, observed in numerical simulations (Sawada et al. 1987) and studied analytically (Spruit 1987). Such shocks can produce accretion at an effective α of 0.01 in hot disks, but are probably not very effective in disks as cool as those in CV's and XRB. A second non-local mechanism is provided by a magnetically accelerated *wind* originating from the disk surface (Blandford 1976; Bisnovatyi-Kogan and Ruzmaikin 1976; Lovelace 1976; Blandford and Payne 1982; for reviews see Blandford 1989; Blandford and Rees 1992; for an introduction see Spruit 1996). In principle, such winds can take care of *all* the angular momentum loss needed to make accretion possible in the absence of a viscosity (Blandford 1976; Königl 1989). The attraction of this idea is that magnetic winds are a strong contender for explaining the strong outflows and jets seen in protostellar objects and AGN. It is not yet clear however if, even in these objects, the wind is actually the main source of angular momentum loss.

In sufficiently cool or massive disks, self-gravitating instabilities of the disk matter can produce internal friction. Paczyński (1978) has proposed that the resulting heating would limit the instability and keep the disk in a well defined moderately unstable state. The angular momentum transport in such a disk has been modeled by several authors (e.g. Ostriker et al. 1999). Disks in XRB are too hot for self-gravity to play a role.

8.1. MAGNETIC VISCOSITY

Magnetic forces can be very effective at transporting angular momentum. If it can be shown that the shear flow in the disk produces some kind of small scale fast dynamo process, that is, some form of magnetic turbulence, an effective $\alpha \sim O(1)$ expected (Shakura and Sunyaev 1973; Eardley and Lightman 1975; Pudritz 1981; Meyer and Meyer-Hofmeister 1982). Numerical simulations of initially weak magnetic fields in accretion disks have now shown that this does indeed happen in sufficiently ionized disks (Hawley et al. 1995; Brandenburg et al. 1995; Armitage 1998). These show a small scale magnetic field with azimuthal component dominating (due to stretching by differential rotation). The effective α's are of the order 0.05. The angular momentum transport is due to magnetic stresses. The fluid motions induced by the magnetic forces contribute only little to the angular momentum transport. In a perfectly conducting plasma this turbulence can develop from an arbitrarily small initial field through magnetic shear instability (also called magnetorotational instability, Velikhov 1959; Chandrasekhar 1961; Balbus and Hawley 1991; 1992). The significance of this instability is that it shows that at large conductivity accretion disks must be magnetic. The actual form of the highly time dependent small scale magnetic field which develops can only be found from numerical simulations.

8.2. VISCOSITY IN RADIATIVELY SUPPORTED DISKS

A disk in which the radiation pressure P_r dominates must be optically thick (otherwise the radiation would escape). The radiation pressure then adds to the total pressure which is larger than it would be, for a given temperature, if only the gas pressure were effective. If the viscosity is then parametrized by (10), it turns out (Lightman and Eardley 1974) that the disk is locally unstable. An increase in temperature increases the radiation pressure, which increases the viscous dissipation and the temperature, leading to a runaway. This has raised the question whether the radiation pressure should be included in the sound speed that enters expression (10). If it is left out, a lower viscosity results, and there is no thermal-viscous runaway. Without knowledge of the process causing the effective viscous stress, this question can not be answered. Sakimoto and Coroniti (1989) have shown, however, that if the stress is due to some form of magnetic turbulence, it most likely scales with the gas pressure alone, rather than the total pressure. Now that it seems likely, from the numerical simulations, that the stress is indeed magnetic, there is reason to believe that in the radiation pressure-dominated case the effective viscosity will scale as $\nu \sim \alpha P_g/(\rho\Omega)$ (this case has not been studied with simulations yet). Nayakshin and Rappaport (1999) show that, depending on how the viscosity scales in the intermediate regime $P_g \approx P_{rad}$, interesting cyclic behavior can occur akin to the 'S-curve' instability in CV disks (section 7.2).

9. Beyond thin disks

Ultimately, much of the progress in developing useful models of accretion disks will depend on detailed numerical simulations in 2 or 3 dimensions. In the disks one is interested in, there is usually a large range in length scales (in LMXB disks, from less than the 10 km neutron star radius to the more than 10^5 km orbital scale). Correspondingly, there is a large range in time scales that have to be followed. This not technically possible at present and in the foreseeable future. In numerical simulations one is therefore limited to studying in an approximate way aspects that are either local or of limited dynamic range in r, t (for examples, see Hawley 1991; Różyczka and Spruit 1993; Armitage 1998). For this reason, there is still a need for approaches that relax the strict thin disk framework somewhat without resorting to full simulations. Due to the thin disk assumptions, the pressure gradient does not contribute to the support in the radial direction and the transport of heat in the radial direction is negligible. Some of the physics of thick disks can be included in a fairly consistent way in the 'slim disk' approximation (Abramowicz et al. 1988). The so-called Advection Dominated Accretion Flows (ADAFs) are related to this approach (for a review see Yi 1998; for an introduction Spruit, this volume).

9.1. BOUNDARY LAYERS

In order to accrete onto a star rotating at the rate Ω_*, the disk matter must dissipate an amount of energy given by

$$\frac{GM\dot{M}}{2R} \left[1 - \Omega_*/\Omega_k(R)\right]^2 . \tag{49}$$

The factor in brackets measures the kinetic energy of the matter at the inner edge of the disk ($r = R$), in the frame of the stellar surface. Due to this dissipation the disk inflates into a 'belt' at the equator of the star, of thickness H and radial extent of the same order. Equating the radiation emitted from the surface of this belt to (49) one gets for the surface temperature T_{sb} of the belt, assuming optically thick conditions and a slowly rotating star ($\Omega_*/\Omega_k \ll 1$):

$$\frac{GM\dot{M}}{8\pi R^2 H} = \sigma_r T_{sb}^4 \tag{50}$$

To find the temperature inside the belt and its thickness, use eq. (29). The value of the surface temperature is higher, by a factor of the order $(R/H)^{1/4}$, than the simplest thin disk estimate (25, ignoring the $(r/r_i)^{1/2}$ factor). In practice, this works out to a factor of a few. The surface of the belt is therefore not very hot. The situation is quite different if the boundary layer is not optically thick (Pringle and Savonije 1979). It then heats up to much higher temperatures. Analytical methods to obtain the boundary layer structure have been used by Regev and Hougerat (1988), numerical solutions of the slim disk type by Narayan and Popham (1993), Popham (1997), 2-D numerical simulations by Kley (1991). These considerations are primarily relevant for CV disks; in accreting neutron stars, the dominant effects of radiation pressure have to be included. More analytic progress on the structure of the boundary layer between a disk and a neutron star and the way in which it spreads over the surface of the star is reported by Inogamov and Sunyaev (1999).

References

Abramowicz, M.A., Czerny B., Lasota J.-P., and Szuszkiewicz E. (1988), *Astrophys. J.* **332**, 646.

Armitage, P.J. (1998), *Astrophys. J.* **501**, L189.

Armitage, P.J. and Livio, M. (1998), *Astrophys. J.* **493**, 898.

Balbus, S.A. and Hawley, J.F. (1991), *Astrophys. J.* **376**, 214.

Balbus, S.A. and Hawley, J.F. (1992), *Astrophys. J.* **400**, 610.

Balbus, S.A. and Hawley, J.F. (1996), *Astrophys. J.* **464**, 690.

Begelman, M.C. (1979, *Mon. Not. R. Astron. Soc.* **187**, 237.

Begelman, M.C., McKee, C.F., and Shields, G.A. (1983), *Astrophys. J.* **271**, 70.

Bisnovatyi-Kogan, G. and Ruzmaikin, A.A. (1976), *Ap. Space Sci.* **42**, 401.

Bisnovatyi-Kogan, G. (1993), *Astron. Astrophys.* **274**, 796.

Blaes, O. and Hawley, J.F. (1988), *Astrophys. J.* **326**, 277.

Blandford, R.D. (1976), *Mon. Not. R. Astron. Soc.* **176**, 465.

Blandford, R.D. and Payne, D.G. (1982) *Mon. Not. R. Astron. Soc.* **199**, 883.

Blandford, R.D. (1989), in F. Meyer, W. Duschl, J. Frank, and E. Meyer-Hofmeister (eds.), *Theory of Accretion Disks*, Kluwer (Dordrecht), p35.

Blandford, R.D. and Rees, M.J. (1992), in S. Holt, S. Neff, and C.M. Urry (eds.), *Testing the AGN paradigm*, Univ. of Maryland, pp3-19.

Bondi, H. (1952), *Mon. Not. R. Astron. Soc.* **112**, 195.

Bradshaw, P. (1969), *J. Fluid Mech.* **36**, 177.

Brandenburg, A., Nordlund, Å., Stein, R.F., and Torkelsson, U. (1995), *Astrophys. J.* **446**, 741.

Cannon, R.C., Eggleton, P.P., Zytkow, A.N., and Podsiadlowski, P. (1992), *Astrophys. J.* **386**, 206.

Cannizzo, J.K., Shafter, A.W., and Wheeler, J.C. (1988), *Astrophys. J.* **333**, 227.

Cordova, F.A. (1995), in W.H.G. Lewin et al. (eds.) *X-ray Binaries*, Cambridge University Press (Cambridge), p331.

Courant, R. and Friedrichs, K.O. (1948), *Supersonic Flow and Shock Waves*, Interscience (New York) and Springer (Berlin) (1976).

Chandrasekhar, C. (1961), *Hydrodynamic and Hydromagnetic Stability*, Oxford Univ. Press (Oxford) and Dover Publications Inc. (1981).

Di Matteo, T., Celotti, A., and Fabian, A.C. (1999), *Mon. Not. R. Astron. Soc.* **304**, 809.

DiPrima, R.C. and Swinney, H.L. (1981), in H.L. Swinney and J.P. Gollub (eds.), *Hydrodynamic Instabilities and the Transition to Turbulence*, Springer, p140.

Eardley, D. and Lightman, A. (1975), *Astrophys. J.* **200**, 187.

Emmering, R.T., Blandford, R.D., and Shlosman, I. (1992), *Astrophys. J.* **385**, 460.

Frank, J., King, A.R., and Raine, D.J. (1992), *Accretion Power in Astrophysics* (2nd edition), Cambridge University Press.

Haardt, F., Maraschi, L., and Ghisellini, G. (1994), *Astrophys. J.* **432**, 95.

Hameury, J.-M., Menou, K., Dubus, G., Lasota, J.-P., and Hure, J.-M. (1998), *Mon. Not. R. Astron. Soc.* **298**, 1048.

Hawley, J.E. (1991) *Astrophys. J.* **381**, 496.

Hawley, J.F., Gammie, C.F., and Balbus, S.A. (1995), *Astrophys. J.* **440**, 742.

Hawley, J.F., Balbus, S.A., and Winters, W.F. (1999), *Astrophys. J.* **518**, 394.

Horne, K. (1985), *Mon. Not. R. Astron. Soc.* **213**, 129.

Horne, K., Marsh, T.R., Cheng, F.H., Hubeny, I., and Lanz, T. (1994), *Astrophys. J.* **426**, 294.

Horne, K. (1993), in J.C. Wheeler (ed.), *Accretion Disks in Compact Stellar Systems*, World Scientific Publishing (Singapore).

Hōshi, R. (1979), *Prog. Theor. Phys.* **61**, 1307.

Inogamov, N.A. and Sunyaev, R.A. (1999), *Astron. Lett.* **25**, 269.

Kippenhahn, R. and Thomas, H.-C. (1981), in D. Sugimoto, D. Lamb, and D.N. Schramm (eds.), *Fundamental Problems in the Theory of Stellar Evolution*, IAU Symp. 93, Reidel (Dordrecht), p237.

King, A.R. (1995) in W.H.G. Lewin, J. van Paradijs, and E.P.J. van den Heuvel (eds.), *X-ray Binaries*, Cambridge University press.

Königl, A. (1989), *Astrophys. J.* **342**, 208.

Königl, A. and Kartje, J.F. (1994), *Astrophys. J.* **434**, 446.

Königl, A. and Ruden, S.P. (1993), in E.H. Levy and J.I. Lunine (eds.), *Protostars and Planets III*, Univ. Arizona Press (Tucson), p641.

Kley, W. (1991), *Astron. Astrophys.* **247**, 95.

Kley,,W., Papaloizou, J.C.B., and Lin, D.N.C (1993), *Astrophys. J.* **416**, 679.

Kuulkers, E., Howell, S.B., and van Paradijs, J. (1996), *Astrophys. J.* **462**, 87.

Landau, L.D. and Lifshitz, E.M. (1959) *Fluid Mechanics*, Pergamon Press (Oxford).

Lewin, W.H.G., van Paradijs, J., and van den Heuvel, E.P.J. (1995), *X-ray Binaries*, Cambridge University press.

Lightman, A.P. and Eardley, D.M. (1974), *Astrophys. J.* **187**, L1.

Lin, D.N.C. and Taam, R.E. (1984), in S.E. Woosley (ed.), *High Energy Transients in Astrophysics*, AIP Conference Procs. 115, p83.

Lovelace, R.V.E. (1976), *Nature* **262**, 649.

Lüst, R. (1952), *Z. Naturforsch.* **7a**, 87.

Lynden-Bell, D. and Pringle, J.E. (1974), *Mon. Not. R. Astron. Soc.* **168**, 603.

Lyutyi, V.M. and Sunyaev, R.A. (1976), *Astron. Zh.* **53**, 511; translation in *Sov. astron.* **20**, 290 (1976).

Massey, B.S. (1968, *Mechanics of Fluids*, Chapman and Hall (London) (6th Ed. 1989).

Meyer, F. (1990), *Rev. Mod. Astron.* **3**, 1.

Meyer, F. and Meyer-Hofmeister, E. (1981), *Astron. Astrophys.* **104**, L10.

Meyer, F. and Meyer-Hofmeister, E. (1982), *Astron. Astrophys.* **106**, 34.

Meyer-Hofmeister, E. and Ritter, H. (1993), in J. Sahade (ed.), *The Realm of Interacting Binary Stars*, Kluwer (Dordrecht), p143.

Meyer-Hofmeister, E., Schandl, S., and Meyer, F. (1997), *Astron. Astrophys.* **318**, 73.

Miller, K.A. and Stone, J.M. (1997), *Astrophys. J.* **489**, 890.

Mineshige, S. and Wheeler, J.C. (1989) *Astrophys. J.* **343**, 241.

Mineshige, S., and Wood, J.A. (1989), in F. Meyer, W. Duschl, J. Frank and E. Meyer-Hofmeister (eds.), *Theory of Accretion Disks*, Kluwer (Dordrecht), p221.

Narayan, R., Goldreich, P., and Goodman, J. (1987), *Mon. Not. R. Astron. Soc.* **228**, 1.

Narayan, R. and Popham, R. (1993), in W. Duschl et al. (eds.), *Theory of Accretion Disks, II*, Kluwer (Dordrecht), p293.

Nayakshin, S., Rappaport, S., and Melia, F. (2000), *Astrophys. J.* **535**, 798.

O'Donoghue, D.O., Chen, A., Marang, F., Mittaz, P.D., Winkler, H., and Warner, B. (1991), *Mon. Not. R. Astron. Soc.* **250**, 363.

Osaki, Y. (1974), *Publ. Astr. Soc. Japan* **26**, 429.

Osaki, Y. (1993) in W. Duschl et al. (eds.), *Theory of Accretion Disks, II*, Kluwer (Dordrecht), p93.

Ostriker, E.C., Gammie, C.F., and Stone, J.M. (1999), *Astrophys. J.* **513**, 259.

Paczyński, B. (1978), *Acta Astron.* **28**, 91.

Paczyński, B. (1991), *Astrophys. J.* **370**, 597.

Piran, T. (1978), *Astrophys. J.* **221**, 652.

Popham, R. and Narayan, R. (1991), *Astrophys. J.* **370**, 604.

Popham, R. (1997), in D.T. Wickramasinghe, G.V. Bicknell, and L. Ferrario (eds.), *Accretion Phenomena and related Outflows* (IAU Colloquium 163), ASP Conference Series 121, p230.

Pringle, J.E. (1981) *Ann. Rev. Astron. Astrophys.* **19**, 137.

Pringle, J.E. and Savonije, G.J. (1979), *Mon. Not. R. Astron. Soc.* **187**, 777.

Pudritz, R. (1981), *Mon. Not. R. Astron. Soc.* **195**, 881.

Różyczka, M. and Spruit, H.C. (1993), *Astrophys. J.* **417**, 677 (with video).

Rees, M.J. (1978), *Physica Scripta* **17**, 193.

Regev, O. and Hougerat, A. (1988), *Mon. Not. R. Astron. Soc.* **232**, 81.

Rybicki, G.R. and Lightman, A.P. (1979), *Radiative Processes in Astrophysics*, Wiley (New York), Ch 1.5.

Rutten R., van Paradijs, J., and Tinbergen, J. (1992), *Astron. Astrophys.* **260**, 213.

Sakimoto, P. and Coroniti, F.V. (1989), *Astrophys. J.* **342**, 49.

Sawada, K., Matsuda, T., Inoue, M., and Hachisu, I. (1987), *Mon. Not. R. Astron. Soc.* **224**, 307.

Schandl, S. and Meyer, F. (1994), *Astron. Astrophys.* **289**, 149.

Schandl, S. and Meyer, F. (1997), *Astron. Astrophys.* **321**, 245.

Shakura, N.I. and Sunyaev, R.A. (1973), *Astron. Astrophys.* **24**, 337.

Smak, J. (1971), *Acta Astron.* **21**, 15.

Smak, J. (1984), *Publ. Astr. Soc. Pac.* **96**, 54.

Spruit, H.C. (1987), *Astron. Astrophys.* **184**, 173.

Spruit, H.C. (1996), in R.A.M.J. Wijers et al. (eds.), *Evolutionary Processes in Binary Stars*, NATO ASI Ser. C477, Kluwer (Dordrecht), p249.

Spruit, H.C. and Rutten, R.G.M. (1998), *Mon. Not. R. Astron. Soc.* **299**, 768.

Spruit, H.C., Matsuda, T., Inoue, M., and Sawada, K. (1987), *Mon. Not. R. Astron. Soc.* **229**, 517.

Spruit, H.C. and Taam, R.E. (1993), *Astrophys. J.* **402**, 593.

Strom, S.E., Edwards, S., and Skrutskie, M.F. (1993), in E.H. Levy and J.I. Lunine (eds.), *Protostars and Planets III*, Univ. Arizona Press (Tucson), p837.

Sunyaev, R.A. and Shakura, N.I. (1977), *PAZh* **3**, 262; *Soviet Astron. Letters* **3**, 138.

Treves, A., Maraschi, L., and Abramowicz, M. (1988) *Publ. Astr. Soc. Pac.* **100**, 427.

Tritton, D.J. (1993), *J. Fluid Mech.* **241**, 503.

Van Paradijs, J. and McClintock, J.E. (1995), in W.H.G. Lewin, J. van Paradijs, and E.P.J. van den Heuvel (eds.), *X-ray Binaries*, Cambridge Univ. Press (Cambridge), p58.

Van Paradijs, J. and Verbunt, F. (1984), in S.E. Woosley (ed.), *High Energy Transients in Astrophysics*, AIP Conference Procs. 115, p49.

Velikhov, V.E. (1959), *J. Expl. Theoret. Phys.* (USSR) **36**, 1398.

Verbunt, F. (1982), *Space Sci. Rev.* **32**, 379.

Wardle, M. and Königl, A. (1993), *Astrophys. J.* **410**, 218.

Warner, B. (1987), *Mon. Not. R. Astron. Soc.* **227**, 23.

Warner, B. (1995), *Cataclysmic Variable Stars*, CUP (Cambridge).

Wood, J.A., Horne, K., Berriman, G., and Wade, R.A. (1989a), *Astrophys. J.* **341**, 974.

Wood, J.A., Marsh, T.R., Robinson, E.L., et al. (1989b), *Mon. Not. R. Astron. Soc.* **239**, 809.

Wood, J.H., Horne, K., and Vennes, S. (1992), *Astrophys. J.* **385**, 294.

Yi, I. (1999), in J.A. Sellwood and J. Goodman (eds.), *Astrophysical Discs*, Astronomical Society of the Pacific Conf. Ser. 160, 279; (astro-ph/9905215).

Zel'dovich Y.B. (1981), *Proc. Roy. Soc. London* **A374**, 299.

RADIATIVELY INEFFICIENT ACCRETION DISKS

H.C. SPRUIT
Max-Planck-Institut für Astrophysik
Postfach 1523, D-85740 Garching, Germany

Abstract. Radiatively inefficient (or advection dominated) disks are discussed at an introductory level. Ion supported and radiation supported flows are discussed, and the different consequences of advection dominated flows onto black holes vs. solid surfaces (neutron stars, white dwarfs), hydrodynamics, the role of the ratio of specific heats, and the possible connection between ADAFs and outflows are described.

1. Introduction

In a thin accretion disk, the time available for the accreting gas to radiate away the energy released by the viscous stress is the accretion time,

$$t_{\mathrm{acc}} \approx \frac{1}{\alpha \Omega_{\mathrm{K}}} \left(\frac{r}{H} \right)^2, \tag{1}$$

where α is the dimensionless viscosity parameter, Ω_{K} the local Keplerian rotation rate, r the distance from the central mass, and H the disk thickness (see Frank et al. 1992 or Spruit in this volume). For a thin disk, $H/r \ll 1$, this time is much longer than the thermal time scale $t_{\mathrm{t}} \approx 1/(\alpha \Omega)$. There is then enough time for a local balance to exist between viscous dissipation and radiative cooling. For the accretion rates implied in observed systems the disk is then rather cool, and the starting assumption $H/r \ll 1$ is justified.

This argument is somewhat circular, however, since the accretion time is long enough for effective cooling only if the disk is assumed to be thin to begin with. Other forms of accretion disks may exist, even at the same accretion rates, in which the cooling is ineffective compared with that of standard (geometrically thin, optically thick) disks.

141

C. Kouveliotou et al. (eds.), The Neutron Star – Black Hole Connection, 141–157.
© 2001 *Kluwer Academic Publishers. Printed in the Netherlands.*

Since radiatively inefficient disks tend to be thick, $H/r \sim O(1)$, they are sometimes called 'quasi-spherical'. However, this does *not* mean that a spherically symmetric accretion model would be a reasonable approximation. The crucial difference is that the flow has angular momentum. The inward flow speed is governed by the rate at which angular momentum can be transferred outwards, rather than by gravity and pressure gradient. The accretion time scale, $t_{\text{acc}} \sim 1/(\alpha\Omega)$ is longer than the accretion time scale in the spherical case (unless the viscosity parameter α is as large as $O(1)$). The dominant velocity component is azimuthal rather than radial, and the density and optical depth are much larger than in the spherical case.

It turns out that there are two kinds of radiatively inefficient disks, the optically thin, and optically thick varieties. A second distinction occurs because accretion flows are different for central objects with a solid surface (neutron stars, white dwarfs, main sequence stars, planets), and those without (i.e. black holes). I start with optically thick flows.

2. Radiation supported advective accretion

If the energy loss by radiation is small, the gravitational energy release $W_{\text{grav}} \approx GM/(2r)$ is converted into enthalpy of the gas and radiation field[1]

$$\frac{1}{2}\frac{GM}{r} = \frac{1}{\rho}[\frac{\gamma}{\gamma-1}P_{\text{g}} + 4P_{\text{r}}], \tag{2}$$

where an ideal gas of constant ratio of specific heats γ has been assumed, and $P_{\text{r}} = \frac{1}{3}aT^4$ is the radiation pressure. In terms of the virial temperature $T_{\text{vir}} = GM/(\mathcal{R}r)$, and assuming $\gamma = 5/3$, appropriate for a fully ionized gas, this can be written as

$$\frac{T}{T_{\text{vir}}} = [5 + 8\frac{P_{\text{r}}}{P_{\text{g}}}]^{-1}. \tag{3}$$

Thus, for radiation pressure dominated accretion, $P_{\text{r}} \gg P_{\text{g}}$, the temperature is much less than the virial temperature. The disk thickness is given by

$$H \approx [(P_{\text{g}} + P_{\text{r}})/\rho]^{1/2}/\Omega, \tag{4}$$

With (3) this yields

$$H/r \sim O(1). \tag{5}$$

In the limit $P_{\text{r}} \gg P_{\text{g}}$, the flow is therefore geometrically thick. This implies that radiation pressure supplies a non-negligible fraction of the support of the gas against gravity (the remainder being provided by rotation).

---

[1]I assume here that a fraction ~ 0.5 of the gravitational potential energy stays in the flow as orbital kinetic energy. See also section 3.

For $P_r \gg P_g$, (2) yields

$$\frac{GM}{2r} = \frac{4}{3}\frac{aT^4}{\rho}. \tag{6}$$

The radiative energy flux, in the diffusion approximation, is

$$F = \frac{4}{3}\frac{d}{d\tau}\sigma T^4 \approx \frac{4}{3}\frac{\sigma T^4}{\tau}, \tag{7}$$

where $\sigma = ac/4$ is Stefan-Boltzmann's radiation constant. Hence

$$F = \frac{1}{8}\frac{GM}{rH}\frac{c}{\kappa} = F_E\frac{r}{8H}, \tag{8}$$

where $F_E = L_E/(4\pi r^2)$ is the local Eddington flux. Since $H/r \approx 1$, a radiatively inefficient, radiation pressure dominated accretion flow has a luminosity of the order of the Eddington luminosity.

The temperature depends on the accretion rate and the viscosity ν assumed. The accretion rate is of the order $\dot{M} \sim 3\pi\nu\Sigma$ (see 'accretion disks' elsewhere in this volume), where $\Sigma = \int \rho dz$ is the surface mass density. In units of the Eddington rate, we get

$$\dot{m} \equiv \dot{M}/\dot{M}_E \approx \nu\rho\kappa/c, \tag{9}$$

where $H/r \approx 1$ has been used, and $\dot{M}_E$ is the Eddington accretion rate onto the central object of size R,

$$\dot{M}_E = \frac{R}{GM}L_E = 4\pi Rc/\kappa. \tag{10}$$

[Note that the definition of $\dot{M}_E$ differs by factors of order unity between different authors. It depends on the assumed efficiency η of conversion of gravitational energy GM/R into radiation. In (10) it is taken to be unity, for accretion onto black holes a more realistic value is $\eta = 0.1$, for accretion onto neutron stars $\eta \approx 0.4$.]

Assume that the viscosity scales with the gas pressure:

$$\nu = \alpha\frac{P_g}{\rho\Omega_K}, \tag{11}$$

instead of the total pressure $P_r + P_g$. This is the form that is likely to hold if the angular momentum transport is due to a small-scale magnetic field (Sakimoto and Coroniti, 1989). Then with (6) we have (up to a factor $2H/r \sim O(1)$)

$$T^5 \approx \frac{(GM)^{3/2}}{r^{5/2}}\frac{\dot{m}c}{\alpha\kappa a\mathcal{R}}, \tag{12}$$

or

$$T \approx 2\,10^8 r_6^{-1/5} (r/r_{\rm g})^{3/10} \dot m^{1/5}, \tag{13}$$

where $r = 10^6 r_6$ and $r_{\rm g} = 2GM/c^2$ is the gravitational radius of the accreting object, and the electron scattering opacity of 0.3 cm^2/g has been assumed. The temperatures expected in radiation supported advection dominated flows are therefore quite low compared with the virial temperature [If the viscosity is assumed to scale with the total pressure instead of $P_{\rm g}$, the temperature is even lower]. The effect of electron-positron pairs can be neglected (Schultz and Price, 1985), since they are present only at temperatures approaching the electron rest mass energy, $T \gtrsim 10^9$K.

In order for the flow to be radiation pressure and advection dominated, the optical depth has to be sufficiently large so the radiation does not leak out. The energy density in the flow, vertically integrated at a distance r, is of the order

$$E \approx aT^4 H, \tag{14}$$

and the energy loss rate per cm^2 of disk surface is given by (8). The cooling time is therefore,

$$t_{\rm c} = E/F = 3\tau H/c. \tag{15}$$

This is to be compared with the accretion time, which can be written in terms of the mass in the disk at radius r, of the order $2\pi r^2 \Sigma$, and the accretion rate:

$$t_{\rm acc} = 2\pi r^2 \Sigma / \dot M. \tag{16}$$

This yields

$$t_{\rm c}/t_{\rm acc} \approx \frac{\kappa}{\pi r c} \dot M = 4\dot m \frac{R}{r}, \tag{17}$$

(where a factor $3/2\,H/r \sim O(1)$ has been neglected). Since $r > R$, this shows that accretion has to be super-Eddington in order to be both radiation- and advection-dominated.

This condition can also be expressed in terms of the so-called *trapping radius* $r_{\rm t}$ (e.g. Rees 1978). Equating $t_{\rm acc}$ and $t_{\rm c}$ yields

$$r_{\rm t}/R \approx 4\dot m. \tag{18}$$

Inside $r_{\rm t}$, the flow is advection dominated: the radiation field produced by viscous dissipation stays trapped inside the flow, instead of being radiated from the disk as happens in a standard thin disk. Outside the trapping radius, the radiation field can not be sufficiently strong to maintain a disk with $H/r \sim 1$, it must be a thin form of disk instead. Such a thin disk can still be radiation-supported (i.e. $P_{\rm r} \gg P_{\rm g}$), but it can not be advection dominated.

Flows of this kind are called 'radiation supported tori' (or radiation tori, for short) by Rees et al. 1982[2]. They must accrete at a rate above the Eddington value to exist. The converse is not quite true: a flow accreting above Eddington is an advection dominated flow, but it need not necessarily be radiation dominated. Advection dominated optically thick acretion flows exist in which radiation does not play a major role (see section 3.1).

That an accretion flow above $\dot{M}_E$ is advection dominated, not a thin disk, follows from the fact that in a thin disk the energy dissipated must be radiated away locally. Since the local radiative flux can not exceed the Eddington energy flux F_E, the mass accretion rate in a thin disk can not significantly exceed the Eddington value (10).

The gravitational energy, dissipated by viscous stress in differential rotation and advected with the flow, ends up at the central object. If this is a black hole, the photons, particles and their thermal energy are conveniently swallowed at the horizon, and do not react back on the flow. Radiation tori are therefore mostly relevant for accretion onto black holes. They are convectively unstable (Bisnovatyi-Kogan and Blinnikov 1977): the way in which energy is dissipated, in the standard α-prescription, is such that the entropy ($\sim T^3/\rho$) decreases with height in the disk. Recent numerical simulations (see section 5) show the effects of this convection.

2.1. SUPER-EDDINGTON ACCRETION ONTO BLACK HOLES

As the accretion rate onto a black hole is increased above $\dot{M}_E$, the trapping radius moves out. The total luminosity increases only slowly, and remains of the order of the Eddington luminosity. Such supercritical accretion has been considered by Begelman and Meier (1982; see also Wang and Zhou 1999); they show that the flow has a radially self-similar structure.

Abramowicz et al. (1988; 1989) studied accretion onto black holes at rates near $\dot{M}_E$. They used a vertically-integrated approximation for the disk, but included the advection terms. The resulting solutions were called 'slim disks'. These models show how with increasing accretion rate, a standard thin Shakura-Sunyaev disk turns into a radiation-supported advection flow. The nature of the transition depends on the viscosity prescription used, and can show a non-monotonic dependence of $\dot{M}$ on surface density Σ (Honma et al. 1991). This suggests the possibility of instability and cyclic behavior of the inner disk near a black hole, at accretion rates near and above $\dot{M}_E$ (for an application to GRS 1915+105 see Nayakshin et al. 2000).

---

[2]Some workers interpret the use of word 'torus' in the context radiatively inefficient accretion as implying a rotating but non-accreting flow. The physics studied by Rees et al. (1982), however, where this name was introduced, explicitly refers to accreting flows such as described here

2.2. SUPER-EDDINGTON ACCRETION ONTO NEUTRON STARS

In the case of accretion onto a neutron star, the energy trapped in the flow, plus the remaining orbital energy, settles onto its surface. If the accretion rate is below $\dot{M}_E$, the energy can be radiated away by the surface, and steady accretion is possible. A secondary star providing the mass may, under some circumstances, transfer more than $\dot{M}_E$, since it does not know about the neutron star's Eddington value. The outcome of this case is still somewhat uncertain; it is generally believed on intuitive grounds that the 'surplus' (the amount above $\dot{M}_E$) somehow gets expelled from the system.

As the transfer rate is increased, the accreting hot gas forms an extended atmosphere around the neutron star, like the envelope of a giant. If it is large enough, the outer parts of this envelope are partially ionized. The opacity in these layers, due to lines of the CNO and heavier elements, is then much higher than the electron scattering opacity. The Eddington luminosity based on the local value of the opacity is then smaller than it is near the neutron star surface. Once an extended atmosphere with a cool surface forms, the accretion luminosity is thus large enough to drive a wind from the envelope (see Kato 1997, where this is discussed in the context of Novae).

This scenario is somewhat dubious however, since it assumes that the mass transferred from the secondary continues to reach the neutron star and generate a high luminosity there. This is not at all obvious, since the mass transfering stream may dissipate in the growing neutron star envelope instead. The result would be a giant (more precisely, a Thorne-Zytkow star), with a steadily increasing envelope mass. Such an envelope is likely to be large enough to envelop the entire binary system, which then develops into a common-envelope (CE) system. The envelope mass is expected to be ejected by CE hydrodynamics (Taam 1994; 2000).

A more speculative proposal, suggested by the properties of SS 433, is that the 'surplus mass' is ejected in the form of jets. The binary parameters of Cyg X-2 are observational evidence for mass ejection in super-Eddington mass transfer phases (King and Ritter 1999; Podsiadlowski and Rappaport 2000; King and Begelman 1999).

3. Hydrodynamics

The hydrodynamics of ADAFs and radiation tori can be studied by starting, at a very simple level, with a generalization of the thin disk equations. Making the assumption that quantities integrated over the height z of the disk give a fair representation (though this is justifiable only for thin disks), and assuming axisymmetry, the problem reduces to a one-dimensional time-dependent one. Further simplifying this by restriction to a steady flow yields the equations

$$2\pi r \Sigma v_r = \dot{M} = \text{cst}, \tag{19}$$

$$r \Sigma v_r \partial_r (\Omega r^2) = \partial_r (\nu \Sigma r^3 \partial_r \Omega), \tag{20}$$

$$v_r \partial_r v_r - (\Omega^2 - \Omega_K^2) r = -\frac{1}{\rho} \partial_r p, \tag{21}$$

$$\Sigma v_r T \partial_r S = q^+ - q^-, \tag{22}$$

where S is the specific entropy of the gas, Ω the local rotation rate, now different from the Keplerian rate $\Omega_K = (GM/r^3)^{1/2}$, while

$$q^+ = \int Q_v dz \qquad q^- = \int \text{div} F_r dz \tag{23}$$

are the height-integrated viscous dissipation rate and radiative loss rate, respectively. In the case of thin disks, equations (19) and (20) are unchanged, but (21) simplifies to $\Omega^2 = \Omega_K^2$, i.e. the rotation is Keplerian, while (22) simplifies to $q^+ = q^-$, expressing local balance between viscous dissipation and cooling. The left hand side of (22) describes the radial advection of heat, and is perhaps the most important deviation from the thin disk equations at this level of approximation (hence the name advection dominated flows). The characteristic properties are seen most cearly when radiative loss is neglected altogether, $q^- = 0$. The equations are supplemented with expressions for ν and q^+:

$$\nu = \alpha c_s^2 / \Omega_K; \qquad q^+ = (r \partial_r \Omega)^2 \nu \Sigma. \tag{24}$$

If α is taken constant, $q^- = 0$, and an ideal gas is assumed with constant ratio of specific heats, so that the entropy is given by

$$S = c_v \ln(p/\rho^\gamma), \tag{25}$$

then equations (19)-(22) have no explicit length scale in them. This means that a special so-called self-similar solution exists, in which all quantities are powers of r. Such self-similar solutions have apparently first been described by Gilham (1981), but have since then been re-invented several times (Spruit et al. 1987; Narayan and Yi 1994). The dependences on r are

$$\Omega \sim r^{-3/2}; \qquad \rho \sim r^{-3/2}, \tag{26}$$

$$H \sim r; \qquad T \sim r^{-1}. \tag{27}$$

In the limit $\alpha \ll 1$, one finds

$$v_r = -\alpha \Omega_K r \left(9 \frac{\gamma - 1}{5 - \gamma} \right), \tag{28}$$

$$\Omega = \Omega_K \left(2\frac{5-3\gamma}{5-\gamma}\right)^{1/2}, \tag{29}$$

$$c_s^2 = \Omega_K^2 r^2 \frac{\gamma-1}{5-\gamma}, \tag{30}$$

$$\frac{H}{r} = \left(\frac{\gamma-1}{5-\gamma}\right)^{1/2}. \tag{31}$$

The precise from of these expressions depends somewhat on the way in which vertical integrations such as in (23) are done (which are only approximate).

The self-similar solution can be compared with numerical solutions of eqs. (19)–(22) with appropriate conditions applied at inner (r_i) and outer (r_o) boundaries (Nakamura et al. 1996; Narayan et al. 1997). The results show that the self-similar solution is valid in an intermediate regime $r_i \ll r \ll r_o$. That is, the solutions of (19)–(22) approach the self-similar solution far from the boundaries, as is characteristic of self-similar solutions.

The solution exists only if $1 < \gamma \le 5/3$, a condition satisfied by all ideal gases. As $\gamma \downarrow 1$, the disk temperature and thickness vanish. This is understandable, since a γ close to 1 means that the gas has a large number of internal degrees of freedom. The accretion energy is shared between all degrees of freedom, so that for a low γ less is available for the kinetic energy (temperature) of the particles.

Second, the *rotation rate vanishes* for $\gamma \to 5/3$. Since a fully ionized gas has $\gamma = 5/3$, it is the most relevant value for optically thin accretion near a black hole or neutron star. Apparently, steady advection dominated accretion can not have angular momentum in this case, and the question arises how an adiabatic flow with angular momentum will behave for $\gamma = 5/3$. In the literature, this problem has been circumvented by arguing that real flows would have magnetic fields in them, which would change the effective compressibility of the gas. Even if a magnetic field of sufficient strength is present, however, (energy density comparable to the gas pressure) the effective γ is not automatically lowered. If the field is compressed mainly perpendicular to the field lines, for example, the effective γ is closer to 2. Also, this does not solve the conceptual problem what would happen to a rotating accretion flow consisting of a more weakly magnetized ionized gas.

This conceptual problem has been solved by Ogilvie (1999), who showed how the low rotation for $\gamma = 5/3$ comes about in a time-dependent manner. He found a similarity solution (depending on distance and time in the combination $r/t^{2/3}$) to the time-dependent version of eqs (19)–(22). This solution describes the asymptotic behavior (in time) of a viscously spreading disk, analogous to the viscous spreading of thin disks (Lynden-Bell and Pringle 1972). As in the thin disk case, all the mass accretes asymptotically onto the central mass, while all the angular momentum travels to infinity together with a vanishing amount of mass. For all $\gamma < 5/3$, the rotation rate at a fixed r tends to a finite value as $t \to \infty$, but for $\gamma = 5/3$ it tends to zero. The slowly-rotating region expands in size as $r \sim t^{2/3}$. It thus seems likely that the typical slow rotation of ADAFs at γ near $5/3$ is a real physical property, and that the angular momentum gets expelled from the inner regions of the flow.

3.1. OTHER OPTICALLY THICK ACCRETION FLOWS

The radiation-dominated flows discussed in section 2 are not the only possible optically thick advection dominated flows. From the discussion of the hydrodynamics, it is clear that disk-like (i.e. rotating) accretion is possible whenever the ratio of specific heats is less than $5/3$. A radiation supported flow satisfies this requirement with $\gamma = 4/3$, but it can also happen in the absence of radiation if energy is taken up in the gas by internal degrees of freedom of the particles. Examples are the rotational and vibrational degrees of freedom in molecules, and the energy associated with dissociation and ionization. If the accreting object has a gravitational potential not much exceeding the $13.6 + 2.2$ eV per proton for dissociation plus ionization, a gas initially consisting of molecular hydrogen can stay bound at arbitrary accretion rates. This translates into a limit $M/M_\odot \, R_\odot/R < 0.01$. This is satisfied approximately by the giant planets, which are believed to have gone through a phase of rapid adiabatic gas accretion (e.g. Podolak et al. 1993).

A more remotely related example is the core-collapse supernova. The accretion energy of the envelope mass falling onto the proto-neutron star is lost mostly through photodisintegration of nuclei, causing the well known problem of explaining how a shock is produced of sufficient energy to unbind the envelope. If the pre-collapse core rotates sufficiently rapidly, the collapse will form an accretion torus (inside the supernova envelope), with properties similar to advection dominated accretion flows (but at extreme densities and accretion rates, by X-ray binary standards). Such objects have been invoked as sources of Gamma-ray bursts (Popham et al. 1999; see also the review by Meszaros, elsewhere in this volume).

A final possibility for optically thick accretion is through *neutrino losses*. If the temperature and density near an accreting neutron star become large enough, additional cooling takes place through neutrinos (as in the cores of giants). This is relevant for the physics of Thorne-Zytkow stars (neutron stars or black holes in massive supergiant envelopes, cf. Bisnovatyi-Kogan and Lamzin 1984, Cannon et al. 1992), and perhaps for the spiral-in of neutron stars into giants (Chevalier 1993; see however Taam 2000).

4. Optically thin advection dominated flows (ADAFs)

The optically thin case has received most attention in recent years, because of the promise it holds for explaining the (radio to X-ray) spectra of X-ray binaries and the central black holes in galaxies, including our own. For a recent review see Yi (1999). This kind of flow occurs if the gas is optically thin, and radiation processes sufficiently weak. The gas then heats up to near the virial temperature. Near the last stable orbit of a black hole, this is of the order 100 MeV, or 10^{12}K. At such temperatures, a gas in thermal equilibrium would radiate at a fantastic rate, even if it were optically thin, because the interaction between electrons and photons becomes very strong already near the electron rest mass of 0.5 MeV. In a remarkable paper, Shapiro, Lightman, and Eardley (1976) noted that this, however, is not what will happen in an optically thin accreting plasma but that, instead, a *two-temperature plasma* forms.

Suppose that the energy released by viscous dissipation is distributed equally among the carriers of mass, i.e. mostly to the ions and $\sim 1/2000$ to the electrons. Most of the energy resides in the ions, which radiate very inefficiently (their high mass prevents the rapid accelerations that are needed to produce electromagnetic radiation). This energy is transferred to the electrons by Coulomb interactions. These interactions are slow, however, under the conditions mentioned. They are slow because of the low density (on account of the assumed optical tickness), and because they decrease with increasing temperature. The electric forces that transfer energy from an ion to an electron act only as long as the ion is within the electron's Debye sphere (e.g. Spitzer 1965). The interaction time between proton and electron, and thus the momentum transfered, therefore decrease as $1/v_{\rm p} \sim T_{\rm p}^{-1/2}$ where $T_{\rm p}$ is the proton temperature.

In this way, an optically thin plasma near a compact object can be in a two-temperature state, with the ions being near the virial temperature, and the electrons, which are doing the radiating, at a much lower temperature around 50–200 keV. The energy transfer from the gravitational field to the ions is fast (by some form of viscous or magnetic dissipation), from the ions to the electrons slow, and finally the energy losses of the electrons fast (by synchrotron radiation in a magnetic field or by inverse Compton scattering of soft photons). Such a flow would be radiatively inefficient since the receivers of the accretion energy, the ions, get swallowed by the hole before getting a chance to transfer their energy to the electrons. The first disk models which take into account the physics of advection and a two-temperature plasma were developed by Ichimaru (1977).

It is clear from this description that both the physics of such flows and the ensuing radiation spectrum depend crucially on the details of the ion-electron interaction and radiation processes assumed. This is unlike the case of the optically thick advection dominated flows, where gas and radiation are in approximate thermodynamic equilibrium. This is a source of uncertainty in the application of the optically thin ADAFs to observed systems, since their radiative properties depend on poorly known quantities, for example, the strength of the magnetic field in the flow.

The various branches of optically thin and thick accretion flows are summarized in Figure 1. Each defines a relation between surface density Σ (or optical depth $\tau = \kappa\Sigma$) and accretion rate. Optically thin ADAFs require low densities, either because of low accretion rates or large values of the viscosity parameter. The condition that the cooling time of the ions by energy transfer to the electrons is longer than the accretion time yields a maximum accretion rate (Rees et al. 1982),

$$\dot{m} \lesssim \alpha^2. \tag{32}$$

If $\alpha \approx 0.05$, as suggested by current simulations of magnetic turbulence, the maximum accretion rate would be a few 10^{-3}. If ADAFs are to be applicable to systems with higher accretion rates, such as Cyg X-1 for example, the viscosity parameter must be larger, on the order of 0.3.

4.1. APPLICATION: HARD SPECTRA IN X-RAY BINARIES

In the hard state, the X-ray spectrum of black hole and neutron star accreters is characterized by a peak in the energy distribution (νF_ν or $E\,F(E)$) at photon energies around 100 keV. This is to be compared with the typical photon energy of ~ 1 keV expected from a standard optically thick thin disk accreting near the Eddington limit. The standard, and by far most likely explanation is that the observed hard photons are softer photons (around 1 keV) that have been up-scattered through inverse Compton scattering on hot electrons. Fits of such Comptonized spectra (e.g. Zdziarski 1998 and references therein) yield an electron scattering optical depth around unity and an electron temperature of 50–100 keV. The scatter in these parameters is rather small between different sources. The reason may lie in part in the physics of Comptonization, but is not fully understood either. Something in the physics of the accretion flow keeps the Comptonization parameters constant as long as it is in the hard state. ADAFs have been applied with some success in interpreting XRB. They can produce reasonable X-ray spectra, and have been used in interpretations of the spectral-state transitions in sources like Cyg X-1 (Esin et al. 1998 and references therein).

An alternative to the ADAF model for the hard state in sources like Cyg X-1 and the black hole X-ray transients is the 'corona' model. A hot corona (Bisnovatyi-Kogan and Blinnikov 1976), heated perhaps by magnetic fields

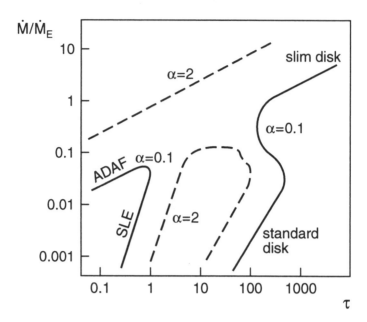

Figure 1. Branches of advection-dominated and thin disks for two values or the viscosity parameter α, as functions of accretion rate and (vertical) optical depth of the flow (schematic, after Chen et al. 1995; Zdziarski 1998). Optically thin branches are the ADAF and SLE (Shapiro-Lightman-Eardley) solutions, optically thick ones the radiation dominated ('slim disk' or 'radiation torus') and SS (Shakura-Sunyaev or standard thin disk). Advection dominated are the ADAF and radiation torus, geometrically thin are the SLE and SS. The SLE solution is a thermally unstable branch.

as in the case of the Sun (Galeev et al. 1979) could be the medium that Comptonizes soft photons radiated from the cool disk underneath. The energy balance in such a model produces a Comptonized spectrum within the observed range (Haardt and Maraschi 1993). This model has received further momentum, especially as a model for AGN, with the discovery of broadened X-ray lines indicative of the presence of a cool disk close to the last stable orbit around a black hole (Lee et al. 2000 and references therein). The very rapid X-ray variability seen in some of these sources is interpreted as due to magnetic flaring in the corona, like in the solar corona (e.g. Di Matteo et al. 1999a).

4.2. TRANSITION FROM THIN DISK TO ADAF

One of the difficulties in applying ADAFs to specific observed systems is the transition from a standard geometrically thin, optically thick disk, which must be the mode of mass transfer at large distances, to an ADAF at closer range. This is shown by Figure 1, which illustrates the situation at some distance close to the central object. The standard disk and the optically thin branches are separated from each other for all values of the viscosity parameter. This separation of the optically thin solutions also holds at larger distances. Thus, there is no plausible continuous path from one to the other, and the transition between the two must be due to additional physics that is not included in diagrams like Figure 1.

A promising possibility is that the transition takes place through a gradual (as a function of radius) *evaporation* (Meyer and Meyer-Hofmeister 1994; Meyer-Hofmeister and Meyer 1999). In this scenario, evaporation initially produces a corona above the disk, which transforms into an ADAF further in.

4.3. QUIESCENT GALACTIC NUCLEI

For very low accretion rates, such as inferred for the black hole in the center of our galaxy (identified with the radio source Sgr A*), the broad band spectral energy distribution of an ADAF is predicted to have two humps (Narayan et al. 1995; Quataert et al. 1999). In the X-ray range, the emission is due to bremsstrahlung. In the radio range, the flow emits synchrotron radiation, provided that the magnetic field in the flow has an energy density order of the gas pressure ('equipartition'). Synthetic ADAF spectra can be fitted to the observed radio and X-ray emission from Sgr A*. In other galaxies where a massive central black hole is inferred, and the center is populated by an X-ray emitting gas of known density, ADAFs would also be natural, and might explain why the observed luminosities are so low compared with the accretion rate expected for a hole embedded in a gas of the measured density. In some of these galaxies, however, the peak in the radio-to-mm range predicted by analogy with Sgr A* is not observed (Di Matteo et al. 1999b). This requires an additional hypothesis, for example that the magnetic field in these cases is much lower, or that the accretion energy is carried away by an outflow.

4.4. TRANSIENTS IN QUIESCENCE

X-ray transients in quiescence (i.e. after an outburst) usually show a very low X-ray luminosity. The mass transfer rate from the secondary in quiescence can be inferred from the optical emission. This shows the characteristic 'hot spot', known from other systems to be the location where the mass transfering stream impacts on the edge of an accretion disk (e.g. van Paradijs and McClintock 1995). These observations thus show that a disk is present in quiescence, while the mass transfer rate can be measured from the brightness of the hot spot. If this disk were to extend to the neutron star with constant mass flux, the predicted X-ray luminosity would be much higher than observed. This has traditionally been interpreted as a consequence of the fact that in transient systems, the accretion is not steady. Mass is stored in the outer parts and released by a disk instability (e.g. King 1995; Meyer-Hofmeister and Meyer 1999) producing the X-ray outburst. During quiescence, the accretion rate onto the compact object is much smaller than the mass transfer from the secondary to the disk.

ADAFs have been invoked as an alternative explanation. The quiescent accretion rate onto the central object is proposed to be higher than in the disk-instability explanation, the greater energy release being hidden on account of the low radiative efficiency of the ADAF. Some transient systems have neutron star primaries, with a hard surface at which the energy accreted by the ADAF must somehow be radiated away. A neutron star, with or without ADAFs, can not accrete in a radiatively inefficient way. In order to make ADAFs applicable, it has been proposed that the neutron stars in these systems have a modest magnetic dipole moment, such that in quiescence the gas in the accretion disk is prevented, by the 'propeller effect' (Illarionov and Sunyaev 1975; Sunyaev and Shakura 1977) from accreting onto the star.

4.5. ADAF-DISK INTERACTION: LITHIUM

One of the strong predictions of ADAF models, whether for black holes or neutron stars, is that the accreting plasma in the inner regions has an ion temperature of 10–100 MeV. Nearby is a cool and dense accretion disk feeding this plasma. If only a small fraction of the hot ion plasma gets in contact with the disk, the intense irradiation by ions will produce nuclear reactions (Aharonian and Sunyaev 1984; Martín et al. 1992). The main effects would be spallation of CNO elements into Li and Be, and the release of neutrons by various reactions. In this context, it is intriguing that the secondaries of both neutron star and black hole accreters have high overabundances of Li compared with other stars of their spectral types (Martín et al. 1992; 1994a). If a fraction of the disk material is carried to the secondary by a disk wind, the observed Li abundances may be accounted for (Martín et al. 1994b).

4.6. ADAF-DISK INTERACTION: HARD X-SPECTRA

The interaction of a hot ion plasma with the cool disk produces a surface layer heated by the incident ions through Coulomb interactions with the electron gas in the disk. Its thickness and temperature turn out to be largely self-regulating: the energy balance is as in the Haardt and Maraschi corona models (in which the interaction is by photons only), while the optical thickness self-regulates through the dependence of the ion penetration depth on the electron temperature. This model (Spruit 1997) produces hard comptonized X-ray spectra whose shape is largely independent of both the energy flux and distance from the central object.

5. Outflows?

The energy density in an advection dominated accretion flow is of the same order as the gravitational binding energy density GM/r, since a significant fraction of that energy went into internal energy of the gas by viscous dissipation, and little of it got lost by radiation. The gas is therefore only marginally bound in the gravitational potential. This suggests that perhaps a part of the accreting gas can escape, producing an outflow or wind. In the case of the ion supported optically thin ADAFs, this wind would be thermally driven by the temperature of the ions, like an 'evaporation' from the accretion torus. In the case of the radiation supported tori, which exist only at a luminosity near the Eddington value, but with much lower temperatures than the ion tori, winds driven by radiation pressure could exist.

The possibility of outflows is enhanced by the viscous energy transport through the disk. In the case of thin accretion disks (not quite appropriate in the present case, but sufficient to demonstrate the effect), the local rate of gravitational energy release (erg cm^{-2}s^{-1}) is $W = \Sigma v_r \partial_r (GM/r)$. The local viscous dissipation rate is $(9/4)\nu\Sigma\Omega^2$. They are related by

$$Q_{\rm v} = 3[1 - (\frac{r_{\rm o}}{r})^{1/2}]W, \tag{33}$$

where $r_{\rm i}$ is the inner edge of the disk (see 'accretion disks' elsewhere in this volume). The viscous dissipation rate is less than the gravitational energy release for $r < (4/9)r_{\rm i}$, and larger outside this radius. Part of the gravitational energy released in the inner disk is transported outward by the viscous stresses, so that the energy deposited in the gas is up to three times larger than expected from a local energy balance argument. The temperatures in an ADAF would be correspondingly larger. Begelman and Blandford (1999) have appealed to this effect to argue that in an ADAF most of the accreting mass of a disk might be expelled through a wind, the energy needed for this being supplied by the viscous energy transport associated with the small amount of mass that actually accretes.

These suggestions are in principle testable, since the arguments are about two-dimensional time dependent flows (axisymmetric), which can be studied fairly well by numerical simulation. Igumenshchev et al. (1996), and Igumenshchev and Abramowicz (1999) present results of such simulations, but unfortunately these give a somewhat ambiguous answer to the question. For large viscosity ($\alpha \sim 0.3$) no outflow is seen, but for small viscosity time dependent flows are seen with outflows in some regions. Some of these flows may be a form of convection and unrelated to systematic outflows.

References

Abramowicz, M.A., Czerny, B., Lasota, J.P., and Szuszkiewicz, E. (1988), *Astrophys. J.* **332**, 646.
Abramowicz, M.A, Kato S., and Matsumoto R. (1989), *Publ. Astr. Soc. Japan* **41**, 1215.
Aharonian, F.A. and Sunyaev, R.A. (1984), *Mon. Not. R. astron. Soc.* **210**, 257.
Begelman, M.C. and Meier, D.L. (1982), *Astrophys. J.* **253**, 873.
Begelman, M.C. and Blandford, R.D. (1999), *Mon. Not. R. astron. Soc.* **303**, L1.
Bisnovatyi-Kogan, G.S. and Lamzin, S.A. (1984), *Astron. Zh* **61**, 323 (translation *Soviet Astron.* **28**, 187).
Bisnovatyi-Kogan, G.S. and Blinnikov, S.I. (1976), *Pi'sma Astron. Zh* **2**, 489 (translation 1977, *Soviet Astron.L* **2**, 191).
Bisnovatyi-Kogan, G.S. and Blinnikov, S.I. (1977), *Astron. Astrophys.* **59**, 111.
Cannon, R.C., Eggleton, P.P., Zytkow, A.N., and Podsiadlowski P. (1992), *Astrophys. J.* **386**, 206.
Chen, X.M., Abramowicz, M.A., Lasota, J.-P., Narayan, R., and Yi, I. (1995), *Astrophys. J.* **443**, L61.
Chevalier, R.A. (1993), *Astrophys. J.* **411**, 33.
Di Matteo, T., Celotti, A., and Fabian, A.C. (1999a), *Mon. Not. R. astron. Soc.* **304**, 809.
Di Matteo, T., Fabian, A. C., Rees, M.J., Carilli, C.L., and Ivison, R.J. (1999b), *Mon. Not. R. astron. Soc.* **305**, 492.
Esin, A.A., Narayan, R., Cui, W., Grove, J.E., and Zhang, S.N. (1998), *Astrophys. J.* **505**, 854.
Frank, J., King, A.R., and Raine, D.J. (1992), *Accretion Power in Astrophysics* (2nd edition), Cambridge University Press.
Galeev, A.A., Rosner, R., and Vaiana, G.S. (1979), *Astrophys. J.* **229**, 318.
Gilham, S. (1981), *Mon. Not. R. astron. Soc.* **195**, 755.
Haardt, F. and Maraschi, L. (1993), *Astrophys. J.* **413**, 507.
Honma, F., Kato, S., Matsumoto, R., and Abramowicz, M. (1991), *Publ. Astr. Soc. Japan* **43**, 261.
Ichimaru, S. (1977), *Astrophys. J.* **241**, 840.
Igumenshchev, I.V., Chen, X.M., and Abramowicz, A. (1996), *Mon. Not. R. astron. Soc.* **278**, 236.
Igumenshchev, I.V. and Abramowicz, M.A. (1999), *Mon. Not. R. astron. Soc.* **303**, 309.
Illarionov, A.F. and Sunyaev, R.A. (1975), *Astron. Astrophys.* **39**, 185.
Kato, M. (1997), *Astrophys. J. Suppl.* **113**, 121.
King, A.R. (1995), in Lewin, W.H.G., van Paradijs, J., and van den Heuvel, E.P.J. (eds.), *X-Ray Binaries*, Cambridge University press, p419.
King, A.R. and Ritter, H. (1999), *Mon. Not. R. astron. Soc.* **309**, 253.
King, A.R., Begelman, M.C. (1999), *Astrophys. J.* **519**, L169.
Lee, J.C., Fabian, W.N., Brandt, W.N., Reynolds, C.S., and Iwasawa, K. (2000), *Mon. Not. R. astron. Soc.* **310**, 973.
Loeb, A. and Laor, A. (1992), *Astrophys. J.* **384**, 115.
Lynden-Bell, D. and Pringle, J.E. (1974), *Mon. Not. R. astron. Soc.* **168**, 603.
Martín, E L., Rebolo, R., Casares, J., and Charles, P.A. (1992), *Nature* **358**, 129.
Martín, E.L., Rebolo, R., Casares, J., and Charles, P.A. (1994a), *Astrophys. J.* **435**, 791.
Martín, E., Spruit, H.C., and van Paradijs, J. (1994b), *Astron. Astrophys.* **291**, L43.

Meyer, F. and Meyer-Hofmeister, E. (1994), *Astron. Astrophys.* **288**, 175.

Meyer-Hofmeister, E. and Meyer, F. (1999), *Astron. Astrophys.* **348**, 154.

Nakamura, K.E., Matsumoto, R., Kusunose, M., and Kato, S. (1996), *Publ. Astr. Soc. Japan* **48**, 769.

Narayan, R. and Yi, I. (1994), *Astrophys. J.* **428**, L13.

Narayan, R. and Yi, I., and Mahadevan, R. (1995), *Nature* **374**, 623.

Narayan, R., Kato, S., and Honma, F. (1997), *Astrophys. J.* **476**, 49.

Nayakshin, S., Rappaport, S., and Melia, F. (2000), *Astrophys. J.* **535**, 798.

Ogilvie, G.I. (1999), *Mon. Not. R. astron. Soc.* **306**, L9.

Podolak, M., Hubbard, W.B., and Pollack, J.B. (1993), *Protostars and Planets II*, University of Arizona Press, p1109.

Podsiadlowski, P. and Rappaport, S. (2000), *Astrophys. J.* **529**, 946.

Popham, R., Woosley, S.E., and Fryer, C. (1999), *Astrophys. J.* **518**, 356.

Quataert, E. and Narayan, R. (1999), *Astrophys. J.* **517**, 101.

Rees, M.J. (1978), *Physica Scripta* **17**, 193.

Rees, M.J., Phinney, E.S., Begelman, M.C., and Blandford, R.D. (1982), *Nature* **295**, 17.

Sakimoto, P. and Coroniti, F.V. (1989), *Astrophys. J.* **342**, 49.

Schultz, A.L. and Price, R.H. (1985), *Astrophys. J.* **291**, 1.

Shapiro, S.L., Lightman, A.P., and Eardley, D.M. (1976), *Astrophys. J.* **204**, 187.

Sunyaev, R.A. and Shakura, N.I. (1975), *Pi'sma Astron. Zh.* **3**, 216 (translation *Soviet Astron. L.* **3**, 114).

Spitzer, L. (1965), *Physics of fully ionized gases*, Interscience Tracts on Physics and Astronomy, Interscience Publication (New York, 1965), 2nd rev.

Spruit, H.C., Matsuda, T., Inoue, M., and Sawada, K. (1987), *Mon. Not. R. astron. Soc.* **229**, 517.

Spruit, H.C. (1997), in E. Meyer-Hofmeister and H.C. Spruit (eds.), *Accretion disks-new aspects, Lecture Notes in Physics* **487**, Springer, p67.

Taam, R.E. (1994), in J. van Paradijs et al. (eds.), *Compact stars in Binaries* IAU Symp 165, Kluwer, p3.

Taam, R.E. (2000), ARAA, in prepapration.

Van Paradijs, J. and McClintock, J.E. (1995), in Lewin, W.H.G., van Paradijs, J., and van den Heuvel, E.P.J. (eds.), *X-Ray Binaries*, Cambridge University press, p58.

Wang, J.M. and Zhou, Y.Y. (1999), *Astrophys. J.* **516**, 420.

Yi, I. (1999), in J.A. Sellwood and J. Goodman (eds.), *Astrophysical Discs*, Astronomical Society of the Pacific Conference series 160, p279 (astro-ph/9905215).

Zdziarski, A.A. (1998), *Mon. Not. R. astron. Soc.* **296**, L51.

Zeldovich, Ya.B., Ivanova, J., and Nadyozhin, D.K. (1972), *Soviet Astron.* **16**, 209.

INNER REGION ACCRETION FLOWS ONTO BLACK HOLES

M. KAFATOS AND P. SUBRAMANIAN
Center for Earth Observing and Space Research
George Mason University
Fairfax, VA 22030, U.S.A.

Abstract. We examine here the inner region accretion flows onto black holes. A variety of models are presented. We also discuss viscosity mechanisms under a variety of circumstances, for standard accretion disks onto galactic black holes and supermassive black holes and hot accretion disks. Relevant work is presented here on unified aspects of disk accretion onto supermassive black holes and the possible coupling of thick disks to beams in the inner regions. We also explore other accretion flow scenarios. We conclude that a variety of scenarios yield high temperatures in the inner flows and that viscosity is likely not higher than alpha ~ 0.01.

1. Introduction

"Accretion is recognized as a phenomenon of fundamental importance in Astrophysics" (Frank, King and Raine 1992). This is indeed the case as gravitational energy released in accretion processes is believed to be the dominant source of energy in a variety of high energy galactic compact sources in binary star systems containing white dwarfs, neutron stars and black holes; as well as extragalactic, supermassive black holes (Shapiro and Teukolsky 1983). Both spherical/quasi-spherical accretion (Bondi 1952; Frank et al. 1992 and references therein; Treves, Maraschi and Abramowicz 1989 and papers therein); and disk accretion (Pringle 1981; Dermott, Hunter, and Wilson 1992; Frank et al. 1992 and references therein; Treves et al. 1989 and papers therein) operate and the type of accretion may depend on boundary conditions such as the motion of gas at infinity, its angular momentum per unit mass, etc. The field of accretion astrophysics is obviously vast. In this paper we concentrate on the inner regions of accretion flows onto galactic (stellar) black hole candidates (GBH) and supermassive

C. Kouveliotou et al. (eds.), The Neutron Star – Black Hole Connection, 159–168.

black holes (SMBH) in active galactic nuclei (AGN). Accretion can be a very efficient process. The luminous energy released as matter accretes with mass accretion rate $\dot{M}$ is

$$L \sim 10^{34} \, (\dot{M}/10^{-9} M_\odot \, \mathrm{yr}^{-1}) \, \mathrm{erg/sec} \quad \text{for a white dwarf}$$
$$L \sim 10^{37} \, (\dot{M}/10^{-9} M_\odot \, \mathrm{yr}^{-1}) \, \mathrm{erg/sec} \quad \text{for a neutron star or a black hole} \quad (1)$$

With corresponding efficiencies $(L = \eta \dot{M} c^2)$

$$\eta \sim 10^{-4} \quad \text{for a white dwarf}$$
$$\eta \sim 0.05 \quad \text{for a neutron star}$$
$$\eta \sim 0.06 \quad \text{for a Schwarzschild black hole}$$
$$\eta \sim 0.42 \quad \text{for a maximally rotating Kerr black hole} \quad (2)$$

In spherical accretion, the specific angular momentum is, by definition zero. High temperatures can be achieved in the inner regions but whether this radiation escapes or not depends on the optical thickness of the accreting gas (Loeb and Laor 1992). Very high temperatures can be achieved in the inner regions of accretion disks: minimum values apply in an optically thick gas where LTE is achieved,

$$\frac{1}{4} a \, T_{\min}^4 = F(r) = \frac{3}{8} \pi \, \frac{GM\dot{M}}{r^3}, \quad (3)$$

whereas when complete internalization of gravitational energy by an optically thin gas is achieved,

$$kT_{\max} \sim \frac{GMm_p}{r_{\mathrm{ms}}} \quad (4)$$

where r_{ms} is the marginally stable radius. $T_{\max} \sim 150$ MeV ($\sim 2 \times 10^{12}$ K) for a Schwarzschild black hole and ~ 750 MeV ($\sim 10^{13}$K) for a Kerr black hole. Section 2 provides a brief summary of a variety of accretion models applicable to BHs. Section 3 covers the physics of accretion as related to the physics of viscous flows. Section 4 covers the topic of outflows, presumably originating close to the central object. Section 5 is discussion and conclusions.

2. Models

Several models pertaining to disk and quasi-spherical accretion flows are presented here.

2.1. STANDARD DISKS

Disk accretion proceeds via the outward transfer of angular momentum of the accreting gas. The seminal work of Shakura and Sunyaev (1973) provided the basic formalism of accretion in what hence came to be known as "standard" disks (see also Novikov and Thorne 1973; Lynden-Bell and Pringle 1974). Such disks are optically thick, geometrically thin, radiate locally as black bodies (BB) and have a radial dependence $T(r) \sim r^{-3/4}$. T is analogous to the effective temperature of a star. The modified BB spectrum has a characteristic broad peak, at low-frequencies $S_\nu \sim \nu^2$ while at high frequencies the spectrum drops exponentially. In between, but over a not too large range of frequencies the spectrum depends on frequency as $\nu^{1/3}$, the characteristic accretion disk law (Pringle 1981; Kafatos 1988). For typical parameters applicable to SMBHs, $10^8 M_\odot$ and near-Eddington accretion, the peak of the modified black body spectrum occurs in the UV, $\log \nu \sim 15.3 - 15.5$, with $T \sim 20,000 - 30,000$ K (Ramos 1997); whereas for GBHs, $\sim 10 M_\odot$, the peak occurs below ~ 1 keV, with $T \sim 1 - a$ few $\times 10^7$ K (Shapiro, Lightman, and Eardley 1976). In reality the disk accretion is more complicated than the above simple expressions. Several regions in the accretion flows have been identified (Shakura and Sunyaev 1973; Novikov and Thorne 1973; Kafatos 1988): i) an outer region where gas pressure dominates over radiation pressure and where the opacity is predominantly free-free; ii) a middle region where gas pressure again dominates but the opacity is primarily due to electron scattering; and iii) an inner region where radiation pressure dominates over gas pressure and the opacity is primarily due to electron scattering. The latter is expected to occur for $r \leq 50 r_g$, where r_g is the gravitational radius of the black hole, GM/c^2. The simple thick disk solution applies to the "outer" disk region (Novikov and Thorne 1973) as well as in accretion disks around white dwarfs.

2.2. MODIFIED DISKS

Further considerations indicate that optically thick disks have to be modified. Modifications include electron scattering and Comptonization. If electron scattering dominates, the emitted flux is lower (Rybicki and Lightman 1979) as follows

$$I_\nu = B_\nu \sqrt{\frac{\kappa_{abs}}{(\kappa_{abs} + \kappa_{es})}} \tag{5}$$

Modifications due to electron scattering are important for $\log \nu > 15$ (Ramos 1997) for SMBHs. Malkan and Sargent (1982), and Ramos (1997) have applied these modifications. At high energies, photons are scattered many times by the electrons before they leave the disk. This was recognized as far back as the original work by Shakura and Sunyaev (1973). The result is that the (relatively) soft photons emanating from the disk are upscattered by the hot electrons and become hardened. This is known as Comptonization (Shapiro, Lightman, and Eardley 1976; Sunyaev and Titarchuk 1980, 1985) and is believed to produce hard, power-law radiation above the usual broad disk peak. Comptonization is important above $\log \nu = 15.8$ for SMBHs (Ramos 1997) and above 10 keV for GBH candidates such as Cygnus X-1 (Shapiro, Lightman, and Eardley 1976).

2.3. TWO-TEMPERATURE DISKS AND ION-SUPPORTED TORI

If $T_e \sim 10^9$ K in the inner portion of the disk, a two-temperature solution is obtained (Shapiro, Lightman, and Eardley 1976; Eilek and Kafatos 1983), where the ions are much hotter than the electrons, $T_i \sim 10^{11} - 10^{13}$K. In such a disk, a puffed-up inner region is formed supported by the ion pressure. Unsaturated Comptonization transfers energy from the electrons to the soft photons emitted in the cooler, underlying disk. The process is described by the dimensionless parameter y (Shapiro, Lightman, and Eardley 1976), where y = < fractional energy change per scattering >< number of scatterings > or

$$y = (\frac{4kT_e}{m_e c^2}) \max(\tau_{es}, \tau_{es}^2) \tag{6}$$

Unsaturated Comptonization occurs for y $\sim$ 1 and is appropriate whenever there is a copious source of soft photons in the inner region (Shapiro, Lightman, and Eardley 1976; Sunyaev and Titarchuk 1985). The y-value is related to the energy flux spectral index A (where the energy flux is measured in keV cm^{-2} s^{-1} keV^{-1}) through A $\sim 0.72\,y^{-0.917}$ (Kafatos 1983) and as such y $\sim$ 1 provides a natural explanation for the spectra of many cosmic sources. The 2T solution is thermally unstable (Piran 1978) and whether it occurs or not depends on a variety of factors, including whether accretion is at near-Eddington rates, where the Eddington luminosity is $L_{Edd} \sim 10^{38}(M/M_\odot)$ erg/sec. Detailed spectra of 2T disks including relativistic effects, ion-ion collisions and resultant radiation spectra at gamma-rays (from pions and relativistic pairs which subsequently radiate via inverse-Compton) have been calculated by Eilek and Kafatos (1983). Both the Shapiro, Lightman, and Eardley (1973) and Eilek and Kafatos (1983) solutions apply to near-Eddington accretion rates. Gamma-gamma scatterings will degrade gamma-rays above $\sim$ MeV as these photons scatter the softer X-rays emerging from the 2T disk. The resultant optical depth for AGNs (Eilek and Kafatos 1983) is

$$\tau_{\gamma\gamma} \sim 5 \times 10^{-2} \, D_{Mpc}^2 \, E_T \, (keV) \, N(2E_T) \, R_\gamma^{-1} \,, \qquad (7)$$

where D_{Mpc} is the distance of the AGN in Mpc, E_T is the relevant threshold X-ray energy for $e^+ e^-$ production and N is the corresponding photon flux at the earth (photons cm^{-2} s^{-1} keV^{-1}) computed at $2E_T$. γ-γ scattering will form a broad shoulder ~ 1 MeV with an exponentially-declining tail and the absence of high-energy radiation in many accreting GBH and SMBH (e.g. Seyferts) sources may be explained by this fundamental physical process (Eilek and Kafatos 1983). High-energy gamma-rays (above 100 MeV − TeV) are probably arising in a jet (see below). A closely-related model to 2T Comptonized disks is the ion-supported torus model (Rees, Begelman, Blandford, and Phinney 1982), proposed as the underlying engine in extragalactic jet sources. They found self-consistent 2T solutions even for $r \gg 2000r_g$ for sub-Eddington rates as low as $10^{-4} \dot{M}_{Edd}$.

2.4. HOT CORONAE

Hot coronae may surround an underlying, cooler disk. Corona models have been proposed for Cygnus X-1 (Liang and Price 1977; Bisnovatyi-Kogan and Blinnikov 1977) and provide a competing model to hot, 2T disks. Coronae provide a natural explanation for unsaturated Comptonization since a corona would envelop an underlying cool disk where the soft photons emanate. Conduction-balanced coronae (see also Rosner, Tucker, and Vaiana 1978) would produce temperatures, $T \sim 10^{11}$K, lower than 2T disks but still higher than standard inner accretion disks. Hot haloes or coronae may also be produced in a bulk motion Comptonization model (see below) and would account for the hard-radiation in such flows.

2.5. BULK COMPTONIZATION

A promising theoretical model proposed for the soft-state GBH candidates is the bulk motion Comptonization model (Chakrabarti and Titarchuk 1995; Titarchuk, Mastichiadis,and Kylafis 1997; Titarchuk and Zanias 1998; Shrader and Titarchuk 1998). In this model (Bautista and Titarchuk 1999), the production of hard photons peaks at $\sim 2r_S$ (where r_S is the Schwarzschild radius, $= 2 \, r_g$). It explains the continuum X-ray spectra of soft-state GBHs in their soft state (Shrader and Titarchuk 1998). Accretion onto the central BH proceeds via a spherically-converging flow at gravitational free-fall speeds. A variant of this model assumes the formation of strong shocks as the convergent inflow speeds become greater than the local sound speed (Chakrabarti and Titarchuk 1995). In the latter model, the disk is divided into two main components, a standard cool disk which extends to the outer boundary and produces soft (UV) photons; and an optically thin sub-Keplerian halo (or corona) which terminates in a standing shock near the black hole. The post-shock region Comptonizes the soft (UV) photons that are subsequently radiated as a hard spectrum with spectral index $\sim$ 1.5.

2.6. THICK DISKS AND ADVECTION-DOMINATED FLOWS

In the above models, it is generally assumed that the disk motion is Keplerian (or quasi-Keplerian). In reality, when radiation pressure is included, the flow becomes non-Keplerian and the disk fattens geometrically (Abramowicz et al. 1978; Jaroszynski et al. 1980; Paczynski and Wiita 1980; Paczynski 1998). A funnel wall is produced as matter cannot reach the axis of rotation. Such thick disk flows may play a role in the production of matter outflows, although the exact mechanism has not been proposed in the above works. These disks are special cases of advective disks (Chakrabarti 1998). Besides pressure effects, radial inflow effects have also to be considered. In normal disks, the radial inflow speed is assumed to be negligibly small. In reality, this speed can be large, particularly in the inner regions. As such, the advection term vdv/dr should be included in the momentum equation. It may also be the case that transonic flows always result in accretion onto black holes (Chakrabarti 1990; Kafatos and Yang 1994) as the inflow speed has to smoothly join the subsonic regime with the supersonic regime near the horizon. It may also be the case that shocks are prevalent (Chakrabarti 1990; Yang and Kafatos 1995; Chakrabarti and Titarchuk 1995; Chakrabarti 1996). Advection-dominated accretion flows (ADAFs) have been widely discussed in the literature (Narayan and Yi 1994; Liang and Narayan 1997; Esin, McClintock, and Narayan 1997; Narayan, Mahadevan, and Quataert 1999). At large or super-Eddington rates (Katz 1977; Abramowicz et al. 1988) optically thick advection solutions are obtained. In these solutions, the large optical depth of the inflowing gas traps or advects it into the central black hole. At low, sub-Eddington accretion rates (Rees et al. 1982; Narayan and Yi 1994), optically-thin advection flows (ADAFs) result. In this model, the accreting gas has low density, is unable to radiate and viscous energy is advected onto the central BH. Optically-thin ADAFs are hot, 2T flows. This model assumes a self-similar solution and can (according to Narayan and co-workers) only operate at low accretion rates and high viscosity values, $\alpha \sim 0.1$. Whether these conditions can be satisfied in realistic flows is another matter.

2.7. ADVECTION DOMINATED INFLOW-OUTFLOW SOLUTIONS

ADAFs have the generic drawback of having a positive Bernoulli parameter in the disk. Such accretion flows thus tend to evaporate, before they accrete. Recently, advection dominated inflow outflow solutions (ADIOS) have been proposed (Blandford & Begelman 1999) which overcome this drawback by postulating a powerful outflow that carries away excess mass, energy and momentum, thus allowing the accretion to proceed. Subramanian et al. (2000) are investigating a scenario where relativistic outflows can be produced in such an advection-dominated accretion flow, as a result of Fermi acceleration due to collisions with kinks in the tangled magnetic field embedded in the accretion flow. This mechanism has been explored in detail by Subramanian et al. (1999), and it is expected that the low-density environment of advection-dominated flows will be ideal sites for the launching of outflows via this mechanism. On the other hand, time-dependent treatments of quasi-spherical accretion (Ogilvie 1999) suggest that advection-dominated accretion can proceed without the need for out-flows.

3. Viscosity Mechanisms

Viscosity in accretion disks has been the subject of investigation for nearly 20 years (for a review, see Pringle 1981). It was recognized early on that ordinary molecular viscosity cannot produce the level of angular momentum transport required to provide accretion rates commensurate with observed levels of emission. In lieu of a detailed model of microphysical viscosity, Shakura & Sunyaev (1973) embodied all the unknown microphysics of viscosity into a single parameter α according to the prescription

$$t_{r\phi} = \alpha P, \qquad (8)$$

where $t_{r\phi}$ denotes the $r\phi$ component of the viscous stress and P is the ambient pressure in the disk. Much of the subsequent developments concentrated on obtaining estimates of the α parameter due to fluid turbulence (Shakura & Sunyaev 1973; Goldman & Wandel 1995) magnetic viscosity (Eardley & Lightman 1975; Balbus & Hawley 1998 and references therein) and radiation viscosity (Loeb & Laor 1992) and ion viscosity (Paczynski 1978; Kafatos 1988). Subramanian et al. (1996) have shown that "hybrid" viscosity (neither pure ion viscosity nor pure magnetic viscosity) due to hot ions scattering off kinks in the tangled magnetic field embedded in accretion disks is the dominant form of viscosity in hot, two-temperature accretion disks. This work assumes the magnetic field embedded in the accretion disk to be isotropically tangled. Recent simulations (see, for instance, Armitage

1998) suggest, however, that the manner in which the magnetic field is tangled might be significantly anisotropic. Based on the detailed calculations presented in Subramanian et al. (1996), we believe that this will merely introduce a direction dependence in the hybrid viscosity, but will not change the overall conclusion that this (hybrid viscosity) is the dominant form of viscosity in hot accretion disks. The viscosity obtained is characterized by a parameter $\alpha \sim 0.01$ (Subramanian et al. (1996). In the turbulent viscosity mechanism characterized by convection (Yang 1999) when gas pressure dominates, an upper limit to α of 0.01 is also obtained. For radiative viscosity (Loeb and Laor 1992) in which radiation pressure dominates, α is greater than 1. These models are, however, producing optically thick, relatively cool inner regions.

4. Jets/Outflows

Several objects for which much of the preceding discussion of accretion disks is relevant also exhibit jets/outflows. Although several mechanisms for producing outflows have been proposed, none of them, with the exception of a few (Das 1998) make the connection between the outflows and the underlying accretion disk. Subramanian et al. (1999) have proposed a model in which the outflow is powered by Fermi acceleration of seed protons due to collisions with magnetic scattering centers embedded in the accretion disk/corona. This is a natural mechanism expected to operate in any accretion disk, and Subramanian et al. (2000) are examining its viability in the context of the ADIOS scenario. The physical scenario in which the Fermi acceleration of protons (which powers the outflow at its base) takes place is the same as that in which hybrid viscosity (Subramanian et al. 1996) is operative. The energy inherent in the shear flow (gravitational potential energy) is dissipated partly by viscous heating of the thermal protons, and partly by Fermi acceleration of the supra-thermal protons, which in turn form a relativistic outflow.

5. Discussion and Conclusions

We have seen that a variety of scenarios predict hot, inner regions, either 2T, hot coronae or ADAFs, where $T_{max} \sim 10^{11} - 10^{13}$K. In most models, these hot inner regions occur for radii not much greater than 10's of gravitational radii. In ADAFs, however, high temperatures persist for hundreds or even 1,000's of radii. Besides spectral observations that would reveal hard X-rays and γ rays, timing observations would be crucial (with characteristic timescales comparable to the light-travel time through the hot region). Several of these models face theoretical difficulties or inconsistencies: 2T disks are unstable (although appropriate viscosity laws may mitigate this difficulty); hot coronae are attractive but the mechanism of heating the corona is unknown. In many (all?) cases, transonic flows and even shocks may be prevalent, breaking up the usual assumptions of quasi-Keplerian or even steady disk structure. ADAFs are particularly problematic: the assumption of self-similarity is probably erroneous. ADAFs assume that the velocity in the θ direction is zero. However, at the boundary between the thin disk and the geometrically thick ADAF solution, v_θ is not zero. Also, at the axis this velocity would be zero but then a funnel would be formed. ADAFs assume no shocks but shocks, transonic flows, etc. are probably prevalent. Although the usual ADAF assumption of bremmstrahlung cooling produces inefficient cooling, there is no reason that other more efficient coolings processes (such as Compton processes) would not be operating. Finally, ADAFs seem to require $\alpha \sim 0.1$ whereas realistic physical calculations for hot, optically thin flows suggest $\alpha \sim 0.01$.

References

Abramowicz, M.A. et al. (1998), *ApJ* **332**, 646.

Armitage, P.J. (1998), *ApJ* **501**, L189.

Balbus, S.A. and Hawley, J.F. (1998), *Rev. Mod. Phys.* **70**, 1.

Bautista, M.A. and Titarchuk, L.G. (1999), *ApJ* **511**, 105.

Bisnovatyi-Kogan, G.S. and Blinnikov, S.I. (1977), *A&A* **59**, 111.

Bondi, H. (1952), *MNRAS* **112**, 195.

Blandford, R.D. and Begelman, M.C. (1999), *MNRAS* **303**, L1.

Chakrabarti, S.K. (1990) *Theory of Transonic Astrophysical Flows*, World Scientific, Singapore.

Chakrabarti, S.K. (1996), *ApJ* **464**, 664.

Chakrabarti, S.K. (1998), in S.K. Chakrabarti (ed.), *Observational Evidence for Black Holes in the Universe*, Kluwer (Dordrecht).

Chakrabarti, S.K. and Titarchuk, L.G. (1995), *ApJ* **455**, 623.

Das, T.K. (1998), in S.K. Chakrabarti (ed.), *Observational Evidence for Black Holes in the Universe*, Kluwer (Dordrecht).

Dermott, S.F., Hunter, J.H.Jr., and Wilson, R.E. (1992), in *Astrophysical Disks, Annals of the New York Academy of Sciences*, **675**, New York Academy of Sciences (New York).

Eardley, D.M. and Lightman, A.P. (1975), *ApJ* **200**, 187.

Eilek, J.A. and Kafatos, M. (1983), *ApJ* **271**, 804.

Esin, A.A., McClintock, J.E., and Narayan, R. (1997), *ApJ* **489**, 865. Frank, J., King, A., and Raine, D. (1992) *Accretion Power in Astrophysics*, Cambridge University Press (Cambridge, U.K).

Goldman, I. and Wandel, A. (1995), *ApJ* **443**, 187.

Jaroszynski, M., Abramowicz, M., and Paczynski, B. (1980), *Acta Astron.* **30**, 1.

Kafatos, M. (1983), in R.M. West (ed.), *Highlights of Astronomy* 6, IAU, p505.

Kafatos, M. (1988), in *Adv. Space Res.*, **8**, COSPAR, Great Britain, pp105-112.

Kafatos, M. and Yang, R. (1994), *MNRAS* **268**, 925.

Katz, J. (1977), *ApJ* **215**, 265.

Liang, E.P.T. and Narayan, R. (1997), in C.D. Dermer, M.S. Strickman, and J.D. Kurfess (eds.), *AIP Sympos. 410, Proccedings of the Fourth Compton Symposium*, AIP (New York), p461.

Liang, E.P.T. and Price, R.H. (1977), *ApJ* **218**, 247.

Loeb, A. and Laor, A. (1992), *ApJ* **384**, 115.

Lynden-Bell, D., and Pringle, J.E. (1974), *MNRAS* **168**, 603.

Malkan, M.A. and Sargent, W.L.W. (1982), *ApJ* **254**, 22.

Narayan, R. and Yi, I. (1994), *ApJ* **428**, L13.

Narayan, R., Mahadevan, R., and Quataert, E. (1999), in M.A. Abramowicz, G. Bjornsson, and J.E. Pringle (eds.), *The Theory of Black Hole Accretion Disks*, Cambridge University Press (Cambridge).

Novikov, I.D. and Thorne, K.S. (1973) in C. DeWitt and B.S. DeWitt (eds.), *Black Holes*, Gordon and Breach (New York), p345.

Ogilvie, G.I. (1999), *MNRAS* **306**, L9.

Paczynski, B. (1978), *Acta Astron.* **28**, 253.

Paczynski, B. (1998), *Acta Astron.* **48**, 667.

Paczynski, B. and Wiita, P.J. (1980), *A&A* **88**, 23.

Pringle, J.E. (1981) *Accretion Discs in Astrophysics, Ann. Rev. Astron. Astrophys.* **19**, pp137-162.

Ramos, E. (1997), *Ph.D Thesis, High-frequency Multifrequency Observations and Analysis of Blazars*, George Mason University.

Ramos, E., Kafatos, M., Fruscione, A., Bruhweiler, F.C., McHardy, I.M., Hartman, R.C., Titarchuk, L.G., and von Montigny, C. (1997), *ApJ* **482**, 167.

Rees, M.J., Begelman, M.C., Blandford, R.D., and Phinney, E.S. (1982), *Nature* **295**, 17.

Rosner, R., Tucker, W.H., and Vaiana, G.S. (1978), *ApJ* **220**, 643.

Rybicki, G.B. and Lightman, A.P. (1979) *Radiative Processes in Astrophysics*, John Wiley & Sons (New York).

Shapiro, S.L., Lightman, A.P., and Eardley, D.M. (1976), *ApJ* **204**, 187.

Shapiro, S.L. and Teukolsky, S.A. (1983) *Black Holes, White Dwarfs and Neutron Stars*, John Wiley & Sons (New York).

Shrader, C. and Titarchuk, L.G. (1998), *ApJ* **499**, L31.

Subramanian, P., Becker, P.A., and Kafatos, M. (1996), *ApJ* **469**, 784.

Subramanian, P., Becker, P.A., and Kazanas, D. (1999), *ApJ* **523**, 203.

Subramanian, P., Becker, P.A., and Kazanas, D., (2000), *ApJ* in preparation.

Sunyaev, R.A. and Titarchuk, L.G. (1980), *A&A* **86**, 121.

Sunyaev, R.A., and Titarchuk, L.G. (1985), *A&A* **143**, 374.

Titarchuk, L.G., Mastichiadis, A., and Kylafis, N.D. (1997), *ApJ* **487**, 834.

Titarchuk, L.G. and Zannias, T. (1998), *ApJ* **493**, 863.

Treves, A., Maraschi, L., and Abramowicz, M. (1989) *Accretion: A Collection of Influential Papers*, World Scientific Publishing (Singapore).

Yang, L.Y. (1999), *Ph.D. Thesis*, George Mason University (in progress).

Yang, R. and Kafatos, M. (1995), *A&A* **295**, 238.

4. Neutron Stars and Black Holes in Binaries

NEUTRON STARS AND BLACK HOLES IN BINARY SYSTEMS: OBSERVATIONAL PROPERTIES

E.P.J. VAN DEN HEUVEL
Astronomical Institute "Anton Pannekoek" and Center for High Energy Astrophysics University of Amsterdam, Kruislaan 403, 1098 SJ Amsterdam, The Netherlands
and
Institute for Theoretical Physics, UC Santa Barbara, CA, USA

Because of his grave illness, Professor van Paradijs was unable to attend the Advanced Study Institute and to present his lectures on the observed properties and mass determinations of neutron stars and black holes in binary systems. At his request, I presented my own review of the same material, in the form of two lectures which were heavily based on van Paradijs' excellent review in the book "The Many Faces of Neutron Stars" (van Paradijs 1998). The reader is referred to this review for an overview of the observed properties of neutron stars and black holes in X-ray binaries, and further to the book "X-ray Binaries" (ed. Lewin et al. 1995). In the present book a very brief summary of the observational material is incorporated in my article on the formation and evolution of neutron stars and black holes in binary systems.

References

Lewin, W.H.G., van Paradijs, J.A., and van den Heuvel, E.P.J. (eds.) (1995), *X-ray Binaries*, Cambridge University Press, pp662.
Van Paradijs, J. (1998), in R. Buccheri, J. van Paradijs, and A. Alpar (eds.), *The Many Faces of Neutron Stars*, NATO ASI Series 515, Kluwer (Dordrecht), pp279-336.

C. Kouveliotou et al. (eds.), The Neutron Star – Black Hole Connection, 171.

FORMATION AND EVOLUTION OF NEUTRON STARS AND BLACK HOLES IN BINARY SYSTEMS

EDWARD VAN DEN HEUVEL

Astronomical Institute "Anton Pannekoek" and Center for High Energy Astrophysics University of Amsterdam, Kruislaan 403, 1098 SJ Amsterdam, The Netherlands and Institute for Theoretical Physics, UC Santa Barbara, CA, USA

1. Introduction

For a general overview of the evolutionary processes in binaries leading to the formation of compact objects I refer to Van den Heuvel (1994a) and for a historical overview of the subject to Van den Heuvel (1994b). In 1962 the first extrasolar X-ray source, Sco X-1 was discovered (Giacconi et al. 1962) and by the end of the sixties several dozen such strong point X-ray sources were known from rocket and balloon experiments.

Their marked concentration in the direction of the galactic center made clear that the distances of a number of them must be of the order of 8 kpc, implying a very large energy output in the form of X-rays, of order 10^{37} to 10^{38} ergs/s (some 10^4 times the total energy output of the sun). What could be the mechanism generating these enormous X-ray luminosities? Largely thanks to the Russian school of Zeldovitch and co-workers since the mid-sixties the idea arose that in a binary system the process of accretion of matter flowing over from a normal companion star to a neutron star or a black hole might power these strong galactic X-ray sources. Indeed, the simple process of accretion of an amount of mass m onto a neutron star (black hole) releases some $0.15\ mc^2$ (0.06 to $0.42\ mc^2$) of gravitational binding energy which, converted into heat, is available for emission in the form of X-rays. (The process of mass accretion onto a super-massive black hole had before that already been suggested as the energy source for quasars and active galaxy nuclei by Salpeter (1964), Zeldovitch (1964), and Zeldovitch and Novikov (1964)).

C. Kouveliotou et al. (eds.), The Neutron Star – Black Hole Connection, 173–243.
© 2001 *Kluwer Academic Publishers. Printed in the Netherlands.*

It should be kept in mind, however, that the existence of neutron stars in nature was not known before the end of 1968 - the year in which the discovery of the radio pulsars was announced (Hewish et al. 1968). The discovery of the Crab Nebula pulsar in November 1968 (Staelin and Reifenstein 1968) and the detection of its large spindown rate made clear that pulsars are neutron stars and that neutron stars are born in a supernova event (Gold 1969, Monaghan 1969) just as had been predicted 34 years earlier by Baade and Zwicky (1934). Before 1968 neutron stars and black holes had been purely theoretical concepts, based on theoretical insights developed in the pioneering studies by Oppenheimer and Volkoff (1938) and Oppenheimer and Snijder (1939), respectively, and subsequently studied by various groups, notably those of J.A. Wheeler in the US and Ya.B. Zeldovitch in the USSR. The first ones to search for black holes in binary systems were Zeldovitch and Guseinov (Guseinov and Zeldovitch 1966; Zeldovitch and Guseinov 1966), who searched for spectroscopic binaries with massive unseen secondary stars. They did, however, not yet mention the possibility that such binaries might emit X-rays, but Novikov and Zeldovitch (1966) did so slightly later. Following the discovery of the faint blue optical counterpart of Sco X-1 (Sandage et al. 1966) this early work culminated in Shklovskii's (1967) neutron-star binary model for Sco X-1. This author showed that the optical light in the system could not arise from the same source as the X-rays, and that the X-ray energy distribution is consistent with thermal bremsstrahlung from an optically thin plasma accreting onto a neutron star. Since the optical spectrum of the source resembles that of a cataclysmic variable (CV: these are binary systems consisting of an accreting white dwarf and a low-mass ordinary star) and since no stellar spectrum is seen (implying that the companion star is faint) he postulated that the neutron star is in a binary and accretes matter from a low-mass companion star. It took another eight years before this "Low-mass X-ray binary" model for Sco X-1 was confirmed by the detection of its 0.86 day orbital period by Gottlieb, Wright and Liller (1975).

Before that, however, the first X-ray satellite UHURU had, in 1971, discovered the existence of the pulsating and eclipsing binary X-ray source Cen X-3, which left no doubt about the existence of neutron stars moving in orbits around massive ordinary stars (Schreier et al. 1972). Shortly before this, earlier in 1971, the UHURU group had discovered rapid X-ray variations in Cyg X-1, with timescales down to 50 msec, with no obvious periodicity (Oda et al. 1971). This rapid variability indicated that the X-ray source in the system cannot be larger than about 10^4 km. Webster and Murdin (1972) and Bolton (1972) independently that same year identified the bright O9.7 supergiant star HD 226868 as the optical counterpart to this source. This identification was an indirect one, through the accurate arcsec position of a radio-source that had been discovered independently by Braes and Miley (1971) and Hjellming and Wade (1971) in the X-ray error box of Cyg X-1. Webster and Murdin and Bolton found the blue supergiant to be a 5.6 day single-lined spectroscopic binary with a large velocity amplitude (72 km/s) indicating, if the O 9.7 supergiant has a normal mass ($\geq$ 15 $M_\odot$) for its spectral type, the presence of a companion of mass > 3 $M_\odot$. As this is above the upper mass limit of a neutron star they suggested Cyg X-1 to be a black hole - a suggestion that nowadays has been fully accepted. However, at the time this was, like that for Sco X-1, a rather indirect indication for the existence of X-ray binaries, and not yet completely convincing. For example, Kristian et al. (1971) dismissed the blue supergiant as the optical counterpart of the X-ray source. On the other hand, there could be little doubt that the X-ray source and the radio source were connected, as the radio source appeared just at the time when a dramatic change in the spectrum of the X-ray source occurred (cf. Tananbaum 1973). Since the radio source coincided within a few arcsec with the blue supergiant star, it seemed quite likely that the star, the radio source and the X-ray source were connected. However, the X-ray source did not eclipse and neither the X-ray source nor the radio source showed a trace of a 5.6 day period variability that might connect them with the 5.6 day blue supergiant binary [only recently such periodic variability has been established at X-ray and radio wavelengths, cf. Hanson et al. 2000].

For these reasons it was only with the discovery of the eclipsing pulsating binary X-ray source Cen X-3 by Schreier et al. (1972) that this collection of observations found a definitive explanation. Here one observed for the first time an X-ray pulsar (neutron star) that shows a beautiful 2.087 day sinusoidal Doppler-modulation of its 4.84 s pulse period and is eclipsed every orbit for about 0.5 days. From the Doppler-effect a projected orbital velocity of 415.1 $\pm$ 0.4 km/sec was derived leading to a minimum companion mass of about 17 $M_\odot$. The presence of this massive companion made it immediately clear that Cyg X-1 might form a similar system with the massive blue supergiant star HD 226868. Thus, by 1972 the existence of both neutron stars and black holes in close orbits around massive companion stars appeared well established. In the same year, UHURU discovered the second pulsating and eclipsing binary X-ray source Her X-1, in which the 1.2 s period X-ray pulsar orbits in 1.7 days a star of rather low mass, about 2.0 $M_\odot$ (Tananbaum et al. 1972). In the seventies and eighties more and more X-ray binaries were discovered. They were found to fall into two broad categories: the High-Mass X-ray Binaries (HMXB) like Cen X-3 and the Low-Mass X-ray Binaries (LMXB) like Her X-1 and Sco X-1.

The recognition that neutron stars and black holes can exist in close binary systems came at first as a surprise, as it did not seem to fit with the then current ideas about the evolution of binary systems. It was known from stellar evolution that the more massive a star is, the shorter its lifetime. Thus, the more massive component of a binary will be the first one to undergo a supernova explosion. If more than half the mass of a circular-orbit binary system is explosively ejected, the orbit of the system becomes hyperbolic and the system is disrupted (Blaauw 1961). This is a simple consequence of the virial theorem (cf. Van den Heuvel 1994a). In a massive binary the mass of the neutron star remnant ($\sim$ 1.4 M$_\odot$) is negligible with respect to the masses of the two components (and to the amount of mass ejected in the supernova), so at first sight one would always expect these systems to be disrupted by the first supernova explosion. Still more puzzling were the almost perfectly circular orbits of Cen X-3 and Her X-1, apparently showing no trace of the effects of the supernova (although Cyg X-1 still has a slightly eccentric orbit). For the HMXBs like Cen X-3, it was soon realized (Van den Heuvel and Heise 1972) that the survival of the systems was due to the effects of large-scale mass transfer that must have occured prior to the supernova explosion. This caused the initially more massive star in the system to have become much less massive than its companion at the time of the explosion, and prevented the system from being disrupted. Tidal effects subsequently circularized the orbits during several millions of years, before the reverse mass transfer began and the system became an X-ray source. (Several other authors somewhat later independently came to similar conclusions: Börner et al. 1972; Tutukov and Yungelson 1973). The model of Van den Heuvel and Heise (1972) built forth on earlier work on close binary evolution developed primarily in the sixties. Morton (1960) had been the first one to show that large-scale mass transfer may explain why the more evolved (sub-giant) component of a close binary system like Algol (β Persei) may be less massive than its brighter and less evolved companion star. The reason is, that in the course of its evolution the envelope of a star gradually expands, and after the exhausion of the hydrogen fuel in the stellar core, swells up to giant dimensions. However, in a close binary system there is no room for a giant star: the sizes of the stars are limited by the dimensions of their so-called Roche lobes (Figure 10). As soon as a star becomes larger than its Roche lobe, the matter outside the Roche lobe will flow over to its companion. The more massive star in a binary will be the first one to overflow its Roche lobe. Evolutionary calculations by Kippenhahn and Weigert (1967), Plavec (cf. Plavec 1968) and Paczynski (cf. Paczynski 1966; 1971a) in the sixties showed that this leads to the transfer of most of the star's hydrogen-rich envelope ($\geq$ 70 % of its mass) to its companion, reversing the mass ratio of the system.

This type of "conservative" evolution (in which mass and orbital angular momentum of the system are conserved) appears to be able to explain the formation of the HMXBs.

However, for the LMXBs it was much harder to understand how the system survived the supernova explosion. The same holds for the close double neutron stars, such as the Hulse-Taylor binary pulsar PSR 1913+16 (Hulse and Taylor 1975), which has an orbital period of only 7.75 hours and an orbital eccentricity of 0.615. These systems must have survived two supernova explosions! In the LMXBs and their close relatives the Cataclysmic Variables the companion of the compact object is a low-mass ordinary star like our sun, and the orbital periods are in general very short: mostly between 11 minutes and about half a day. All these systems must in the course of their lives (and in the case of LMXBs and binary pulsars: before the last supernova in the system) have lost a very large amount of mass and orbital angular momentum. The evolution of these systems was therefore much more complicated than that of the HMXBs. The first models for the evolution of binaries with a large loss of mass and orbital angular momentum were made by Van den Heuvel and De Loore (1973) who showed that a HMXB may later in life turn into a very close binary system consisting of a helium star (the helium core of the massive star) and a compact star. They suggested that the 4.8 hour X-ray binary Cyg X-3 is such a helium-star system, which was confirmed almost 20 years later (Van Kerkwijk et al. 1992). The Hulse-Taylor binary pulsar is a logical later evolutionary product of such a system (cf. Flannery and Van den Heuvel 1975; De Loore et al. 1975). The first model for explaining the origin of an LMXB was that of Sutantyo (1975a) for the origin of Her X-1. He showed that in order to obtain such a system one should start out with a binary with components that differ very much in mass. In this case, due to the large difference between the thermal timescales of the envelopes of the stars, the low-mass component can hardly accept any mass from its more massive companion, once that star begins to overflow its Roche lobe. As a result most of the overflowing matter is lost from the systems, together with its orbital angular momentum, and only the helium core of the more evolved star is left, together with the practically unchanged low-mass companion. Still, in order to not disrupt the system when this helium star explodes, the initial conditions must have been very fine-tuned. Therefore, the formation of a LMXB is a much rarer event than the formation of HMXBs (cf Van den Heuvel 1983; 1994b; Webbink and Kalogera 1994; Kalogera and Webbink 1998), and the same holds for double neutron stars.

An important new ingredient that has been introduced in these binary evolution models already since 1975 (Flannery and Van den Heuvel 1975) is the occurrence of velocity kicks of order a few hundred km/s that are imparted to the neutron stars in their birth events. Ample evidence for the occurrence of such kicks has been inferred in the last decade from the space and velocity distribution of radio pulsars and from a variety of other observational facts (cf. Dewey and Cordes 1987; Tauris and Van den Heuvel 2000; Hartman 1996; 1997; Van den Heuvel and Van Paradijs 1997). Without the occurrence of these kicks a variety of properties of the X-ray binaries and the binary radio pulsars, including their birth rate in the galaxy, would be difficult to understand (Verbunt and Van den Heuvel 1995; Kalogera 1998a, b; Kalogera and Webbink 1998).

As mentioned above, already in the mid-seventies a link was suggested between the massive X-ray binaries and the Hulse-Taylor binary pulsar. The magnetic field of this pulsar is relatively weak ($\sim 10^{10}$ G, some two orders of magnitude lower than the average of pulsar magnetic fields) and its spin period is very short (0.059 sec). As it was believed at the time that neutron star magnetic fields decay spontaneously on a relatively short timescale ($\sim 10^7$ yrs) the weak field suggested that this neutron star is old. Since in X-ray binaries with accretion disks angular momentum is fed to the neutron star, one observes the X-ray pulsars in these systems to show a gradual decrease of their pulse periods in the course of time (see section 2.4), so-called "spin-up". Bisnovatyi-Kogan and Komberg (1975) suggested that if the binary system is disrupted in the second supernova, this spun-up neutron star may again become observable as a radio pulsar. It was suggested by Smarr and Blandford (1976) that PSR 1913+16 is the old "spun-up" neutron star in the system (due to its weak magnetic field it now spins down only very slowly, on a timescale of several 10^8 yrs). The other neutron star in the system was then produced by the second supernova explosion, and must be a young strong-field neutron star. It is not observable, either because it has already rapidly spun down (Srinivasan and Van den Heuvel 1982), or because the Earth is outside of the pulsar beam. Old neutron stars spun-up by accretion in X-ray binaries, which are now observable as radio pulsars are called "recycled pulsars" (Radhakrishnan and Srinivasan 1981, 1982). In order to become observable as a radio pulsar, there should no longer be gas in the system, so accretion should have terminated. The companions of recycled pulsars should therefore either be white dwarfs or neutron stars. The system may also have been disrupted, resulting in a single recycled pulsar. The discovery of the first millisecond radio pulsar in 1982 gave the recycling idea an enormous boost: Alpar et al. (1982) and Radhakrishnan and Srinivasan (1982) suggested that millisecond pulsars are very old neutron stars which were spun-up by accretion in Low-mass X-ray Binaries (LMXBs). Because of the very long duration the accretion phase in LMXBs ($\geq 10^8$ yrs), very much angular momentum can be fed to these neutron stars, leading to spin-up to millisecond periods. This recycling model has recently been beautifully confirmed by the discovery of the first millisecond X-ray pulsar in the LMXB-system SAX 1808.4-3658 (Wijnands and Van der Klis 1998). In order to spin up a neutron star to a millisecond period, its magnetic field should have decayed to below 10^9 G (see section 3.6). The general idea now is that the decay of the magnetic field in accreting neutron stars is somehow related to the accretion process (Taam and Van den Heuvel 1986), although the precise mechanisms for field decay are still being debated (cf. Konar and Bhattacharya, 1999).

2. Types of binaries with compact objects

2.1. INTRODUCTION

The binaries with compact objects can be divided into the categories and types listed in Table 1. Basically they fall into the categories "compact star plus ordinary star" and "two compact stars". The first category consists of the X-ray binaries and the Cataclysmic Variables, the second category of the binary radio pulsars and the double white dwarfs. Each of these categories can be divided into a few main types, that can be further divided into sub-types, as indicated in the table, where also examples of the different sub-types are given. A few binaries with compact objects do not fit into the above categories, notably the so-called "ante-deluvian" binary radio pulsars, which are young pulsars in an eccentric orbit around a massive star, which presumably are the progenitors of HMXBs. Four such systems are presently known: PSR 1259-63 (P_{orb}=3.4yr), PSR 1820-11 ($P_{orb} = 357{,}8^d$), PSR J1740-3052 (P_{orb}=231^d) and PSR J0045-7319 (P_{orb}=51^d) in the Large Magellanic Cloud. In the first system and the last-mentioned two systems the companion of the young pulsar is a B-type star, in PSR 1820-11 the companion is not known. Furthermore, there are the two peculiar X-ray binaries with relativistic jets, SS433 and Cyg X-3. In SS433 the companion of the jet-producing compact object is probably an early-type hydrogen-rich star (cf. Margon 1983). In Cyg X-3 it is a Wolf-Rayet star (helium star, cf. Van Kerkwijk et al. 1992).

I now briefly describe the main characteristics of each of the various types of neutron-star and black-hole binary systems listed in the table. In this review I will concentrate mainly on these systems as these are the ones that have primarily been discovered thanks to observations from space. Only where this is appropriate, I will also mention the CV-systems and double white dwarfs.

TABLE 1. Main Categories and types of binaries with compact objects

Category	Main Types	Sub-Types	Example (NS/BH)
X-ray Binaries	High-Mass Donor $(M_d \geq 10 M_\odot)$	"Standard" HMXB	Cen X-3, P_{orb} $= 2^d.087$ (NS) Cyg X-1, P_{orb} $= 5^d.60$ (BH)
		"Be"-HMXB	A0535+26, P_{orb} $= 104^d$ (NS)
	Low-Mass Donor $(M_d < M_\odot)$	Galactic Disk	Sco X-1, P_{orb} $= 0^d.86$ (NS) A0620-00, P_{orb} $= 7^h.75$ (BH)
		Globular Cluster	X 1820-30, P_{orb} $= 11^m$ (NS)
	Intermed. Mass Donor $1 \leq M_d/M_\odot < 10$		Her X-1, P_{orb} $= 1^d.7$ (NS) Cyg X-2, P_{orb} $= 9.8^d$ (NS) V 404 Cyg, P_{orb} $= 6^d.5$ (BH)
Binary Radio Pulsars with unevolved companion star	Main-seq. star plus young pulsars in eccentric orbit	B-type companion	PSR 1259-63, $P_{orb} = 3^{yr}.4$
		Low-mass companion	PSR 1820-11, $P_{orb} = 357^d.8$
Binary Radio Pulsars with compact companion	"High-mass" companion $(M_c = 0.5 - 1.4 \ M_\odot)$	Double NS, (recycled pulsar)	PSR 1913+16 $(P_{orb} = 7^h.75)$
		Massive WD companion circular orbit (recycl. PSR)	PSR 0655+64 $(P_{orb} = 1.03^d)$
		Massive WD companion eccentr. orbit (non-recycl. PSR)	PSR 2303+46 $(P_{orb} = 12^d.34)$
	"Low-Mass" companion $(M_c \leq 0.45 M_\odot)$	Circular orbits (recycled PSR)	PSR 1855+09 $(P_{orb} = 12^d.33)$ PSR 0820+02 $(P_{orb} = 1232^d.4)$
CV-like binaries	Novae and related systems	$M_{donor} \leq M_{WD}$	DQ Her (P_{orb} $= 0^d.194)$ SS Cyg (P_{orb} $= 0^d.275)$
	Super Soft X-ray sources	$M_{donor} > M_{wd}$	CAL 83 (P_{orb} $= 1^d.04)$ CAL 87 (P_{orb} $= 10^h.6)$
Double White Dwarfs	AMCVn Systems	He WD + CO WD (He WD fills R. Lobe)	AMCVn $(P_{orb} = 22^m)$
		CO WD + CO WD	WD1204+450 $(P_{orb} = 1^d.603)$

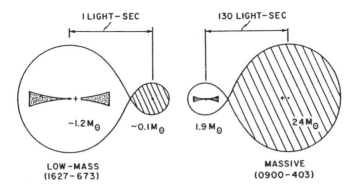

Figure 1. Comparison of the very different dimensions of a low mass X-ray binary and a high mass X-ray binary. The optical stars are depicted to fill their Roche lobes (hatched). The radial extent of the accretion disks (stippled) around the neutron stars is sketched nearly to scale. The dots and the crosses (+) mark the centers of mass of the individual stars and the system, respectively (after Bradt and McClintock 1983).

2.2. X-RAY BINARIES

a) High- and Low-Mass X-ray Binaries

The X-ray binaries can, broadly speaking, be divided into two main groups, the HMXBs and LMXBs, which differ in a number of important characteristics, listed in Table 2, and graphically depicted in Figure 1. (The characteristics listed and depicted are for systems containing neutron stars but most of them also hold for the X-ray binaries that contain black holes; also these can be divided into HMXBs and LMXBs, see Figure 3). For further details we refer here to the reviews of these two types of systems in the book "X-ray Binaries" (Lewin et al. 1995). In the HMXBs the companion of the X-ray source is a luminous early-type star of spectral type O or B, like in the Cen X-3 system, with a mass typically between 10 and 40 $M_\odot$. In the LMXBs it is a faint star of mass $\leq M_\odot$; in most cases the stellar spectrum is not even visible as the light of the systems is dominated by that of the accretion disk around the compact star. The orbital periods of the LMXBs are generally short (mostly $\leq 0.5\ ^d$), though on average somewhat longer than those of the Cataclysmic Variables.

b) Intermediate Mass X-ray Binaries

Only recently it was realized that apart from these two main groups of X-ray binaries, which each contain some 10^2 known systems in our Galaxy (cf. Van Paradijs and McClintock 1995), there are a few X-ray binaries in which

TABLE 2. The two main types of strong Galactic Binary X-ray Sources.

HMXB	LMXB
- Optical counterparts massive and luminous early type stars, spectrum O and early B; $L_{opt}/L_x > 1$	Faint blue optical counterparts $L_{opt}/L_x < 0.1$
- Concentrated in space towards the galactic plane: young stellar population, age $< 10^7$ years	Concentrated in space towards the galactic center; fairly wide spread around the galactic plane: old stellar population, age $(5 - 15) \times 10^9$ years
- Type of time variability: Regular X-ray pulsations: no X-ray outbursts	Type of time-variability: often X-ray outbursts; only in 3 cases regular X-ray pulsations.
- Relatively hard X-ray spectra $kT \geq 15$ KeV	Softer X-ray spectra: $kT \leq 10$ KeV

the companion is or has been a star of "intermediate mass", i.e. between 1 $M_\odot$ and 10 $M_\odot$. In fact, Her X-1 is a system of this type as well as Cyg X-2. In the latter system the companion presently has a mass $< M_\odot$, but is highly overluminous for this mass, which indicates that it is an evolved star that started out with a mass between 3 and 4 $M_\odot$ (cf. Podsiadlowski and Rappaport 2000; King and Ritter 1999; Tauris et al. 2000; Kolb et al. 2000).

c) Spin and Magnetic Field in neutron-star X-ray binaries

¿From the characteristics of the HMXBs and LMXBs listed in Table 2 and graphically depicted in Figure 1 it will be clear that the HMXBs belong to a very young stellar population, as OB stars do not live longer than about 10^7 yrs. This implies that the neutron stars in these systems in general have ages of at most only a few million years. On the other hand, the characteristics of the LMXBs show that they tend to belong to a much older stellar population, with ages ranging from a few hunderd million years to over 10^{10} years. (The last-mentioned age is typical for the about a dozen known globular cluster X-ray sources.) Hence, the neutron stars in these systems tend to be old. This age difference is strikingly clear from their different galactic distribution depicted in Figure 2. A further difference is the presence of strong magnetic fields (B $\geq 10^{12}$G) in the neutron stars in

HMXBs, as evidenced by their regular X-ray pulsations, and the absence of such regular pulsations and thus of strong fields in the bulk of the neutron stars in LMXBs. The occurrence of thermonuclear X-ray bursts in many LMXBs confirms that their magnetic fields must be weak, as fields stronger than $10^{10} - 10^{11}$ G suppress such bursts (cf. Lewin et al. 1995). This difference in magnetic field strength between the two groups is most probably due to field decay related to evolution in a binary system (cf. Taam and Van den Heuvel 1986). The neutron stars in the LMXBs have been living next to their companions and spinning down as well as accreting on average for a much longer period ($10^8 - 10^9$ yrs) than those in the HMXBs (not more than a few times 10^5 yrs, see below in section 3.2). The various ways in which this large difference in timescales for evolution may affect the surface magnetic field strength are discussed in section 3.6.4 (see Bhattacharya and Srinivasan 1995; Ruderman 1998).

With the long-lasting accretion of matter through a disk, as occurs in the LMXBs (see sections 3.4 to 3.6), the neutron star will be spun up to millisecond periods, as mentioned in the Introduction. The relation between X-ray binaries and binary and millisecond radio pulsars will be considered in section 3.6.

d) The Black Hole X-ray Binaries

Here a great breakthrough came with the discovery by McClintock and Remillard (1986) that the K5V companion of the "X-ray NOVA Monoceros 1975" – the source A0620-00 – is a spectroscopic binary with an orbital period of 7.75 hours and a velocity amplitude > 470 km/s. This large orbital velocity indicates that even if the K-dwarf would have zero mass, the compact object has a mass > 3 $M_\odot$, and therefore must be a black hole. Since then eight such systems consisting of a black hole with a low-mass donor star been discovered (see the review by Tanaka, this volume). Figure 3 depicts the 3 types of black hole X-ray binaries presently known (from McClintock 1992). Also here there is an "intermediate-mass-companion" system, that of LMC X-3, and there may be several more, notably V404 Cygni and X-ray Nova Sco 1994 (cf. Tanaka and Lewin, 1995, Nelemans et al. 1999).

e) The High Mass X-ray Binaries in more detail

It appears that the HMXBs fall into two sub-types which differ in a number of important characteristics:

(i) the "standard" HMXBs in which the massive OB-type companion of the X-ray source is close to filling its Roche lobe, as is evidenced from its double-wave optical lightcurve, that shows that the star is tidally deformed

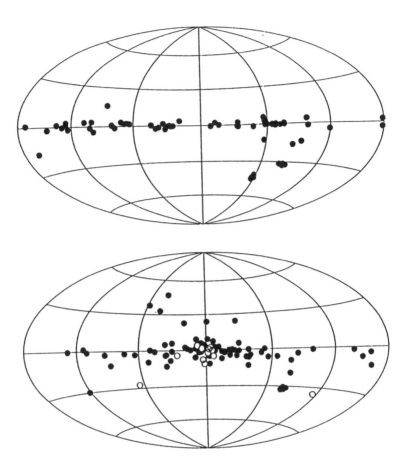

Figure 2. Sky maps (in galactic coordinates) of the high-mass X-ray binaries (top panel) and low-mass X-ray binaries (bottom panel); the latter also include the globular-cluster sources (indicated by open circles). The 27 LMXB within 2° of the Galactic center have not been included to avoid congestion of the map. These maps are based on the catalogue of Van Paradijs (1995).

("pear shaped"; cf. Van Paradijs and McClintock 1995). These are the systems of the type depicted in the left part of Figure 1. They are persistent X-ray sources and except for one, they have orbital periods ≤ 11 days.

(ii) The B-emission X-ray binaries, in which the B-type companion is a rapidly rotating unevolved star that is deep inside its Roche lobe. These systems have orbital periods between 15^d and over one year, and in most cases are "transient" X-ray sources that may occasionally turn on for a few weeks to several months, with long "off" periods in between (see Figure 4). Table 3 lists a number of characteristic examples of both types of systems (after Rappaport and Van den Heuvel 1982 and Van den Heuvel and Rap-

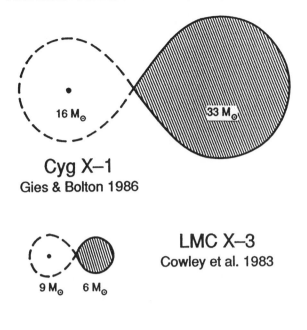

Cyg X–1
Gies & Bolton 1986

LMC X–3
Cowley et al. 1983

A0620-00
McClintock & Remillard 1986

Figure 3.   Schematic sketch, to scale, of plausible models for three dynamical black hole candidates. The optical companions (shaded regions) are shown filling their critical Roche equipotential lobes. The value of the mass function and the masses given in the figure determine the following values of the inclination angle: $i(CygX - 1) = 32^{\circ}$ $i(LMCX - 3) = 63^{\circ}$ and $i(A0620 - 00) = 45^{\circ}$. (Adapted from McClintock [1992]).

paport 1987, see also Apparao 1994). The Be/X-ray binaries are by far the most numerous group of HMXBs: over 50 such systems are already known in our Galaxy and a dozen of them is known in the Small Magellanic Cloud (SMC - a small satellite galaxy of our Galaxy, with a mass of only 1 % of the latter). Judging from these numbers, the total number of Be/X-ray binaries in our Galaxy may easily be as large as 10^3. On the other hand, not more than ten "standard" HMXBs are known in our Galaxy, and only one in the SMC and a few in the LMC. Extrapolating from the about ten known "standard" HMXBs in our sector of the galaxy, it seems likely that there are not more than some 50 to 100 such systems in the entire galaxy.

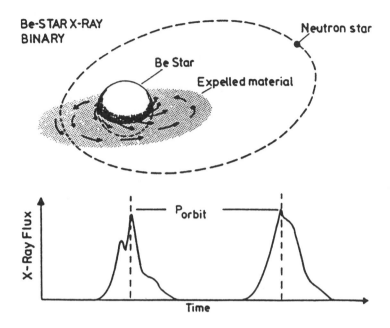

Figure 4. Schematic model of a Be star X-ray binary system such as X0535+26 and X0332+53. The neutron star moves in a moderately eccentric orbit around the Be star, which is much smaller than its own critical equipotential lobe. The rapidly rotating Be star is temporary surrounded by matter expelled in its equatorial plane. Near its periastron passage the neutron star enters this circumstellar matter and the resultant accretion produces an X-ray outburst lasting several days to weeks.

TABLE 3. The "standard" massive X-ray binaries (top) and some examples of Be X-ray binaries (bottom).

Source	Optical counterpart	Spectral type	Pulse period (s)	Orbital period (d)	Eccentricity
LMCX-4	Sk-Ph	O7 III-V	13.5	1.408	$e = 0.00$
Cen X-3	Krz's star	O6.5 II-III	4.84	2.087	$e = 0.00$
4U 1700-37	HD 153919	O6.5 f	-	3.412	$e \sim 0$
SMC X-1	Sk 160	B0Ib	0.717	3.89	$e = 0.00$
4U 1538-52	QV Nor	B0Iab	529	3.73	$e = 0.00$
Cyg X-1	HD 226868	O9.7Iab	-	5.60	$e = 0.05$
4U 0900-40	HD 77581	B0.5Ib	283	8.965	$e = 0.09$
GX 301-2	Wra 977	B1.5Ia	696	41.5	$e = 0.47$
4U 0115+63	John's star	Be	3.61	24.3	$e = 0.34$, transient
4U 0352+30	X Per	O9.5(III-V)e	835	250	$e \sim 0$, very weak,
A 0535+26	HD 245770	B0Ve	104	111	steady transient
4U 1145-61	Hen 715	B1Vne	292	188	highly variable
4U 1258-61	MMV star	B2Vne	272	133	highly variable

2.3. THE BINARY AND MILLISECOND RADIO PULSARS

a) Spin and magnetic field

The binary radio pulsars are characterized by in general much shorter spin periods than the ordinary single radio pulsars. This can be clearly observed in the $\dot{P}$ vs P diagram of radio pulsars, where P is the pulse period and $\dot{P}$ is the period derivative. Figure 5 shows this diagram for the about 1000 pulsars that were known in the galactic disk early in 2000 (Camilo 2000). Dots are single pulsars, circles are binary pulsars.

The surface dipole magnetic field strength B_s of pulsars is related to P and $\dot{P}$ by the equation (cf. Manchester and Taylor 1977; Lyne & Smith 1990)).

$$B_s = (\frac{3c^3 I}{8\pi^2 R^6} P\dot{P}) = 3.2 \times 10^{19}(P\dot{P})^{1/2}G \qquad (1)$$

where I and R are the moment of inertia and radius of the neutron star, respectively (this follows from equating the rotational energy loss to the electromagnetic energy loss from the spinning magnetized neutron star, cf. Manchester and Taylor 1977). In Figure 5 lines of constant B_s are indicated. The figure shows that a large fraction of the binary pulsars is millisecond pulsars, i.e. has a spin period shorter than 0.01 s. In the figure also lines are drawn of constant spin-down age, as defined by (cf. Manchester and Taylor 1977):

$$t_{sd} = P/2\dot{P} \qquad (2)$$

The figure shows that the binary pulsars typically have magnetic field strengths in the range $10^8 - 10^{10}$ G and "ages" of order $(1\text{-}10)\times10^9$ yrs, whereas the "garden variety"pulsars that make up the cloud of "ordinary pulsars" typically have magnetic field strengths of $10^{12} - 10^{13}$G and ages $\leq 10^7$ years.

In the figure a number of well-known binary radio pulsars is indicated, such as the three close double neutron stars PSR 1913+16, (Hulse-Taylor pulsar), PSR 1534+12 (Wolszczan's pulsar) and PSR 2127+11c (in the globular cluster M15) and the eccentric-orbit neutron-star plus white dwarf system PSR 2303+46 (cf. Van Kerkwijk and Kulkarni 1999). In the latter system the pulsar is an ordinary strong magnetic-field pulsar. Table 4 lists some vital data of these binary pulsars and of some other representative binary pulsars. Many pulsars are known also in globular clusters. Most of these are millisecond pulsars and some 40 % - and in some clusters, like 47 Tuc, over 60 % (Camilo et al. 2000) - of them is in binaries (against some 7 % of the total pulsar population).

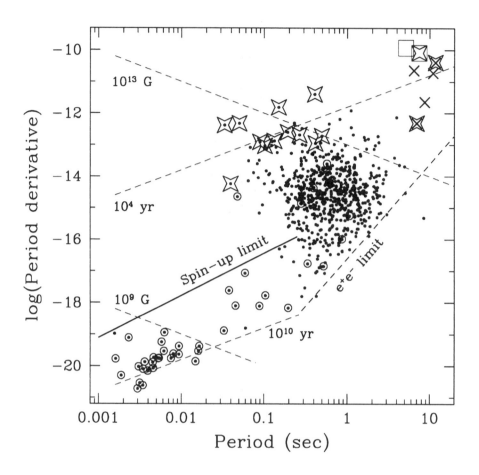

Figure 5. $P - \dot{P}$ diagram for radio pulsars. Circles indicate binary pulsars, stars indicate pulsars associated with supernova remnants, crosses are "Anomalous X-ray Pulsars" and squares are Soft Gamma-Ray Repeaters. (The latter two classes of objects are young single neutron stars with extremely strong magnetic fields (10^{14} to 10^{15} Gauss). Lines of constant surface dipole magnetic field and of constant spin-down age are indicated, as well as the "spin-up limit" and the "death-line" (e^+e^- limit). Diagram courtesy F. Camilo (2000).

b) Orbits and companion stars

With the exception of the three "ante-deluvian" systems mentioned in Table 1, the companion stars in radio pulsar binaries are themselves also dead stars: neutron stars or white dwarfs. They can be divided into the following categories (see also Table 1), as depicted in Figure 6:

(1) "The PSR 1913+16 Class", in which the companion to the neutron star is another neutron star or a massive white dwarf.

TABLE 4.   Some vital data of representative binary radio pulsars with neutron-star and white-dwarf companions. They fall into four categories (see also
Table 1): (NS, NS)$_e$: eccentric orbit double neutron star (NS, CO)$_e$: eccentric
orbit neutron star + CO-white dwarf (NS, CO)$_c$: circular orbit neutron star +
CO-white dwarf (NS, He)$_c$: circular orbit neutron star + He-white dwarf. Except
in the (NS, CO)$_e$ systems the observed pulsars in these neutron-star binaries are
"recycled": they were the first-born compact stars in their systems. (Data partly
after Brown, Lee, Portegies Zwart and Bethe 2000 and references therein, and
after Van den Heuvel 1994a and references therein).

Pulsar	P_{orb} [days]	P_{spin} ms	M_p [$M_\odot$]	M_c [$M_\odot$]	e	B [G]
(ns, ns)						
B1534+12	0.421	37.9	1.339	1.339	0.274	10^{10}
B1913+16	0.323	59.0	1.441	1.387	0.617	2.3×10^{10}
B2127+11C	0.335	30.5	1.349	1.363	0.681	1.2×10^{10}
(ns, co)$_e$						
B2303+46	12.34	1066.	< 1.44	> 1.20	0.658	7.9×10^{11}
J1141-6545	0.198	394.		> 0.97	0.172	
(ns, co)$_c$						
J2145-0750	6.839	16.1		> 0.43	2.1×10^{-5}	6×10^8
J1022+1001	7.805	16.5		> 0.73	9.8×10^{-5}	8.4×10^8
J0621+1002	8.319	28.9		> 0.45	0.00245	1.6×10^9
B0655+64	1.029	195.7		> 0.7	0.75×10^{-5}	1.26×10^{10}
1957+20	0.38	1.6		0.02	$< 10^{-3}$	1.6×10^8
1855+09	12.33	5.4		0.20	2.1×10^{-5}	3×10^8
1953+29	117.35	6.1		0.30	3.3×10^{-4}	4×10^8
0820+02	1232.4	864.9		0.30	1.2×10^{-2}	3×10^{11}

(2) "The PSR 1953+29 Class", in which the companion to the neutron star
is a low-mass helium white dwarf (M< 0.45 $M_\odot$). The latter systems always
have circular orbits, with periods covering a large range, from less than one
day to 1232 days. In systems with orbital periods shorter than ~ 150 days
the pulsars tend to be millisecond pulsars, with very weak magnetic fields,
in the range 10^8 -10^9 G.

Category (1) can be further subdivided into
(a) Systems in which the companion is itself also a neutron star. These have
very eccentric orbits and their orbital periods tend to be very short. The
magnetic fields of these pulsars are relatively weak, in the range 10^9-10^{10}G.
(b) Systems in which the companion is a massive white dwarf (M_{wd} = 0.5
- 1.4 $M_\odot$). These fall into two groups:

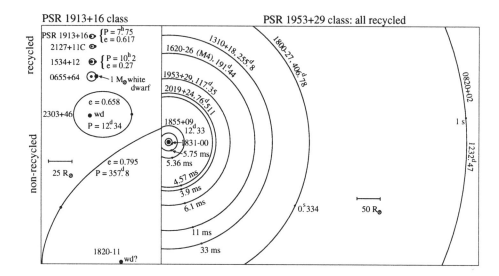

Figure 6. The main classes of binary radio pulsars (orbits drawn to scale). Left: The PSR 1913+16-class of systems tend to have narrow and very eccentric orbits; the companion of the pulsar is itself a neutron star or a massive white dwarf. When, in the case of white dwarf companions, the orbit is circular, the neutron star is recycled; if the orbit is eccentric, the neutron star is not recycled - this is the case in 2303+46 (see text). Right: The PSR 1953+29-class systems tend to have wide and circular orbits; here the companion stars have a low mass, in the range 0.2-0.4 $M_\odot$, or even smaller - and most probably are helium white dwarfs (see text).

(i) Those with eccentric orbits, such as PSR 2303+46, in which the pulsar is a young strong-magnetic field neutron star.
(ii) Those with circular orbits, such as PSR 0655+64, in which the pulsar is an old neutron star with a relatively weak magnetic field, in the range 10^9-10^{10}G.

2.4. SPIN EVOLUTION OF ACCRETING NEUTRON STARS; THE CONCEPTS OF ALFVEN RADIUS, MAGNETOSPHERE, EQUILIBRIUM SPIN, SPIN-UP AND SPIN-DOWN

a) Observed spin behaviour of accreting neutron stars
Figure 7 shows the pulse-period vs. orbital period relation of the accreting X-ray pulsars (Liu 2001). Also the three binary radio pulsars with B-type companion stars are indicated in the diagram. One observes in this figure that many of the X-ray pulsars have long pulse periods between 10 and 1000

s. In persistent sources, spin periods of order of seconds are found only in systems where from UV and optical observations we have clear evidence for the presence of an accretion disk. These are systems where (a large part of) the mass transfer is due to Roche-lobe overflow: the low-mass X-ray binary Her X-1 and the high-mass systems in which the supergiant donors are just beginning to overflow their Roche lobes: Cen X-3, SMC X-1 and LMC X-4. These are indicated by the crosses in the diagram. These sources are expected to be powered by a combination of stellar wind and beginning Roche-lobe overflow (Savonije 1978, 1983). In most of these systems the X-ray pulsars show a secular decrease of the pulse period (spin-up) on a relatively short timescale: a few thousand years in the massive systems and $\sim 10^5$ yrs in the case of Her X-1 (see Bildsten et al. 1997; Finger 1998). Figure 8 shows for example the observed spin evolution of Cen X-3. Although on short timescales episodes of spin-up and spin-down alternate, the average trend is that of spin-up. On the other hand in persistent systems that are purely wind-fed, such as the HMXBs with blue supergiant companions that do not yet fill their Roche lobes, the pulse periods are very long, and they vary erratically in time showing no clear secular trends. This can be explained by the fact that the amount of angular momentum carried by the supersonic winds is negligible, and eddies form in the wind downstream of the neutron star, which alternatingly may feed corotating and counter-rotating angular momentum to it (Taam and Fryxell 1988; Fryxell and Taam 1988). The B-emission (Be) X-ray binaries show an again different spin behaviour: these sources are transients, in many cases recurrent. During a transient outburst they in general show rapid spin-up, indicating that a disk has been formed. However, between outbursts they spin down, as at the beginning of the next outburst the spin period is generally observed to be much longer than at the end of the previous one. An important clue to what drives the spin-up and -down in the Be-X-ray systems is given by the correlation between spin periods and orbital periods in these systems (Figure 7), discovered by Corbet (1984). In order to understand this we first have to consider the concept of equilibrium spin period, which is the topic of the next paragraph.

b) The concepts of Alfven radius, magnetosphere and equilibrium spin
These concepts were introduced by Davidson and Ostriker (1973) and Lamb et al. (1973). The Alfven radius r_A is the distance from the neutron star where the kinetic energy density 0.5. ρ v^2 of the inflowing matter equals the magnetic energy density of $\frac{\mu B^2}{2}$ of the neutron-star's dipole magnetic field, so:

$$0.5\rho v(r)^2 = 0.5\mu B(r)^2 \tag{3}$$

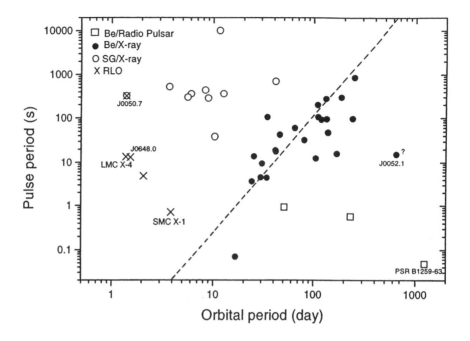

Figure 7.   The pulse period versus the orbital period for the accretion powered pulsars with known orbital period (after Liu 2001). Crosses indicate persistent sources with accretion disks (Roche lobe overflow), open circles are persistent sources with blue supergiant companion (wind accretors) and filled circles are the B-emission X-ray binaries, which are mostly transient sources. Squares are radio pulsars with Be companions. The dashed line indicates Corbet's (1984) relation between pulse period and orbital period for the Be-X systems.

where

$$B(r) = B_{\mathrm{s}}(R/r)^3 \tag{4}$$

in which B_{s} is the magnetic field strength at the neutron-star radius R and μ is the permeability of the vacuum. Here $v(r)$ is either the free-fall velocity or the Kepler-velocity in the disk (which are the same, except for a factor $\sqrt{2}$), and $\rho(r)$ can be expressed in terms of the accretion rate $\dot{M}$, $v(r)$ and r (e.g.: see Bhattacharya and van den Heuvel 1991, van den Heuvel 1994a). This leads to

$$r_{\mathrm{A}} = (\mu B_{\mathrm{s}}^2 R^6 / \dot{M}\sqrt{2GM})^{2/7} \tag{5}$$

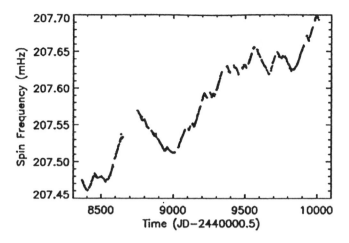

Figure 8.   Cen X-3 spin frequency measurements with BATSE (after Finger 1998).

For r > r_A the flow of the matter is not influenced by the neutron star's magnetic field. On the other hand, for distances smaller than the Alfven radius, the magnetic field forces the matter to flow in along the field lines. The region around the neutron star closer than the Alfven-radius is called the Magnetosphere. For r < r_A the inflowing matter is forced to corotate with the magnetosphere of the neutron star, for r > r_A matter can freely orbit the star.

A second important radius is the corotation radius r_{co}. This is the distance where the Kepler velocity of matter around the neutron star just equals the rotational velocity of the magnetosphere ωr. The value of r_{co} is given by:
$\Omega^2 r = GM/(r^2)$, where $\Omega = 2\pi/P$, P being the rotation period of the star. Thus:

$$r_{co} = (GMP^2/4\pi^2)^{1/3} \qquad (6)$$

There are now two possibilities for the accretion, illustrated by Figure 9:

(i) if $r_A > r_{co}$, matter at the magnetospheric boundary cannot flow in: as soon as the matter enters the magnetosphere it is forced to corotate; however, it then rotates faster than the Keplerian velocity and is centrifuged out of the magnetosphere again. In this situation the accretion is therefore shut off.
(ii) if $r_A < r_{co}$, accretion can take place. In the latter situation, matter with angular momentum will enter the magnetosphere and flow to the neutron

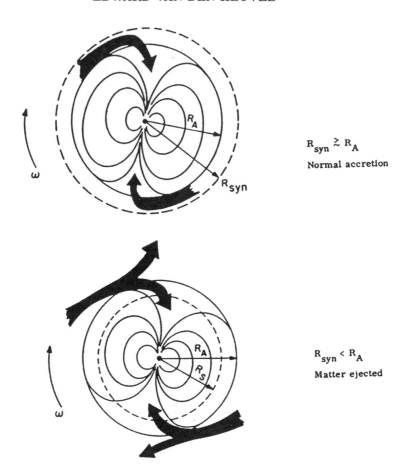

Figure 9. Schematic representation of Alfven-radius R_A and corotation radius R_{syn} of a rotating magnetized neutron star. R_A depends on accretion rate $\dot{M}$, mass M and dipole magnetic field strength B_9 at stellar surface); R_{syn} depends on rotation period P and mass of the neutron star. When $R_A = R_{syn}$ the neutron star spins at its equilibrium spin period P_{eq}. If it rotates slower, accretion is possible, if it rotates faster, centrifugal forces on matter entering the magnetosphere will swing this matter out and accretion is impossible (after Schreier 1977, Ann. N.Y. Acad.Sci. 402, 445).

star surface, causing the angular velocity of rotation to increase, i.e. the spin period to decrease. This will go on until $r_A = r_{co}$, after which further accretion and spin-up is impossible. The star will therefore settle at a spin period for which these two radii are just equal to each other. This is called the equilibrium spin period P_{eq}, which is given by:

$$P_{eq} = (2.4ms)(B_9)^{6/7}(R_6)^{16/7}(M)^{-5/7}(\dot{M}/\dot{M}_{\rm Edd})^{-3/7} \qquad (7)$$

(cf.van den Heuvel 1994a), where B_9 is the surface dipole magnetic field strength in units of 10^9 Gauss, R_6 is the neutron star radius in units 10^6cm, M is the mass of the neutron star in units $M_\odot$, and $\dot{M}_{Edd}$ is the Eddington accretion rate.

The X-ray pulsars with accretion disks seem all to be spinning near to their equilibrium spin periods. Their magnetic field strengths, inferred from X-ray cyclotron lines, are typically of order 10^{12} G (Trümper et al. 1978), yielding equilibrium spin periods of order of seconds, just as observed in these sources. The fact that their spin periods still show a secularly decreasing trend (spin-up) may be due either to a secular decrease of the surface dipole magnetic field strength, or a secular increase of the accretion rate, or both (Finger 1998). A decrease of the surface dipole strength might be due to temporary "burying" of the field due to the accretion process (see section 3.6.4).

c) Possible explanation for the "Corbet-relation" for Be/X-ray binaries
As shown by Waters and van Kerkwijk (1989) the Corbet relation between the spin periods and orbital periods of the Be/X-ray binaries indicates that the Be stars are surrounded by a disk of matter, ejected from its equatorial regions at relatively low velocities (one may also call this a slow "wind", that carries with it the angular momentum it had at the moment the wind particles left the very rapidly rotating star). When the wind density is low, the neutron star spins faster than its equilibrium spin period (see above) and accretion is not possible. The neutron star now spins down due to the torques that the surrounding accretion disk exerts on the magnetosphere (cf.Ghosh and Lamb 1979a, b). When the wind density increases, the magnetosphere is compressed, and the equilibrium spin period becomes shorter than the spin period of the neutron star, such that accretion becomes possible (see Figure 9). At this moment the star becomes an X-ray source, and spin-up will occur due to the accretion of disk angular momentum. On average one expects the spin period of the neutron star to hover around a value equal to the equilibrium period corresponding to the mean density in the wind of the Be star near the orbit of the neutron star. The wider the orbit, the less dense the wind, and thus the lower the mean accretion rate, and the longer the corresponding equilibrium spin period.

3. Formation and evolution of compact stars in binary systems

3.1. THE CONCEPT OF "ROCHE LOBE" AND THE EFFECTS OF MASS TRANSFER AND MASS LOSS ON BINARY ORBITS

a) The concept of "Roche-lobe"

Assuming that the two stars move in circular orbits and corotate with

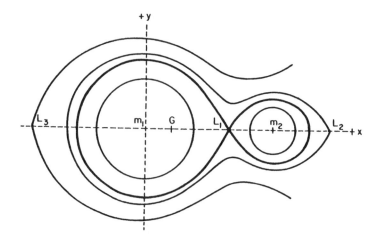

Figure 10. The cross section with the orbital plane of some special Roche equipotentials. The surface through the inner Lagrangian point L_1 forms the two Roche lobes around the stars. The point G denotes the binary's center of mass from Plavec 1968; (further explanations in section 3).

the binary system, one can use the corotating coordinate system as the reference system. The equipotential surfaces in this system are stationary and are due to a combination of the gravitational attraction of the components and the centrifugal acceleration produced by the rotation of the system. These are the so-called Lagrangian equipotential surfaces, depicted in Figure 10.

Close to the stars they are nearly circular; at greater distances from the stellar centers they deform and ultimately touch in what is called the first Lagrangian point L_1. For still higher values of the potential the surfaces envelope both stars, as illustrated in the figure.

The equipotential surface passing through L_1 is called the critical Roche surface, and the pear-shaped part of this surface around a star is called "Roche-lobe". One defines the radius R_L of the Roche lobe as that of a sphere with the same volume as the lobe. R_L is a function only of the orbital radius a and the mass ratio $q = m_2/m_1$ of the binary components. Various approximation formulas for R_L exist. The most precise one is that due to Eggleton (1983):

$$R_L/a = 0.49q^{2/3}/(0.6q^{2/3} + ln(1 + q^{1/3})) \qquad (8)$$

which is accurate within a few percent for all values of q. A very convenient alternative expression for q > 1.25 is

$$R_L/a = 0.462(m_1/m_1 + m_2)^{1/3} \qquad (9)$$

b. Orbital changes due to "conservative" mass transfer

When a star fills its Roche lobe, matter in its outer layers can freely flow over towards the companion star along L_1.

If one assumes that all the matter lost by the star is captured by the other star, and that the rotational angular momentum of the stars is negligible with respect to the orbital angular momentum, one will have conservation of both total mass and total orbital angular momentum of the system, which is called "conservative evolution" of the binary. In this case it is easy to calculate how the orbital radius a changes during the mass transfer, as one has the equations

$$(m_1 + m_2) = M = const. \qquad (10)$$

and

$$J^2 = G\frac{m_1^2 m_2^2}{(m_1 + m_2)}a = const. \qquad (11)$$

where J is the orbital angular momentum.
Eqs. (10) and (11) yield

$$a(m_1^2 m_2^2) = const. \qquad (12)$$

leading to the following equation for the change in a:

$$a/a_o = (\frac{m_1^o m_2^o}{m_1 m_2})^2 \qquad (13)$$

where index zero indicates the initial situation. Assuming m_1 to be the star that is losing mass along L_1, the mass of this star will decrease. It is easy to see from eq. (13) that if $m_1^0 > m_2^0$, the orbital radius a will *decrease* due to the mass transfer. On the other hand, if $m_1^0 < m_2^0$ (mass-transfer from the less massive to the more massive star) leads to an *increase* of a.

c. Orbital changes due to loss of mass and orbital angular momentum from the system

In the case of mass loss from the system we are no longer dealing with a closed "conservative" system and there is, in principle, an infinite range of possibilities for the changes of the orbit depending on the amounts of mass lost and of the angular momentum lost with this mass. For example, if a fraction α of the matter exchanged between the components leaves the system, one finds from eq. (11) that the rate of change of the orbital separation can be written as:

$$\dot{a}/a = -2(1 + (\alpha - 1)(m_1/m_2) - \frac{\alpha}{2}\frac{m_1}{m})(\dot{m}/m_1) + 2\dot{J}_{orb}/J_{orb} \qquad (14)$$

It is clear that this equation can be solved only if $\dot{J}_{orb}/J_{orb}$ is known.

In binary systems with very short orbital periods, angular momentum loss will occur due to the emission of gravitational waves (cf. Shapiro and Teukolsky 1983):

$$\dot{J}_{orb}/J_{orb} = \frac{-32G^3}{5c^5}\frac{m_1 m_2(m_1 + m_2)}{a^4}s^{-1} \qquad (15)$$

For sufficiently narrow orbits, eq. (15) becomes the dominant term in eq. (14) and will cause a to decrease. Therefore, the orbits of very narrow binaries will tend to continuously shrink, forcing the components to transfer mass (if a component fills the Roche Lobe). (If mass is transferred from the less massive to the more massive component the net result of gravitational radiation losses plus mass transfer may also be a widening of the orbit). Gravitational radiation losses are therefore a major force driving the mass transfer in very narrow binaries, such as Cataclysmic Variables and low-mass X-ray binaries (Faulkner 1971). For a review of a variety of "modes" of mass angular momentum losses from binaries I refer to Van den Heuvel (1994a) and Soberman et al. (1997). Of particular importance for CVs and LMXBs is also: "magnetic braking" which may cause an enhanced loss of orbital angular momentum. We come back to this in section 3.4.

d. Effects of sudden (explosive) mass loss on the orbits

We consider systems with initially circular orbits and we neglect the effects of the impact of the ejected shell on the companion star. We assume the explosion to be instantaneous (infinitely short duration). In this case the orbital changes can be simply expressed in terms of the ratio $\mu_f = \frac{m_1^f + m_2^f}{m_1^o + m_2^o}$ of the total mass of the system after and before the explosion. We express the orbital semi-major axis a and orbital period P in units of the initial orbital radius a_0 and initial orbital period P_0, respectively; zero and f indicate quantities before and after the explosion. One then obtains the following simple expressions for the orbital parameters after the explosion (cf. Flannery and Van den Heuvel 1975):

$$a_f = \mu_f/(2\mu_f - 1) \qquad (16)$$

$$e_f = (a_f - 1)/a^f = (1 - \mu_f)/\mu_f \qquad (17)$$

$$p_f = \mu_f/(2\mu_f - 1)^{3/2} \qquad (18)$$

One thus observes that the system is disrupted (hyperbolic orbit) if $\mu_f < 0.5$. This is a consequence of the Virial Theorem.

The runaway velocity of the center of gravity of bound systems after star 2 (with orbital velocity V_2) ejected an amount Δm is given by

$$V_g = \Delta m V_2/(m_1^f + m_2^f) \qquad (19)$$

Hence, the largest velocity that a post-explosion system can attain is V_2.

3.2. INTERMEZZO: REASONS FOR THE EXISTENCE OF THE CLASSES OF HIGH- AND LOW-MASS X-RAY BINARIES: THE CONCEPTS OF EDDINGTON LIMIT, ROCHE-LOBE OVERFLOW AND STELLAR WINDS

a) Eddington limit

The typical X-ray luminosities of the High- and Low-Mass X-ray Binaries are in the range $10^{35} - 10^{38}$ ergs/s, corresponding to mass accretion rates onto a neutron star in the range 10^{-11} to 10^{-8} $M_\odot$/yr. When the mass-transfer rate exceeds a few times 10^{-8} $M_\odot$/yr, the X-ray luminosity exceeds the so-called Eddington limit of $10^{4.5}$ $(M/M_\odot)$ $L_\odot$ ($\sim 10^{38}$ ergs/s for a 1.4 $M_\odot$ neutron star), at which the radiation pressure force on the accreting matter exceeds the gravitational attraction force of the compact star (Davidson and Ostriker 1973). This places a natural upper limit to the accretion rate of

$$\dot{M}_{Edd} \simeq 1.5 R_6 \times 10^{-8} M_\odot/\text{yr} \qquad (20)$$

where R_6 is the radius of the compact star in units of 10^6 cm. At accretion rates $> \dot{M}_{Edd}$ the excess accreting matter will pile up around the compact object and form a cloud optically thick to X-rays, thus quenching the source. Therefore, in the observed "persistent" HMXBs and LMXBs the accretion rates clearly must be in the range 10^{-11} and 10^{-8} $M_\odot$/yr. One may now ask the following question: assuming that neutron stars are born as companions of stars of any kind of mass, in which mass ranges do we then expect to observe (reasonably) long-lived persistent X-ray sources? In other words, in what mass range do we expect companions to lose mass in such a way that the neutron star will have an accretion rate between 10^{-11} and 10^{-8} $M_\odot$/yr? It appears that the answer to this question is that one then expects just the two observed groups of persistent X-ray binaries to occur: one group with companion masses $\leq M_\odot$ and another group with companion masses ≥ 15 $M_\odot$. The reasons for this are simple (Van den Heuvel 1975) and have to do with the ways in which companion stars can lose mass, as follows.

b) Modes of mass transfer and types of binary X-ray sources
Basically, the companion can lose mass in two ways: (1) by Roche lobe overflow: in this case the donor star fills its Roche lobe and transfers mass along the first Lagrangian point L_1 to its companion; (2) by a high-velocity stellar wind; in this case the star needs not fill its Roche lobe, and the neutron star captures some matter from the wind. The accretion rates predicted from these two types of mass loss are as follows:
(1) Roche lobe overflow. In this case the mass loss takes place at low velocities and practically all mass lost by the star will be accreted by the companion. The mass-transfer from a donor star that is more massive than its companion takes place on the thermal time-scale of the donor (cf. Kippenhahn and Weigert 1967; Paczynski 1971a), leading to a mass-transfer rate (see below, in section 3.3.5; cf. Van den Heuvel 1994a):

$$\dot{M}_{\rm Roche} \simeq 3.10^{-8} M^3 [M_\odot/\rm yr] \qquad (21)$$

where M is the donor mass in solar masses.

Equation (21) shows that only for $M \leq M_\odot$ mass-transfer rates below the Eddington rate can be obtained. However, for $M \leq M_\odot$ the mass-losing star is no longer more massive than the compact star, and equation (21) will no longer be valid. For a donor that is less massive than the compact star, mass transfer will take place on the nuclear timescale of the donor star (10^{10} yrs), or due to narrowing of the orbit by gravitational radiation losses and/or "magnetic braking" (see section 3.4). In both cases a mass transfer rate of order 10^{-10} $M_\odot$/yr will ensue (cf. Van den Heuvel 1994a).
(2) Accretion from a stellar wind. Strong stellar winds with (total) mass loss rates $\geq 10^{-9}$ $M_\odot$/yr are found only in stars more massive than about 15 $M_\odot$ as was first discovered in a rocket experiment by Morton

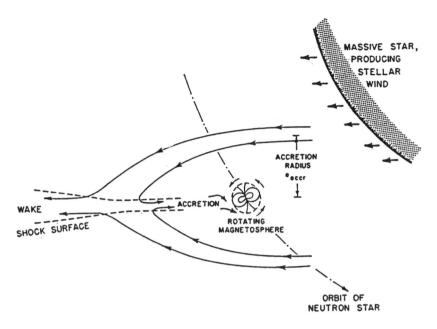

Figure 11. Streamlines of stellar-wind material, in the frame of an accreting neutron star. Relative dimensions are not to scale (from Davidson and Ostriker 1973).

(1967). Blue supergiants like the companions of Cyg X-1 and Cen X-3 typically have wind mass loss rates of a few times 10^{-6} $M_\odot$/yr with outflow velocities of order (1-2) $\times 10^3$ km/s. It was realized by Davidson and Ostriker (1973) that a neutron star passing through such a wind will capture only a tiny fraction of the wind matter, of order 10^{-11} to 10^{-10} $M_\odot$/yr, precisely enough to power a $10^{35} - 10^{36}$ ergs/s X-ray source. Figure 11, after Davidson and Ostriker (1973) depicts this situation. The neutron star captures only the matter that passes it within its gravitational capture radius r_{acc} given by

$$r_{acc} = \sqrt{2GM_n/V_w^2} \qquad (22)$$

where V_w is the velocity of the wind and M_n is the mass of the neutron star. The amount of the wind matter captured is $\dot{M}_{acc} = \frac{\pi r_{acc}^2}{4\pi a^2} \dot{M}_w$ where a is the orbital radius and $\dot{M}_w$ is the wind mass-loss rate. For M_n=1.4 $M_\odot$, V_w=10^3 km/s, a=50 $R_\odot$, as in a typical "standard" HMXB, one has $\dot{M}_{acc}$ = $5.10^{-5} \dot{M}_w$.

The conclusion from the above is that only companion stars with M $\leq$ $M_\odot$ and $\geq$ 15 $M_\odot$ are able to turn a compact companion star into a steady X-ray source. This explains the existence of the

two main observed classes of binary X-ray sources, the LMXBs and HMXBs and the virtual absence of persistent binary X-ray sources with donor stars with $M_\odot \le M \le 15\ M_\odot$ (Van den Heuvel 1975).

b) Mass accretion in transient sources

For the B-emission X-ray transients the cause for the "turn-on" as an X-ray source is clearly to be sought in the unstable outer layers of the rapidly rotating B-emission star (Maraschi et al. 1976). Be stars are known to go with irregular time intervals through "emission" phases, in which, for unknown reasons, the star ejects a disk of gas in its equatorial plane. The passage of the neutron star through this disk will then cause a recurrent X-ray outburst, as depicted in Figure 4.

On the other hand, for the black-hole "X-ray novae", like NOVA Monoceros 1975 (A0620-00) (see Figure 3), the cause of the outburst is probably to be sought in an instability in the accretion disk around the black hole (cf. King et al. 1997; King 1998; 2000 (and references therein); Belloni et al. 1997).

3.3. EVOLUTION OF SINGLE STARS AND BINARIES LEADING TO THE FORMATION OF COMPACT OBJECTS

3.3.1. *Introduction*

Stars start out their evolution as homogeneous globes of gas consisting roughly of 70 % hydrogen, 28 % helium and a few % heavier elements. They go through a series of phases of nuclear fusion in their central regions and at the end of their lives leave compact objects. Single stars less massive than about 8 $M_\odot$ are expected to leave white dwarfs as remnants, with masses $\le 1.4\ M_\odot$ and in most cases consisting of carbon, oxygen and some helium (more rarely they may also contain magnesium and neon). They have radii of order 5.10^3 km and are supported against gravitational collapse by the huge Fermi-pressure exerted by the degenerate electron gas in their interior (cf. Srinivasan 1997, Kawaler 1997). In single stars more massive than about 8 $M_\odot$ the burned-out core reaches a mass larger than the Chandrasekhar limit, the maximum mass possible for an electron-degenerate configuration ($\sim 1.4\ M_\odot$, see Srinivasan 1998). Therefore, at the end of life the cores of such stars collapse and, if they are not too heavy, will be able to reach a stable configuration as a neutron star. During this collapse some 3×10^{53} ergs of gravitational binding energy is released ($0.15\ M_n\ c^2$, where M_n is the mass of the neutron star), far more than the binding energy of the stellar envelope, causing this envelope to be violently ejected at velocities $\sim 10^4$ km/s. This is observable as a supernova event.

Neutron stars typically have masses between 1.2 and about 2.0 $M_\odot$ (see Van Paradijs 1998), and a radius of about 10 km. Evidence is mounting (based on a variety of observational and theoretical considerations involving the black-hole X-ray binaries and Gamma-Ray Bursts (cf. Ergma and Van den Heuvel 1998; Nelemans et al. 1999; Sollerman et al. 2000) indicating that single stars more massive than about 20 to 25 $M_\odot$ produce burned-out cores too massive to leave stable neutron stars, causing them to collapse to black holes. The positional and temporal coincidence of about one quarter of the optically identified Gamma Ray Bursts with a peculiar new type of supernova (cf. the review by Van Paradijs 1999; Van Paradijs et al. 2000) suggests that for the first time the birth events of black holes as the end-product of massive evolution have been observed (cf. Iwamoto et al. 1998; Woosley 2001).

In binary systems the above-indicated mass ranges for progenitors of compact stars will be somewhat modified by the effects of mass transfer and mass loss, but they still can be used as a rough first approximation (cf. Van den Heuvel 1994a). Table 5 summarizes the various types of remnants expected to result as a function of initial stellar mass.

TABLE 5. Types of final evolutionary products expected as a function of stellar mass (explanation in text)

Main-sequence mass	He core mass	Final product
3-8 $M_\odot$	1.4-1.9 $M_\odot$	CO white dwarf
8-10 (12) $M_\odot$	1.9-2.2 (3.0) $M_\odot$	Degen. O-Ne-Mg core → collapse to NS
10(12) $M_\odot$-25 $M_\odot$	2.2 (3.0)-8 $M_\odot$	Collapsing iron core → collapse to NS
$\gtrsim$ 25 $M_\odot$	8 $M_\odot$	Collapsing iron core → collapse to BH

3.3.2. The three basic stellar timescales

Three timescales are of fundamental importance in stellar evolution (cf. Cox and Giuli 1968). When the hydrostatic equilibrium of a star is disturbed (e.g. because of sudden mass loss), the star will restore this equilibrium on a so-called dynamical timescale (also called pulsational timescale), which is of the order of the time it takes a sound wave to cross one stellar radius. Numerically the dynamical timescale is given by

$$\tau_d = 50(\overline{\rho}_\odot / \overline{\rho})^{1/2} \ [min] \tag{23}$$

where $\bar{\rho}$ and $\bar{\rho}_\odot$ denote the mean density of the star and the sun, respectively ($\bar{\rho}_\odot = 1.4$ g/cm^3). When the thermal equilibrium of the star is disturbed, it will restore this equilibrium on a thermal timescale τ_{th} (also called the Kelvin-Helmholtz timescale), which is the time it takes to emit its thermal energy content GM^2/R at its present luminosity L.

Thus

$$\tau_{\text{th}} = GM^2/RL \qquad (24)$$

Stars with nuclear burning obey the so-called mass-luminosity relation, which for $M > M_\odot$ is well approximated by $L \propto M^{3.5}$, and for $M < M_\odot$ by $L \propto M^3$. Main-sequence stars, furthermore, obey a mass-radius relation, which for $M > M_\odot$ follows $R \propto M^{0.5}$, and for $M \leq M_\odot R \propto M$. Substituting these expressions in eq. (24) one finds that for $M > M_\odot \tau_{\text{th}}$ can be approximated numerically as

$$\tau_{\text{th}} = 3 \times 10^7 (M/M_\odot)^{-2} [yr] \qquad (25)$$

The third stellar timescale is the nuclear one, which is the time needed for the star to exhaust its nuclear fuel reserve (which is proportional to M), at its present fuel consumption rate (which is proportional to L), so

$$\tau_{\text{nuc}} = 10^{10} (M/L)(L_\odot/M_\odot) [yr] \qquad (26)$$

With the above expressions for the mass-luminosity relation one finds that numerically τ_{nuc} can be approximated as

$$\tau_{\text{nuc}} = 10^{10} (M/M_\odot)^{-2.5} [yr] \qquad (27)$$

for $M > M_\odot$, and

$$= 10^{10} (M/M_\odot)^{-2} [yr] \qquad (28)$$

for $M \leq M_\odot$.

3.3.3. *The variation of the outer radius and luminosity during stellar evolution*

Figure 12a depicts how the observable stellar parameters – luminosity (L), radius (R) and effective surface temperature (T_e) – change during the evolution of the stellar interior. Since $L = 4\pi R^2 \sigma T_e^4$, only two of these parameters are independent of one another. The figure depicts the evolutionary tracks in the Hertzsprung-Russell diagram of a "massive" star (15 M$_\odot$), an "intermediate-mass" star (5 M$_\odot$) and two "low-mass" stars (2.25 M$_\odot$ and 1.0 M$_\odot$). Figure 12b also separately depicts the evolution of the radius of the 5 $M_\odot$ star. Important evolutionary stages are indicated in the figures. Between points 1 and 2 the star is in the long-lasting phase of core hydrogen burning (nuclear timescale). At point 3 hydrogen ignites in a shell around the helium core. For stars more massive than 1.2 $M_\odot$ the entire star briefly contracts between the points 2 and 3, causing its central temperature to rise. In stars more massive than about 2.3 $M_\odot$ the core continues to contract after hydrogen-shell ignition, upon which the outer layers begin to expand. When the central temperature reaches $T \sim 10^8 K$, helium ignites (point 4). At this moment the star has become a red giant, with a very dense core and a very large radius. During helium burning it describes a loop in the Hertzsprung-Russell diagram. The star with M $\geq$ 2.3 $M_\odot$ moves from the point 2 to 4 on a thermal timescale (stage II in Figure 12b) and describes the helium-burning loop on a nuclear timescale (from 4 to 5). During helium shell burning the outer radius expands again and at carbon ignition the star has become a red supergiant, on the so called Asymptotic Giant Branch (AGB). The evolution of *low-mass* stars (M $\lesssim$ 2.3 $M_\odot$) takes a somewhat different course. After hydrogen-shell ignition the helium core becomes degenerate and the hydrogen burning shell generates the entire stellar luminosity. While its core mass grows, the star gradually climbs upwards along the giant branch until it reaches helium ignition with a flash (point 4). For all stars less massive than about 2.3 $M_\odot$ the helium core at helium-flash ignition has a mass of about 0.45 $M_\odot$.

The evolution described above holds for an initial chemical composition of 70% hydrogen, 28% helium and neglecting the effects of convective overshooting (for details see Maeder and Meynet (1989)).

3.3.4. *Types of close binary evolution in terms of the evolutionary state of the companion star at onset of mass transfer: cases A, B and C*

When a star is born as a member of a binary system (with a radius smaller than that of its Roche lobe), it may, due to evolutionary expansion of the envelope, after some time begin to overflow this lobe. When this happens, the matter flowing out along the first Lagrangian point L_1 will fall towards the companion. The further evolution of the system will now depend on: (i) the evolutionary state of the star at the onset of the overflow, which is determined by the mass of the primary star, the orbital separation a and the mass ratio of the systems. (ii) whether the envelope of the star at the onset of the transfer is in radiative or convective equilibrium. As to the factor (i), Kippenhahn and Weigert (1967) defined three types of close binary evolution, called cases A, B and C, depicted in Figure 12b. In case A, the system is so close that the primary star already begins to overflow its Roche lobe after the end of core-hydrogen burning but before helium ignition (stage II) and in case it overflows its Roche lobe during helium-

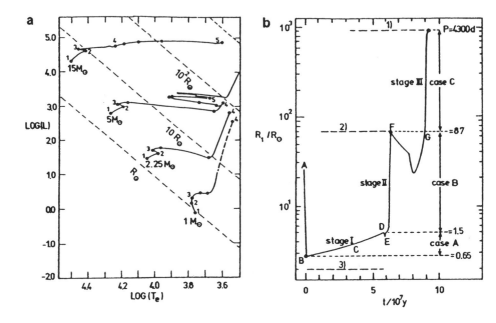

Figure 12. Evolutionary tracks in the Hertzsprung-Russell diagram of stars of 1 $M_\odot$, 2.25 $M_\odot$, 5 $M_\odot$ and 15$M_\odot$, after Iben (1967). b) Secular change of the radius of a star of 5$M_\odot$, from pre-main-sequence contraction (A) to carbon ignition. Also indicated are the orbital periods (in days) of close binaries in which the star just fills its Roche lobe at various evolutionary stages, for binaries with mass ratio $q = M_2/M_1 = 0.5$. The ranges of binary periods for evolution according to the cases A, B and C (see section 5) are indicated. 1) carbon ignition, 2) helium ignition, 3) main-sequence model (after Paczynski (1971a)).

shell burning or beyond (stage III in Figure 12b). It is clear from the figure that cases B and C occur over a wide range of orbital periods, case C even up to periods ~ 10 years. The precise orbital period ranges for the cases A, B and C depend on the initial primary mass M_1° and the mass ratio. Figure 13 shows for stars with masses between 1 and 16 $M_\odot$ the orbital periods at which they fill their Roche lobes, for an assumed mass ratio q=0.5.

3.3.5. *Evolution of the systems after the onset of Roche-Lobe overflow; the concepts of "conservative" vs. "common-envelope" evolution*

a) Contact star with a radiative envelope

When the contact star loses an amount of mass ΔM to its companion it will restore its hydrostatic equilibrium on a dynamical timescale, i.e. almost instantaneously. In stars with radiative envelopes the new hydrostatic equilibrium radius is smaller than its radius before the onset of the mass

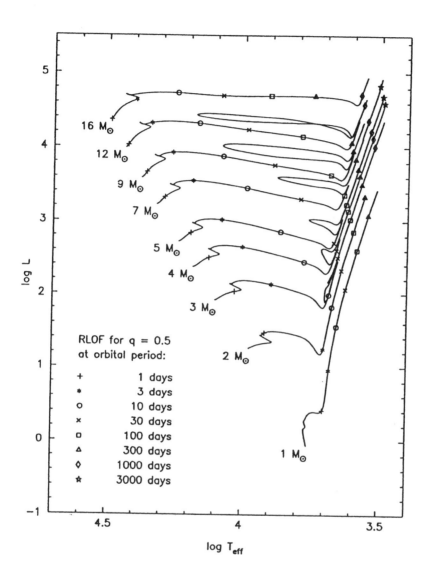

Figure 13. Evolutionary tracks in the HR diagram of 1 $M_\odot$ to 16 $M_\odot$ calculated by O. Pols using the evolutionary code of P.P. Eggleton. On the tracks are indicated the radii at which the stars fill their Roche lobes in binaries with the indicated orbital periods, for $q = M_2/M_1 = 0.5$.

transfer, but as the star is now out of thermal equilibrium, its radius will begin to expand on a thermal timescale to restore this equilibrium. If the mass M_1 of the mass-losing star is larger than that (M_2) of its companion, the transfer will make the orbit - and with it the Roche lobe radius R_L of

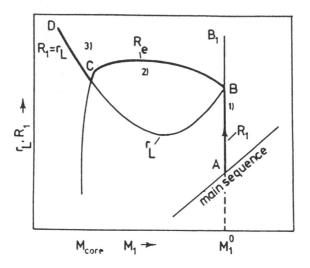

Figure 14.   Variation of the thermal-equilibrium radius of the primary component R_e and the radius of its Roche lobe $r_L(= R_L)$ during the mass transfer from the primary to the secondary. Only after the primary has become less massive than the secondary will r_L begin to increase. Consequently, equilibrium is not possible before the primary has lost so much matter than it has become the less massive component.

the mass-losing star - shrink (see equations (8), (9) and (13)). If the radius of the star after readjustment to hydrostatic equilibrium is still larger than that of its Roche lobe, the mass loss will continue until so much mass has been lost that it fits within its new Roche lobe. The subsequent expansion of the star on a thermal timescale will, however, cause the mass transfer to resume, causing the Roche lobe to shrink further, and so on. As a result the star continues to transfer mass until it has itself become the less massive component of the system. Its further expansion and mass loss will now cause, according to equation (13) the orbit to expand such that finally the thermal equilibrium radius of the star may be able to fit within its Roche lobe. Figure 14 depicts schematically the change of the thermal equilibrium radius R_e and the Roche-lobe radius R_L of the mass-losing star as a function of its decreasing mass. It is assumed here that the star reaches its Roche lobe after the end of core-hydrogen burning (i.e. in stage II in Figure 12b: this is Case B evolution). The figure shows that only after almost the entire hydrogen-rich envelope of the star has been lost, R_e becomes smaller than R_L again. Hence, in systems evolving according to Case B, practically only the helium core of the mass-losing star remains afer the mass transfer, surrounded by a hydrogen-rich outer layer of only very small mass.

Calculations show that also in Case A the thermal equilibrium radius can only become smaller than the Roche-lobe radius when the mass-losing star has become the less massive component of the system. The star may now, during its further nuclear evolution, either shrink or slowly expand on a nuclear timescale. The latter is the case in Case A systems and in Case B systems with original primary masses smaller than about 2.3 $M_\odot$. After the mass transfer these systems are still in the phase of hydrogen burning: in Case A systems hydrogen still burns in the core, in the low-mass Case B systems it burns in a shell surrounding a degenerate helium core of low mass (< 0.45 $M_\odot$). The radii of these stars gradually increase when hydrogen burning advances. Therefore they continue to slowly transfer mass on a nuclear timescale. The systems in this slow mass-transfer phase are the Algol-type binaries. In the more massive Case B systems the helium core has a mass ≥ 0.45 $M_\odot$ and contracts further and ignites helium burning, upon which the star expels the last part of its hydrogen-rich envelope and becomes a helium-burning pure helium star. Such stars have a very small radius, and in a binary are therefore deep inside their Roche lobes. Figure 15 schematically depicts the subsequent evolutionary phases of a close binary in which the more massive star at the onset of the mass transfer has a radiative envelope. The order of magnitude timescale of the first phase of mass-transfer is simply the thermal timescale of the (initial) primary star, given by equation (25). This leads to an order of magnitude mass-transfer rate

$$\dot{M}_1 = -M_1/\tau_{th} = -3.10^{-8}(M_1/M_\odot)^3[M_\odot/yr] \qquad (29)$$

After the reversal of the mass ratio and stabilization of the system, the mass transfer takes place on a nuclear timescale, leading with equation (26) to a mass-transfer rate of order

$$\dot{M}_1 = -10^{-10}(L_1/L_\odot)[M_\odot/yr] \qquad (30)$$

which according to eq (27) for $M > M_\odot$ leads to

$$\dot{M}_1 = 10^{-10}(M_1/M_\odot)^{3.5}[M_\odot/yr] \qquad (31)$$

b) Response of the accreting companion
Being the less massive component, it has a longer thermal timescale than its companion. Therefore the rapid transfer of mass to it causes its thermal equilibrium to be strongly perturbed, especially because the accreting material carries considerable kinetic energy which is dissipated into heat when the matter accretes on the surface. The matter also carries a fraction of the orbital angular momentum which will strongly spin up the accretion's rotation. The details of the accretion process are still poorly known, but it is likely that if the mass ratio of the system is small ($\lesssim 0.3$) the star will

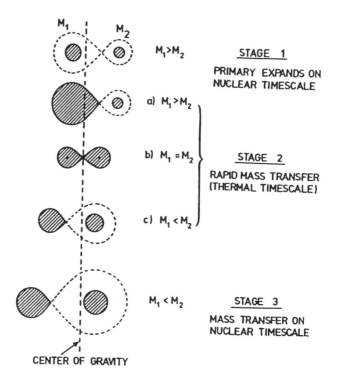

Figure 15. Subsequent evolutionary stages and orbital dimensions of a close binary system in which the primary star has a radiative envelope at the time it reaches its Roche lobe.

swell up to overflow its own Roche lobe. This occurs for radiative as well as convective envelopes of the mass-losing star. A common-envelope then forms around the binary. In the next paragraph we consider the evolution which then ensues.

c) Contact star with a convective envelope: Common-Envelope Evolution

A star with a deep convective envelope has the characteristic that upon losing mass its radius expands on a dynamical timescale (cf. Bhattacharya and van den Heuvel 1991), which is only of the order of hours to days. If this star is the more massive component of a close binary, its Roche lobe will shrink as a result of the mass transfer. Therefore the result is a violently unstable phase of mass transfer in which theoretically mass-transfer rates of order 10^{-2} to 10^{-1} $M_{\odot}$/yr could result. However, the companion will never in such a short time be able to accomodate these large amounts of mass dumped onto it. Therefore, as first suggested by Paczynski (1976) and

Ostriker (1973), in this case the expanding envelope of the primary star will completely engulf the secondary star, leading to the formation of a common envelope, inside which the secondary star and the core of the primary star orbit each other. Due to the large frictional drag on this orbital motion inside the common envelope, the secondary star will rapidly spiral inwards. In this process a large amount of (orbital) potential energy is converted into heat, causing the envelope to be expelled. As a result, one expects after the spiral in a very close binary to remain, consisting of the secondary and the evolved core of the primary star. Paczynski and Ostriker proposed this type of evolution to explain the existence of the Cataclysmic Variable binaries. These have extremely short orbital periods, in general of order of only a few hours, but they often contain massive white dwarfs, which can only be produced when stars are on the Asymptotic Giant Branch, i.e. have radii of hundreds of solar radii. The original orbits of these systems, before the onset of the mass transfer, must therefore have been hundreds of days up to a number of years. An alternative outcome of the Common-Envelope evolution (in case there is not enough orbital energy available to expell the envelope, or if the core of the primary star is not sufficiently compact) is that the stars will completely merge, resulting in the formation of a rapidly rotating single star. Numerical Hydrodynamic calculations of Common-Envelope (further abbreviated here as CE-evolution) have shown that the CE-process proceeds very rapidly, on a timescale of order only 10^2 to 10^3 years (Taam and Bodenheimer 1991; Taam and Sandquist 2000). CE-evolution is not only important for understanding the formation of the Cataclysmic Variables, but also for that of the Low-Mass X-ray Binaries and the double compact objects with short orbital periods such as the close double neutron stars like the Hulse-Taylor binary pulsar and the close double white dwarfs such as the AM CVn systems. In Figure 13 the vertical parts of the tracks where the stars ascend the giant branch indicate where the envelope of the mass-losing star is convective and CE-formation is expected to occur.

d) Further reasons why a Common Envelope might form
Apart from mass transfer from a convective envelope, formation of a common envelope may ensue also for other reasons, as follows:

(i) Binary mass ratio far from unity
If the two stars in a binary differ very much in mass, their thermal timescales will be very different. The timescale on which the secondary can accomodate the large amounts of mass dumped onto it without getting far out of thermal equilibrium is its own thermal timescale. However, the primary dumps the matter onto the secondary on the primary's thermal timescale

which, in case the mass ratio of the system is smaller than 0.3, differs from that of the secondary by an order of magnitude or more. Therefore, if the mass ratio of secondary and primary is below 0.3 one expects the secondary to rapidly swell up as a result of the mass transfer and to begin to overflow its own Roche lobe after a few tens of percents of the envelope mass of the primary has been transferred. Further mass transfer then leads to the formation of a common envelope around the two stars.

(ii) Tidal instability

Tidal forces will attempt to keep the two components of a close binary in corotation with the orbital motion. However, Darwin (1879) showed that if the orbital angular momentum of a system, in case the components would be corotating, is less than three times the rotational angular momentum of these (corotating) components, corotation can never be achieved, and the attempt of the tidal forces to drive the system towards corotation will drain the orbital angular momentum away, converting it continuously into rotational angular momentum of the components without ever reaching corotation. The system will therefore be tidally unstable and will evolve towards a tidal catastrophe in which the two stars will inevidably spiral inwards and merge with each other. Darwin discovered this instability for the Earth-Moon system, but it equally holds for binary stellar systems. Counselman (1973) rediscovered Darwin's theorem when he studied the expected secular evolution of the X-ray binary Centaurus X-3. In practice this tidal instability occurs only for systems like the HMXBs, with mass ratio very far from unity: in the Centaurus X-3 system this ratio is about 14, and the neutron star can be considered a point mass, with negligible rotational angular momentum.

e) A simplified approach for calculating the outcome of Common-Envelope Evolution

It will be clear from the above that knowledge of CE-evolution is essential for understanding the final evolution and fate of the High-Mass X-ray Binaries and the formation of a large variety of close binaries that contain compact components, such as the CVs.

In calculating the order of magnitude outcome of the spiral-in process we will follow here an approach, based on rather simple energy considerations, introduced by Webbink (1984). The numerical hydrodynamical computations by various groups (e.g.Bodenheimer and Taam 1984, Taam and Bodenheimer 1991, Rasio and Livio 1996, cf. Taam and Sandquist 2000) lend support to this formalism for making a fair order of magnitude estimate of the final orbital parameters of the system after spiral in, given the initial system configuration. It is assumed here that the companion has a mass M_2 and that the primary star has a compact core (e.g.: helium core or CO core) of mass M_{1f} and an extended envelope of mass M_{1e}. To calculate the orbital period with which the system may terminate after spiral in we assume that a fraction α of the loss of the orbital energy is used to expell the envelope. α is called the "efficiency factor" of CE evolution. The above assumptions lead to the following equation for the relation between the original and the final orbital radii a1 and a_2 (Webbink 1984):

$$G(M_{1f}+M_{1e})M_{1e}/\lambda a_1 R_{L1} = \alpha[GM_2.M_1/2a_2-G(M_{1f}+M_{1e})M_2/2a_1] \quad (32)$$

where R_{L1} a1 denotes the Roche lobe radius of the primary star at the onset of the mass transfer and λ is a weighting factor (< 1) for the gravitational binding energy of the core and envelope. Equation (32) yields

$$a_2/a_1 = (M_2.M_{1f})/[(M_{1e} + M_{1f}).(M_2 + 2M_{1e}/\alpha.\lambda.R_{L1})] \quad (33)$$

3.3.6. Conservative evolution and the formation of the high-mass X-ray binaries

a) Formation of the "standard" High Mass X-ray Binaries The formation of the HMXBs can be understood in terms of the "conservative" evolution of normal massive close binary systems in the way described above in paragraph a of section 3.3.5. We consider as an example the evolution of a massive Case B system as depicted in Figure 16 (after van den Heuvel 1977), which started out with components of 20 and 8 $M_\odot$ and an initial orbital period of 4.7 days. The subsequent evolutionary phases are described in the figure caption and are as follows: After 6.17×10^6 yrs the primary star has terminated core-hydrogen burning and overflows its Roche lobe. In only 3×10^4 yrs it transfers 14.66 $M_\odot$ to its companion and leaves behind a 5.34 $M_\odot$ helium core, which becomes a helium-burning pure helium star. The orbital period is now 10.86 days and the companion has become a 22.66 $M_\odot$ star which is still practically unevolved as it has been rejuvenated by the 14.66 $M_\odot$ of unprocessed primordial hydrogen-rich matter from its companion. As first noticed by Paczynski (1967) binary systems consisting of a helium-burning helium star plus a massive main-sequence star exactly resemble the Wolf-Rayet binaries, of which hundreds are known in our Galaxy and the Magellanic Clouds. Wolf-Rayet stars are hydrogen-deficient stars of very high luminosity and relatively low mass: in the Wolf-Rayet binaries they are practically always some 2 to 4 times less massive than their O- or early B-type companion stars (cf. van den Heuvel 1994a). Just as expected for helium burning helium stars in this mass range, they have roughly the same luminosity as their massive main-sequence companions and have a remarkable emission-line spectrum indicating large mass loss by stellar winds at a rate typically of order 10^{-5} $M_{\odot}$/yr, and at velocities of order several thousands of Km/s.

Some 60 % of all WR-stars are found in close binaries of this type. The further evolution of helium stars has been studied for example by Paczynski (1971b), Arnett (1973), Nomoto (1984), Habets (1985, 1986), and Pols (2000). These studies have shown that the radii of helium stars more massive than 3.5 $M_{\odot}$ do not expand very much before the termination of their evolution with the collapse of their burnt-out core to a neutron star or a black hole. For the 5.34 $M_{\odot}$ helium star of Figure 16 this occurs 0.69 × 10^6 yrs after the mass transfer. Due to the explosive mass ejection in the supernova (assuming symetric mass ejection) the orbital period increases to 12.63 days and a small orbital eccentricity is induced, and the center of gravity of the system is accelerated to a runaway velocity of about 35 km/s. At age 10.41 × 10^6 yrs the 22.66 $M_{\odot}$ companion has terminated its core-hydrogen burning and has become a blue supergiant with a strong stellar wind. This induces the system to become a High-Mass X-ray binary, resembling Vela X-1. The HMXB phase, until the star overflows its Roche lobe, probably lasts only between 10^4 and 10^5 yrs. After reaching its Roche lobe a Common Envelope will form. Assuming the system to survive this phase as a binary, the outcome will be a very close binary consisting of a helium star (Wolf-Rayet star) and the compact companion in a very narrow orbit. It is now known that the 4.8-hour X-ray binary Cyg X-3 is such a system, consisting of a helium star (Wolf-Rayet star) and a compact object (neutron star or black hole), cf. van Kerkwijk et al. (1992); Hanson et al. (2000); Ergma and Yungelson (1998).

b) Formation of the B-emission X-ray Binaries

The existence of the Be/X-ray binaries can also be simply explained in terms of conservative evolution, in this case: of binaries that started out with somewhat lower initial masses of the components, i.e.: primary stars typically in the mass range 8 to about 15 $M_{\odot}$. In this case, due to the dependence of helium-core mass on initial stellar mass, the mass exchange will produce systematically longer orbital periods of the post-mass-exchange binaries (van den Heuvel 1983, van den Heuvel and Rappaport, 1987; van den Heuvel et al. 2001). The generally high orbital eccentricities of the Be/X-ray binaries cannot be explained with symmetric supernova mass ejection (van den Heuvel 1994a; Verbunt and van den Heuvel 1995), and therefore are a clear indication that velocity kicks are imparted to neutron stars in their birth events (van den Heuvel and van Paradijs 1997).

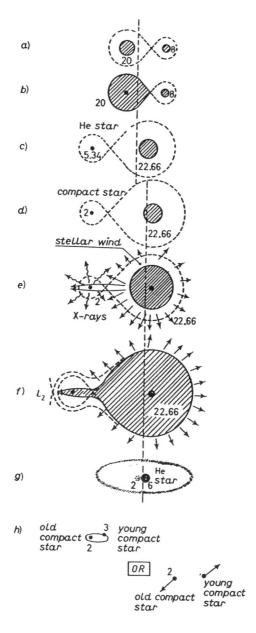

Figure 16.   Subsequent stages in the evolution of a massive close binary that started out
with components of 20 $M_\odot$ and 8 $M_\odot$ in a circular orbit with P=4.70d (for details see
Van den Heuvel 1994a). It is assumed that the supernova explosion of the primary leaves
a 2 $M_\odot$ compact star (neutron star or black hole). a), b) $t = 0, P = 4.70d$, onset of the
first stage of mass exchange; c) $t = 6.20 \times 10^6y, P = 12.63d$, He star (=Wolf-Rayet star)
has exploded as supernova; e) $t = 10.41 \times 10^6y, P = 12.63d$, the normal star becomes a
supergiant; its strong stellar winds turns the compact star into powerful X-ray source; f)
$t = 10.45 \times 10^6y, P = 12.63d$, onset of second stage of mass exchange, the X-ray source
is extinguished and large mass loss from the system begins; g) $t \sim 10.47 \times 10^6y, P \sim 4h$,
onset of second Wolf-Rayet stage; h) $t \sim 11 \times 10^6y$, the second He star has exploded as
a supernova, survival or disruption of the system depends on the mass of the remnant.

3.3.7. *Formation of double neutron stars: descendants of Be/X-ray binaries*

Equation (33) predicts that due to their small orbital separations, the standard HMXBs are unlikely to survive the second mass-transfer phase and spiral-in. On the other hand, the wide Be/X-ray binaries such as X Per ($P_{orb} = 250$ days) are expected to survive spiral-in as systems depicted in Figure 16g: short-period binaries consisting of a helium star together with a compact star. When the helium star terminates its evolution with a supernova explosion, the system either is disrupted or a double neutron star with a very eccentric orbit is formed, closely resembling the Hulse-Taylor and Wolszczan binary radio pulsars (see Figure 6). It thus seems most likely that these binary pulsars are the descendants of Be/X-ray binaries (van den Heuvel 1992, 1994a). Recently some doubts have been expressed as to whether or not the neutron star can survive the spiral-in process through the envelope of a massive companion, without turning into a black hole due to highly super-Eddington accretion (Chevalier 1993; Brown 1995). However, there are various arguments indicating that the neutron star can indeed survive this spiral in. First of all, in order that highly super-Eddington accretion can occur, neutrino-cooling must become the dominant cooling process of the accreting neutron star. This will occur only after the accretion rate has risen to 10^4 times the Eddington rate. Before that, photon cooling will dominate. However, already when the accretion rate slightly exceeds the Eddington rate, the radiation pressure will prevent it from growing further. If the accretion is non-spherical, through a disk, one may perhaps exceed the Eddington rate by an order of magnitude or so, as is observed sometimes in some of the massive neutron-star X-ray binaries such as SMC X-1 and A0538-66 in the Large Magellanic Cloud. However, there is to my knowledge no conceivable physically realistic way in which the gap between exceeding the Eddington rate by one order of magnitude (due to non-spherical accretion) and by four orders of magnitude, as required for neutrino cooling to take over, can be bridged: radiation pressure will prevent it. The recent discovery of two close binary pulsars with eccentric orbits, consisting of a young pulsar and a massive white dwarf (section 2.3) indicates that indeed a compact star (in this case a massive white dwarf) can survive the spiral-in through the envelope of a companion star of about $10\ M_\odot$ (Tauris and Sennels 2000). The core of that companion became a helium star of about $2.5\ M_\odot$ which subsequently exploded as a supernova, leaving behind the young neutron star in the system and inducing the eccentricity of the orbit. Furthermore, taking into account the angular momentum of the accreted matter, Chevalier (1996) and Armitage and Livio (2000) have expressed doubts as to whether the neutron star spiralling-in through the envelope of a wide binary companion (i.e.: a later evolutionary state of a Be/X-ray binary) will indeed be able to accrete much matter. A strong argument against the neutron stars in the Be/X-ray binaries turning into black holes during spiral-in is that in that case the formation rate of close binary pulsars consisting of a black hole and a young neutron star would be of similar order of magnitude as the formation rate of the Be/X-ray binaries (as a sizeable fraction of these will survive spiral in). (The binaries become very close if the neutron star can accrete the entire envelope of its companion). The estimate is that there are several thousands of Be/X-ray systems in the galaxy (van den Heuvel 1992, 1994a). These systems live shorter than 10^7 years, so their formation rate is of order 10^{-4} to 10^{-3} per year in the galaxy. This implies that the formation rate of the close binaries consisting of a black hole and a young neutron star should be at least 10^{-4} per year, i.e. about 1 % of the pulsar birth rate. Due to the large gravitational attraction of the black hole disruption of the systems in the second supernova is unlikely, even if the neutron star receives a sizeable kick velocity when it is born. One would therefore expect some ten close black-hole/neutron star systems to be present among the about 1000 presently known young radio pulsars, while there is none. This is a strong argument against the scenario in which the neutron stars in HMXBs would not survive spiral-in as neutron stars.

Figure 17 shows the various ways in which the persistent ("standard") and Be/X-ray binaries are expected to terminate their evolution. If the Be star is less massive than 8 - 10 $M_\odot$ (the precise value depends on the orbital period) its burned-out core is not massive enough to leave a neutron star and becomes a CO- or O-Ne-Mg white dwarf. In that case a very close binary radio pulsar with a circular orbit will remain, consisting of a neutron star and a massive white dwarf. PSR 0655+64 (P_{orb}=1.04) is thought to be such a system (see section 3.4.2.C).

3.3.8. Common-envelope evolution and the formation of cataclysmic variables and low-mass X-ray binaries

Cataclysmic Variables (CVs) and Low-Mass X-ray Binaries (LMXBs) both consist of a compact stellar remnant and a low-mass donor star in a very narrow orbit, with orbital periods typically of order of about one hour to a few days. As mentioned in section 3.3.5 the concept of CE-evolution was put forward to explain the existence of the CV-binaries: their progenitors were wide binaries consisting of a red giant with a degenerate CO or O-Ne-Mg core and a low-mass companion star with a mass typically < $M_\odot$. The red giant can have any mass between about 1.5 $M_\odot$ and 10 $M_\odot$. The envelope of the red giant engulfed the companion star, causing it to spiral inwards, driving off the giant's envelope in the process. The final outcome

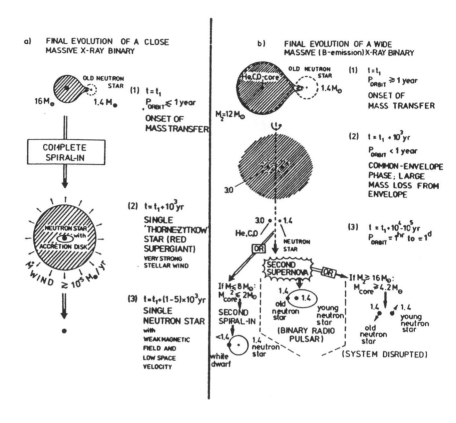

Figure 17. The various possibilities for the final evolution of a High-Mass X-ray Binary. In all cases the onset of Roche-lobe overflow leads to formation of a common envelope and the occurrence of spiral-in. (a) In systems with orbital periods less than about 1 year there is most probably not enough energy available in the orbit to eject the common envelope, and the neutron star spirals down into the core of its companion. Subsequently, the envelope is ejected by the liberated accretion energy flux, leaving a single recycled radio pulsar. (b) In systems with orbital periods longer than about a year the common envelope is ejected during spiral-in, and a close binary can be left, consisting of the neutron star and the core, consisting of helium and heavier elements, of the companion. Companions initially more massive than 8-12 $M_{\odot}$ leave cores that will explode as a supernova, leaving an eccentric-orbit binary pulsar or two runaway pulsars. Systems with companions less massive than $\sim 8 - 12\, M_{\odot}$ leave close binaries with a circular orbit and a massive white dwarf companion, similar to PSR 0655+64.

is a very close binary consisting of the degenerate core of the giant plus the unchanged low-mass companion star. The first detailed calculation of the formation of a CV- binary through this process was carried out by Meyer and Meyer-Hofmeister (1978) which started out from a system consisting of a 5 $M_{\odot}$ red giant with a companion of one solar mass. The LMXBs must have originated through a similar type of process, the only difference being that the red giant in the progenitor system was more massive than 8-10 $M_\odot$, such that it produced a heavy-element core too massive to finish as a white dwarf. The first model put forward for such evolution was that of Sutantyo (1975a) for the formation of the Her X-1 system, which is presented as an example in Figure 18. The progenitor was assumed to be a binary system with component masses of 15 $M_\odot$ + 2 $M_\odot$, evolving according to Case B or C. The 15 $M_\odot$ star produces a 4 $M_\odot$ helium core. Using equation (33) with $\lambda = R_{L_1} = 0.5$ and $\alpha = 1$ one finds the orbital radius to shrink by a factor $a_1/a_2 = 168$ during spiral-in. To keep the 2 $M_\odot$ star inside its Roche lobe after spiral in a_2 should be larger than 4.5 $R_\odot$, leading to $a_1 > 3.5$ AU (756 $R_\odot$), and a pre-spiral-in orbital period of 1.6 yrs. Comparison between the outcome of equation (33) and of detailed calculations of CE-evolution mentioned above shows that α-values between 1 and 4 are possible. With $\alpha = 2$ one finds for the system of Figure 18 that $a_1/a_2 = 86.5$, implying an initial orbital separation > 390 $R_\odot$ ($P_{orb} > 0.6$ yrs).

3.3.9. *Direct core collapse vs. accretion-induced collapse*

a) Direct core collapse In the model of Figure 18 the neutron star formed directly from the core collapse of the evolved 4 $M_\odot$ helium star. This type of neutron-star formation is expected for all helium stars more massive than 2.2 to 2.5 $M_\odot$ (Habets 1985, 1986, cf. van den Heuvel 1994a). Helium stars in the mass range 2.2 to 3.5 $M_\odot$ evolve during helium-shell burning towards a giant phase. They will therefore in a close binary go through a second mass-transfer phase, after which only a relatively small-mass core is left ($<$ 2 $M_\odot$). Relatively little mass is therefore ejected in the supernova explosion. However, one has to take into account that in neutron-star formation a considerable velocity kick can be imparted to the neutron star. Typical mean kick velocities inferred from pulsar observations are of order 200 km/s with a spread of up to 500 km/s. Using typical kick-velocity distributions like that of Lyne and Lorimer (1994) or Hartman 1996; Hartman et al. (1997), Kalogera and Webbink (1998) made extensive binary population synthesis evolution calculations. They found that without velocity kicks the LMXBs with short orbital periods (< 0.5 days) cannot be formed. Therefore, kicks are absolutely required for forming the observed LMXB population.

b) Accretion-induced collapse; relation to Super Soft Binary X-ray Sources In CV-like binaries in which the white dwarf is composed of O-Ne-Mg, mass transfer can induce the white dwarf to collapse to a neutron star (Nomoto and Kondo 1991). This is most likely to occur if the mass-transfer rate is in the range $1 - 4 \times 10^{-7}$ $M_\odot$/yr, as in this range the accreted matter will start steady hydrogen burning on the surface of the white dwarf, without much radius expansion. This leads to a continuous growth of the mass of

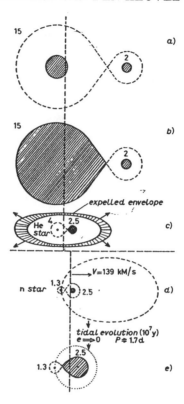

Figure 18. Scenario for the formation of the Her X-1 system (a Low-Mass X-ray Binary) out of an initially massive close binary with an extreme mass ratio. When the 15 $M_\odot$ primary begins to overflow its Roche lobe, the 2 $M_\odot$ secondary spirals in and causes the hydrogen-rich envelope of the primary to be expelled, leaving a 4 $M_\odot$ helium star plus the 2 $M_\odot$ secondary, which is expected to have accreted only 0.5 $M_\odot$ in this process. The supernova mass ejection and impact effects imparted a 139 km s^{-1} velocity to the center of mass of the system, and induced a large orbital eccentricity. After the supernova explosion of the helium star the system is 'quiet' for about 5 $\times 10^8$ yr and the orbit is circularized by tidal friction. When the star begins to transfer mass to the neutron star the system becomes an X-ray source (after Sutantyo [1975a]; explanation in the text).

the white dwarf. CV-like systems with mass-transfer rates in this range are the so-called Super Soft Binary X-ray Sources (SSS, cf. Kahabka and van den Heuvel 1997). The mass donors have here (initial) masses larger than that of the white dwarf, i.e. in the range 1.2 $M_\odot$ to 3 $M_\odot$. Therefore the mass transfer proceeds on a thermal timescale of the donor star, and is in the above-mentioned range. In this way a SSS-binary like CAL 83 (M_d = 2 $M_\odot$, P_{orb}=1.04 d) might also produce a system like Her X-1, although the progenitor of the latter system must have started out with a somewhat longer orbital period. As, however, the orbital periods of the "classical" SSS-binaries range up to several days, this can easily be accomodated. To obtain the large distance of Her X-1 from the galactic plane (2 kpc), in this case a considerable velocity kick must be imparted to the neutron star at its birth. For detailed model calculations for the origin of the Her X-1 system, including kicks, I refer to Sutantyo (1992; 1999).

3.4. MECHANISMS DRIVING THE MASS TRANSFER IN LOW-MASS X-RAY BINARIES AND CATACLYSMIC VARIABLES

3.4.1. *Gravitational radiation vs. "magnetic braking"*
A. Introduction

This topic was already covered order-of-magnitude wise in section 3.2, but a number of important refinements should be added, which is the subject of this section. We consider the case in which the donor star is less massive than the compact star, which is the case in most CVs and LMXBs. In this case the mass transfer is stable. The only exception is Her X-1. Here, however, beginning atmospheric Roche-lobe overflow can power the source for several times 10^5 yrs before the Eddington limit is exceeded (Savonije 1978). Assuming the donor star to fill its Roche lobe, the orbital period is related to its average density (Faulkner 1971) as follows from combining Kepler's third law with equation (9) for calculating the volume of the Roche lobe:

$$(\frac{(2\pi)^2}{P}) = G(M_1 + M_2)/a^3 \tag{34}$$

$$(R_L/a)^3 = (0.46)^3 M_2/(M_1 + M_2) \tag{35}$$

yielding:

$$P = 9h(R_2/R_\odot)^{1.5}(M_\odot/M_2)^{1/2} = 9h(\overline{\rho}_\odot/\overline{\rho})^{1/2} \tag{36}$$

TABLE 6. Approximate mass-radius relations for Roche-lobe-filling low-mass ($\leq 1 M_\odot$) stars in thermal equilibrium (during the mass transfer the thermal equilibrium may gradually be disturbed such that deviations from these relations become possible, after Verbunt 1990.

Main sequence	$R_2/R_\odot = M/M_\odot$	$P = 9^h(M_2/M_\odot)$
Helium main sequence	$R_2/R_\odot = 0.2 M_2/M_\odot$	$P = 0.^h9(M_2/M_\odot)$
Degenerate star	$R_2/R_\odot = 0.013(1+X)^{5/3} \times$	$P = 48^s(1+X)^{5/2} \times$
	$(M_\odot/M_2)^{1/3}$	$M_\odot/M_2$ [a]

[a] Here X is the fractional hydrogen abundance.

Table 6 gives relation (36) for main-sequence stars, helium stars and degenerate stars. Equation (36) therefore directly indicates what kind of

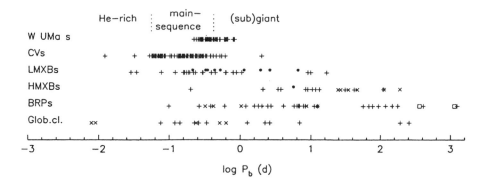

Figure 19. Orbital periods of X-ray binaries and binary radio pulsars in our Galaxy, and, for comparison, cataclysmic variables and contact binaries (also called W UMa binaries). Each symbol indicates one system. For the X-ray binaries • indicates a system with a low-mass white dwarf companion, × a system with a high-mass white dwarf or neutron star companion, and □ a main-sequence companion. For globular clusters a × indicates an X-ray binary, a + a radio pulsar binary (Courtesy: F.W.M. Verbunt).

stars fit in a system, once the orbital period is known. Figure 19 shows the orbital period distributions of the CVs and LMXBs (after Kolb and Ritter 1998, courtesy F. Verbunt 2001). The figure shows that all three types of mass donors listed in Table 6 are possible in short-period systems. Figure 19 also shows that for the CVs there is a clear "period gap" between 2 and 3 hours, where very few systems are present, while the LMXBs do not show such a gap. Table 6 shows that systems with orbital periods shorter than 80 minutes most probably have helium stars or degenerate stars as donors. Binaries with orbital periods between 80 minutes and 8h are expected to contain normal hydrogen-rich unevolved main-sequence stars with masses not larger than about one solar mass. We will now consider the evolution of these systems in more detail. In section 3.4.2 we will consider the evolution of systems with evolved donor stars.

B. Effects of orbital angular momentum losses by gravitational radiation and "magnetic braking"

We will largely follow here the analysis by Verbunt(1990). For simplicity we will tentatively assume that the total mass of the binary system is conserved, i.e. $\dot{M}_1 = -\dot{M}_2$ (for a straightforward generalization to the case with mass loss from the system: see Rappaport et al. 1982). Equation (14) with $\alpha = 0$ yields:

$$\dot{a}/a = 2\dot{J}_{\text{orb}}/J_{\text{orb}} - 2(1 - M_2/M_1)(\dot{M}_2/M_2) \qquad (37)$$

Since we have designated the donor as star 2, $\dot{M}_2 < 0$. A necessary condition for stable mass transfer is $M_2 < M_1$ (see above). Substituting the logarithmic derivative of equation (9) into equation (37) one obtains:

$$\dot{R}_{\text{L}}/R_{\text{L}} = 2\dot{J}_{\text{orb}}/J_{\text{orb}} - 2(1 - M_2/M_1)(\dot{M}_2/M_2) + \dot{M}_2/3M_2 \qquad (38)$$

which gives the change of the Roche lobe radius in terms of the angular momentum loss rate from the system. If one writes the mass-radius relation of the donor star as $R_2 = M_2^n$ one will have:

$$\dot{R}_2/R_2 = n\dot{M}_2/M_2 \qquad (39)$$

In order to have stable mass transfer R_2 should remain equal to R_{L}. In that case the combination of equations (38) and (39) yields:

$$\dot{J}_{\text{orb}}/J_{\text{orb}} = (5/6 + n/2 - M_2/M_1)\dot{M}_2/M_2 \qquad (40)$$

For stable mass transfer both sides of equation (40) should be negative which implies:

$$M_2/M_1 < 5/6 + n/2 \qquad (41)$$

For a given value of n this sets a limit to the mass ratio above which the transfer is unstable. For example for a non-relativistically degenerate donor star $n = -1/3$ and $M_2/M_1 < 2/3$ is required for stable mass transfer. On the other hand, for a low-mass main-sequence donor $n = 1$ and M_2/M_1 should be $< 4/3$ for stable mass transfer. If the mechanism of angular momentum loss is prescribed, the evolution of the mass-transfer rate can be calculated with equation (40). The two mechanisms that are generally accepted to drive the mass transfer in close (P $<$ 12 h) systems are: angular momentum loss by: (i) gravitational radiation (see equation (15) and (ii) "magnetic braking". The latter process, first introduced for these systems by Verbunt and Zwaan (1981), takes place when the Roche-lobe filling donor is a solar-type dwarf star with a convective envelope. Such stars have hot coronae, surface magnetic fields and stellar winds. The observed rotational slow-down of solar-type stars is ascribed to the fact that the magnetic field keeps the winds in corotation out to at least 5 to 10 stellar radii, such that these winds continuously carry off significant amounts of rotational angular momentum from the stars, even though the wind-mass-loss rates are negligible on a stellar evolutionary timescale ($\sim 10^{-13}$ $M_\odot$/yr). These magnetically-coupled winds thus produce the "magnetic braking" of the star's rotation. In a close binary the donor will be kept in corotation due to tidal forces. This means that when its rotation slows down due to magnetic braking, it will be spun-up back into corotation by tidal forces. This goes at the expense of the orbital angular momentum of the system, which is continuously converted by the tidal forces into rotational angular momentum of the donor, and then lost from the system by magnetic braking. Thus magnetic braking is a drain on the orbital angular momentum of the system, causing the two stars to spiral towards each other. It appears that in CVs and LMXBs the angular momentum losses by magnetic braking may exceed those by gravitational radiation losses by an order of magnitude, such that according to equation (40) they may drive mass-transfer rates an order of magnitude larger than gravitational radiation (Verbunt and Zwaan 1981).

C. Evolution of CVs and LMXBs with a main-sequence donor: the period minimum

Combining equations (40) and (15) and eliminating a by using equation (9) together with the mass-radius relation given by equation (39), one obtains the mass-transfer rate $\dot{M}_2$ as a function of M_1 and M_2 for the case the transfer is driven by gravitational radiation losses. For a main-sequence donor n = 1. Assuming M_1 =1.4 $M_\odot$, one then obtains for M_2 in the mass range 0.1 to 1.0 $M_\odot$, a mass-transfer rate $\dot{M}_2 = - 10^{-10}$ $M_\odot$/yr, as depicted in Figure 20b (after Verbunt 1990), which in an LMXB will yield an X-ray luminosity of $\sim 10^{36}$ ergs/s. In many CVs mass-transfer rates of $\sim 10^{-9}$ $M_\odot$/yr are observed and a sizeable fraction of the LMXBs with short orbital periods has X-ray luminosities in the range 10^{37} to 10^{38} ergs/s, indicating mass-transfer rates in the range 10^{-9} to 10^{-8} $M_\odot$/yr. The work of Verbunt and collaborators (e.g. Rappaport, Verbunt and Joss 1983) has shown that angular momentum losses by magnetic braking can well explain these much higher mass-transfer rates.Figure 20a shows how the orbital period decreases when the donor mass decreases from 1.0 $M_\odot$ to 0.2 $M_\odot$, both for main-sequence stars and helium stars. The absence of CVs and LMXBs with H-rich donors and orbital periods below 80 minutes is due to the reversal of the mass-radius relation around P=80 minutes at which point a hydrogen-rich donor, with its reduced mass becomes degenerate and begins to follow a reverse mass-radius relation $R_2 \propto M_2^{-1/3}$. For this reason for hydrogen-rich donors the minimum possible orbital period is about 80 minutes, as was first pointed out by Paczynski and Sienkiewicz (1981).

This minimum period can be reached only if the donor is not driven far out of thermal equilibrium due to the preceding mass loss, as in that case the star will swell up, as it cannot radiate out its excess heat content fast enough. The latter may occur if the timescale for the mass loss $M_2/\dot{M}$

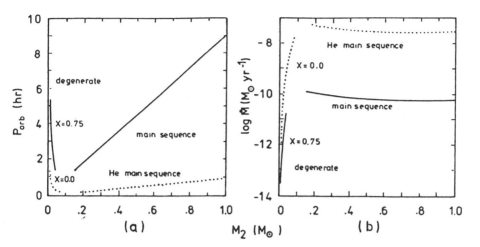

Figure 20. (a) Orbital period P_{orb} and (b) mass transfer rate $\dot{M}$ in low-mass X-ray binaries that evolve by means of angular momentum loss by gravitational radiation, as a function of the mass of the donor star, M_2. Donors are assumed to be in thermal equilibrium. Helium stars have shorter orbital periods and higher $\dot{M}$ than hydrogen-rich stars (after Verbunt 1990).

becomes shorter than the thermal timescale of the donor. For donors with $M_2 < 0.3$ $M_\odot$ the mass-loss timescale by magnetic braking becomes much smaller than the thermal timescale, whereas that by gravitational radiation is comparable or larger than the thermal timescale. The fact that the observed period minimum is around 80 minutes indicates that for systems with $M_2 < 0.3$ $M_\odot$ apparently magnetic braking no longer plays an important role.

D. Systems below the period minimum: hydrogen-poor donor stars

There are a few CVs and LMXBs with periods below the period minimum. Among the LMXBs these are for example 1626-67 (41 min), and 1820-30 (11 min). Among the CVs these are for example the AM CVn systems (P ~ 20 minutes), which have no hydrogen in their spectrum. Clearly, the donors in these systems must all be hydrogen-poor stars (see Figure 20a).

E. Reasons for the period-gap of CV binaries

A plausible explanation for the presence of the period gap between 2 and 3 hours for the CV binaries (Figure 19) was first put forward by Spruit and Ritter (1983). They noticed that with magnetic braking, a mass-losing hydrogen-rich donor star will, when it approaches 0.3 $M_\odot$, begin to deviate substantially from thermal equilibrium. Such a star therefore has an inflated radius, considerably larger than its thermal-equilibrium radius. Approaching 0.3 $M_\odot$ the orbital period of the system has decreased to about 3 h. Spruit and Ritter noticed that at about 0.3 $M_\odot$ a hydrogen-rich star becomes completely convective: the radiative core in the mass-losing star becomes smaller and smaller and completely disappears when M $\sim$ 0.3 $M_\odot$. Stars without convective cores are not expected to have magnetic activity any more, so at 0.3 $M_\odot$, magnetic braking is expected to switch off. This causes a sudden decrease in the orbital angular-momentum-loss rate by at least an order of magnitude, and thus also of the star's mass-loss rate to its companion by a similar factor. The result is that the star now has time to relax back towards thermal equilium, and thus to shrink: it will therefore detach itself from the Roche lobe, causing the mass transfer to stop completely and the star to shrink to its thermal equilibrium radius. Orbital angular momentum losses by gravitational radiation will continue, so the orbit keeps gradually shrinking, such that after a long time the donor star again fills its Roche lobe and the mass transfer resumes, now on a GR timescale. By that time the orbital period is about 2 h. This model explains the virtual absence of CVs with orbital periods between 2 and 3 h.

F. Why there is no period gap for LMXBs

In LMXBs during the long-lasting mass-transfer phase ($> 10^8$ yrs) driven by magnetic braking and gravitational radiation, the neutron star will have been spun up to a very short spin period, of order of milliseconds (see section 3.6). Thus, when the evolving system enters the period gap and the mass transfer shuts off, this rapidly spinning magnetized neutron star becomes a millisecond radio pulsar. This pulsar is so powerful an energy source that it is expected, with its relativistic pulsar wind (composed of a relativistic $e^- + e^+$ plasma), to be able to substantially ablate or evaporate the companion star (Ruderman et al. 1989), as is indeed observed to be taking place in some millesecond binary radio pulsar systems such as PSR 1957+20 (Fruchter et al 1988). This "evaporation" may be one viable way to get rid of companion stars and to explain the absence of LMXB-systems below the period gap (van den Heuvel and van Paradijs 1988). Furthermore, once, due to the termination of the mass transfer, the rapidly spinning neutron star has become active as a pulsar, the high pressure exerted by the relativistic pulsar wind onto the companion will prevent further accretion onto the neutron star, even if the companion would resume the mass transfer: the matter spilling over the Roche lobe would simply be blown out of the system. Therefore, LMXB systems entering the period gap will never become X-ray binaries again. This explains the absence of LMXBs below the period gap (with the exception of some systems with helium star or degenerate companions, which have had a different evolution).

3.4.2. *Mass transfer in LMXBs AND CVs driven by internal nuclear evolution of the donor star*

A. Introduction

Low-Mass X-Ray Binaries and CVs with orbital periods longer than about 10 hours cannot contain Roche-Lobe filling main-sequence stars with masses $< M_\odot$. If the donor is a low-mass star, as clearly is the case in Sco X-1 (P=0.86d) or Cyg X-2 (P= 9.8d) it must be an evolved star, i.e. a (sub-) giant. Recently it has been argued by various authors (King and Ritter 1999; Podsiadlowski and Rappaport 2000; Tauris et al. 2000) that the donor star in Cyg X-2, which has a mass of 0.5 $M_\odot$ but a luminosity of some 30 to 50 $L_\odot$, must be the evolved remnant of an "intermediate mass" star, in this case with an initial mass of some 2 $M_\odot$. [Such an evolution of this system had been suggested earlier by van den Heuvel (1981; 1994a; 1995; 1996) as a later evolutionary phase of a system like Her X-1]. The systems with evolved donors can therefore be divided into two groups: those with evolved "low-mass" donors, i.e. stars that started out less massive than their neutron-star or white-dwarf companion, and "intermediate mass" donors, i.e that started out more massive than the compact star in the system, up to about a mass $\sim$ 4 to 5 $M_\odot$. We will consider these two groups separately.

B. Evolved Low-mass donors: < 1.4 $M_\odot$

After leaving the main sequence, stars less massive than about 2.3 $M_\odot$ become subgiants powered by hydrogen burning in a shell around a degenerate helium core. The outer radius R and luminosity L of these stars are uniquely determined by the mass M_c of the degenerate core (Webbink et al 1983; Taam 1983; Rappaport et al. 1995; Joss et al. 1987). As the star evolves the core mass grows and the outer radius and luminosity increase, making the star ascend the giant branch in the Hertzsprung-Russell diagram. In a close binary in which this star is the less massive component, the growth of its outer radius will drive continuous mass transfer, which causes the orbit to widen and therefore is stable. Figure 21 shows as an example the evolution of a binary system (calculated by Joss and Rappaport 1983) that started out with a donor star and compact star both of one $M_\odot$, and an initial orbital period of 12.5 days. After the end of the mass transfer phase, which lasts 8.1×10^7 yrs, the donor has transferred its entire hydrogen-rich envelope to the neutron star and leaves behind a 0.31 $M_\odot$ degenerate helium white dwarf which is in a 117-day orbit around a 1.69 $M_\odot$ neutron star. The evolution of this system was tuned to model the formation of the first-discovered millisecond radio-pulsar binary PSR 1953+29 (Boriakoff et al. 1983), which consists of a 6 msec pulsar and a low-mass white dwarf companion. As the outer radius of the giant is completely determined by its degenerate helium core mass, the final Roche-lobe radius, just before the end of the mass transfer, is determined by this core mass, and thus is the

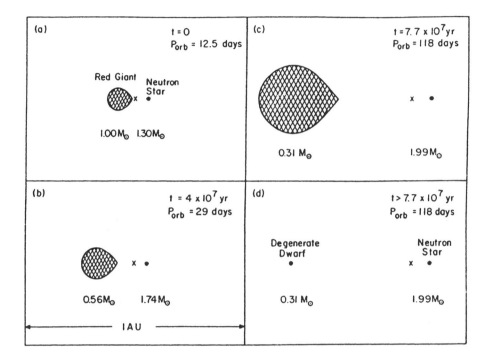

Figure 21. Evolution of a wide low-mass X-ray binary into a wide radio-pulsar binary with a circular orbit and a low-mass helium white-dwarf companion, such as PSR
1953+29. At the onset of the mass transfer, the low-mass companion is a (sub-)giant
with a degenerate helium core of $0.24 M_\odot$. Its light is generated by hydrogen fusion in
a shell around the core. The mass transfer from the giant to the neutron star is due to
the slow expansion of the giant, driven by this hydrogen-shell burning. During the mass
transfer the orbit gradually expands (due to angular-momentum conservation) and after
8×10^7 yr the system terminates as a wide radio-pulsar binary (after Joss and Rappaport
[1983]).

final orbital period. In most of the millisecond binary radio pulsars (with
circular orbits) the mass functions are so small that the companions must
be similar helium white dwarfs less massive than 0.45 $M_\odot$ (the maximum
mass that such a white dwarf can have before igniting helium burning).

**C. Evolved "Intermediate-mass" donors ($> 1.4 M_\odot$) and the
formation of the radio pulsars with massive white dwarf companions**

Here two types of evolution are possible as recently described in detail
by Tauris et al (2000), depending on whether or not the envelope of the
star at the onset of the transfer is radiative or convective.

a. Radiative envelope. In this case the mass transfer can keep the star inside its shrinking Roche lobe (see section 3.3.5), such that the formation of a common envelope is avoided (see for example also King and Begelman 1999; King, Taam, and Begelman 2000). The mass transfer will proceed on the thermal timescale of the donor, leading to highly super-Eddington mass-transfer rates. It seems reasonable to assume that the excess (super-Eddington) amount of transferred mass is ejected from the neutron star, carrying the specific orbital angular momentum of that star (this is the so-called "isotropic re-emission mode" of mass transfer, cf. Bhattacharya and van den Heuvel 1991; Soberman et al. 1997). With this assumption Tauris et al. (2000) find that the systems evolve at first on a thermal timescale of the donor star, followed after reversal of the mass ratio by a phase of evolution on a nuclear timescale, during which for donors with initial masses < 2 $M_\odot$ still a sub-Eddington phase of evolution on a nuclear timescale may occur. Presumably Cyg X-2 is presently in such a state (cf. Kolb et al. 2000, and references therein). For the higher donor masses there are no sub-Eddington phases and is unlikely therefore to observe these systems as X-ray binaries. Nevertheless, they have relatively long mass-transfer phases (10^6 to 10^7 yrs) during which they can accrete between 0.01 and 0.1 $M_\odot$, enough to spin them up to (very) short spin periods, as the accretion will take place through a disk. In the end a system remains with an orbital period between about 2 days and a few weeks, consisting of a recycled neutron star and a CO white dwarf with a mass between 0.5 and 1.1 $M_\odot$. Due to the long-lasting phase of Roche-lobe overflow one expects the orbit of these systems to have been circularized. This type of evolution explains the existence of the short-period binary pulsars with massive white dwarf companions in circular orbits with periods > 2-3 days (see Table 1).

b. Convective envelope. In this case, which concerns systems with wider initial orbits (initial orbital period longer than a few days, the lower limit depending on the donor mass, cf.Tauris et al. 2000) the formation of a common envelope cannot be avoided. The neutron star will be engulfed by the common envelope and spiral down into it. The result will be a very close system consisting of a neutron star and a massive white dwarf in a narrow orbit. The prime example of such a system is PSR 0655+64 (P= 1.04d). Tauris et al (2000) found that this system and several others of this type cannot have formed in any other way (see also Figure 17).

3.5. THE GLOBULAR CLUSTER X-RAY BINARIES AND THEIR FORMATION AND EVOLUTION

The discovery in 1973 by Gursky (see van den Heuvel 1994b) that in three of the error boxes of UHURU X-ray sources a globular cluster is present, came as a great surprise, since already this observation alone would indicate that X-ray sources are much more abundant (per unit mass) in globular clusters than in the Galaxy as a whole (Gursky 1973, private communication; Katz 1975). This is because the total mass of all globular clusters together is not more than 10^{-4} of the mass of the galaxy, and there are no more than some 100 strong UHURU sources in the galaxy. Finding 3 UHURU sources in such clusters then already implies that such sources are 300 times more abundant in globular clusters than in the rest of the Galaxy as Gursky (1973) immediately realized. This point was made in writing for the first time by Katz (1975), after Clark (1975) independently discovered the globular cluster sources and had obtained much more accurate positions of these sources with the SAS-C satellite, showing them to be inside their clusters. There are now a dozen strong globular cluster X-ray sources known, indicating a rougly thousand times higher incidence of strong sources in globular clusters than in the rest of the galaxy (cf. Lewin and Joss 1983). The globular cluster sources are all Type I X-ray bursters, indicating that they are neutron stars (van Paradijs 1978, cf. Lewin, van Paradijs and Taam 1995) and binary periods have been detected in a number of them, indicating that they are definitely Low-Mass X-ray Binaries. The abnormally high incidence of X-ray binaries in globular clusters indicated that here a different formation mechanism must be operating than for the X-ray sources in the galactic disk, and it was soon suggested that the formation of these LMXBs was due to close encounters of neutron stars and ordinary stars in the dense central regions of the clusters, leading to binary formation by gravitational "collision" processes. These were suggested to be either : tidal capture (Fabian et al. 1975) or real collisions, between giant stars and neutron stars (Sutantyo 1975b). Later, particularly through numerical stellar dynamics calculations, it became clear that binary-binary collisions or binary-single star collisions followed by an exchange of an ordinary star by a neutron star, are the most likely processes for the formation of these systems (cf. Hut 1983; 1984; Hut and Bahcall 1983; Rappaport et al. 1989; Romani et al. 1987; Verbunt et al. 1987). It will be clear that these processes may lead to X-ray binaries that may be quite different from those that formed through the ordinary processes of close binary evolution.

Indeed, the remnants of these X-ray binaries, the globular cluster binary radio pulsars, have distributions of orbital characteristics different from those of the binary radio pulsars in the galactic disk (e.g.: see Camilo et al. 2000). See further section 3.6.3.

3.6. FROM X-RAY BINARIES TO BINARY RADIO PULSARS: THE CONCEPT OF "RECYCLING"

3.6.1. *Introduction*
In the foregoing chapters we have seen that the double neutron stars and neutron-star binaries with massive white dwarf companions are the rem-

nants of HMXBs and intermediate-mass X-ray binaries, respectively, and that the binary pulsars with low-mass helium white dwarf companions (and other low- mass companions such as ablated hydrogen dwarfs) are the remnants of LMXBs. Evolutionary schemes leading to these different types of binary pulsars are given in: (i) Figure 17 for the double neutron stars and the close systems consisting of a neutron star and a massive white dwarf in a circular orbit, and (ii) Figure 21 for the neutron star plus low-mass helium white dwarf systems. The neutron stars in all these types of systems have gone through an extensive accretion phase, which particularly in the LMXBs was very long lasting ($\sim 10^8$ yrs or more). As this was disk accretion the neutron stars in these systems will have acquired a large amount of angular momentum. The observed spin-up behaviour of the X-ray pulsars that are accreting from a disk (see section 2.4 and Figure 8) shows indeed that one expects such neutron stars to evolve towards shorter and shorter spin periods. If at the same time the magnetic field has weakened to the values observed in the double neutron stars ($\sim 10^{10}$G) or the binary pulsars with white dwarf companions (10^8 to 10^{10} G) (see Figure 5) then, as equation (7) shows, these neutron stars can be spun up to periods of the order of 20 ms in the case of double neutron stars, and 1 millisecond in the case of sytems with helium white dwarf companions. There is overwhelming evidence that the magnetic fields in these binary neutron stars have decayed to the values quoted above. Possible reasons for this decay will be discussed in section 3.6.4.

3.6.2. *Evolution of a neutron star that is born in a close binary: Recycling*

The ingredients discussed in the last section allow us to describe the evolution in the surface dipole magnetic field strength (B_s) versus spin period (P) diagram of a neutron star that is born in a close binary, next to an unevolved stellar companion. This is depicted in Figure 22 (the diagram is in fact a different representation of the $\dot{P}$ vs. P diagram of Figure 5, as $B_s = $ const $(P\dot{P})^{0.5}$): The neutron star is born in the upper left of the diagram as a strongly magnetized rapidly spinning pulsar. It will within a few million years spin down along a horizontal track through the diagram and cross the "deathline". To the right of this line no radio pulsars can exist as the polar cap electric field has become too weak to create pairs, and no more pulsar particle wind can be produced, causing the pulsar process to stop (cf. Chen and Ruderman 1993; Björnsson 1996). The region to the right of the deathline is the "graveyard". During its stay in the graveyard its further spindown in the weak stellar wind of its companion (or other processes) will make its magnetic dipole moment decay (see below), causing it to move downwards in the B_s vs. P diagram. When finally the companion begins to overflow its Roche Lobe and begins to transfer matter through a disk, the rotation rate of the neutron star will rapidly increase, causing it to move towards shorter and shorter spin periods, i.e.: towards the left in the B_s vs. P diagram. Thus, as sketched in Figure 22, it will cross the deathline from the right and move again into the region of the "living" radio pulsars in the diagram. However, in this phase it is still an X-ray source, and will not generate radio pulses. It can at maximum be spun-up to the spin period corresponding to the "spin-up line" indicated in Figure 22, which corresponds to the shortest possible spin period which an accreting neutron star with a given dipole field strength B_s can attain, i.e.: for the maximum possible accretion rate, which is the Eddington rate. Thus, according to equation (7) the equation for the spin-up line is:

$$P_{min} = (2.4\,ms)(B_9)^{6/7} R_6^{16/7} M^{-5/7} \qquad (42)$$

Assuming R and M to be the same for all neutron stars, this gives the Spin-up line depicted in Figure 22. The importance of the spin-up line was first realized by Radhakrishnan and Srinivasan (1981; 1982; 1984), [see also Srinivasan and van den Heuvel 1982] and Alpar et al. (1982). Only after the accretion phase is terminated and the donor has itself become a compact object (neutron star or white dwarf), [or if the accretion process is otherwise interrupted, e.g. by the entering of the system into the "period gap", see section 3.4.1.F] the rapidly spinning magnetized neutron star will become observable as a radio pulsar again. Because of their return from the graveyard, such pulsars are called "recycled". [The term "recycled" was coined by Radhakrishnan]. The recycling model has been very successful in explaining the peculiar position of the bulk of the binary and millisecond radio pulsars in the $\dot{P}$ vs. P diagram (see Figure 5). The model predicts that the recycled pulsars can only be found between the spin-up line and the deathline, and this is indeed exactly the region in the $\dot{P}$ vs. P diagam where they are found. It also predicts that the companions should be dead stars (neutron stars or white dwarfs), which - with the exception of a few "evaporating" companions - is indeed the case. It furthermore predicts that the orbits of the double neutron stars must be eccentric (because of the second supernova in the system) whereas those of the recycled pulsars with white dwarf companions should be circular, because of the tidal circularization of the orbits during the long mass transfer phase. All of these predictions fit excellently with the observations, confirming "recycling" to be the cause of the formation of these two classes of binary radio pulsars.

3.6.3. *Further confirmation of the recycling model: the presence of the binary and millisecond radio pulsars in globular clusters*
As noticed in section 3.5, the incidence of Low-Mass X-ray Binaries in globular clusters is about 1000 times higher (per unit stellar mass) than

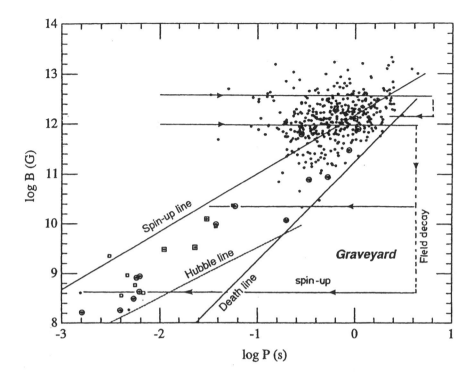

Figure 22.   Magnetic Field Strength ($\propto \sqrt{P.\dot{P}}$) versus pulse period diagram of radio pulsars with possible evolutionary tracks of pulsars in binaries indicated. Pulsars are born in the left upper part of the diagram and - if no field decay occurs - move towards the right along horizontal tracks (fully drawn). In the graveyard the field of a single pulsar probably does not decay (see text), but that of a binary neutron star does decay, presumably due to external circumstances (accretion, spin down). Furthermore, the accretion of matter with angular momentum causes these neutron stars to be 'spun up' towards the left in the diagram along the indicated lines until they reach the Spin-up line. After the companion has itself become a compact star (neutron star of white dwarf) or has disappeared, the spun-up neutron star becomes observable as a radio pulsar. It will then slowly spin-down, i.e. move towards the right again in the diagram. Dots indicate normal ("garden variety") non-recycled pulsars. Circles indicate some well-known radio pulsars in binaries, squares with dots are globular cluster binary pulsars, open squares are some single pulsars in globular clusters.

in the galaxy as a whole, indicating that the situation in globular clusters is extremely favourable for the formation of these objects. If the recycling picture is correct, one would therefore expect also many binary and millisec-ond pulsars to exist in globular clusters. With this idea in mind, A.G.Lyne started to survey globular clusters for such objects and found the first globular-cluster millisecond pulsar: a 3 ms pulsar in the cluster M28 (Lyne et al. 1987). Radio surveys have since turned up dozens of globular cluster radio pulsars, the majority of which are millisecond pulsars. Many are in close binary systems, just as predicted by the "recycling" picture. The globular clusters richest in radio pulsars are 47 Tuc and M15, in which 20 and 8 such objects have been detected, respectively (Anderson 1992, Manchester et al. 1991, Robinson et al. 1995, Camilo et al. 2000). In 47 Tuc they are all millisecond pulsars, in M15 the population is more varied, including slowly-spinning pulsars and a double neutron star (Prince et al. 1991), closely resembling the Hulse-Taylor binary pulsar. In 47 Tuc at least 13 of the 20 millisecond pulsars are in binaries. Calculations of globular cluster evolution, including dynamics, predict that the total neutron star pupulation of 47 Tuc is at least of order 1000, many of which are expected to be recycled (Rasio et al. 2000). Detection of millisecond pulsars in close binaries in clusters is difficult, due to the Doppler-variations of the pulse periods, and also since most clusters are distant and many of the pulsars are weak (Camilo et al. 2000). The observed sample is therefore likely to be just the "tip of the iceberg". Indeed, recent Chandra X-ray observations of 47 Tuc have revealed the presence of over 100 weak X-ray sources (10^{30}-10^{33} ergs/s) among which are all of the known radio pulsars (Grindlay 2000). As millisecond pulsars are known to be X-ray sources with these luminosities (cf. Becker and Truemper 1998), it is likely that most of the over one hundred faint X-ray sources in 47 Tuc are also millisecond radio pulsars. Also in several other globular clusters similar faint Chandra-X-ray sources have been found: 19 in NGC 6752 and 14 in NGC 6121 (B. Gaensler, priv. comm.). From their spectra, which are similar in hardness to those of the millisecond pulsars, the majority of them is also expected to be millisecond pulsars. Thus, the globular clusters have provided a most beautiful confirmation of the recycling piture, even though a few of the globular cluster pulsars may have had a more complex history, involving multiple stellar encounters.

3.6.4. *Why did the magnetic fields of the recycled neutron stars decay?*

We know that all the millisecond pulsars and most of the binary pulsars were recycled in binaries (the few single millisecond pulsars in the galactic disk as well as in globular clusters are also thought to have been recycled in binaries; for a review of possible formation models see Bhattacharya and Van den Heuvel 1991; van den Heuvel 1994a). Phenomenologically one may therefore either think that the field decay is somehow related either (i) to the accretion process that took place, or (ii) to the spindown to very long spin periods that presumably took place when the companion star was still unevolved and deeply inside its Roche lobe, and had only a very weak stellar wind. For both these processes models have been proposed. Space does not permit to discuss here the merits of these different models. We restrict ourselves therefore to some general statements.

As to field decay by accretion: models have been proposed in which the field is only confined to the neutron star crust, and there is no magnetic field in the superconducting interior; in that case, accretion can under certain conditions, lead the crustal field to decay (Bhattacharya 1995; Urpin et al. 1998 and references therein; Konar and Bhattacharya 1999). However, the discovery of the "magnetars", neutron stars with dipole fields of order 10^{15} gauss (Kouveliotou et al 1998; predicted by Thompson and Duncan 1995; see Thompson, this volume) makes it impossible to have the fields only in the crusts: they must thread the entire superconducting interior, as the pressures of these fields are too large to be solely sustained by the crust (Duncan, priv.comm). Therefore, it seems hard to make the total field decay by accretion onto the crust. If the field threads the superconducting interior the weakening of the outside field by accretion can only be achieved by "burying" the crustal field and hoping that it never comes out again. However, it seems very hard to "bury" the magnetic field by accretion for a long time (Cummings et al. 2001). This is because during the accretion new crust of the neutron star is continuously added, and the accreted hydrogen burns to helium, and then later, in thermonuclear flashes, to carbon and oxygen, which still later burns further to heavier elements. This creates a neutron star crust with many impurities, which causes its electrical conductivity to be low (Cummings et al. 2001; Bildsten, priv. Comm). This causes any buried field to re-emerge above the stellar surface within a few million years after the termination of the accretion phase. Also, due to the relatively short timescale of the field diffusion process through the crust, one can only successfully "bury" the crustal field temporarily if the accretion rate is not too low: at rates below 10^{-10} $M_{\odot}$/yr one expects there will be observable surface magnetic fields (Ruderman, priv.comm.). These considerations lead to two conclusions: (1) the internal (core) magnetic dipole moment of the millisecond pulsars must have been the internal (core) dipole moments of the neutron stars in LMXBs. (2) The internal (core) dipole moments of these stars must in some way have been weakened by a factor of order 10^4 during their lifetimes.

The most promising model for achieving this is the one proposed by Ruderman (e.g. see Ruderman 1998), which is based on the fact that in the superfluid and superconducting interior the magnetic flux is concentrated in quantized flux tubes which are strongly pinned to the quantized vortices in the rotating neutron superfluid. The density of fluxtubes per m^2 is proportional to the magnetic field strength and the density of vortex tubes is proportional to the neutron star's angular velocity of rotation. When the rotation of the neutron star slows down, the density of the quantized vortex tubes in the interior decreases, forcing the quantized vortex tubes to move outwards, carrying with them the magnetic flux tubes. The outwards moving vortices dissappear when they reach the crust, and the magnetic flux tubes they carried along will pile up below the inner crust. Because of the finite conductivity of the crust, the currents associated with the field will gradually dissipate, causing the field to decay. Thus, basically, the spindown of the star is the cause of the decay of the core field. Spindown to long rotation periods is expected in the progenitors of the LMXBs during the very long time interval before the donor star fills its Roche lobe (cf. Bhattacharya 1995), and thus the magnetic fields of these neutron stars are expected to have decayed very much. In the HMXBs the neutron stars had no more than about 10^7 years to spin down, such that given the diffusion timescale of several million years of the field through the crust, the field cannot have weakened much more than an order of magnitude, before spin-up occurred and the field-lines were dragged in again. Thus, in the pulsars recycled in HMXBs one does not expect the fields to have weakened to the same low values as in the LMXBs.

3.7. ORIGINS OF THE BLACK HOLES IN X-RAY BINARIES AND OF THE "MAGNETARS"

3.7.1. *Observational Evidence on Black hole formation*

The formation of the black-hole X-ray binaries is, in principle, similar to that of the neutron star X-ray binaries, the only difference being that the progenitor of the compact star left a black hole. The mass range in which stars leave black holes as remnants is not well known, but the X-ray binaries give some interesting information on this. Especially from the properties of the Low-Mass Black hole X-ray binaries (i.e. the ones with low-mass donor stars) one has been able to derive that the progenitors of the black holes in these systems cannot have been more massive than 20 to 25 $M_\odot$ (see the arguments given by Ergma and van den Heuvel 1998). Also, it has become clear that the formation of the black holes in these systems has been accompanied by considerable explosive mas ejection. This is inferred from: (i) the observed runaway velocities of the black hole X-ray binaries, which indicate that in the formation of the black holes in Cygnus X-1 and X-ray NOVA Sco 1994 at least several solar masses of matter must have been explosively ejected (Nelemans et al 1999); (ii) the observation of a large overabundance of alpha-process elements in the companion star of X-ray NOVA Sco 1994 (Israelian et al. 1999) which indicates that the progenitor of the black hole in that system was a Wolf-Rayet-like star that ejected at least several solar masses of elements ranging from C and O to S, Si and Ca (but no iron-group elements) during its formation.

In addition, the Gamma-Ray Bursts (GRB) have also given us important information on the formation of black holes. This holds particularly for the GRB of 25 April 1998, which appeared to coincide with a very peculiar supernova SN 1998bw in the spiral galaxy ESO 184-G82 at a distance of some 45 Mpc (Galama et al 1998). This supernova, of type Ic, turned out to be special in many ways: (i) it was the brightest radio-supernova ever seen, with a synchrotron radio spectrum, indicating the ejection of matter with mildly relativistic velocities (Lorentz factor of about 2), (ii) its lightcurve, energy and spectrum indicated that the exploding star must have been a C-O star with a mass larger than at least 6 solar masses, and most likely about 12 solar masses, with a collapsing core larger than 3 solar masses (Iwamoto et al. 1998). As the latter is above the upper mass limit for a stable neutron star, (Nauenberg and Chapline 1973, Rhoades and Ruffini 1973, Kalogera and Baym 1996) this indicates that this supernova was the birth event of a black hole, the first one ever observed.

The fact that the collapsing star here was a C-O star indicates that this must have been a so-called WC star (Carbon Wolf-Rayet star). To produce a C-O star of about 10 solar masses, as was the case here, taking into account that Wolf-Rayet stars lose mass in the form of a stellar wind at a rate 10^{-5} to 10^{-4} $M_\odot$/yr, implies that it must have started out as a helium star of at least 20 and more likely 25 $M_\odot$, which implies an original mass of the progenitor on the main sequence of at least 40 to 50 $M_\odot$.

We thus see that thanks to X- and Gamma-ray astronomy we now begin to get an idea of the mass range in which stars leave black holes as remnants: from some 20 to 25 up to at least 40 to 50 $M_\odot$.

3.7.2. *Fate of the most massive stars: formation of "magnetars"?*

It has been argued that due to the enormous stellar wind mass loss rates of the most massive stars, stars with masses above 50 to 60 $M_\odot$ have at the end of their evolution masses smaller than the stars from the "middle" group, in the range 20 to 50 $M_\odot$, and that therefore the stars more massive than about 60 $M_\odot$ might again leave neutron stars as remnants (cf. Wellstein and Langer 1999; Woosley et al. 1993; 1995).

This is an interesting proposal, that possibly might receive support from the discovery of the "magnetars": neutron stars with magnetic fields in the range 10^{14} to 10^{15} G (Kouveliotou et al. 1998). These objects were discovered as so-called Soft Gamma Ray Repeaters (SGRs), soft gamma ray sources that give recurrent outbursts. For a review see Thompson (this volume). These are objects in our galaxy, very close to the galactic plane, indicating that they are very young, and related to massive stars. It has been found that two of these objects coincide with very young clusters of very massive stars, in one case containing a Luminous Blue Variable (LBV) with a luminosity of 5×10^6 L$_\odot$ (Fuchs et al. 1999; Vrba et al. 2000; Eikenberry and Dror 2000). To have such a luminosity the star must have had an initial mass of > 100 M$_\odot$. The fact that the SGR is within one parsec from the cluster center indicates that it belongs to this cluster. As it is already in a more advanced evolutionary phase than the LBV, it must have started out as an even more massive star, which implies that it then is the remnant of a star more massive than 100 M$_\odot$. This seems to confirm the idea that the most massive stars again leave neutron stars as remnants, though neutron stars of a very special type, i.e with extremely large magnetic fields. It seems too early to draw definitive conclusions about this all, but this evidence is at least suggestive.

Acknowledgements

I wish to thank Jan van Paradijs, Chryssa Kouveliotou, Chris Thompson, Robert Duncan, Lars Bildsten, Thomas Tauris and Mal Ruderman for many enlightening discussions. This research was supported by the Netherlands Research Organization NWO through a Spinoza Grant to the author and by Grant Nr. PHY 99-07949 to the ITP-UCSB, Santa Barbara.

References

Alpar, M.A., Cheng, A.F., Ruderman, M.A., and Shaham, J. (1982), *Nature* **300**, 728.
Anderson, S.B. (1992), *Ph.D. Thesis*, California Institute of Technology.
Apparao, K.M.V. (1994), *Space Science Reviews* **69**, 255.
Armitage, P.J. and Livio, M. (2000), *ApJ* **532**, 540.
Arnett, W.D. (1973), *ApJ* **179**, 249.
Baade, W. and Zwicky, F. (1934), *Phys. Rev.* **45**, 138.
Becker, W. and Truemper, J. (1998), in R. Buccheri, J.A. van Paradijs, and A. Alpar (eds.), *The Many Faces of Neutron Stars*, Kluwer (Dordrecht), pp525-537.
Belloni, T., Mendez, M., King, A.R., van der Klis, M., and van Paradijs, J.A. (1997), *ApJ* **479**, L145.
Bhattacharya, D. and van den Heuvel, E.P.J. (1991), *Phys.Rep.* **203**, 1.
Bhattacharya, D. (1995), *J.Astroph.& Astron.* **16**, 217.
Bhattacharya, D. and Srinivasan, G. (1995), in W.H.G. Lewin, J.A. van Paradijs, and E.P.J. van den Heuvel (eds.), *X-Ray Binaries*, Cambridge Univ. Press, p495.
Bildsten, L. et al. (1997), *ApJ Supp.* **113**, 367.
Bisnovatyi-Kogan, G.S. and Komberg, B.V. (1974), *Astron. Zh.* **51**, 373.
Bisnovatyi-Kogan, G.S. and Komberg, B.V. (1975), *Sov. Astron.* (English translation) **18**, 217.
Björnsson, C.-I. (1996), *ApJ* **471**, 321.
Blaauw, A. (1961), *Bull. Astron. Inst. Netherlands* **15**, 265.
Bodenheimer, P. and Taam, R.E. (1984), *ApJ* **280**, 771.
Bolton, C.T. (1972), *Nature* **235**, 271.
Boriakoff, V., Buccheri, R., and Fauci, F. (1983), *Nature* **304**, 417.
Bradt, H.V.D. and McClintock, J.E. (1983), *Ann.Rev.Astr.Ap.* **21**, 13.
Braes, L. and Miley, G. (1971), *Nature* **232**, 246.
Brown, G.E. (1995), *ApJ* **440**, 270.
Brown, G.E., Lee, C.-H., Portegies Zwart, S.F., and Bethe, H. (2000), (preprint).
Börner, G., Meyer, F., Schmidt, H.U., and Thomas, H.-C. (1972), *paper presented at the*

meeting of the Astron. Gesellschaft, Wien, Austria (abstract)

Camilo, F. (2000), *private comm.*

Camilo, F., Lorimer, D.R., Freire, P., Lyne, A.G., and Manchester, R.N. (2000), *ApJ* **535**, 975.

Chen, K.-Y. and Ruderman, M.A. (1993), *ApJ* **402**, 264.

Chevalier, R.A. (1993), *ApJ* **411**, L33.

Chevalier, R.A. (1996), *ApJ* **459**, 322.

Clark, G.W. (1975), *ApJ* **199**, L143.

Corbet, R.H.D. (1984), *A&A* **141**, 91.

Counselman, C.C. (1973), *ApJ* **180**, 307.

Cox, J.P. and Giuli, R.T. (1968), *Principles of Stellar Structure I & II* Gordon and Breach, pp1327.

Cummings, A., Zweibel, E., and Bildsten, L. (2001), *ApJ* (in press).

Darwin, G.H. (1879), *Proc. R. Soc. London* **29**, 168.

Davidson, K. and Ostriker, J.P. (1973), *ApJ* **179**, 585.

De Loore, C., de Greve, J.P., and de Cuyper, J.P. (1975), *Ap. Sp. Sci.* **36**, 219.

Dewey, R.J. and Cordes, J.M. (1987), *ApJ* **321**, 780.

Eggleton, P.P. (1983), *ApJ* **268**, 368.

Eikenberry, S.S. and Dror, D.H. (2000), *ApJ* **537**, 429.

Ergma, E. and van den Heuvel, E.P.J. (1998), *A&A* **331**, L29.

Ergma, E. and Yungelson, L.R. (1998), *A&A* **333**, 151.

Fabian, A.C., Pringle, J.E., and Rees, M.J. (1975), *MNRAS* **172**,

Faulkner, J. (1971), *ApJ* **170**, L99.

Finger, M. (1998), in R. Buccheri, J.A. van Paradijs, and A. Alpar (eds.), *The Many Faces of Neutron Stars*, Kluwer (Dordrecht), pp369-384.

Flannery, B.P. and van den Heuvel, E.P.J. (1975), *A&A* **39**, 61.

Fruchter, A.S., Stinebring, D.R., and Taylor, J.H. (1988), *Nature* **333**, 237.

Fryxell, B. and Taam, R.E. (1988), *ApJ* **335**, 862.

Fuchs, Y., Mirabel, F., Chaty, S., Claret, A., Cesarsky, C.J., and Cesarsky, D.A. (1999), *A&A* **350**, 891.

Galama, T.J., Vreeswijk, P.M., van Paradijs, J.A. et al. (1998), *Nature* **395**, 670.

Ghosh, P. and Lamb, F.K. (1979), *ApJ* **232**, 259.

Ghosh, P. and Lamb, F.K. (1979), *ApJ* **234**, 296.

Giacconi, R., Gursky, H., Paolini, F.R., and Rossi, B.B. (1962), *Phys. Rev. Letters* **9**, 439.

Gold, T. (1969), *Nature* **221**, 25.

Gottlieb, E.W., Wright, E.L., and Liller, W. (1975), *ApJ* **195**, L33.

Grindlay, J. (2000), *paper presented at ITP conference on Spin and Magnetism in Young Neutron Stars, UC Santa Barbara.*

Gursky, H. (1973), *presented at NATO Adv. Study Inst. on Neutron Stars and Black Holes, Cambridge UK (unpublished).*

Guseinov, O. and Zeldovitch, Ya.B. (1966), *Astr. Zh.* **43**, 313.

Habets, G.M.H.J. (1985), *Ph.D. Thesis, Advanced Evolution of Helium Stars and Massive Close Binaries*, University of Amsterdam.

Habets, G.M.H.J. (1986), *A&A* **187**, 209.

Hanson, M.M., Still, M.D., and Fender, R.P. (2000), *ApJ* **541**, 308.

Hartman, J.W. (1996), in S. Johnston, M.A. Walker, and M. Bailes (eds.), *Pulsars, Problems and Progress* ASP Conf. series 105, p53.

Hartman, J.W. (1997), *A&A* **322**, 127.

Hewish, A., Bell, S.J., Pilkington, J.D., Scott, P.F., and Collins, R.A. (1968), *Nature* **217**, 709.

Hjellming, R.M. and Wade, C. (1971), *ApJ* **168**, L21.

Hulse, A.R. and Taylor, J.H. (1975), *ApJ* **191**, L51.

Hut, P. (1983), *Astron. J.* **88**, 1549.

Hut, P. and Bahcall, J. N. (1983), *ApJ* **268**, 319.

Hut, P. (1984), *ApJ Supp.* **55**, 301.

Iben, I. (1967), *Ann.Rev.Astron.Ap.* **5**, 571.

Israelian, G., Rebolo, R., Basri, G., Casares, J., and Martin, E.L. (1999), *Nature* **401**, 142.

Iwamoto, K., Mazzali, P. A., Nomoto, K., Umeda, H., Nakamura, T. et al. (1998), *Nature* **395**, 672.

Joss, P.C. and Rappaport, S.A. (1983), *Nature* **304**, 419.

Joss, P.C., Rappaport, S.A., and Lewin, W.G.H. (1987), *ApJ* **319**, 180.

Kahabka, P. and van den Heuvel, E.P.J. (1997), *Ann. Rev. Astron. Ap.* **35**, 69.

Kalogera, V. and Baym, G. (1996), *ApJ* **470**, L61.

Kalogera, V. (1998), in R. Buccheri, J.A. van Paradijs, and A. Alpar (eds.), *The Many Faces of Neutron Stars*, Kluwer (Dordrecht), pp505-551.

Kalogera, V. (1998), *ApJ* **493**, 368.

Kalogera, V. and Webbink, R.F. (1998), *ApJ* **493**, 351.

Katz, J.I. (1975), *Nature* **253**, 698.

Kawaler, S.D. (1998), in S.D. Kawaler, I. Novikov, and G. Srinivasan (eds.), *Stellar Remnants*, Springer (Heidelberg), pp1-95.

King, A.R., Kolb, U., and Suszkewicz, E. (1997), *ApJ* **488**, 89.

King, A.R. (1998), *MNRAS* **269**, L45.

King, A.R. and Begelman, M.C. (1999), *ApJ* **519**, 169.

King, A.R. and Ritter, H. (1999), *MNRAS* **309**, 253.

King, A.R. (2000), *MNRAS* **312**, L39.

King, A.R., Taam, R.E., and Begelman, M.C. (2000), *ApJ* **530**, L25.

Kippenhahn, R. and Weigert, A. (1967), *Zeitsch. Ap.* **65**, 251.

Kolb, U., Davies, M.B., King, A.R., and Ritter, H. (2000), *MNRAS* **317**, 438.

Konar, S. and Bhattacharya, D. (1999), *MNRAS* **308**, 795.

Kouveliotou, C., Dieters, S., Strohmayer, T., van Paradijs, J.A., Fishman, G.J., Meegan, C. A., Hurley, K., Kommers, J., Smith, I., Frail, D., and Murakani, T. (1998), *Nature* **393**, 235.

Kristian, J., Brucato, R., Visvanatan, N., Lanning, H., and Sandage, A. (1971), *ApJ* **168**, L91.

Lamb, F.K., Pethick, C.J., and Pines, D. (1973), *ApJ* **182**, 271.

Lewin, W.H.G. and Joss, P.C. (1983), in W.H.G. Lewin and E.P.J. van den Heuvel (eds.), *Accretion Driven Stellar X-Ray Sources*, Cambridge Univ. Press, pp41-115.

Lewin, W.H.G., van Paradijs, J.A., and Taam, R.E. (1995), in W.H.G. Lewin, J.A. van Paradijs, and E.P.J. van den Heuvel (eds.), *X-Ray Binaries*, Cambridge Univ. Press, p175.

Liu, Q.Z. (2001), *private communication*.

Lyne, A.G., Brinklow, A., Middleditch, J., Kulkarni, S.R., Backer, D.C., and Clifton, T.R. (1987), *Nature* **328**, 399.

Lyne, A.G. and Smith, F.G. (1990), *Pulsar Astronomy*, Cambridge Univ. Press, p274.

Lyne, A.G. and Lorimer, D.R. (1994), *Nature* **369**, 127.

Maeder, A. and Meynet, G. (1989), *A&A* **210**, 155.

Manchester, R.N. and Taylor, J.H. (1977), *Pulsars*, Freeman (San Francisco), p281.

Manchester, R.N., Lyne, A.G., Robinson, C., D'Amico, N., Bailes, M., and Lim, J. (1991), *Nature* **352**, 2191.

Maraschi, L., Treves, A., and van den Heuvel, E.P.J. (1976), *Nature* **259**, 292.

Margon, B. (1983), in W.H.G. Lewin and E.P.J. van den Heuvel (eds.), *Accretion Driven Stellar X-Ray Sources*, Cambridge Univ. Press, p287.

McClintock, J.E. and Remillard, R.A. (1986), *ApJ* **308**, 110.

McClintock, J.E. (1992), in E.P.J. van den Heuvel and S.A. Rappaport (eds.), *X-ray Binaries and Recycled Pulsars*, Kluwer, (Dordrecht), p27.

Meyer, F. and Meyer-Hofmeister, E. (1978), *A&A* **78**, 167.

Monaghan, J. (1969), unpublished (manuscript submitted to Nature, but not accepted).

Morton, D.C. (1960), *ApJ* **132**, 146.

Morton, D.C. (1967), *ApJ* **147**, 1017.
Nauenberg, M. and Chapline, G. (1973), *ApJ* **197**, 277.
Nelemans, G., Tauris, T.M., and van den Heuvel, E.P.J. (1999), *A&A* **352**, L87.
Nomoto, K. (1984), *ApJ* **277**, 791.
Nomoto, K. and Kondo, Y. (1991), *ApJ* **367**, L19.
Novikov, I.D. and Zeldovitch, Ya.B. (1966), *Nuova Cim. Sup.* **4**, 827.
Oda, M., Gorenstein, P., Gursky, H., Kellogg, E., Schreier, E., Tananbaum, H., and
    Giacconi, R. (1971), *ApJ* **166**, L1.
Oppenheimer, J. and Volkoff, G. (1938), *Phys. Rev.* **55**, 374.
Oppenheimer, J. and Snijder, H. (1939), *Phys. Rev.* **56**, 455.
Ostriker, J.P. (1973), *private communication to B. Paczynski*
Paczynski, B. (1966), *Acta Astron.* **16**, 231.
Paczynski, B. (1967), *Acta Astron.* **17**, 355.
Paczynski, B. (1971), *Acta Astron.* **21**, 1.
Paczynski, B. (1971), *Ann. Rev. Astron. Ap.* **9**, 183.
Paczynski, B. (1976), in P. Eggleton et al. (eds.), *Structure and Evolution of Close Binary
    Systems*, Reidel (Dordrecht), pp75-80.
Paczynski, B. and Sienkiewicz, R. (1981), *ApJ* **248**, L27.
Plavec, M.J. (1967), *Commun. Obs. R. Belgique, Uccle* **B17**, 83.
Podsiadlowski, P. and Rappaport, S. (2000), *ApJ* **529**, 946.
Pols, O. (2000), *MNRAS* (in press).
Pooley, G.G., Fender, R.P., and Brocksopp, C. (1999), *MNRAS* **302**, L1.
Prince, T.A., Anderson, S.B., Kulkarni, S.R., and Wolszczan, A. (1991), *ApJ* **374**, L41.
Radhakrishnan, V. and Srinivasan, G. (1981), in B. Hidayat and W. Feast (eds.), *Proc.
    2nd Asia-Pacific Regional Meeting of the IAU Bandung Tira Pustaka*, (Jakarta 1984),
    p423.
Radhakrishnan, V. and Srinivasan, G. (1982), *Current Science* **51**, 1096.
Rappaport S. and van den Heuvel E.P.J. (1982), *IAU Symp.* **98**, 327.
Rappaport, S.A., Joss, P.C., and Webbink, R.F. (1982), *ApJ* **254**, 616.
Rappaport, S.A., Verbunt, F., and Joss, P.C. (1983), *ApJ* **275**, 713.
Rappaport, S.A., Putney, A., and Verbunt, F. (1989), *ApJ* **345**, 210.
Rappaport, S.A. et al. (1995), *MNRAS* **273**, 731.
Rasio, F.A. and Livio, M. (1996), *ApJ* **471**, 366.
Rasio, F.A., Pfahl, E.D., and Rappaport, S.A. (2000), *ApJ* **532**, L47.
Rhoades, C.E. and Ruffini, R. (1974), *Phys.Rev.Lett.* **32**, 324.
Ritter, H. and Kolb, U. (1998), *A&A Supp.* **129**, 83.
Robinson, C.R., Lyne, A.G., Manchester, R.N., Bailes, M., D'Amico, N., and Johnston,
    S. (1995), *MNRAS* **274**, 547.
Romani, R.W., Kulkarni, S.R., and Blandford, R.D. (1987), *Nature* **329**, 309.
Ruderman, M., Shaham, J., Tavani, M., and Eichler, D. (1989), *ApJ* **343**, 292.
Ruderman, M. (1998), in R. Buccheri, J.A. van Paradijs, and A. Alpar (eds.), *The Many
    Faces of Neutron Stars*, Kluwer (Dordrecht), p77.
Salpeter, E.E. (1964), *ApJ* **140**, 796.
Sandage, A.R., Osmer, P., Giacconi, R. et al. (1966), *ApJ* **146**, 316.
Savonije, G.J. (1978), *A&A* **62**, 317.
Savonije, G.J. (1983), in W.H.G. Lewin and E.P.J. van den Heuvel (eds.), *Accretion
    Driven Stellar X-Ray Sources*, Cambridge Univ. Press, pp343-366.
Schreier, E., Levinson, R., Gursky, H., Kellogg, E., Tananbaum, H., and Giacconi, R.
    (1972), *ApJ* **172**, L79.
Shapiro, S.L. and Teukjolsky, S.A. (1983), *Black Holes, White Dwarfs and Neutron Stars*,
    John Wiley & Sons (New York), p645.
Shklovski, I. (1967), *ApJ* **148**, L1.
Smarr, L.L. and Blandford, R.D. (1975), *ApJ* **207**, 574.
Sobermann, G.E., Phinney, E.S., and van den Heuvel, E.P.J. (1997), *A&A* **327**, 620.
Sollerman, J., Kozma, C., Fransson, C., Leibuntgut, B., Lundquist, P., Ryde, F., and

Woudt, P. (2000), *ApJ* **537**, L127.

Spruit, H.C. and Ritter, H. (1983), *A&A* **124**, 267.

Srinivasan, G. and van den Heuvel, E.P.J. (1982) *A&A* **108**, 143.

Srinivasan, G. (1998), in S.D. Kawaler, I. Novikov, and G. Srinivasan (eds.), *Stellar Remnants*, Springer (Heidelberg), pp97-235.

Staelin, D.H. and Reifenstein, E.C. III (1968), *IAU Circ. 2110*.

Sutantyo, W. (1975a), *A&A* **41**, 47.

Sutantyo, W. (1975b), *A&A* **44**, 227.

Sutantyo, W. (1992), in E.P.J. van den Heuvel and S.A. Rappaport (eds.), *X-Ray Binaries and Recycled Pulsars*, Kluwer (Dordrecht), p293.

Sutantyo, W. (1999), *A&A* **344**, 505.

Taam, R.E. (1983), *ApJ* **270**, 694.

Taam, R.E. and van den Heuvel, E.P.J. (1986), *ApJ* **305**, 235.

Taam, R.E. and Fryxell, B.A. (1988), *ApJ* **327**, L73.

Taam, R.E. and Bodenheimer, P. (1991), *ApJ* **373**, 246.

Taam, R.E. and Sandquist, E.L. (2000), *Annual Rev. Astron. Ap.*, 113.

Tanaka, Y. and Lewin W.H.G. (1995), in W.H.G. Lewin, J.A. van Paradijs, and E.P.J. van den Heuvel (eds.), *X-Ray Binaries*, Cambridge Univ. Press, p126.

Tananbaum, H., Gursky, H., Kellogg, E., Levinson, R., Schreier, E., and Giacconi, R. (1972), *ApJ* **174**, L143.

Tananbaum, H. (1973), in H. Bradt and R. Giacconi (eds.), *X- and Gamma-Ray Astronomy*, Reidel (Dordrecht), p9.

Tauris, T. and Savonije, G.J. (1999), *A&A* **350**, 928.

Tauris, T. and van den Heuvel, E.P.J. (2000), in M. Kramer et al. (eds.), *Pulsar Astronomy - 2000 and beyond*, IAU Coll.177, ASP CS 202, pp595.

Tauris, T., van den Heuvel, E.P.J., and Savonije, G.J. (2000), *ApJ* **530**, L93.

Tauris, T.M. and Sennels, T. (2000), *A&A* **355**, 236.

Thompson, C. and Duncan, R. (1995), *MNRAS* **275**, 255.

Trümper, J. et al. (1978), *ApJ* **219**, L105.

Tutukov, A.V. and Yungelson, L.R. (1973), *Nautsnie Informatsie* **27**, 58.

Urpin, V., Konenkov, D., and Geppert, U. (1998), *MNRAS* **299**, 73.

Van den Heuvel, E.P.J. and Heise, J. (1972), *Nature* **239**, 67.

Van den Heuvel, E.P.J. and de Loore, C. (1973), *A&A* **25**, 387.

Van den Heuvel, E.P.J. (1975), *ApJ* **198**, L109.

Van den Heuvel, E.P.J. (1977), in *Proc. 7th Texas Symp. Rel. Ap.* Annals N.Y. Acad. Sci. **302**, p14.

Van den Heuvel, E.P.J. (1981), in D. Sugimoto et al. (eds.), *Fundamental Problems in the Theory of Stellar Evolution*, Reidel (Dordrecht), p155.

Van den Heuvel, E.P.J. (1983), in W.H.G. Lewin and E.P.J. van den Heuvel (eds.), *Accretion Driven Stellar X-Ray Sources*, Cambridge Univ. Press.

Van den Heuvel, E.P.J. and Rappaport, S. (1987), *IAU Colloq. 92*, 291.

Van den Heuvel, E.P.J. and van Paradijs, J.A. (1988), *Nature* **334**, 227.

Van den Heuvel, E.P.J. (1992), in E.P.J. van den Heuvel and S.A. Rappaport (eds.), *X-Ray Binaries and Recycled Pulsars*, Kluwer (Dordrecht), pp233-56.

Van den Heuvel, E.P.J. (1994a), in H. Nussbaumer and A. Orr (eds.), *Interacting Binaries*, Springer (Heidelberg), pp263-474.

Van den Heuvel, E.P.J. (1994b), in S.S. Holt and C.S. Day (eds.), *The Evolution of X-ray Binaries*, AIP Conf.Proc. 308, p18.

Van den Heuvel, E.P.J. (1995), *J. of Astroph.& Astron.* **16**, 255

Van den Heuvel, E.P.J. (1996), in S. Johnston, M.A. Walker, and M. Bailes (eds.), *Pulsars, Problems and Progress* ASP Conf. series 105, pp557-559.

Van den Heuvel, E.P.J. and van Paradijs, J.A. (1997), *ApJ* **483**, 399.

Van den Heuvel, E.P.J., Portegies Zwart, S., Bhattacharya, D., and Kaper, L. (2000), *A&A* **364**, 563.

Van den Heuvel, E.P.J. (2001), in J. Ventura et al. (eds.) *Proc. NATO, ASI on Neutron*

Stars and Black Holes (in press), Kluwer (Dordrecht).

Van der Hucht, K.A. (1999), in K.A. van der Hucht et al. (eds.), *Wolf-Rayet Phenomena in Massive Stars and Starburst*, ASP. Conf. series (San Fransisco), p13.

Van Kerkwijk, M.H., Charles, P.H., Geballe, T.R., King, D.L., Miley, G., Molnar, L.A., and van den Heuvel, E.P.J. (1972), *Nature* **355**, 703.

Van Kerkwijk, M.H. and Kulkarni, S.R. (1999), *ApJ* **516**, L25.

Van Paradijs, J.A. (1978), *Nature* **274**, 650.

Van Paradijs, J. (1995), W.H.G. Lewin et al. (eds.), *X-ray Binaries*, Cambridge Univ. Press, pp536-577.

Van Paradijs, J.A. and McClintock, J.E. (1995), in W.H.G. Lewin, J.A. van Paradijs, and E.P.J. van den Heuvel (eds.), *X-Ray Binaries*, Cambridge Univ. Press, p58.

Van Paradijs, J.A. (1998), R. Buccheri, J.A. van Paradijs, and A. Alpar (eds.), *The Many Faces of Neutron Stars*, Kluwer (Dordrecht), pp279-336.

Van Paradijs, J.A. (1999), *Science* **286**, 693.

Van Paradijs, J.A. et al. (2000), *Ann. Rev. Astron. Ap.* **38**, 379.

Verbunt, F. and Zwaan, C. (1981), *A&A* **100**, L7.

Verbunt, F., van den Heuvel, E.P.J., van Paradijs, J.A., and Rappaport, S.A. (1987), *Nature* **329**, 312.

Verbunt, F. (1990), *Neutron Stars and Their Birth Events* Kluwer (Dordrecht), p179.

Verbunt, F. and van den Heuvel, E.P.J. (1995), in W.H.G. Lewin, J.A. van Paradijs, and E.P.J. van den Heuvel (eds.), *X-Ray Binaries*, Cambridge Univ. Press, 457.

Verbunt, F. (2001), *private comm.*

Vrba, F.J., Henden, A.A., Luginbuhl, C.B., Guetter, H.H., Hartmann, D.H., and Klose, S. (2000), *ApJ* **533**, L17.

Waters, L.B.F.M. and van Kerkwijk, M.H. (1989), *A&A* **223**, 196.

Webbink, R.F., Rappaport, S.A., and Savonije G.J. (1983), *ApJ* **270**, 678.

Webbink, R.F. (1984), *ApJ* **277**, 355.

Webbink, R.F. and Kalogera, V. (1994), in S.S. Holt and C.S. Day (eds.), *The Evolution of X-ray Binaries*, AIP Conf.Proc. 308, p321.

Webster, B.L. and Murdin, P. (1972), *Nature* **235**, 37.

Wellstein, S. and Langer, N. (1999), *A&A* **350**, 148.

Wijnands, R. and van der Klis, M. (1998), *Nature* **394**, 344.

Woosley, S.E., Langer, N., and Weaver, T.A. (1993), *ApJ* **823**, 411.

Woosley, S.E., Langer N., and Weaver, T.A. (1995), *ApJ* **448**, 315.

Woosley, S. (2001), in E.P.J. van den Heuvel, L. Kaper, and P.A. Woudt (eds.), *Black Holes in Binaries and Galactic Nuclei*, Springer-Verlag (Heidelberg).

Zeldovitch, Ya.B. (1964), *Sov.Phys.Dokl.* **9**, p195 and p246

Zeldovitch, Ya.B. and Novikov, I.D. (1964), *Dokl.Acad.Nauk.USSR* **155**, 678.

Zeldovitch, Ya.B. and Novikov, I.D. (1964), *Dokl.Acad.Nauk.USSR* **158**, 811.

Zeldovitch, Ya.B. and Guseinov, O. (1966), *ApJ* **144**, 840.

TRANSIENT LOW-MASS X-RAY BINARIES IN QUIESCENCE

L. BILDSTEN
Institute for Theoretical Physics and Department of Physics
Kohn Hall, University of California, Santa Barbara
Santa Barbara, CA 93106

AND

R.E. RUTLEDGE
Space Radiation Laboratory, MS 220-47
Caltech, Pasadena, CA 91125

Abstract. We summarize the quiescent X-ray observations of transient low-mass X-ray binaries. These observations show that, in quiescence, binaries containing black holes are fainter than those containing neutron stars. This has triggered a number of theoretical ideas about what causes the quiescent X-ray emission. For black hole binaries, the options are accretion onto the black hole or coronal emission from the rapidly rotating stellar companion. There are more possibilities for the neutron stars; accretion, thermal emission from the surface or non-thermal emission from a "turned-on" radio pulsar. We review recent theoretical work on these mechanisms and note where current observations can distinguish between them. We highlight the re-analysis of the quiescent neutron star emission by Rutledge and collaborators that showed thermal emission to be a predominant contributor in many of these systems. Our knowledge of these binaries is bound to dramatically improve now that the *Chandra* and *XMM-Newton* satellites are operating successfully,

1. Introduction

Many black holes and neutron stars are in binaries where a steady-state accretion disk (one that supplies matter to the compact object at the same rate as mass is donated from the Roche-lobe filling companion) is thermally unstable (Van Paradijs 1996; King et al. 1996). This instability results in a

C. Kouveliotou et al. (eds.), The Neutron Star – Black Hole Connection, 245–259.

limit cycle – as in dwarf novae (where the compact object is a white dwarf) – with matter accumulating in the outer disk for months to decades until a thermal instability is reached (Huang and Wheeler 1989; Mineshige and Wheeler 1989) that triggers rapid accretion onto the compact object. The substantial brightening in the X-rays (typically to levels near the Eddington limit, $10^{38} - 10^{39}$ erg s^{-1} for $M = 1 - 10 M_\odot$ stars) brings attention to these otherwise previously unknown binaries. Both neutron stars (NS) and black holes (BH) exhibit these X-ray outbursts, separated by periods ($\sim$ months to decades) of relative quiescence (for recent reviews of the outburst properties, see Tanaka and Lewin 1995; Tanaka and Shibazaki 1996; Chen et al. 1997). The neutron stars are identified by Type I X-ray bursts from unstable thermonuclear burning on their surfaces. For those that "appear" to be black holes (based on their spectral and/or timing properties and lack of Type I bursts), detailed optical spectroscopy in quiescence is undertaken. Many of the measured optical mass functions are in excess of the maximum possible neutron star mass ($\approx 3 M_\odot$), making these binaries an excellent hunting ground for black holes (see McClintock 1998 for a summary).

Our purpose is to discuss the X-ray emission from these binaries when they are in their faint "quiescent" state between outbursts. X-ray observations with sensitive pointed instruments (ROSAT and ASCA) of these transients in quiescence have detected all of those harboring neutron stars and some that contain black holes. It is clear that, on average, the binaries containing black holes are less luminous than those with neutron stars (Barret et al. 1996; Narayan et al. 1997a; Asai et al. 1998). It is still a mystery as to what powers the very faint X-ray emission ($L_x \ll 10^{35}$ erg s^{-1}) from these binaries when in quiescence, and we will review the possibilities here.

Accretion is the most often discussed energy source and clearly powers the dwarf novae (the analogous systems that contain white dwarfs in systems with orbital periods typically less than three hours) in quiescence. These were found by *Einstein* to be faint X-ray sources ($10^{30} - 10^{32}$ erg s^{-1}) when in their quiescent state (Cordova and Mason 1984; Patterson and Raymond 1985). The inferred accretion rate onto the white dwarf is a few percent of the rate being transferred within the binary, and the X-rays originate from the boundary layer near the white dwarf (Patterson and Raymond 1985). This was confirmed via eclipse observations with ROSAT of three short orbital period DN in quiescence (Mukai et al. 1997; Van Teeseling 1997; Pratt et al. 1999). In all of these systems, the X-ray emission was eclipsed when the white dwarf was behind the companion. The physics that sets this low inflow rate towards the white dwarf is not clear and might well be different than in the binaries containing neutron stars and black holes, which are the focus of this review.

2. Black Hole Transients in Quiescence: Advection Dominated Accretion Flows or Coronal Emission?

At this time, three BHs have been detected in quiescence: A0620−00 (Mc-Clintock et al. 1995), GS 2023+33 (Verbunt et al. 1994; Wagner et al. 1994), and GRO J1655−40 (Hameury et al. 1997). The puzzle of the emission mechanism began when ROSAT/PSPC detected X-rays from A0620−00 at a level $L_x \approx 6 \times 10^{30}$ erg s^{-1} (McClintock et al. 1995). McClintock et al. (1995) made it clear that this X-ray emission could not be due to a steady-state accretion disk around the black hole, as if so, there would be a production of optical and UV photons from the outer parts of the disk that would far exceed that observed.

To solve this puzzle and to explain the higher quiescent luminosities ($10^{32} - 10^{33}$ erg s^{-1}) of the transient NSs, Narayan et al. (1997b) invoked an advection-dominated accretion flow (ADAF) onto the compact object at a rate $\dot{M}_q$ in quiescence. For the black holes, the X-rays are produced via Compton up-scattering of the optical/UV synchrotron emission from the inner parts of the flow. The model predicts X-ray emission as a fraction of the optical/UV emission (in excess of that from the stellar companion) – an observable ratio which is used to evaluate the model's success. Current ADAF spectral modeling of the X-ray detected BH's requires that $\dot{M}_q$ be $\sim 1/3$ of the total mass transfer rate in the binary (Narayan et al. 1997a). Accretion rates this high are required because of the relative inefficiency of the flow at producing X-rays (Narayan et al. 1997b; Hameury et al. 1997). The much higher efficiency (by 3-4 orders of magnitude) of X-ray production from accretion onto a neutron star forces the quiescent accretion rate onto these objects to be much lower than for the black holes. In other words, if the implied $\dot{M}_q$ from BHs was landing on a NS, it would shine in quiescence at about 10^{36} erg s^{-1}, a factor of 1000 brighter than observed. A solution is to just dial $\dot{M}_q$ to be uniformly lower in the NS systems than in the BH systems; though this is not easy to do (Menou et al. 1999).

Another possible mechanism for the faint emission from black hole binaries is coronal X-ray emission from the tidally locked companion star (Verbunt 1996; Bildsten and Rutledge 2000). The analogous systems are tidally locked stars in tight binaries, such as the RS CVn systems. These have X-ray luminosities from coronal activity that reach the level observed from the black hole transients. For a convective star that is rotating rapidly, most X-ray observations point to $L_x/L_{bol} \approx 10^{-3}$ as a "saturation limit" in coronal X-ray emission (Vilhu and Walter 1987; Singh et al. 1999).

In Figure 1, we display the X-ray detections and upper-limits for several observed BH systems, with their optically-derived bolometric flux, along with this saturation limit in L_x/L_{bol}, and the prospects of detecting coronal X-rays from undetected systems with *Chandra* (X-ray flux limits for *XMM-*

Newton are about a factor of three lower). Of the previously undetected sources, only GS 2000+25 stands out as a possible new detection of stellar coronal emission (the companion of 4U 1543−47 is not convective, and thus no coronal emission is expected).

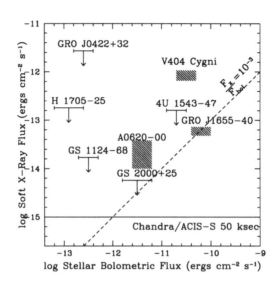

Figure 1. X-ray flux vs. stellar bolometric flux for X-ray detected black hole binaries (shaded regions) and 2σ upper-limits. The dashed line shows a typical coronal value $F_x/F_{bol}=10^{-3}$. The bottom line is the *Chandra*/ACIS-S 50 ksec detection limits for a $N_H=0.2\times10^{22}$ cm^{-2}, Raymond-Smith model at $kT=1.0$ keV. Detection of X-rays from GRO J0422+32, GS 1124-68, and H 1705−25 with *Chandra* would be well above that expected from coronal emission. GS 2000+25 is close to the $L_x/L_{bol}=10^{-3}$ limit and may be detected. While 4U 1543−47 is apparently well within X-ray detectable range, the early-type companion (A2V) is non-convecting, which makes it a "clean" system to study X-ray emission which is not coronal. From Bildsten & Rutledge (2000).

Distinguishing between these two competing mechanisms – ADAFs and coronal emission – can be done with high S/N X-ray spectroscopy across the 0.1-4.0 keV energy range, where the X-ray emission is detected. Stellar coronal X-ray spectra are measured in low-mass stars and are reasonably well described as a Raymond-Smith plasma (Raymond and Smith 1977). These X-ray spectra are distinct from those calculated from ADAFs, which consist of a featureless continuum from the Compton-scattered optical/UV emission and weak line emission that does not appear detectable with the current generation of detectors (Narayan and Raymond 1999).

3. X-Ray Spectra of Quiescent Neutron Stars

Centaurus X-4 was the first NS transient detected in quiescence (Van Paradijs et al. 1987). More recently, quiescent X-ray spectral measurements have been made of Aql X−1 (Verbunt et al. 1994) and 4U 2129+47 (Garcia and Callanan 1999) with the *ROSAT*/PSPC; of EXO 0748−676 with Einstein IPC (Garcia and Callanan 1999); and of Cen X−4 and 4U 1608−522 with *ASCA* (Asai et al. 1996b). The X-ray spectrum of Aql X−1 (0.4–2.4 keV) was consistent with a blackbody (BB), a bremsstrahlung spectrum, or a pure power-law (Verbunt et al. 1994). For 4U 1608−522, the spectrum (0.5–10.0 keV) was consistent with a BB ($kT_{BB} \approx 0.2$–0.3 keV), a thermal Raymond-Smith model ($kT = 0.32^{+0.18}_{-0.5}$ keV), or a very steep power-law (photon index 6^{+1}_{-2}). Similar observations of Cen X−4 with *ASCA* found its X-ray spectrum consistent with these same models, but with an additional power-law component (photon index ≈ 2.0) above 5.0 keV (recent observations with *BeppoSAX* of Aql X−1 in quiescence also revealed a power-law tail; Campana et al. 1998b). These high energy power-law components are not fully understood, and their relationship to the thermal component is unclear. We discuss this more in §4.3.

In four of these five sources (the exception being EXO 0748−676), BB fits implied an emission area with a radius ≈ 1 km, much smaller than a NS. This has little physical meaning however, as the emitted spectrum from a quiescent NS atmosphere with light elements at the photosphere is far from a blackbody (Romani 1987). For a weakly-magnetic ($B \leq 10^{10}$ G) pure hydrogen or helium [1] atmosphere at effective temperatures $kT_{eff} \lesssim$ 0.5 keV the opacity is dominated by free-free transitions (Rajagopal and Romani 1996; Zavlin et al. 1996). Because of the opacity's strong frequency dependence ($\propto \nu^{-3}$), higher energy photons escape from deeper in the photosphere, where $T > T_{eff}$ (Pavlov and Shibanov 1978; Romani 1987; Zampieri et al. 1995). Spectral fits near the peak and into the Wien tail (which is the only part of the spectrum sampled with current instruments) with a BB curve then overestimate T_{eff} and underestimate the emitting area, by as much as two orders of magnitude (Rajagopal and Romani 1996; Zavlin et al. 1996).

Rutledge et al. (1999; 2000) showed that fitting the spectra of quiescent NS transients with realistic atmospheric models yielded emitting areas consistent with a 10 km radius NS. In Figure 2, we compare the measured H atmosphere and blackbody spectral parameters for the quiescent NSs. The emission area radii are larger from the H atmosphere spectra by a factor of a few to ten, and are consistent with the canonical radius of a NS. There is thus observational evidence that thermal emission from a pure hydrogen

---

[1] The strong surface gravity will stratify the atmosphere within ~ 10 s (Alcock & Illarionov 1980; Romani 1987). Hence, for accretion rates $\lesssim 2 \times 10^{-13} M_\odot$ yr^{-1} (corresponding to an accretion luminosity $\lesssim 2 \times 10^{33}$ erg s^{-1}), metals will settle out of the photosphere faster than the accretion flow can supply them (Bildsten, Salpeter, & Wasserman (1992)). As a result, the photosphere should be nearly pure hydrogen if $\dot{M}_q$ is small.

photosphere contributes to – and perhaps dominates – the NS luminosity at photon energies of 0.1–1 keV. This will be tested much better with upcoming *Chandra* and *XMM-Newton* observations.

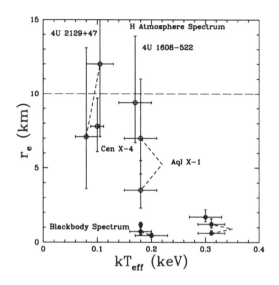

Figure 2. Comparison between the spectral parameters r_e and kT_{eff}, derived from spectral fits of the quiescent X-ray emission from Aql X−1, Cen X−4, 4U 1608−522 and 4U 2129+47. The *open points* are for the H atmosphere spectrum and the *solid points* are from a black-body spectrum. The two points connected for Aql X−1 correspond to the upper- and lower- distance limits for that source. The two connected points for 4U 2129+47 are for two different distance/N_H estimates. The H atmosphere fits produce values of r_e, consistent with a 10 km NS. From Rutledge et al. (2000).

4. What Powers the Quiescent Emission from the Neutron Stars?

In our earlier discussion of black holes, we pointed out that the expected coronal X-ray emission from the tidally locked and rapidly rotating stellar companions is at a level consistent with that observed from the black hole transients A 0620-00 and GRO J1655-40 (though see Lasota 2000 for an argument against this). The BH transient V404 Cygni is too bright to be explained this way, as are all binaries containing NSs. Several energy sources for the quiescent NS emission have been discussed and developed (Stella et al. 1994). These include late-time thermal emission from heat released deep in the NS crust during outbursts (Brown et al. 1998), accretion (Van Paradijs et al. 1987; Menou et al. 1999), and non-thermal emission from a turned-on radio pulsar (Campana et al. 1998a).

4.1. THERMAL EMISSION FROM DEEP NUCLEAR ENERGY RELEASE

Brown, Bildsten and Rutledge (1998) showed that the "rock-bottom" emission from these systems is set by thermal emission from the neutron star. This minimum luminosity comes from nuclear energy deposited in the inner crust (at a depth of ≈ 300 m) during the large accretion events. The freshly accreted material compresses the inner crust and triggers nuclear reactions that deposit about an MeV per accreted baryon there (Haensel and Zdunik 1990). This heats the NS core on a 10^4-10^5 yr timescale, until it reaches a steady-state temperature $\approx 4 \times 10^7 (\langle \dot{M} \rangle / 10^{-11} \, M_\odot \, \mathrm{yr}^{-1})^{0.4}$ K (Bildsten and Brown 1997), where $\langle \dot{M} \rangle$ is the time-averaged accretion rate in the binary. A core this hot makes the NS "glow" at a luminosity

$$L_q \approx \frac{1 \, \mathrm{MeV} \langle \dot{M} \rangle}{m_p} \approx 6 \times 10^{32} \frac{\mathrm{erg}}{\mathrm{s}} \left(\frac{\langle \dot{M} \rangle}{10^{-11} M_\odot \, \mathrm{yr}^{-1}} \right), \tag{1}$$

even after accretion halts (Brown et al. 1998). The NS is then a thermal emitter in quiescence, consistent with the inferrences from the quiescent spectroscopy noted earlier. This quiescent emission is inevitable (unless accelerated core cooling mechanisms are active in these stars that do not seem to occur in young neutron stars) and provides a "floor" for the quiescent luminosity. Additional emission mechanisms can add to this, increasing the luminosity and modifying the spectral shape.

In exploring this scenario Rutledge et al. (2000) analysed only observations made during periods of the lowest observed flux, to minimize contributions from accretion. They calculated the bolometric luminosity from the H atmosphere fits. Using these new bolometric quiescent luminosities for Aql X-1, Cen X-4, and 4U 1608−522, they plotted (Figure 3) L_q/L_o as a function of t_r/t_o (Brown et al. 1998). Here L_q and L_o are the observed quiescent and average outburst luminosities, and t_r and t_o are the recurrence interval and outburst duration. We show this relation for the NSs (*open circles*) Aql X-1, Cen X-4, 4U 1608−522, and EXO 0748−676 and the BHs (*filled circles*) H 1705−250, 4U 1543−47, Tra X-1, V 404 Cyg (GS 2023+33), GS 2000+25, and A 0620−00. We denote with an arrow those BHs for which only an upper limit on L_q is known.

The expected incandescent luminosity is plotted for two amounts of heat per accreted baryon deposited in the inner crust during an outburst: 1 MeV (*solid line*) and 0.1 MeV (*dotted line*). With the exception of Aql X-1, Cen X-4, 4U 1608−522, and the Rapid Burster, the data from this plot are taken from Chen et al. (1997). For Aql X-1 and the Rapid Burster, L_o and t_o are accurately known (*RXTE*/All-Sky Monitor public data); for the remaining sources L_o and t_o are estimated from the peak luminosities and the rise and decay timescales.

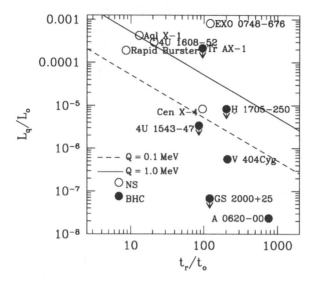

Figure 3. The ratio of quiescent luminosity L_q to outburst luminosity L_o as a function of the ratio of recurrence interval t_r to outburst duration t_o. The lines are for different amounts of heat, 0.1 MeV (*dashed line*) and 1.0 MeV (*solid line*), per accreted nucleon deposited at depths where the thermal time is longer than the outburst recurrence time. Also plotted are the observed ratios for several NSs (*open circles*) and BHs (*filled circles*). For most of the BHs, only an upper limit (*arrow*) to L_q is known. Data is from Chen et al. (1997), with the exception of L_q for the Rapid Burster (Asai et al. 1996b). For Aql X-1 and the Rapid Burster, L_o and t_o are accurately known (*RXTE*/All-Sky Monitor public data); for the remaining sources L_o and t_o are estimated from the peak luminosities and the rise and decay timescales. From Rutledge et al. (2000).

Four of the five NSs are within the band where the quiescent luminosity is that expected when the emitted heat is between 0.1-1.0 MeV per accreted baryon. The fifth NS (EXO 0748−676), has a higher quiescent luminosity (by a factor of 10), which we interpret as being due to continued accretion, an interpretation which is reinforced by the observation of spectral variability during the quiescent observations with *ASCA* (Corbet et al. 1994; Thomas et al. 1997), on timescales of ∼1000 sec and longer. (Garcia and Callanan 1999 measured $L_x=1\times10^{34}$ erg s^{-1} from Einstein/IPC observations of this source.) The BHs on this figure are more spread out across the parameter-space, qualitatively indicating a statistical difference – although not one which is particular for each object – between the two classes of objects. This suggests the NS quiescent luminosity is more strongly related to the accreted energy than the BH quiescent luminosities.

4.2. ACCRETION ONTO THE NEUTRON STAR DURING QUIESCENCE

Accretion was initially suggested (Van Paradijs et al. 1987) as the energy source of quiescent emission from transiently accreting NSs, partially because few other emission mechanisms were known at the time. Thermal emission was not considered, as it was presumed that these neutron stars with low-mass companions were clearly older than the NS core cooling timescale (10^6 yr) and would have cold cores. The work of Brown et al. (1998) changed that.

There are presently no observational results which exclude the possibility that part of the quiescent luminosity of these NSs is due to accretion. Indeed, some observational evidence suggests that accretion occurs onto the NS surface during quiescence; long-term (months-years) variability in the observed flux has been reported (a factor of 4.2 ± 0.5 in 8 days from Cen X-4; Campana et al. 1997) and in 4U 2129+47 (by a factor of 3.4 ± 0.6 between Nov-Dec 1992 and March 1994; Garcia and Callanan 1999; Rutledge et al. 2000). If so, then the required rate is about $\dot{M}_q = 10^{-14}$–10^{-15} $M_\odot$ yr^{-1}, a factor of 10^{-4} below the time-averaged rate. While this intensity variability can be explained by a variable absorption column depth, active accretion during quiescence is also a possibility.

Recent observations of Aql X-1 at the end of an outburst showed an abrupt fading into quiescence (Campana et al. 1998b) associated with a sudden spectral hardening (Zhang et al. 1998a). This was followed by a period of ~ 15 days, over which the source was observed (three times) with a constant flux level (Campana et al. 1998b). This behavior was interpreted as the onset of the "propellor effect" (Illarionov and Sunyaev 1975; Stella et al. 1986) in this object, which would inhibit – perhaps completely – accretion onto the NS. The energy source for the long-term nearly constant flux is most likely thermal emission (Brown et al. 1998).

A thermal spectrum alone cannot distinguish between accretion and a hot NS core as the energy source. This is because the accretion energy is likely deposited deep beneath the photosphere and is re-radiated as thermal emission (Zampieri et al. 1995). This emission is nearly identical to that expected from the hot NS core. The only possible difference would be if the accretion rate is high enough (about $> 10^{-13} M_\odot$ yr^{-1}, see footnote in § 3), to constantly replenish metals in the photosphere and if spallation of these elements is not too strong (Bildsten et al. 1992). These metals, particularly Oxygen, will imprint photoabsorption edges in the emergent spectrum (Rajagopal and Romani 1996). The presence of such metallic absorption in the NS quiescent emission spectra – aside from being astrophysically important – would clearly indicate active accretion onto the NS.

4.3. MAGNETOSPHERES AND SPINS

Evolutionary scenarios that connect the accreting neutron stars in Low-Mass X-ray Binaries to millisecond radio pulsars predict that these neutron stars should be rapidly rotating (at a few milliseconds) and magnetized at $10^8 - 10^9$ G. There is only one transiently accreting neutron star that unambiguously looks like this, SAX J1808.4-3658, at $\nu_s = 401$ Hz (Wijnands and Van der Klis 1998; Chakrabarty and Morgan 1998), and $B = 10^8 - 10^9$ G (Psaltis and Chakrabarty 1999). We know little about the magnetic fields and spins of the neutron stars in the other transients. The only one for which we know the spin is Aql X-1, where nearly coherent oscillations during Type I bursts imply a rotation rate of 550 Hz (Zhang et al. 1998b).

The neutron star's spin and magnetic field are important for two reasons. The first is the distinct possibility of shutting off accretion onto the neutron star from the "propellor" effect. This can happen when the magnetospheric radius exceeds the co-rotation radius (where the Kepler period equals the spin period), which for a neutron star spinning at ν_s with magnetic moment μ, will happen when the accretion rate is below $\dot{M}_p \approx 7 \times 10^{-11} M_\odot$ yr$^{-1}(\mu/10^{26}$G cm$^3)^2(\nu_s/300$ Hz$)^{7/3}$, suggesting a minimum accretion luminosity of $\approx 10^{36}$ erg s^{-1} for the fiducial parameters in the accretion rate equation. Even once in this regime, Campana et al. (1998a) discussed the possibility of emission from the gravitational energy release of matter striking the magnetosphere itself, which would yield a lesser amount of energy per gram and thus a lower total luminosity. Finally, if a magnetic field plays an important role in the geometry of quiescent accretion, one might expect some asymmetries that would produce X-ray pulsations. This was not observed in the recent fading of Aql X-1 (Campana et al. 1998b; Zhang et al. 1998a), where stringent limits on the pulsed fraction of the emission were placed ($\leq 1.2\%$ rms variability, 95% confidence; Chandler and Rutledge 2000) at a time believed to be just at the onset of the propellor for a 10^8 G field. What this implies about the magnetic field strength of Aql X-1 is still unknown.

The second place where the magnetic field and spin matter is when the magnetosphere becomes larger than the light cylinder. One might imagine the neutron star turning into a millisecond radio pulsar at this stage (see Stella et al. 1994 for an overview). However, a millisecond radio pulsar at the position of a transient X-ray binary has never been observed, even for SAX J1808.4-3658. Perhaps it is difficult for the accretion rates in these systems to become low enough to allow pulsar activity.

Campana et al. (1998b) conjectured that the hard X-rays sometimes seen in quiescence might be non-thermal emission from an active pulsar. If so, then the energy source is a fraction of the spin-down luminosity. As

Stella et al. (1994) noted, if the fraction of spin-down energy going into X-rays in transiently accreting binaries is similar to that observed from millisecond radio pulsars ($L_x \approx 10^{-3}\dot{E}$; Becker & Trümper 1999), and the magnetic field strengths are sufficient to "turn on" a millisecond pulsar, then the predicted X-ray luminosities are close to those observed ($\sim 10^{32}$-10^{33} erg s^{-1}). Indeed, a few of the X-ray detected millisecond pulsars have X-ray luminosities in this range (Becker & Trümper 1999). Brown et al. (1998) noted the same possibility for the neutron star in SAX J1808.4-3658. The X-ray spectral energy distribution of such a non-thermal component is hard and power-law like (Becker & Trümper 1999) similar to the hard power-law tails observed in Cen X-4 and Aql X-1 (Asai et al. 1996b; Campana et al. 1998b).

5. Conclusions and The Future

It is only recently that the focus of studying quiescent NSs has changed from parameterizing the phenomenology, to measuring the physics behind the emission. Higher quality X-ray data from the X-ray spectroscopy missions *Chandra* and *XMM-Newton* will provide much better data than any of the previous missions. These will also provide the means to account for possible contributions due to a hard-power law component in the black holes, as well as the neutron stars.

Bildsten and Rutledge (2000) have recently argued that two (A0620−00 and GRO J1655−40) of the three X-ray detected black hole binaries exhibit X-ray fluxes entirely consistent with coronal emission from the companion star. The current upper limits on the remaining BHs are also consistent with production via chromospheric activity in the secondary. All four NSs (Aql X−1, Cen X−4, 4U 1608−522, 4U 2129+47) have quiescent X-ray luminosities which are at least ten times greater than expected from chromospheric emission alone.

This suggests that a viable hypothesis for the majority of the transient NSs and BHs is that little accretion occurs in quiescence. Though mass is continuously transferred from the companion to the outer accretion disk (as is clear from the Hα line emission), accretion onto the compact object appears to be small. In the absence of accretion, the quiescent X-rays from a NS would then be dominated by thermal flux from a hot NS core (Brown et al. 1998), while for BHs, the quiescent X-rays come from the chromospheric activity of the secondary. The advantage of this hypothesis is that it explains the X-ray luminosities of BHs and (separately) NSs, without having to invoke dramatically different quiescent accretion rates that depend on the type of compact object (Menou et al. 1999).

At odds with this simple scenario is the detection of X-rays from V404

Cygni at a level which is a factor of ten brighter than can be explained as coronal emission. In addition, the observed variability in the NS quiescent luminosity can not be explained easily without some accretion. Both of these observations point to the possibility that the transferred matter sometimes can make it down to the central compact object. The fraction of the time this occurs and the reason why still needs to be better understood.

One of the most important areas of research in the coming years will be the search for and exploitation of photospheric absorption edges in NS quiescent spectra (§ 4.2). These edges are a kind of "holy grail" of NS spectroscopy as the known energy permits us to measure the gravitational redshift. We can also use realistic atmospheric spectra to derive the emission area radius divided by the distance to the NS. This radius, combined with the photospheric redshift, will provide an independent measure of the NS mass and radius, and thus its equation of state. The major uncertain parameter in these systems – the source distance – can be measured with the Space Interferometric Mission, set for launch in 2006, which will measure parallactic distances to objects as faint as 20th magnitude to 4μarcsec; which can find the majority of systems in this review.

In addition to better understanding of known sources, we hope that the new satellites will also probe quiescent emission from other populations of transient accretors. A likely place for progress with *Chandra* are the low-luminosity X-ray sources observed in globular clusters (Hertz and Grindlay 1983) which are either cataclysmic variables (Cool et al. 1995; Grindlay et al. 1995) or transient neutron stars in quiescence (Verbunt et al. 1984). X-ray spectroscopy can identify these objects as NSs radiating thermal emission from the atmosphere, or imply a different origin for the emission. As discussed above and elsewhere (Brown et al. 1998), the quiescent luminosities of these sources are set by the time averaged accretion rate. Thus, the low luminosity (10^{31} erg s^{-1}) X-ray sources in globular clusters, if they were transient neutron stars in quiescence, would have $\langle \dot{M} \rangle \approx 2\times10^{-13}$ $M_\odot$ yr^{-1}. The advantage to the cluster work will be the prior knowledge of the distance and reddening to the sources. *Simultaneous thermal spectroscopy of multiple sources in the same globular cluster at a known distance might well provide the first unambiguous and simultaneous measurements of many neutron star radii!*

We also expect progress in quiescent observations of transiently accreting X-ray pulsars. These systems are typically in high mass ($\gtrsim 10 M_\odot$) X-ray binaries, where the companion can contribute a significant fraction of the expected persistent X-ray luminosity, forcing us to depend on a pulse for secure detection of thermal emission in quiescence. The high magnetic fields (10^{12}–10^{13} G) perturb the opacity of the NS atmosphere and produce a pulse even if the underlying flux is uniform. Pulsations at the same luminosity level (10^{32} erg s^{-1}; cf. Eq. 1) as observed from the low-magnetic field systems was recently seen from A 0535+26 at a time when the circumstellar disc was absent (Negueruela et al. 2000). After excluding a magnetospheric origin, this pulsed emission was interpreted as due either to matter leaking onto the polar caps or to thermal emission from the NS core (Negueruela et al. 2000) – heated from nuclear emission deposited in the inner crust during the accretion outbursts, as described by Brown et al. (1998).

In binaries containing black holes, the major observational challenge is to distinguish between the ADAF and stellar coronal emission models. The best test is likely to be high S/N X-ray spectroscopy in the 0.1-4 keV range, where spectral lines contribute significantly in Raymond-Smith plasma (coronal) models, but not in ADAFs (Narayan and Raymond 1999). Once this is done, more focused studies of the accretion flows around black holes can be carried out.

We are clearly at the forefront of discovery regarding the physics of the quiescent emission of neutron stars and black holes. We have moved beyond initial detection, and at present a variety of mechanisms have been proposed to explain emission from both black holes and neutron stars. In the present era of *Chandra* and *XMM-Newton*, we will study these emission mechanisms in detail for the brightest of sources. The opportunity to detect neutron star photospheric absorption edges, and the ability to measure the neutron star radius from the broad-band spectroscopy may well constrain the neutron star equation of state.

Acknowledgements

We thank Ed Brown, George Pavlov and Slava Zavlin for the collaboration on much of this work. We are grateful to Ed Brown for preparing Figure 3. This work was supported in part by the National Science Foundation through Grant NSF94-0174 and NASA via grant NAG5-3239. L.B. is a Cottrell Scholar of the Research Corporation.

References

Alcock, C. and Illarionov, A. (1980), *ApJ* **235**, 534.
Asai, K., Dotani, T., Hoshi, R., Tanaka, Y., Robinson, C.R., and Terada, K. (1998), *Publ.Astron.Soc.Japan* **50**, 611.
Asai, K., Dotani, T., Mitsuda, K., Hoshi, R., Vaughan, B., Tanaka, Y., and Inoue, H. (1996b), *Publ.Astron.Soc.Japan* **48**, 257.
Barret, D., McClintock, J.E., and Grindlay, J.E. (1996), *ApJ* **463**, 963.
Becker, W. and Trümper, J. (1999), A&A, **341**, 803.
Bildsten, L. and Brown, E.F. (1997), *ApJ* **477**, 897.
Bildsten, L. and Rutledge, R.E. (2000), *ApJ* **541**, 908.
Bildsten, L., Salpeter, E.E., and Wasserman, I. (1992), *ApJ* **384**, 143.
Brown, E.F., Bildsten, L., and Rutledge, R.E. (1998), *ApJ* **504**, L95.

Campana, S., Colpi, M., Mereghetti, S., Stella, L., and Tavani, M. (1998a), *A&A Reviews* **8**, 279.

Campana, S., Mereghetti, S., Stella, L., and Colpi, M. (1997), A&A **324**, 941.

Campana, S., Stella, L., Mereghetti, S., Colpi, M., Tavani, M., Ricci, D., Fiume, D.D., and Belloni, T. (1998b), *ApJ* **499**, L65.

Chakrabarty, D. and Morgan, E.H. (1998), *Nature* **394**, 346.

Chandler, A. and Rutledge, R.E. (2000), *ApJ* **545**, 1000.

Chen, W., Shrader, C.R., and Livio, M. (1997), *ApJ* **491**, 312.

Cool, A.M., Grindlay, J.E., Cohn, H.N., Lugger, P.M., and Slavin, S.D. (1995), *ApJ* **439**, 695.

Corbet, R. H.D., Asai, K., Dotani, T., and Nagase, F. (1994), *ApJ* **436**, L15.

Cordova, F.A. and Mason, K.O. (1984), *MNRAS* **206**, 879.

Garcia, M.R. and Callanan, P.J. (1999), *AJ* **118**, 1390.

Grindlay, J.E., Cool, A.M., Callanan, P.J., Bailyn, C.D., Cohn, H.N., and Lugger, P.M. (1995), *ApJ* **455**, L47.

Haensel, P. and Zdunik, J.L. (1990), A&A **227**, 431.

Hameury, J.-M., Lasota, J.-P., McClintock, J.E., and Narayan, R. (1997), *ApJ* **489**, 234.

Hertz, P. and Grindlay, J.E. (1983), *ApJ* **267**, L83.

Huang, M. and Wheeler, J.C. (1989), *ApJ* **343**, 229.

Illarionov, A.F. and Sunyaev, R.A. (1975), A&A **39**, 185.

King, A.R., Kolb, U., and Burderi, L. (1996), *ApJ* **464**, L127.

Lasota, J.-P. (2000), *A&A* **360**, 575.

McClintock, J.E. (1998), in S.S. Holt and T.R. Kallman (eds.), *Accretion Processes in Astrophysical Systems: Some Like it Hot! Eighth Astrophysics Conference, College Park, MD, October 1997.*, AIP Conf. Proc. 431, American Institute of Physics, p290.

McClintock, J.E., Horne, K., and Remillard, R.A. (1995), *ApJ* **442**, 358.

Menou, K., Esin, A.A., Narayan, R., Garcia, M.R., Lasota, J.P., and McClintock, J.E. (1999), *ApJ* **520**, 276.

Mineshige, S. and Wheeler, J.C. (1989), *ApJ* **343**, 241.

Mukai, K., Wood, J.H., Naylor, T., Schlegel, E.M., and Swank, J.H. (1997), *ApJ* **475**, 812.

Narayan, R., Barret, D., and McClintock, J.E. (1997a), *ApJ* **482**, 448.

Narayan, R., Garcia, M.R., and McClintock, J.E. (1997b), *ApJ* **478**, L79.

Narayan, R. and Raymond, J. (1999), *ApJ* **515**, L69.

Negueruela, I., Reig, P., Finger, M.H., and Roche, P. (2000), *A&A* **356**, 1003.

Patterson, J. and Raymond, J.C. (1985), *ApJ* **292**, 550.

Pavlov, G.G. and Shibanov, I.A. (1978), *Soviet Astronomy* **22**, 214.

Pratt, G.W., Hassall, B. J.M., Naylor, T., and Wood, J.H. (1999), *MNRAS* **307**, 413.

Psaltis, D. and Chakrabarty, D. (1999), *ApJ* **521**, 332.

Rajagopal, M. and Romani, R.W. (1996), *ApJ* **461**, 327.

Raymond, J.C. and Smith, B.W. (1977), *ApJ (Suppl)* **35**, 419.

Romani, R.W. (1987), *ApJ* **313**, 718.

Rutledge, R.E., Bildsten, L., Brown, E.F., Pavlov, G.G., and Zavlin, V.E. (1999), *ApJ* **514**, 945.

Rutledge, R.E., Bildsten, L., Brown, E.F., Pavlov, G.G., and Zavlin, V.E. (2000), *ApJ* **529**, 985.

Singh, K.P., Drake, S.A., Gotthelf, E.V., and White, N.E. (1999), *ApJ* **512**, 874.

Stella, L., Campana, S., Colpi, M., Mereghetti, S., and Tavani, M. (1994), *ApJ* **423**, L47.

Stella, L., White, N.E., and Rosner, R. (1986), *ApJ* **308**, 669.

Tanaka, Y. and Lewin, W. (1995), in W. Lewin, J. Van Paradijs, and E. Van Den Heuvel (eds.), *X-Ray Binaries*, Cambridge University Press, p126.

Tanaka, Y. and Shibazaki, N. (1996), *Ann.Rev.Astr.Ap.* **34**, 607.

Thomas, B., Corbet, R., Smale, A.P., Asai, K., and Dotani, T. (1997), *ApJ* **480**, L21.

Van Paradijs, J. (1996), *ApJ* **464**, L139.

Van Paradijs, J., Verbunt, F., Shafer, R.A., and Arnaud, K.A. (1987), A&A **182**, 47.

Van Teeseling, A. (1997), A&A **319**, L25.

Verbunt, F. (1996), *IAU Symposia* **165**, 333.

Verbunt, F., Belloni, T., Johnston, H.M., Van der Klis, M., and Lewin, W.H.G. (1994), A&A **285**, 903.

Verbunt, F., Elson, R., and Van Paradijs, J. (1984), *MNRAS* **210**, 899.

Vilhu, O. and Walter, F.M. (1987), *ApJ* **321**, 958.

Wagner, R.M., Starrfield, S.G., Hjellming, R.M., Howell, S.B., and Kreidl, T.J. (1994), *ApJ* **429**, L25.

Wijnands, R. and Van der Klis, M. (1998), *Nature* **394**, 344.

Zampieri, L., Turolla, R., Zane, S., and Treves, A. (1995), *ApJ* **439**, 849.

Zavlin, V.E., Pavlov, G.G., and Shibanov, Y.A. (1996), A&A **315**, 141.

Zhang, S.N., Yu, W., and Zhang, W. (1998a), *ApJ* **494**, L71.

Zhang, W., Jahoda, K., Kelley, R.L., Strohmayer, T.E., Swank, J.H., and Zhang, S.N. (1998b), *ApJ* **495**, L9.

A COMPARISON OF RADIO EMISSION FROM NEUTRON STAR AND BLACK HOLE X-RAY BINARIES

R.P. FENDER
Astronomical Institute 'Anton Pannekoek'
and Center for High Energy Astrophysics,
University of Amsterdam, Kruislaan 403,
1098 SJ Amsterdam, The Netherlands

1. Introduction : radio emission from X-ray binaries

Radio emission has now been detected from $\sim 20\%$ of all X-ray binaries. In nearly all cases it has been found to be variable and to display a nonthermal[1] spectrum, and synchrotron emission has been established as the most likely emission mechanism.

Radio outbursts from X-ray binaries generally follow a pattern of fast rise and power law and/or exponential decay, with an evolution of the spectral index $\alpha = \Delta \log S_\nu / \Delta \log \nu$ from 'inverted' ($\alpha \geq 0$), probably arising in (partially) optically thick emission, to optically thin ($\alpha \leq -0.5$), with a corresponding shift in the peak of emission to lower frequencies. This is in qualitative agreement with models of an expanding, synchrotron-emitting cloud, as proposed by van der Laan (1966) for outbursts of active galactic nuclei. Furthermore, in recent years the time evolution of several outbursts have been imaged at high angular resolution with arrays such as VLA, MERLIN and VLBA, and these images reveal the outflow of radio-emitting matter along more or less collimated paths at relativistic velocities. More recently relatively stable radio emission from at least one persistent X-ray binary has also been resolved into a jet-like structure. As a result it has become widely, if not universally, accepted that radio emission from X-ray binaries arises in synchrotron-emitting jets. For detailed reviews and references, see e.g. Hjellming & Han (1995), Mirabel & Rodríguez (2000), Fender (2000a).

---

[1]Here 'nonthermal' is taken to mean arising from a non-Maxwellian particle distribution which cannot be described by a single temperature.

C. Kouveliotou et al. (eds.), The Neutron Star – Black Hole Connection, 261–266.
© 2001 *Kluwer Academic Publishers. Printed in the Netherlands.*

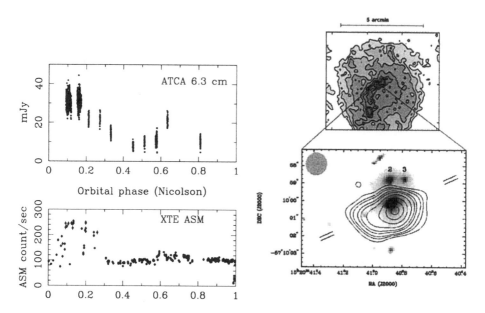

Figure 1.   Radio outbursts from the neutron star X-ray binary Cir X−1. Left panel shows XTE ASM and radio monitoring around one 16.6-day orbit; flaring presumably occurring during periastron passage of the neutron star in an elliptical orbit (from Fender 1997). Right panel shows arcmin- (top) and arcsec-scale radio jets (bottom) from the system (from Fender et al. 1998).

In this paper I compare, briefly, the properties of radio emission from both transient and persistent neutron-star and black-hole(-candidate) X-ray binaries.

2.   Transients

Bright X-ray transients are generally accompanied by transient radio emission which follows the pattern outlined in the introduction. Bright transients can contain both neutron stars (e.g. Aql X−1) and black holes (e.g. GS 1124−684); do the radio properties of these two populations differ?

The answer is yes and no. Firstly, it is clear that, excluding bright, exotic objects for which classification of the compact object type has proved impossible to date (e.g. Cyg X−3, SS 433, LSI +61° 303), the black hole transients are, at the peak of outburst, the brightest radio sources associated with X-ray binaries. However, this is also the case for their X-ray emission – i.e. the brightest transients in the X-ray band are also the black holes. It is unclear at present whether this simply reflects the larger average masses of the black holes, or differences in the accretion flows onto the two types of compact accretor. On the other hand, the ratio of radio to X-ray peak fluxes is comparable for both neutron star and black hole X-ray binaries. As an example, the neutron star transient Cen X-4 reached peak fluxes of ~ 4 Crab and ~ 10 mJy at soft X-ray and radio wavelengths respectively during its 1979 outburst. For comparison, the black hole transient A 0620−00 reached peak fluxes of ~ 45 Crab and ~ 200 mJy during its 1975 outburst. While the ratios are not exactly the same (but bear in mind there is likely to be a large scatter in observed radio fluxes due to beaming – see Kuulkers et al. 1999), their order-of-magnitude correspondence indicates that in both neutron star and black hole systems the ratio of X-ray to radio luminosities is comparable. Furthermore, the ratio is similar for most other systems (Fender & Kuulkers 2000). This in turn implies that the accretion and jet formation mechanisms are broadly the same, during outburst, for both types of system.

Cir X−1 stands out as an example of a neutron-star X-ray binary which undergoes transient radio-bright outbursts every 16.6 days (Figure 1). The outbursts are believed to occur at periastron passage of a neutron star in a highly eccentric orbit with a main sequence or slightly evolved mass donor. The system displays a number of unique characteristics including radio jets which connect to a radio-bright synchrotron nebula (Fender et al. 1998 and references therein), rapidly evolving X-ray and radio light curves, and the highest measured radial velocity (~ 400 km s^{-1}) of any X-ray binary (Johnston, Fender & Wu 1999). Given its relative brightness (core + jets are typically ≥ 10 mJy) and the predictability of its radio outbursts, Cir X−1 may be the best source in which to study the formation of jets by a neutron-star accretor.

3. Persistent sources

Only four persistently bright X-ray sources in our Galaxy are believed to contain black holes : Cyg X−1, GX 339−9, 1E 1740.7−2942 and GRS 1758−258. The latter two are too faint for regular monitoring, however for Cyg X−1 and, especially, GX 339−4, we have a good idea of how radio emission is related to X-ray state. This is explored in detail in Fender (2000b), and is summarised in Table 1. Discrete, bright, outbursts correspond to major state transitions and/or the Very High State; the formation of a steady outflow seems to occur in the Low/Hard and, more weakly, Off states.

Neutron star systems do not display analogs of the black hole states, but can nonetheless be classified into different groups. These are the 'Z' and 'atoll' X-ray binaries (believed to contain low magnetic field neutron stars), and the high-field X-ray pulsars. Thus the population of neutron star

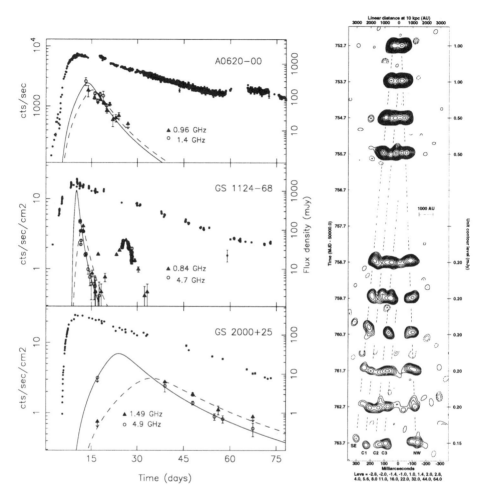

Figure 2.   Radio outbursts from black hole X-ray binaries. Left panel compares X-ray
and radio lightcurves of three black hole candidate X-ray transients (from Kuulkers et al.
1999). Note the similarity of the X-ray lightcurves (black dots) while radio light curves
are clearly different. Right panel illustrates clearly resolved relativistic ejections from the
black hole candidate GRS 1915+105 (from Fender et al. 1999).

X-ray binaries allows us to explore the effects of mass accretion rate and
accretor magnetic field on the production of radio jets. The Z sources (and
the unusual atoll source GX 13+1) are all regularly detected as variable
radio sources.

Penninx et al. (1988), in observations of the Z-source GX 17+2 established a link between radio emission and location on the Z track (corresponding to the soft X-ray colours, and hence presumably the state of the
accretion disc, at the time of observation – see e.g. van der Klis 1995),

TABLE 1. The relation of radio emission to black hole X-ray state. From Fender (2000b).

State	Radio properties
Very High	Bright ejections with spectral evolution from absorbed $\rightarrow$ optically thin
High/Soft	Radio suppressed by factor ≥ 25
Intermediate	Weak ?
Low/Hard	Low level, steady, flat spectrum extending to at least sub-mm
Off	Weak; similar to Low/Hard but reduced by a factor ≥ 10

TABLE 2. Comparison of derived mean intrinsic radio luminosities for the BHC/Z, Atoll and X-ray pulsar classes of persistent X-ray binary, plus simple interpretations of their physical differences. From Fender & Hendry (2000).

Source type	$S_\nu/(\mathrm{kpc}^2)$ (mJy)	Inferred physical characteristics $(\dot{m}/\dot{m}_{\mathrm{Edd}})$	B (Gauss)	inner disc radius (km)
BHC (low/hard state)	55 ± 13	≤ 0.1	–	few $\times 100$
Z (horizontal branch)	"	$0.1 - 1.0$	$10^9 - 10^{10}$	few $\times 10$
Atoll	$\leq 10 \pm 2$	$0.01 - 0.1$	$10^9 - 10^{10}$	few $\times 10$
X-ray pulsar	$\leq 6 \pm 2$	≤ 1.0	$\geq 10^{12}$	≥ 1000

such that radio emission is strongest on the 'horizontal branch', weak on the 'normal branch' and absent on the 'flaring branch'. This relation is in agreement with (most) subsequent studies, and at face value appears to demonstrate an anti-correlation between radio emission and accretion rate in these sources, but this is almost certainly an oversimplification.

The atoll sources are in general not detected except during outbursts (e.g. Aql X−1 is an atoll source and a transient, which had a radio outburst) and/or at very high mass accretion rates (which may be the cause in the case of GX 13+1). No X-ray pulsar has ever been detected as a radio source.

These results, and their inferences, are summarised in Table 2 (adapted from Fender & Hendry 2000). It seems clear that an inner (≤ 1000 km) accretion disc, as a result of a low ($\leq 10^{10}$ G) accretor magnetic field and, most importantly and intuitively, a high mass accretion rate, are required for jet formation. In addition, it is demonstrated in Table 1 that the Z sources (plus GX 13+1) have approximately the same radio luminosity as the persistent black hole candidates when in the Low/Hard X-ray state. Once again we are forced to conclude that the accretion and jet formation processes in neutron star and black hole X-ray binaries are similar (although it is noted that the Z sources are apparently a little more luminous in X-rays on average than the black holes in the Low/Hard state).

4. Conclusions

This comparison of the radio properties of the neutron-star and black-hole X-ray binaries has revealed that whether the systems are *transient*, as a result (in most cases) of low average accretion rates and disc instability mechanisms, or *persistent*, as a result of high average accretion rates, the coupling between radio jet formation and accretion luminosity is similar for both classes of accretor. It appears that the disc-jet coupling does not really care too much about the nature of the accretor (as long as the magnetic field is not strong enough to disrupt the accretion disc, as in the case of the X-ray pulsars).

However, this is an oversimplification of the situation, and many important questions remain. For example, it is the atoll sources, amongst the neutron star X-ray binaries which appear to show the strongest X-ray spectral evidence for Comptonising coronae, yet they are weak radio sources ... while in the black holes the presence of the corona is (nearly) always associated with observable radio emission. Is it just a question of accretion rate, or is another factor allowing only one of the classes to form jets ? This and other questions will only be resolved by future coordinated radio and X-ray observations of neutron-star, as well as black-hole, X-ray binaries.

References

Fender, R.P. (1997), in C.D. Dermer, M.S. Strickman, and J.D. Kurfess (eds.), *Proceedings of 4th Compton Symposium*, AIP Conf. Proc. **410**, p798.

Fender, R.P. (2000a), in *Astrophysics and Cosmology : A collection of critical thoughts*, Springer Lecture Notes in Physics, in press (astro-ph/9907050).

Fender, R.P. (2000b), in L. Kaper, E.P.J. van den Heuvel, P.A. Woudt (eds.), *Black holes in binaries and galactic nuclei, ESO workshop*, Springer-Verlag, in press.

Fender, R.P. and Hendry, M.A. (2000), *MNRAS* **317**, 1.

Fender, R., Spencer, R., Tzioumis, T., Wu, K., van der Klis, M., van Paradijs J., and Johnston, H. (1998), *ApJ* **506**, L21.

Fender, R.P., Garrington, S.T., McKay, D.J., Muxlow, T.W.B., Pooley, G.G., Spencer, R.E., Stirling, A.M., and Waltman, E.B. (1999), *MNRAS* **304**, 865.

Hjellming, R.M. and Han, X.H. (1995), in W.H.G. Lewin, J. van Paradijs, E.P.J. van den Heuvel (eds.), *X-ray binaries*, Cambridge Univ. Press, p308.

Johnston, H.M., Fender, R.P., and Wu, K. (1999), *MNRAS* **308**, 415.

Kuulkers, E., Fender, R.P., Spencer, R.E., Davis, R.J., and Morison, I. (1999), *MNRAS* **306**, 919.

Mirabel, I.F. and Rodríguez, L.F. (1999), *Annual Rev. Astron. & Astrophys.* **37**, in press (astro-ph/9902062).

Penninx, W., Lewin, W.H.G., Zijlstra, A.A., Mitsuda, K., van Paradijs, J., and van der Klis, M. (1988), *Nature* **336**, 146.

Van der Klis, M. (1995), in W.H.G. Lewin, J. van Paradijs, and E.P.J. van den Heuvel (eds.), *X-ray binaries*, Cambridge Univ. Press, p252.

Van der Laan, H. (1966), *Nature* **211**, 1131.

BLACK HOLE AND NEUTRON STAR JET SOURCES

I.F. MIRABEL

Centre d'Etudes de Saclay/ CEA/DSM/DAPNIA/SAP
91911 Gif/Yvette, France &
Intituto de Astronomía y Física del Espacio. Buenos Aires,
Argentina

Abstract. Black holes of stellar mass and neutron stars in binary systems are first detected as hard X-ray sources using high-energy space telescopes. Relativistic jets in some of these compact sources are found by means of multiwavelength observations with ground-based telescopes. The X-ray emission probes the inner accretion disk and immediate surroundings of the compact object, whereas the synchrotron emission from the jets is observed in the radio and infrared bands, and in the future could be detected at even shorter wavelengths. Black-hole X-ray binaries with relativistic jets mimic, on a much smaller scale, many of the phenomena seen in quasars and are thus called microquasars. Because of their proximity, their study opens the way for a better understanding of the relativistic jets seen elsewhere in the Universe. From the observation of two-sided moving jets it is inferred that the ejecta in microquasars move with relativistic speeds similar to those believed to be present in quasars. The simultaneous multiwavelength approach to microquasars reveals on short time-scales the close connection between instabilities in the accretion disk seen in the X-rays, and the ejection of relativistic clouds of plasma observed as synchrotron emission at longer wavelengths. Besides contributing to a deeper comprehension of accretion disks and jets, microquasars may serve in the future to determine the distances of jet sources using constraints from special relativity, and the spin of black holes using general relativity.

1. Jets in astrophysics

While the first evidence of jet-like features emanating from the nuclei of galaxies goes back to the discovery by Curtis (1918) of the optical jet from

C. Kouveliotou et al. (eds.), The Neutron Star – Black Hole Connection, 267–282.
© *2001 Kluwer Academic Publishers. Printed in the Netherlands.*

the elliptical galaxy M87 in the Virgo cluster, the finding that jets can also be produced in smaller scale by binary stellar systems is much more recent. The detection by Margon et al. (1979) of large, periodic Doppler drifts in the optical lines of SS 433 resulted in the proposition of a kinematic model (Fabian & Rees 1979; Milgrom 1979) consisting of two precessing jets of collimated matter with velocity 0.26c. High-resolution angular radio imaging as a function of time showed the presence of outflowing radio jets and fully confirmed the kinematic model. The early history of SS 433 has been reviewed by Margon (1984).

Since the detection of Sco X-1 at radio wavelengths (Ables 1969), some X-ray binaries had been known to be strong, time-variable non-thermal emitters. Ejection of synchrotron-emitting clouds was suspected, but the actual confirmation of radio jets came only with the observations of SS 433. At present, there are about 200 known galactic X-ray binaries (van Paradijs 1995), of which about 10 % are radio-loud (Hjellming & Han 1995). Of these radio-emitting X-ray binaries, 10 have shown evidence of relativistic jets of synchrotron emission, and this review focuses on this set of objects. I use the term "jets" to designate collimated ejecta that have opening angles $\leq 15°$.

In the last years it has become clear that collimated ejecta can be produced in several stellar environments when an accretion disk is present. Jets with terminal velocities of the order of a few hundred to a few thousand km s^{-1} are now known to emanate from objects as diverse as very young stars (Reipurth & Bertout 1997), nuclei of planetary nebulae (López 1997), and accreting white dwarfs that appear as supersoft X-ray sources (Rappaport et al. 1994). These types of stellar jets have, however, non-relativistic velocities ($\sim$100–10000 km s^{-1}) and their associated emission is dominantly thermal (i.e. free-free continuum emission in the radio as well as characteristic near-IR, optical and UV lines). Interestingly, in all known types of jet sources a disk is believed to be present. Here I concentrate on synchrotron jets with velocities that can be considered relativistic ($v \geq 0.1c$), which are observed in X-ray binaries that contain a compact object, that is, a neutron star or a black hole.

2. Microquasars

Black holes were first predicted by John Michell (1783) in the context of Newtonian physics and the corpuscular theory of light. He was the first to suggest that they could be detected by the motion of nearby luminous objects. In the fourth year of the french revolution, Pierre-Simon Laplace speculated on the possible existence of both, stellar-mass and supermassive black holes. In the "Exposition du Syteme du Monde" Laplace proposed:

1) that stellar-mass black holes could be as numerous as stars - "en aussi grand nombre que les étoiles"-, and 2) that the most massive objects of the universe could be black holes- "il est donc possible que les plus grands...corps de l'univers, soient par cela même, invisibles". In the nineteenth century the ondulatory conception of light became predominant and the idea of black hole was forgotten for more than a century, until it became a natural consequence of general relativity.

The recent finding in our own galaxy of *microquasars* (Margon, 1994; Mirabel et al. 1992; Mirabel & Rodríguez 1994; 1998) has opened new perspectives for the astrophysics of black holes (see Figure 1). These scaled-down versions of quasars are believed to be powered by spinning black holes but with masses of up to a few tens that of the Sun. The word *microquasar* was chosen to suggest that the analogy with quasars is more than morphological, and that there is an underlying unity in the physics of accreting black holes over an enormous range of scales, from stellar-mass black holes in binary stellar systems, to supermassive black holes at the centre of distant galaxies (Rees 1998).

At first glance it may seem paradoxical that relativistic jets were first discovered in the nuclei of galaxies and distant quasars and that for more than a decade SS 433 was the only known object of its class in our Galaxy (Margon 1984). The reason for this is that disks around supermassive black holes emit strongly at optical and UV wavelengths. Indeed, the more massive the black hole, the cooler the surrounding accretion disk is. For a black hole accreting at the Eddington limit, the characteristic black body temperature at the last stable orbit in the surrounding accretion disk will be given approximately by $T \sim 2 \times 10^7 \, M^{-1/4}$ (Rees 1984), with T in K and the mass of the black hole, M, in solar masses. Then, while accretion disks in AGNs have strong emission in the optical and ultraviolet with distinct broad emission lines, black hole and neutron star binaries are usually identified for the first time by their X-ray emission. Among these sources, SS 433 is unusual given its broad optical emission lines and its brightness in the visible. Therefore, it is understandable that there was an impasse in the discovery of new stellar sources of relativistic jets until the recent developments in X-ray astronomy. Strictly speaking and if it had not been for the historical circumstances described above, the acronym *quasar* ("quasi-stellar-radio-source") would have suited better the stellar mass versions rather than their super-massive analogs at the centers of galaxies.

Since the characteristic times in the flow of matter onto a black hole are proportional to its mass, variations with intervals of minutes in a microquasar correspond to analogous phenomena with durations of thousands of years in a quasar of $10^9 \, M_\odot$, which is much longer than a human life-time. Therefore, variations with minutes of duration in microquasars could be

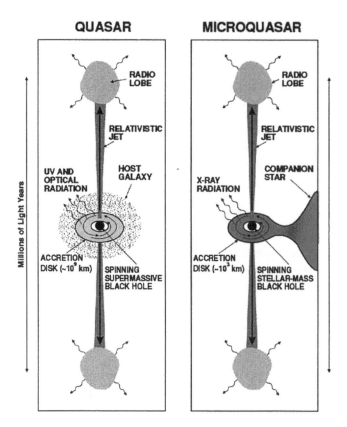

Figure 1. Diagram illustrating current ideas concerning quasars and microquasars (not to scale). As in quasars, in microquasars the following three basic ingredients are found: 1) a spinning black hole, 2) an accretion disk heated by viscous dissipation, and 3) collimated jets of relativistic particles. However, in microquasars the black hole is only a few solar masses instead of several millon solar masses; the accretion disk has mean thermal temperatures of several millon degrees instead of several thousand degrees; and the particles ejected at relativistic speeds can travel up to distances of a few light-years only, instead of the several millon light-years in some giant radio galaxies. In quasars matter can be drawn into the accretion disk from disrupted stars or from the interstellar medium of the host galaxy, whereas in microquasars the material is being drawn from the companion star in the binary system. In quasars the accretion disk has sizes of $\sim 10^9$ km and radiates mostly in the ultraviolet and optical wavelengths, whereas in microquasars the accretion disk has sizes of $\sim 10^3$ km and the bulk of the radiation comes out in the X-rays. It is believed that part of the spin energy of the black hole can be tapped to power the collimated ejection of magnetized plasma at relativistic speeds. This analogy between quasars and microquasars resides in the fact that in black holes the physics is essentially the same irrespective of the mass, except that the distance and time-scales of phenomena are proportional to the black hole mass. Because of the relative proximity and shorter time scales, in microquasars it is possible to firmly establish the relativistic motion of the sources of radiation, and to better study the physics of accretion flows and jet formation near the horizon of black holes.

sampling phenomena that we have not been able to study in quasars. The repeated observation of two-sided moving jets in a microquasar (Rodríguez & Mirabel 1999) has led to a much greater acceptance of the idea that the emission from quasar jets is associated with material moving at speeds close to that of light. Furthermore, simultaneous multiwavelength observations of this microquasar (Mirabel et al. 1998; Eikenberry et al. 1998) are revealing the connection between the sudden disappearance of matter through the horizon of the black hole and the ejection of expanding clouds of relativistic plasma.

3. Superluminal sources

Expansions at up to ten or more times the speed of light have been observed in quasars for more than 20 years (Pearson & Zensus 1987; Zensus 1997). At first these superluminal motions provoked concern because they appeared to violate relativity, but they were soon interpreted as illusions due to relativistic aberration (Rees 1966). However, the ultimate physical interpretation had remained uncertain. In the extragalactic case the moving jets are observed as one-sided (because strong Doppler favoritism renders the approaching ejecta detectable) and it is not possible to know if superluminal motions represent the propagation of waves through a slowly moving jet, or if they reflect the actual bulk motion of the sources of radiation.

In the context of the microquasar analogy, one may ask if superluminal motions could be observed from sources known to be in our own Galaxy. Among the handful of black holes of stellar mass known so far, three transient X-ray sources have indeed been identified at radio wavelengths as sporadic sources of superluminal jets. The first superluminal source to be discovered (Mirabel & Rodríguez 1994) was GRS 1915+105, a recurrent transient source of hard X-rays first found and studied with the satellite GRANAT (Castro-Tirado et al. 1994). The discovery of superluminal motions in GRS 1915+105 stimulated a search for similar relativistic ejecta in other transient hard X-ray sources. Soon after, the same phenomenon was observed by two different groups (Tingay et al. 1995; Hjellming & Rupen 1995) in GRO J1655–40, a hard X-ray nova found with the Compton Gamma Ray Observatory (Zhang et al. 1994). A third superluminal source may be XTE J1748–288 (Hjellming et al. 1998), a transient source with a hard X-ray spectrum recently found with XTE.

King (1998) proposes that the superluminal sources are black hole binaries with the secondary in the Hertzsprung-Russell gap, which provides super-Eddington accretion into the black hole. In the Galaxy there would have been $\geq 10^3$ systems of this class with a lifetime for the jet phase of $\leq 10^7$ years, which is the spin-down phase of the black hole.

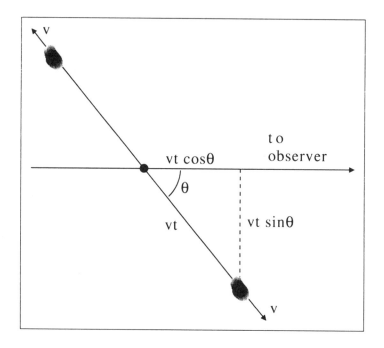

Figure 2.   Geometry of the two-sided ejection. The emission is symmetric, but when the emitting clouds move at relativistic speeds the approaching component of the pair appears to move faster and to be brighter than the receding component.

4. Special relativity effects

The main characteristics of the superluminal ejections can be understood in terms of the simultaneous ejection of a pair of twin condensations moving at velocity β ($\beta = v/c$), with v being the velocity of the condensations and c the speed of light), with the axis of the flow making an angle θ ($0° \leq \theta \leq 90°$) with respect to the line of sight of a distant observer (Rees 1966; see Figure 2). The apparent proper motions in the sky of the approaching and receding condensations, μ_a and μ_r, are given by:

$$\mu_a = \frac{\beta \; sin \; \theta}{(1 - \beta \; cos \; \theta)} \frac{c}{D}, \tag{1}$$

$$\mu_r = \frac{\beta \; sin \; \theta}{(1 + \beta \; cos \; \theta)} \frac{c}{D}, \tag{2}$$

where D is the distance from the observer to the source. These two equations can be transformed to the equivalent pair of equations:

$$\beta \cos \theta = \frac{\mu_a - \mu_r}{\mu_a + \mu_r},\tag{3}$$

$$D = \frac{c \, \tan \theta \, (\mu_a - \mu_r)}{2} \frac{}{\mu_a \mu_r}.\tag{4}$$

If only the proper motions are known, an interesting upper limit for the distance can be obtained from eqns. (3) and (4):

$$D \le \frac{c}{\sqrt{\mu_a \mu_r}}.\tag{5}$$

In all equations we use cgs units and the proper motions are in radians s^{-1}. In the case of the bright ejection event of March 19, 1994 for GRS 1915+105, the proper motions measured were $\mu_a = 17.6 \pm 0.4$ mas day^{-1} and $\mu_r = 9.0 \pm 0.1$ mas day^{-1}. Using eqn. (5), we derive an upper limit for the distance, $D \le 13.7$ kpc, confirming the galactic nature of the source.

The distance to GRS 1915+105 is found to be, from HI absorption studies, 12.5±1.5 kpc (Rodríguez et al. 1995; Chaty et al. 1996). Then, the proper motions of the approaching and receding condensations measured with the VLA in 1994 and 1995 imply apparent velocities on the plane of the sky of $v_a = 1.25c$ and $v_r = 0.65c$ for the approaching and receding components respectively. The ejecta move with a true speed of $v = 0.92c$ at an angle $\theta = 70°$ with respect to the line of sight (Mirabel & Rodríguez 1994). The faster proper motions of 24 mas/day measured with MERLIN (Fender et al. 1999) and the VLBA (Dhawan et al. 1999) in 1997 would imply a true speed of 0.98c at an angle of 66° to the line of sight.

The ratios of observed to emitted flux density S_o, from a twin pair of optically-thin, isotropically emitting jets are:

$$\frac{S_a}{S_o} = \delta_a^{k-\alpha},\tag{6}$$

$$\frac{S_r}{S_o} = \delta_r^{k-\alpha},\tag{7}$$

where α is the spectral index of the emission ($S_\nu \propto \nu^\alpha$), and k is a parameter that accounts for the geometry of the ejecta, with k = 2 for a continuous jet and k = 3 for discrete condensations. Then, the ratio of observed flux densities (measured at equal separations from the core) will be given by

$$\frac{S_a}{S_r} = \left(\frac{1 + \beta \cos \theta}{1 - \beta \cos \theta}\right)^{k-\alpha},\tag{8}$$

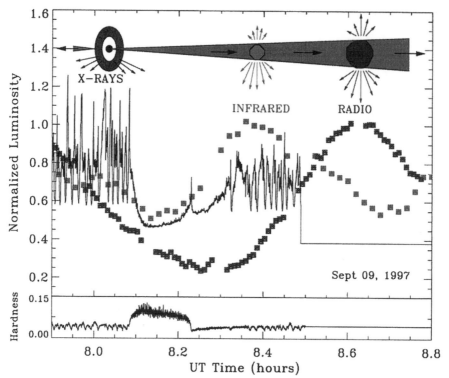

Figure 3. Radio, infrared, and X-ray light curves for GRS 1915+105 at the time of quasi-periodic oscillations on September 9, 1997 (Mirabel et al. 1998). The infrared flare starts during the recovery from the X-ray dip, when a sharp, isolated X-ray spike is observed. These observations show the connection between the rapid disappearance and follow-up replenishment of the inner accretion disk seen in the X-rays (Belloni et al. 1997), and the ejection of relativistic plasma clouds observed as synchrotron emission at infrared wavelengths first and later at radio wavelengths. A scheme of the relative positions where the different emissions originate is shown in the top part of the figure. The hardness ratio (13-60 keV)/(2-13 keV) is shown at the bottom of the figure.

Since for the March 19, 1994 event $\beta \cos \theta = 0.323$ and $\alpha = -0.8$ the flux ratio in the case of discrete condensations would be 12, whereas for a continuous jet it would be 6. For a given angular separation it was found that the observed flux ratio between the approaching and receding condensations is 8 ± 1. Similar results were found using the MERLIN observations by Fender et al. (1999). Therefore, irrespective of the distance to the source, the flux ratios for equal angular separations from the core are consistent with the assumption of a twin ejection at relativistic velocities.

5. Accretion disk instabilities and jet formation

Collimated jets seem to be systematically associated with the presence of an accretion disk around a star or a collapsed object. In the case of black holes, the characteristic dynamical times in the flow of matter are proportional to the black hole's mass, and the events with intervals of minutes in a microquasar could correspond to analogous phenomena with duration of thousands of years in a quasar of 10^9 $M_\odot$ (Sams et al. 1996; Rees 1998). Therefore, the variations with minutes of duration observed in a microquasar in the radio, IR, optical, and X rays could sample phenomena that we have not been able to observe in quasars.

X-rays probe the inner accretion disk region, radio waves the synchrotron emission from the relativistic jets. The long-term multiwavelength light curves of the superluminal sources show that the hard X-ray emission is a necessary but not sufficient condition for the formation of collimated jets of synchrotron radio emission. In GRS 1915+105 the relativistic ejection of pairs of plasma clouds have always been preceded by unusual activity in the hard X-rays (Harmon et al. 1997), more specifically, the onset of major ejection events seems to be simultaneous to a sudden drop from a luminous state in the hard X-rays (Foster et al. 1996; Mirabel et al. 1996a). However, not all unusual activity and sudden drops in the hard X-ray flux appear to be associated with radio emission from relativistic jets. In fact, in GRO J1655–40 there have been several hard X-ray outbursts without subsequent radio flare/ejection events. A more detailed summary of the long-term multifrequency studies of black hole binaries can be found in Zhang et al. (1997).

The episodes of large-amplitude X-ray flux variations on time-scales of seconds and minutes, and in particular, the abrupt dips observed in GRS 1915+105 are believed to be evidence for the presence of a black hole, as discussed below. These variations could be explained if the inner (≤ 200 km) part of the accretion disk goes temporarily into an advection-dominated mode (Abramowicz et al. 1995). In this mode, the time for the energy transfer from ions (which get most of the energy from viscosity) to electrons (which are responsible for the radiation) is larger than the time of infall to the compact object. Then, the bulk of the energy produced by viscous dissipation in the disk is not radiated (as happens in standard disk models), but instead is stored in the gas as thermal energy. This gas, with large amounts of locked energy, is advected (transported) to the compact object. If the compact object is a black hole, the energy quietly disappears through the horizon. In constrast, if the compact object is a neutron star, the thermal energy in the superheated gas is released as radiation when it collides with the surface of the neutron star and heats it up. The cooling time of the neutron star photosphere is relatively long, and in this case a slow decay in the X-ray flux is observed. Thus, one would expect the luminosity of black hole binaries to vary over a much wider range than that of neutron star binaries. The idea of advection-dominated flow has also been proposed to explain the X-ray delay in an optical outburst of GRO J1655-40.

Figure 3 shows that during large-amplitude variations in the X-ray flux of GRS 1915+105, remarkable flux variations on time-scales of minutes have also been reported at radio (Pooley & Fender 1997; Rodríguez & Mirabel 1997; Mirabel et al. 1998) and near-infrared wavelengths (Fender et al. 1997; Fender and Pooley 1998; Eikenberry et al. 1998; Mirabel et al. 1998). The rapid flares at radio and infrared wavelengths are thought to come from expanding magnetized clouds of relativistic particles. This idea is supported by the observed time shift of the emission at radio wavelengths as a function of wavelength and the finding of infrared synchrotron precursors to the follow-up radio flares (Mirabel et al. 1998). Sometimes the oscillations of radio waves appear as isolated events composed of twin flares with characteristic time shifts of 70±20 minutes (e.g. Pooley & Fender 1997; Dhawan, Mirabel & Rodríguez 1999). The time shift between the twin peaks seems to be independent of wavelength (Mirabel et al. 1998), and no Doppler boosting is observed. This suggests that these quasiperiodic flares may come from expanding clouds moving in opposite directions with non-relativistic bulk motions.

Mirabel et al. (1998) have estimated that the minimum mass of the clouds that are ejected every few tens of minutes is $\sim 10^{19}$ g. On the other hand, the estimated total mass that is removed from the inner accretion disk in one cycle of a few tens of minutes is of the order of $\sim 10^{21}$ g (Belloni et al. 1997). Given the uncertainties in the estimation of these masses, it is still unclear what fraction of mass of the inner accretion disk disappears through the horizon of the black hole. Anyway, it seems plausible that during accretion disk instabilities consisting of the sudden disappearance of its inner part, most of it is advected into the black hole, and only some fraction is propelled into synchrotron-emitting clouds of plasma.

6. Other sources of relativistic jets in the galaxy

In X-ray binaries there is a general correlation between the X-ray properties and the jet properties. The time interval and flux amplitude of the variations in radio waves seem to correspond to the time and amplitude variations in the X-ray flux. More specifically, persistent X-ray sources are also persistent radio sources, and the transient X-ray sources produce at radio wavelengths sporadic outburst/ejection events. Persistent sources of hard X-rays (e.g. 1E1740.7–2942, GRS 1758–258) are usually associated with faint, double-sided radio structures that have sizes of several arcmin (parsec scales). The radio cores of these two persistent sources are weak (≤ 1 mJy) and do not exhibit high amplitude variability. On the contrary, rapidly variable hard X-ray transients (e.g. GRS 1915+105, GRO J1655–40, XTE J1748–288) may exhibit variations in the X-ray and radio fluxes of several orders of magnitude in short intervals of time. Because these black-hole X-ray transients produce sporadic ejections of discrete, bright plasma clouds, the proper motions of the ejecta can be measured.

Probably all hard X-ray sources that accrete at super-Eddington rates produce relativistic jets. However, the observational study of these jets presents in practice several difficulties. Persistent hard X-ray sources such as Cygnus X-1 are surrounded by faint non-thermal radio features extending several arcmin (Martí et al. 1996), and even in the cases where they are well-aligned with the variable compact radio counterpart it is very difficult to prove conclusively that the faint and extended radio features are actually associated with the X-ray source. This was the case with Sco X-1, where possible large-scale radio "lobes" were found to be extragalactic sources symmetrically located in the plane of the sky with respect to Sco X-1 (Fomalont & Geldzahler 1991). On the other hand, in transient black hole binaries one may observe transient sub-arcsec jets, but unless the interferometric observations are conveniently scheduled, the evolution is too rapid and it may not be possible to follow up the proper motions of discrete clouds. This may have been the case in the radio observations of the X-ray sources Nova Oph 93 (Dela Valle, Mirabel, & Rodríguez 1994) and Nova Muscae (Ball et al. 1995), among others. In Table 1 of Mirabel & Rodríguez (1999) can be found the list of known sources of relativistic jets in the Galaxy.

7. Microblazars and gamma-ray bursts

It is interesting that in all three sources where θ (the angle between the line of sight and the axis of ejection) has been determined, a large value is found (that is, the axis of ejection is close to the plane of the sky). These values are $\theta \simeq 79°$ (SS 433; Margon 1984), $\theta \simeq 66° - 70°$ (GRS 1915+105; Mirabel & Rodríguez 1994; Fender et al. 1999), $\theta \simeq 85°$ (GRO J1655–40; Hjellming & Rupen 1995), and $\theta \geq 70°$ for the remaining sources. This result is not inconsistent with the statistical expectation since the probability of finding a source with a given θ is proportional to $\sin \theta$. We then expect to find as many objects in the $60° \leq \theta \leq 90°$ range as in the $0° \leq \theta \leq 60°$ range. However, this argument suggests that we should eventually detect objects with a small θ. For objects with $\theta \leq 10°$ we expect the time-scales to be shortened by 2γ and the flux densities to be boosted by $8\gamma^3$ with respect to the values in the rest frame of the condensation. For instance, for motions with $v = 0.98c$ ($\gamma = 5$), the time-scale will shorten by a factor of ~10 and the flux densities will be boosted by a factor of ~ 10^3. Then, for a galactic source with relativistic jets and small θ we expect fast and intense variations in the observed flux. These microblazars may be quite hard to detect in practice, both because of the low probability of small θ values and because of the fast decline in the flux. Gamma-ray bursts are at cosmological distances and ultra-relativistic bulk motion and beaming appear as essential ingredients to solve the enormous energy requirements (e.g. Kulkarni et al. 1999; Castro-Tirado et al. 1999). Beaming reduces the energy release by the beaming factor $f = \Delta\Omega/4\pi$, where $\Delta\Omega$ is the solid angle of the beamed emission. Additionally, the photon energies can be boosted to higher values. Extreme flows from collapsars with bulk Lorentz factors > 100 have been proposed as sources of γ-ray bursts (Mészáros & Rees 1997). High collimation (Dar 1998; Pugliese et al. 1999) can be tested observationally (Rhoads 1997), since the statistical properties of the bursts will depend on the viewing angle relative to the jet axis.

Recent studies of gamma-ray burst afterglows suggest that they are highly collimated jets. The brightness of the optical transient associated with GRB 990123 showed a break (Kulkarni et al. 1999), and a steepening from a power law in time t proportional to $t^{-1.2}$, ultimately approaching a slope $t^{-2.5}$ (Castro-Tirado et al. 1999). Furthermore, the achromatic steepening of the optical light curve and early radio flux decay of GRB 990510 are inconsistent with simple spherical expansion, and well fit by jet evolution. It is interesting that the power laws that describe the light curves of the ejecta in microquasars show similar breaks and steepening of the radio flux density (Rodríguez & Mirabel 1999). In microquasars, these breaks and steepenings have been interpreted (Hjellming & Johnston 1988) as a transition from slow intrinsic expansion followed by free expansion in two dimensions. Linear polarizations of about 2% were recently measured in the optical afterglow of GRB 990510 (Covino et al. 1999), providing strong evidence that the afterglow radiation from gamma-ray bursters is, at least in part, produced by synchrotron processes. Linear polarizations in the range of 2-10% have been measured in microquasars at radio (Rodríguez et al. 1995; Hannikainen et al. 1999), and optical (Scaltriti et al. 1997) wavelengths.

In this context, microquasars in our own Galaxy seem to be less extreme local analogs of the super-relativistic jets associated to the more distant γ-ray bursters. However, γ-ray bursters are different to the microquasars found so far in our own Galaxy. The former do not repeat and seem to be related to catastrophic events, and have much larger super-Eddington luminosities. Therefore, the scaling laws in terms of the black hole mass that are valid in the analogy between microquasars and quasars do not seem to apply in the case of γ-ray bursters.

8. Future perspectives

The study of relativistic jets from X-ray binaries in our own galaxy sets on a firmer basis the relativistic ejections seen elsewhere in the Universe. The analogy between quasars and microquasars led to the discovery of superluminal sources in our own galaxy, where it is possible to follow the motions of the two-sided ejecta. This permits astronomers to overcome the ambiguities that had dominated the physical interpretation of one-sided moving jets in quasars, and conclude that the ejecta consist mainly of matter moving with relativistic bulk motions, rather than waves propagating through a slowly moving jet. The Lorentz factors of the bulk motions in the jets from microquasars seem to be similar to those believed to be common in quasars. From the study of the two-sided moving jets in one microquasar, an upper limit for the distance to the source was derived, using constraints from special relativity.

Because of the relatively short time-scales of the phenomena associated with the flows of matter around stellar mass black holes, one can sample phenomena which we have not been able to observe in quasars. Of particular importance is to understand the connection between accretion flow instabilities observed in the X-rays and the ejection of relativistic clouds of plasma observed in the radio, infrared, and possibly in the optical. The detection of synchrotron infrared flares implies that the ejecta in microquasars contain very energetic particules with particle Lorentz factors of at least 10^3.

The discovery of microquasars opens several new perspectives which could prove to be particularly productive:

1. They provide a new method to determine distances using special relativity constraints. If the proper motions of the two-sided ejecta and the Doppler factor of a spectral line from one of the ejecta are measured, the distance to the source can be derived. With the rapid advance of technological capabilities in astronomy, this relativistic method to determine distances may be applied first to black hole jet sources in galactic binaries, and in the decades to come to quasars.

2. Microquasars are nearby laboratories that can be used to gain a general understanding of the mechanism of ejection of relativistic jets. The multiwavelength observations of GRS 1915+105 during large-amplitude oscillations suggest that the clouds are ejected during the replenishment of the inner accretion disk that follows its sudden disappearance beyond the

last stable orbit around the black hole. In the context of these new data, the time seems to be ripe for new theoretical advances on the models of formation of relativistic jets.

3. High sensitivity X-ray spectroscopy of jet sources with future X-ray space observatories may clarify the phenomena in accretion disks that are associated with the formation of jets.

4. More microquasars will be discovered in the future. Among them, microblazars should appear as sources with fast and large-amplitude variations in the observed flux. Depending on the beaming angle and bulk Lorentz factor they will be observed up to very high photon energies. Microquasars in our own Galaxy may be less extreme local analogs of the super-relativistic jets that seem to be associated with distant gamma-ray bursters.

5. The spin of stellar mass black holes could be derived from the observed maximum stable frequency of the QPOs observed in the X-rays, provided the mass has been independently determined. However, theoretical work is needed to distinguish between the alternative interpretations which have been proposed in the context of general relativity for the maximum stable frequency of QPOs.

6. Finally, microquasars could be test grounds for general relativity theory in the strong field limit. General relativity theory in weak gravitational fields has been successfully tested by observing in the radio wavelengths the expected decay in the orbit of a binary pulsar, an effect produced by gravitational radiation damping. We expect that phenomena observed in microquasars could be used in the future to investigate the physics of strong field relativistic gravity near the horizon of black holes.

Acknowledgments: I thank L.F. Rodríguez for permission to reproduce here results from our joint research.

References

Ables, J.G. (1969), *Proc. Astron. Soc. Australia* **1**, 237.

Abramowicz, M.A., Chen, X., Kato, S., Lasota, J.P., and Reguev, O. (1995), *ApJ* **438**, L37.

Ball, L., Kesteven, M.J., Campbell-Wilson, D., Turtle, A.J., and Hjellming, R.M. (1995), *MNRAS* **273**, 722.

Belloni, T., Méndez, M., King, A.R., van der Klis, M., and van Paradijs, J. (1997), *ApJ* **479**, L145.

Castro-Tirado, A.J., Geballe, T.R., and Lund, N. (1996), *ApJ* **461**, L99.

Castro-Tirado, A.J. et al. (1999), *Science* **283**, 2069.

Chaty, S., Mirabel, I.F., Duc, P.A., Wink, J.E., and Rodríguez, L.F. (1996), *Astron. Astrophys.* **310**, 825.

Covino, S. et al. (1999), *IAU Circular 7172*.

Curtis, H.D. (1918), *Publ. Lick Obs.* **13**, 9.

Dar, A. (1998), *ApJ* **500**, L93.

Dela Valle, M., Mirabel, I.F., Rodríguez, L.F. (1994), *Astron. Astrophys.* **290**, 803.

Dhawan, V., Mirabel, I.F., and Rodríguez, L.F. (1999), in preparation

Eikenberry, S.S., Matthews, K, Morgan, E.H., Remillard, R.A., and Nelson, R.W. (1998), *ApJ* **494**, L61.

Fabian, A.C. and Rees, M.J. (1979), *MNRAS* **187**, 13.

Fender, R.P., Garrington, S.T., McKay, D.J., Muxlow, T.W.B., Pooley, G.G., Spencer, R.E., Stirling, A.M., and Waltman, E.B. (1999), *MNRAS* **304**, 865.

Fender, R.P. and Pooley, G.G.. (1998), *MNRAS* **300**, 573.

Fender, R.P., Pooley, G.G., Brocksopp, C., and Newell, S.J. (1997), *MNRAS* **290**, L65.

Fomalont, E.B. and Geldzahler, B.J. (1991), *ApJ* **383**, 289.

Foster, R.S., Waltman, E.B., Tavani, M., Harmon, B.A., Zhang, S.N. et al. (1996), *ApJ* **467**, L81.

Hannikainen, D.C., Hunstead, R.W., Sault, R.J., McKay, and D.J. (1999), in J. Paul, T. Montmerle, and E. Auburg (eds.), *Proc. 19th Texas Symposium on Relativistic Astrophysics and Cosmology*, Paris, December (1998).

Harmon, B.A., Deal, K.J., Paciesas, W.S., Zhang, S.N., Gerard, E., Rodríguez, L.F., and Mirabel, I.F. (1997), *ApJ* **477**, L85.

Hjellming, R.M. and Han, X. (1995), in W.H.G. Lewin, J.A. van Paradijs, and E.P.J. van den Heuvel (eds.), *X-Ray Binaries*, Cambridge Univ. Press, p308.

Hjellming, R.M. and Johnston, K.J. (1988), *ApJ* **328**, 600.

Hjellming, R.M. and Rupen, M.P. (1995), *Nature* **375**, 464.

Hjellming, R.M., Rupen, M.P., Mioduszewski, A.M. et al. (1998), in *Workshop on Relativistic Jet Sources in the Galaxy*, Paris, December 12-13 (1998).

King, A. (1998), in *Workshop on Relativistic Jet Sources in the Galaxy* Paris, December 12-13 (1998).

Laplace, P.-S. (1795), *Exposition du Systeme du Monde* Volume II, 2nd edition.

López, J.A. (1997), in H.J. Habing and H.J.G.L.M. Lamers (eds.), *Planetary Nebulae*, IAU Symposium 180, Kluwer, pp197-203.

Margon, B.A., Stone, R.P.S., Klemola, A., Ford, H.C., Katz, J.I. et al. (1979), *ApJ* **230**, L41.

Margon, B.A. (1984), *Annu. Rev. Astr. Astrophys.* **22**, 507.

Martí, J., Rodríguez, L.F., Mirabel, I.F., and Paredes, J.M. (1996), *Astron. Astrophys.* **306**, 449.

Mészáros, P. and Rees, M.J. (1997), *ApJ* **482**, L29.

Michell, J. (1784), *Philosophical Transactions of the Royal Society*, pp35-57.

Milgrom, M. (1979), *Astron. Astrophys.* **79**, L3.

Mirabel, I.F., Dhawan, V., Chaty, S., Rodríguez, L.F., Robinson, C., Swank, J., and Geballe, T. (1998), *Astron. Astrophys.* **330**, L9.

Mirabel, I.F. and Rodríguez, L.F. (1994), *Nature* **371**, 46.

Mirabel, I.F. and Rodríguez, L.F. (1998), *Nature* **392**, 673.

Mirabel, I.F. and Rodríguez, L.F. (1999), *Annu. Rev. Astr. Astrophys.* **37**, in press.

Mirabel, I.F., Rodríguez, L.F., Chaty, S., Sauvage, M., Gerard, E. et al. (1996), *ApJ* **472**, L111.

Mirabel, I.F., Rodríguez, L.F., Cordier, B., Paul, J., and Lebrun, F. (1992), *Nature* **358**, 215.

Pearson, T.J. and Zensus, J.A. (1987), in J.A. Zensus and T.J. Pearson (eds.), *Superluminal Radio Sources*, Cambridge University Press, p1.

Pooley, G.G. and Fender, R.P. (1997), *MNRAS* **292**, 925.

Pugliese, G., Falcke, H., and Biermann, P.L. (1999), *Astron. Astrophys.* **344**, L37.

Rappaport, S. et al. (1994), *ApJ* **469**, 255.

Rees, M.J. (1966), *Nature* **211**, 468.

Rees, M.J. (1984), *Annu. Rev. Astr. Astrophys.* **22**, 471.

Rees, M.J. (1998), in R.M. Wald (ed.), *Black Holes and Relativistic Stars*, University of Chicago, pp79-101.

Reipurth, B. and Bertout, C. (1997), *Herbig-Haro Flows and the Birth of Stars*, IAU

Symposium 182, Kluwer.

Rhoads, J.E. (1997), *ApJ* **487**, L1.

Rodríguez, L.F., Gerard, E. Mirabel, I.F., Gómez, Y., and Velázquez, A. (1995), *ApJ Suppl.* **101**, 173.

Rodríguez, L.F. and Mirabel, I.F. (1997), *ApJ* **474**, L123.

Rodríguez, L.F. and Mirabel, I.F. (1999), *ApJ* **511**, 398.

Sams, B.J., Eckart, A., and Sunyaev, R. (1996) *Nature* **382**, 47.

Scaltriti, F., Bodo, G., Ghisellini, G., Gliozzi, M., and Trussoni, E. (1997), *Astron. Astrophys.* **327**, L29.

Tingay, S.J., Jauncey, D.L., Preston, R.A., Reynolds, J.E., Meier, D.L. et al. (1995), *Nature* **374**, 141.

Van Paradijs, J. (1995), in W.H.G. Lewin, J.A. van Paradijs, and E.P.J. van den Heuvel (eds.), *X-Ray Binaries*, Cambridge Univ. Press, p536.

Zensus, J.A. (1997), *Annu. Rev. Astr. Astrophys.* **35**, 607.

Zhang, S.N., Mirabel, I.F., Harmon, B.A., Kroeger, R.A., Rodríguez, L.F. et al. (1997), in C.D. Dermer, M.S. Strickman, and J.D. Kurfess (eds.), *Proceedings of the Fourth Compton Symposium*, AIP (New York), pp141-62.

BULK-FLOW COMPTONIZATION AND TIME LAGS DUE TO COMPTONIZATION

N. D. KYLAFIS AND P. REIG
University of Crete
Physics Department
P. O. Box 2208
710 03 Heraklion, Crete
Greece and
Foundation for Research and Technology-Hellas
P. O. Box 1527
711 10 Heraklion, Crete
Greece

1. Introduction

Compton up-scattering of low-energy photons by a *thermal* distribution of electrons is a well-known mechanism for producing power-law, high-energy, X-ray spectra. The presentation of Rashid Sunyaev in this conference covered this topic in great detail. We will demonstrate here that Compton up-scattering of low-energy photons by the *bulk motion* of electrons is an equally good mechanism for producing power-law, high-energy, X-ray spectra.

Matter accreting into a black hole acquires speeds close to the speed of light c near the black-hole horizon. There, the kinetic energy of the electrons is comparable to their rest-mass energy. Thus, by inverse Compton scattering the electrons can raise the energy of the photons to a few hundred keV.

To make the analogy between thermal Comptonization and bulk-motion Comptonization clearer, let us consider the following idealized example. Consider a box whose walls are partly reflective and partly transmissive. Inside the box there are cold ($T_e = 0$) electrons circling in a torus at speed close to the speed of light and a source of low-energy ($E_0 \ll m_e c^2$) photons. A fraction of these photons leak out instantly and the rest are scattered off the electrons several times before they escape. Since the photons hit the

C. Kouveliotou et al. (eds.), The Neutron Star – Black Hole Connection, 283–294.
© 2001 *Kluwer Academic Publishers. Printed in the Netherlands.*

electrons from every possible direction, they sample electron velocities from $\sim -c$ to $\sim c$, much like in the thermal Comptonization. The distribution of electron velocities is not Maxwellian, but this has only quantitative effects. The qualitative effects are the same. Thus, thermal Comptonization and bulk-motion Comptonization are expected to be similar.

In realistic situations the role of the leaky box is played by an accretion disk. We will demonstrate that, for *unsaturated* Comptonization in an accretion disk around a black hole, the escaping spectrum is a power law at high energies, $dN/dE \propto E^{-\alpha}$, and we will derive the photon-number index α.

The motion of the electrons need not be circular for the bulk-motion Comptonization to work. In all cases, in which the photons can on average extract energy from the kinetic energy of the electrons, the process works. The simplest such case is spherical accretion into black holes. The problem is one-dimensional, relatively simple mathematically and approximate analytic solutions have been found. We will therefore start with this problem first.

2. Comptonization in a converging fluid flow

Let's consider the problem of steady state, spherically symmetric accretion of plasma onto a compact object. If $\dot{M}$ is the mass accretion rate and

$$v(r) = c \left(\frac{r_s}{r} \right)^{1/2} \tag{1}$$

is the radial, free-fall, inward speed, then the Thomson optical depth of the flow from some radius r to infinity is given by

$$\tau_{\mathrm{T}}(r) = \int_r^\infty dr \, n_e(r) \, \sigma_{\mathrm{T}} = \dot{m} \left(\frac{r_s}{r} \right)^{1/2}, \tag{2}$$

where r_s is the Schwarzschild radius, $n_e(r)$ is the electron number density, σ_{T} is the Thomson cross section and $\dot{m}$ is the mass accretion rate in units of the Eddington accretion rate, which is defined by

$$\dot{M}_{\mathrm{E}} \equiv \frac{L_{\mathrm{E}}}{c^2} = \frac{4\pi G M m_{\mathrm{p}}}{\sigma_{\mathrm{T}} c}. \tag{3}$$

Here L_{E} is the Eddington luminosity, M is the mass of the compact object, m_{p} is the proton mass, and G is the gravitational constant. We remark here that for spherically accreting black holes, $\dot{M}$ can be significantly larger than $\dot{M}_{\mathrm{E}}$, while the escaping luminosity L remains much less than L_{E}, because the efficiency of converting accretion energy into luminosity is significantly less than unity in this case. Thus, our assumption in equation (1) of a free-fall velocity profile is justified for black holes but not for neutron stars.

For super-critical accretion into black holes ($\dot{m} > 1$), it is evident from equation (2) that there should be regions in the flow from which the photons escape diffusively. Of *qualitative* importance in our discussion below is the trapping radius r_{tr} (Rees 1978; Begelman 1979) defined by the relation

$$3\frac{v(r_{\mathrm{tr}})}{c}\tau_{\mathrm{T}}(r_{\mathrm{tr}}) \equiv 1. \tag{4}$$

No *quantitative* importance should be given to the trapping radius, because the mean number of scatterings that photons undergo depends on the photon source distribution.

For our problem, it is convenient to use not the Thomson optical depth τ_{T} but the effective optical depth

$$\tau'(r) \equiv 3\frac{v(r)}{c}\tau_{\mathrm{T}}(r) = 3\dot{m}\frac{r_{\mathrm{s}}}{r} , \tag{5}$$

defined such that $\tau'(r_{\mathrm{tr}}) = 1$.

The physical picture here is the following: Consider a low-energy photon that finds itself in the accretion flow. As the photon diffuses outward, it scatters off the inflowing electrons. Since this, in most cases, is an almost head-on collision between a fast moving electron and a low energy photon, the photon gains energy on average. The problem to be solved can then be described qualitatively as follows: Consider a source of photons with energy E_0, or a distribution around this energy, such that $E_0 \ll m_{\mathrm{e}}c^2$. The source of photons is placed anywhere in the accretion flow between the inner and outer boundaries of the flow and with any spatial distribution. For large optical depths in the accretion flow, the photons diffuse in the flow and either they are absorbed by the compact object or escape from the flow. For small optical depths (say of order unity), the majority of the input photons escape after a few scatterings without changing their energies significantly (almost coherently scattered). However, a small fraction of the input photons undergo effective scatterings with significant energy change (Hua & Titarchuk 1995). In all cases, the diffusing photons gain energy from the bulk and the thermal motion of the electrons. The objective is to determine the emergent spectrum.

Since the bulk motion involves high speeds, the thermal motion of the electrons has little effect on the emergent spectrum if $kT_{\mathrm{e}} \ll m_{\mathrm{e}}c^2$. In such a case, the asymptotic behavior of the emergent spectrum is a power-law. The exact value of the power-law exponent depends on the condition at the inner boundary (the type of compact object) and the accretion rate.

If the inner boundary is fully reflective, all input photons escape. Furthermore, the repeated reflection of the photons at the inner boundary causes them to scatter many times with the inflowing electrons and the emergent power law is as flat as it can be.

In the opposite case, where the inner boundary is fully absorptive, as is the case of a black-hole horizon, the emergent spectrum consists only of the photons that *did not* reach the inner boundary. These photons have had fewer chances to collide with the inflowing electrons and thus the emergent power law is steeper than in the fully reflective case.

In all cases, the power-law spectrum produced by the bulk motion of the electrons has a cutoff at high energies due to Compton recoil.

The problem of Compton upscattering of low-frequency photons in an optically thick, converging flow has been studied by several researchers. Blandford and Payne were the first to address this problem in a series of three papers (Blandford & Payne 1981a,b; Payne & Blandford 1981).

In the first paper of the series they wrote a photon kinetic equation, which took into account both thermal and bulk motions of electrons. This equation describes how the spectrum of the photons evolves as they diffuse through the accreting matter. It is a second-order partial differential equation for which two boundary conditions are needed. For different boundary conditions, different analytic solutions have been found. Below we describe each of these solutions separately.

It is important to remark that in the derivation of the photon kinetic equation of Blandford & Payne (1981a) terms of order $E/m_e c^2$, $kT_e/m_e c^2$, and v/c were not kept in a consistent way. This was done properly by Psaltis & Lamb (1997). They also showed that terms of order $(v/c)^2$ *must* be included in the equation even if $v/c \ll 1$. Such terms were not included in the equation of Blandford & Payne (1981a).

The various inner boundary conditions examined are treated as subsections in the rest of this section. In all cases, the outer boundary condition is that of no incoming flux.

2.1. THE BLACK HOLE IS A POINT

Payne & Blandford (1981) assumed that the accretion flow extends inwards all the way to $r = 0$ and that the flow-velocity profile is of the form $v(r) \propto r^{-\beta}$. For the inner boundary condition they assumed adiabatic compression of photons as $r \to 0$. By neglecting thermal Comptonization they showed that for $\tau' \gg 1$ all emergent spectra have a high-energy, power-law tail with photon-number index $\alpha = 1 + 3/(2 - \beta)$ (for free fall $\beta = 1/2$ and $\alpha = 3$), which is independent of the low-frequency source distribution. This result has been verified by Monte Carlo calculations (Kylafis, unpublished).

The same problem, but with thermal Comptonization included, was solved by Colpi (1988). As expected, at low values of the electron temperature ($kT_e/m_ec^2 \ll 1$) bulk Comptonization dominates. At higher temperatures thermal Comptonization dominates.

2.2. REFLECTIVE INNER BOUNDARY AT A FIXED RADIUS

Mastichiadis & Kylafis (1992) assumed that the accretion flow extends inwards to $r = R$, where R is, e.g., the radius of a neutron star. The flow-velocity profile was taken to be that of equation (1) and the surface $r = R$ was assumed to be reflective. Using the photon kinetic equation of Blandford & Payne (1981a) and neglecting thermal Comptonization, they showed that for $\tau' \gg 1$ the emergent photon-number spectrum is the sum of the photon-number spectrum derived by Payne & Blandford (1981) plus a term proportional to E^{-1}, which dominates at high energies.

This result can be understood as follows: Photons that escape without ever reaching the surface $r = R$ do not feel the inner boundary condition. In other words, these photons never learn whether there is a black hole inside the surface $r = R$ or a neutron-star surface at $r = R$. These photons will therefore emerge with a spectrum that is identical to that of accretion into black holes. On the other hand, the photons that reach the surface $r = R$ are reflected there and experience nearly head-on collisions with the inflowing electrons. Thus, for many photons repeated reflections at $r = R$ may occur. It is these photons that acquire high energy and give rise to the power law E^{-1}. A qualitative understanding of the index $\alpha = 1$ is given in Mastichiadis &Kylafis (1992).

2.3. SCHWARZSCHILD HORIZON AT THE INNER BOUNDARY

Turolla et al. (1996) and Zane et al. (1996) studied Comptonization in spherically accreting matter into a Schwarzschild black hole. With a relativistic treatment, but neglecting thermal Comptonization, they found that for $\tau' \gg 1$ the emergent photon-number spectrum is at high energies proportional to $E^{-(3-40/3\tau')}$. For moderate optical depths, their high-energy spectrum is flatter than that found by Payne & Blandford (1981).

Laurent & Titarchuk (1999; see also Titarchuk & Zannias 1998) extended the above work to include thermal Comptonization. With their Monte Carlo calculations they found that the high-energy photon-number power-law index α is a function of the accretion rate $\dot{M}$ and the electron temperature T_e. For a range of values of $\dot{M}$ and T_e they find α comparable to what is observed in black-hole candidate X-ray sources.

2.4. SURFACE AT INNER BOUNDARY HAS FIXED ALBEDO

Titarchuk, Mastichiadis & Kylafis (1997) extended the work of Mastichiadis & Kylafis (1992) to include thermal Comptonization and a surface at $r = R$ which has albedo A. For $A = 1$ the surface is reflective and for $A = 0$ it is absorptive. For a non-relativistic treatment, the case of $A = 0$ corresponds to a black hole as the central object with the horizon at $r = R$. For such a case, they found that the high-energy photon-number spectrum is a power law with index α, which depends on M and T_e. For $\dot{m}(kT_e/10 \text{ keV}) < 50$, bulk-motion Comptonization dominates the thermal one. For values of T_e satisfying this relation, the index α varies between 2 and 2.5. This has been verified by Monte Carlo calculations (Kylafis, Litchfield & Messaritaki 2000).

3. Comptonization in an accretion disk

Contrary to spherical accretion onto a compact object, which is simple mathematically but rather unphysical, it is widely accepted that in most cases matter falls onto the compact object through an accretion disk. It is therefore important to examine the effect of Compton upscattering of low-energy photons in an accretion disk.

We have considered a very simple, but nevertheless illustrative model of Comptonization in an accretion disk (for more details see Reig, Kylafis & Spruit 2000). We assume that the matter in the disk rotates with Keplerian speed from some outer radius to the black-hole horizon, namely

$$v_K(r) = c \left(\frac{r_h}{r} \right)^{1/2}, \qquad (6)$$

where r_h is the radius of the black-hole horizon and r is the radial distance in cylindrical coordinates. No radial component of the velocity is taken into account and thus the disk is more like a system of Keplerian rings than a real accretion disk. In this simple picture, we take the density n_e and the height H of the disk to be constant. Thus, the relevant parameter is the vertical Thomson optical depth $\tau_z = n_e \sigma_T H$.

Bulk-motion Comptonization that produces hard X-rays is effective only near the black-hole horizon, where the speed is very high. Thus, only those photons that find themselves in the inner part of the disk, and manage to escape eventually, will have their energy significantly increased. If the disk is geometrically thin ($H/r_h \ll 1$), but optically thick ($\tau_z \gtrsim 1$), the photons sample before escape a radial distance in the disk of order H from the point of emission (in case of emission in the disk) or from the point of first scattering (if the photons come into the disk from outside). In such a case, the soft-photon source must be near the horizon or, if it is outside the disk, to send some of its photons there. However, if the disk is geometrically and optically thick ($H/r_h \gg 1$, $\tau_z \gtrsim 1$), the photons sample before escape a radial distance which is a significant fraction of the radial extent of the disk. Thus, no matter where the source of photons is, some of the photons that diffuse in the disk will experience scatterings by the fast moving electrons. It is these photons that produce the high-energy power-law spectra.

In what follows, we will consider a geometrically thick disk ($H = 30\,r_h$) and we will neglect thermal Comptonization ($kT_e \ll m_e c^2$). Also for simplicity we will assume a monochromatic source of soft photons of energy $E_0 \ll m_e c^2$.

3.1. SOURCE OF SOFT PHOTONS INSIDE THE DISK

Consider soft photons emitted by the disk itself. We assume monochromatic point sources in the mid-plane of the disk at radii $r = 1.2\,r_h$, $r = 7\,r_h$, and $r = 25\,r_h$. Each source emits isotropically in its rest frame photons of energy $E_0 = 1$ keV. In the lab frame, the emission is forwardly (i.e., in the direction of the flow at the point of emission) peaked and the photons have a distribution of energies. The emergent spectra from the disk have been calculated by Monte Carlo simulations. For the three point sources considered above, the emergent spectra are given respectively in Figures 1a, 1b and 1c for various values of the optical depth τ_z.

Due to the high speed at $r = 1.2\,r_h$ ($v_K = 0.9c$), the distribution of source-photon energies in the lab frame is very wide. This distribution becomes narrower as we consider sources further out in the disk. In all cases however, the emergent spectrum has a high-energy power-law tail that extends to a few hundred keV. The photon-number spectral index α is in the range 2.3–3.

Compton upscattering of soft photons in the boundary layer between a neutron star and an accretion disk has been studied by Hanawa (1991).

3.2. SOURCE OF SOFT PHOTONS OUTSIDE THE DISK

Consider now soft photons of energy $E_0 = 1$ keV impinging normally on the disk at $r = 7\,r_h$. The emergent spectra from the disk have been calculated by Monte Carlo for the same values of τ_z as in Figure 1 and are given in Figure 2. It is evident that in all cases a high-energy power-law spectrum emerges from the disk. The photon-number spectral index α is the same as in the corresponding cases of Figure 1b.

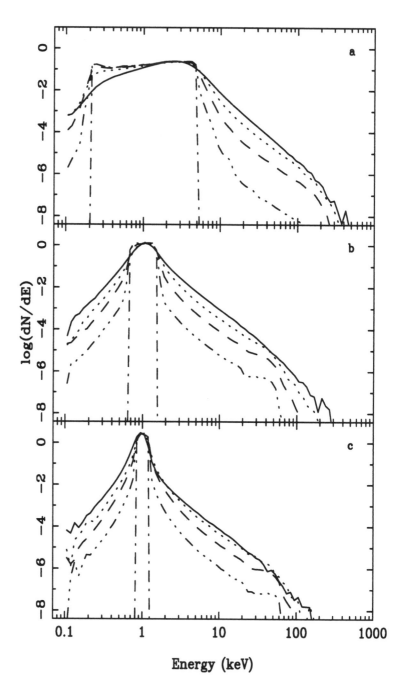

Figure 1. Photon flux as a function of energy. The spectra in each panel correspond to the values $\tau_z = 0$ (dot-dashed), 0.1 (triple dot-dashed), 1 (dashed), 3 (dotted), and 10 (solid) of the vertical optical depth of the disk. The three panels correspond to the positions of the soft-photon source in the midplane of the disk, namely $r = 1.2\, r_h$ (top), $r = 7\, r_h$ (middle) and $r = 25\, r_h$ (bottom).

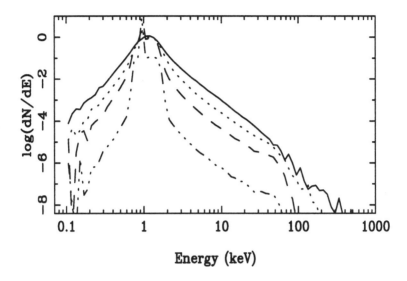

Figure 2.   Same as Fig 1, but the photons of energy $E_0 = 1$ keV impinge normally on the disk at $r = 7\,r_{\rm h}$.

4. Time lags due to Compton scattering

Let us examine again a simple but illustrative example. Consider a point source of photons of energy E_0 at the center of a homogeneous spherical cloud of plasma of radius R and Thomson optical depth τ. For $\tau \gtrsim 3$, so that the diffusion approximation is valid, the mean travel time of the photons before escape is

$$\bar{t} \approx \bar{n}_{\rm sc} \frac{l}{c} \approx \tau^2 \frac{R/\tau}{c} = \tau \frac{R}{c}, \tag{7}$$

where $\bar{n}_{\rm sc} \approx \tau^2$ is the mean number of scatterings and $l \equiv R/\tau$ is the mean free path. Thus, due to diffusion, the mean travel time of the photons in the cloud is τ times the crossing time R/c.

   If the scattering cloud is hot $(kT_{\rm e} \gg E_0)$, then on average the more times a photon is scattered in the cloud the more energy it acquires from the hot electrons. Therefore, the hard photons *lag* the soft photons by a time interval (Sunyaev & Titarchuk 1980)

$$\Delta t \sim \frac{1}{c} \frac{R}{\tau} \ln \left(\frac{E_{\rm hard}}{E_{\rm soft}} \right). \tag{8}$$

A time lag of photons of energy E_{hard}, relative to photons of energy E_{soft}, proportional to $\ln(E_{\text{hard}}/E_{\text{soft}})$ has been observed in Cygnus X-1 (Nowak et al. 1999).

Suppose now that the point source of photons of energy E_0 is varying periodically with period P. If the mean travel time $\bar{t}$ of the photons in the cloud is much greater than P, the variability of the source will not be detected by a distant observer, because it will be washed out by the distribution of photon-escape times (Kylafis & Klimis 1987; Kylafis &Phinney 1989). In other words, Fourier frequencies ν satisfying the relation

$$\nu \gg \frac{1}{\tau}\frac{c}{R} \tag{9}$$

will not be seen in observations of the source. All variability with frequencies ν satisfying inequality (9) will be washed out in the cloud.

5. Time lags in X-ray observations

Following Vaughan & Nowak (1997), we consider simultaneous X-ray observations of a source in two energy bands. Let $f_1(t_k)$ be the X-ray flux observed at times t_k in the energy band $[E_1, E_1 + \Delta E_1]$ and $f_2(t_k)$ be the simultaneous flux in the energy band $[E_2, E_2 + \Delta E_2]$. The Fourier transforms of $f_1(t_k)$ and $f_2(t_k)$ at frequency ν_j are

$$F_1(\nu_j) = a_1(\nu_j)e^{i\phi_1(\nu_j)} \tag{10}$$

and

$$F_2(\nu_j) = a_2(\nu_j)e^{i\phi_2(\nu_j)} \tag{11}$$

respectively, where $P_1(\nu_j) = |a_1(\nu_j)|^2$ and $P_2(\nu_j) = |a_2(\nu_j)|^2$ are the corresponding power spectra.

Consider now the cross spectrum

$$F_1^*(\nu_j)F_2(\nu_j) = a_1(\nu_j)a_2(\nu_j)e^{i[\phi_2(\nu_j)-\phi_1(\nu_j)]} \equiv A(\nu_j)e^{i\phi(\nu_j)}. \tag{12}$$

For better statistics consider the average

$$C(\nu_j) = < F_1^*(\nu_j)F_2(\nu_j) > \tag{13}$$

of several such cross spectra, which is computed for different time intervals of the same lightcurve. The Fourier *phase lag* $\phi(\nu_j)$ is the phase of the average cross power spectrum, i.e.,

$$\phi(\nu_j) = \arg[C(\nu_j)]. \tag{14}$$

If $\phi(\nu_j)$ remains constant from observation to observation, then we say that the coherence function of the X-ray source is equal to 1 (Vaughan & Nowak 1997). For the X-ray sources Cyg X-1, GX 339-4, 4U 1705-44, and 4U 0614+091 it has been shown by Ford et al. (1999) that the coherence function is close to 1 for a wide range of Fourier frequencies.

The Fourier *time lag* is constructed from $\phi(\nu_j)$ by dividing through by $2\pi\nu_j$, i.e.,

$$t_{\text{lag}}(\nu_j) \equiv \frac{\phi(\nu_j)}{2\pi\nu_j}. \tag{15}$$

The Fourier time lag can be either positive or negative. If the energies $[E_1, E_1 + \Delta E_1]$ are smaller than the energies $[E_2, E_2 + \Delta E_2]$, then a positive time lag indicates that the light curve of the harder photons lags the light curve of the softer photons. For simple Comptonization models (such as the one discussed above of a point source at the center of a uniform, hot cloud), a constant time lag is expected at all Fourier frequencies. According to equation (15), this means that $\phi(\nu_j) \propto \nu_j$. Instead, what has been found by Ford et al. (1999) is that $\phi(\nu_j)$ is nearly constant at all Fourier frequencies. According to equation (15) this means that $t_{\text{lag}}(\nu_j) \propto 1/\nu_j$.

Now the question arises: How can this happen if we want to save Comptonization as a model for the energy spectra? An answer was provided by Kazanas, Hua & Titarchuk (1997) and Hua, Kazanas & Titarchuk (1997) (see also the contribution of D. Kazanas in this book).

Consider a non-uniform, spherical cloud of plasma with temperature T_e and electron density

$$n_e(r) = n_0 \frac{r_0}{r}, \tag{16}$$

where n_0 is a constant density and r_0 is the inner radius of the cloud (assumed much smaller than the outer radius). Then the optical depth to electron scattering from radius r_0 to radius r is

$$\tau_T(r) = \int_{r_0}^{r} n_e(r) \, \sigma_T \, dr = n_0 \, r_0 \, \sigma_T \, \ln\frac{r}{r_0}. \tag{17}$$

In other words, there is equal optical depth per decade of radius. For simplicity, let's assume that the total optical depth of the cloud is equal to 1. If we now consider a source of photons of energy $E_0 \ll kT_e$ at $r = 0$, then the emitted photons are scattered once (on average) in the cloud and *this scattering occurs with equal probability in all decades of radius*. The energy gain of the scattered photon is independent of where the scattering took place. Let us visualize the decades of radius as "zones" and consider a source which is variable at all frequencies. "Zones" with large radii produce large time lags between the scattered and the unscattered photons, but allow small Fourier frequencies to pass (see equation 9). "Zones" with small radii produce small time lags with scattering, but allow large Fourier frequencies to pass. Thus, if a photon is scattered at $r_{\rm sc}$, a time lag $t_{\rm lag} \sim r_{\rm sc}/c$ will occur and this will be seen at Fourier frequency $\nu \sim c/r_{\rm sc} \sim 1/t_{\rm lag}$.

6. Summary and conclusions

Power-law, high-energy, X-ray spectra like the ones observed in black-hole X-ray sources can be produced either with thermal Comptonization or with bulk-motion Comptonization. However, the observed time lags, which reach ~ 1 s, cannot be explained by bulk-motion Comptonization. This is because the Comptonization occurs near the black hole (i.e., at small radii) and the resulting time lags are much smaller than the ones observed.

Thermal Comptonization in a hot, uniform cloud cannot explain the observed correlation between time lag and Fourier frequency. This correlation can be explained with a hot, non-uniform cloud (Kazanas et al. 1997), but this cloud of temperature $kT_{\rm e} \sim 50$ keV must extend out to $\sim 3 \times 10^{10}$ cm.

References

Begelman, M.C. (1979), *MNRAS* **187**, 237.
Blandford, R.D. and Payne, D.G. (1981a), *MNRAS* **194**, 1033.
Blandford, R.D. and Payne, D.G. (1981b), *MNRAS* **194**, 1041.
Colpi, M. (1988), *ApJ* **326**, 223.
Ford, E.C., van der Klis, M., Méndez, M., van Paradijs, J., and Kaaret, P. (1999), *ApJ* **512**, L31.
Hanawa, T. (1991), *ApJ* **373**, 222.
Hua, X.-M., Kazanas, D., and Titarchuk, L. (1997), *ApJ* **482**, L57.
Hua, X.-M. and Titarchuk, L.G. (1995), *ApJ* **449**, 188.
Kazanas, D., Hua, X.-M., and Titarchuk, L. (1997), *ApJ* **480**, 735.
Kylafis, N.D. and Klimis, G.S. (1987), *ApJ* **323**, 678.
Kylafis, N.D., Litchfield, S.J., and Messaritaki, E.I. (2001), *A&A*, to be submitted.
Kylafis, N.D. and Phinney, E.S. (1989), in H. Ögelman and E.P.J. van den Heuvel (eds.), *Timing Neutron Stars*, NATO ASI Ser C262 (Kluwer, Dordrecht), p731.
Laurent, P. and Titarchuk, L. (1999), *ApJ* **511**, 289.
Mastichiadis, A. and Kylafis, N.D. (1992), *ApJ* **384**, 136.
Nowak, M.A., Vaughan, B.A., Wilms, J., Dove, J.B., and Begelman, M.C. (1999), *ApJ* **510**, 874.
Payne, D.G. and Blandford, R.D. (1981), *MNRAS* **196**, 781.
Psaltis, D. and Lamb, F.K. (1997), *ApJ* **488**, 881.
Rees, M.J. (1978), *Phys. Scr.* **17**, 193.
Reig, P., Kylafis, N.D., and Spruit, H. (2001), *A&A*, to be submitted.
Sunyaev, R. and Titarchuk, L. (1980), *A&A* **86**, 121.
Titarchuk, L., Mastichiadis, A., and Kylafis, N.D. (1997), *ApJ* **487**, 834.
Titarchuk, L. and Zannias, T. (1998), *ApJ* **493**, 863.
Turolla, R., Zane, S., Zampieri, L., and Nobili, L. (1996), *MNRAS* **283**, 881.
Vaughan, B.A. and Nowak, M.A. (1997), *ApJ* **474**, L43.
Zane, S., Turolla, R., Nobili, L., and Erna, M. (1996), *ApJ* **466**, 871.

GRS 1915+105 AS A BLACK HOLE ACCRETION DISK LABORATORY

T. BELLONI

Osservatorio astronomico di Brera
Via E. Bianchi 46, I-23807 Merate, Italy

Abstract. GRS 1915+105 is a unique source among black hole candidates because of its extreme and complex variability. The major cause for these variations has been attributed to the effects of a thermal-viscous instability in the inner accretion disk. Here I compare the basic properties of this source with those of other systems and show how this peculiar system can be used as a laboratory for testing and extending accretion disk models.

1. Introduction

Unlike the case of Low-Mass X-ray Binaries (LMXBs) containing a neutron star (see van der Klis 1995), the X-ray properties of black-hole candidates (BHC) have so far escaped a tight classification. In the recent years, thanks to the observation of a number of black-hole transients, it has been possible to isolate four characteristic (or "canonical") states of BHCs as a function of accretion rate. As a function of decreasing accretion rate, a "canonical" BHC is expected to go through: a Very High State (VHS), a High State (HS), an Intermediate State (IS) and a Low State (LS) (see van der Klis 1995; Méndez & van der Klis 1998 for a more detailed description of the four states). Unfortunately, not all sources are strictly "canonical", in the sense that some objects do not change state despite obvious large changes in luminosity (and therefore presumably in accretion rate, see Tanaka & Lewin 1995), and some others show mixed or anomalous features that make it difficult to clearly classify them (see Oosterbroek et al. 1997, Tanaka & Lewin 1995). However, the state classification is a good starting point for the understanding of the X-ray properties of BHCs.

C. Kouveliotou et al. (eds.), The Neutron Star – Black Hole Connection, 295–300.

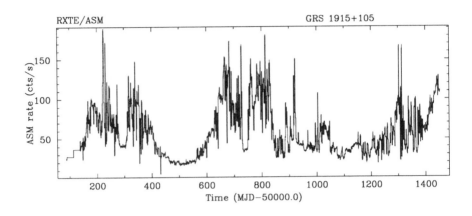

Figure 1. RXTE/ASM light curve of GRS 1915+105. Bin size is 1 day.

2. Where do we go from here?

In order to assess whether this classification is meaningful in terms of the
physics of accretion disks, there are essentially two main ways. The first is
to sit on a single persistent source, like Cyg X−1 or GX 339−4, and wait for
a rare state transition. Since the launch of the Rossi X-ray Timing Explorer
(RXTE), these two sources, which are usually observed in their LS, showed
a transition, to the IS for Cyg X−1 (Belloni et al. 1996) and to the HS
for GX 339−4 (Belloni et al. 1999). The problem with this approach is the
relatively long waiting times. The second is to wait for new or recurrent
transients and follow their evolution. Quite a few have indeed been observed
by RXTE since its launch. This approach is complicated by the necessity
of comparing different systems. However, there is a third possibility: find a
single source that does more and faster. Luckily, we do have such a source.

3. GRS 1915+105

The X-ray transient GRS 1915+105, discovered in 1992 with the Granat
satellite (Castro Tirado, Brandt & Lund 1992), did not stand out among
the others until 1994, when superluminal radio jets were discovered from
the source (Mirabel & Rodríguez 1994). With the launch of RXTE, thanks
to a very extensive campaign of observations, what appeared to be yet another X-ray transient turned out to be one of the most extraordinary X-ray
sources in the sky because of its extreme variability, which is accompanied by strong spectral changes. Examples of light curves can be seen in
Greiner, Morgan & Remillard (1996), Belloni et al. (1997a,b), Taam, Chen

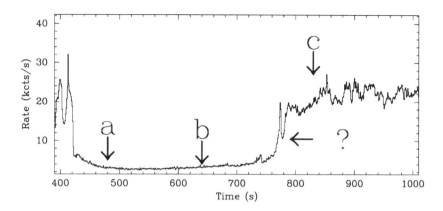

Figure 2. Ten minutes of PCA light curve of GRS 1915+105. (a): the inner portion of the disk is not observable; (b): refilling is under way; (c): the disk is all observable.

& Swank (1997), Belloni et al. (2000). With more than 300 observations made with RXTE, we can say that GRS 1915+105 already showed more "state transitions" than all other X-ray binaries together in almost 40 years of X-ray astronomy. But: are they state transitions as the ones observed from more "well behaved" sources? In this paper I will try to summarize some results on the spectral variability of this source and to compare and link them to those from "canonical" black hole candidates.

3.1. THE UNSTABLE DISK

At first sight, it does not seem likely that anything "canonical" could be applied to GRS 1915+105. Figure 1 shows the light curve of the source obtained with the All-sky Monitor (ASM) on board RXTE since the beginning of 1996: how would state transitions be defined here? Indeed the variability seen here has been explained in a completely different way. Belloni et al. (1997a,b) proposed a model for the interpretation of the time variability of GRS 1915+105 involving the appearing/disappearing of the inner region of the accretion disk, driven by a thermal-viscous instability. In this framework (see Figure 2) the drops in count rate correspond to the disappearing from view of a portion of the disk (a), the low-rate intervals to its refilling (b), the turn-on to the re-observability of that portion (c) and to the subsequent shining of the whole disk. This model allowed Belloni et al. (1997b) to map the radial dependence of the viscous time scale (measured as the length of the refill period): the results agree well with the predictions of the standard thin accretion disk theory.

3.2. TEMPERATURE OSCILLATIONS

The instability model could qualitatively explain all observations but one. This led Belloni et al. (2000), after the analysis of a large number of observations, to the identification of an additional feature in the light curves. An example of such a feature is the sharp (and soft) dip marked by a question mark in Figure 2. Overall the picture that emerges is that of an accretion disk that, besides undergoing rather dramatic episodes of instability, has a varying inner temperature. These temperature variations, which must be related to changes in the local accretion rate, can be very fast; in some cases, the temperature can vary substantially on a time scale well below a second (Belloni et al. 2000).

3.3. HOW CANONICAL IS GRS 1915+105?

When the instability is not at work, i.e. when the disk is observable in its entirety, the energy spectrum of GRS 1915+105 is not unlike that of a typical BHC in its VHS: a combination of a bright disk component and a rather steep power law. Also its power spectrum resembles that of a VHS: a band-limited noise component with $\sim$10% fractional rms (see Belloni 1998). Interestingly, the VHS as observed in GS 1124−68 (Miyamoto et al. 1994) and GX 339−4 (Miyamoto et al. 1991) seems to have a different 'timing mode', when the variability is much lower and the transitions to which can be fast. This could correspond to the "soft-disk" intervals mentioned in the previous section. The conclusion is that GRS 1915+105 could be a source in the VHS (which is associated to a high accretion rate), undergoing some serious instability.

During the instability intervals, the energy spectrum changes in the following way: the disk component becomes softer, due to the non-observability of the inner portion of the disk; the power law flattens. During extremely long intervals of instability, the disk component becomes so soft that it is no longer observable, and the power law flattens down a photon index of 1.6 (Belloni et al. 2000). In the power spectrum, a flat-top noise component appears, whose break frequency moves to lower and lower frequencies as the power law flattens (Belloni 1998, Trudolyubov, Churazov & Gilfanov, 1999). This power spectrum is similar to the Intermediate/Low state and for low break frequencies approaches the Low state. The identification of the intervals with the canonical IS/LS is however complicated by the fact that such an instability is not observed in any other source. It seems more reasonable to argue that the instability *mimics* some of the features of these states, causing the power law to flatten and the characteristic time scale in the power spectrum to shorten.

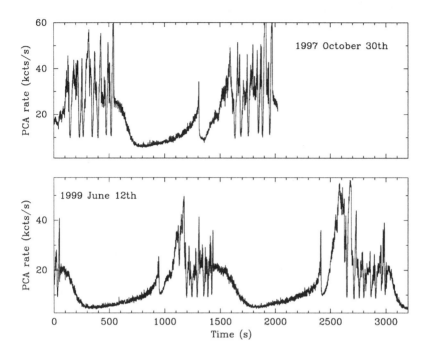

Figure 3.   Examples of two RXTE/PCA light curves of GRS 1915+105.

4. GRS 1915+105 as a laboratory

As we have seen, GRS 1915+105 provides us with a challenging set of features. A few are similar to those of standard BHCs, but many are absolutely peculiar to this source. What is important to note is that it is precisely the combination of peculiar and normal aspects that makes this source a laboratory for the study of accretion disks. As an example, the viscous time scale *vs* radius dependence found by Belloni et al. (1997b) is a fundamental dependence which applies to all thin accretion disks, and not only to that of GRS 1915+105. More is there to be learned. The soft-hard oscillations of the disk described above are independent of the instability events, since they are also observed when the disk appears to be stable (Belloni et al. 2000). However, whenever the inner disk is unstable and appears again after having been unobservable, the disk *always* goes to its soft state (see Figure 2), even if only for a few seconds. This provides a link between the two otherwise unrelated phenomena, link that can be exploited for the understanding of both.

An important point to remark about the variability of GRS 1915+105 is that, although a number of remarkable patterns have been observed with RXTE in the past four years, these patterns tend to repeat, to the extent that they could be classified into a small number of classes by Belloni et al. (2000). The similarity between the structure of the variability in different observations can be striking, as shown in the two examples in Figure 3, where two observations more than one and a half years apart are shown. Not only the overall shape, but even small details are similar in both observations. This of course cannot be a coincidence. What is observed here must be something very fundamental in the accretion disk of this source. Not only the disk becomes unstable, but the instability triggers some very specific and reproducible response.

In conclusion, modeling the observed characteristics of GRS 1915+105 is a mighty challenge, but it is the best laboratory we have to test and understand accretion models. It is not clear why no other source behaves like this one. It might be an effect of the high accretion rate or of the high spin of the black hole, or a combination of both. Whatever the reason, the perturbations induced on the (otherwise normal) accretion disk allow us to study its response to them, thus working for us as a probe into the elusive inner disk.

References

Belloni, T. et al. (1996), *ApJ* **472**, L107.
Belloni, T. et al. (1997a), *ApJ* **479**, L145.
Belloni, T. et al. (1997b), *ApJ* **488**, L109.
Belloni, T. (1998), *New. Astr. Rev.* **42**, 585.
Belloni, T. et al. (1999), *ApJ* **519**, L159.
Belloni, T. et al. (2000), in preparation.
Castro-Tirado, A.J., Brandt, S., and Lund, N. (1992), *IAUC No. 5990*.
Greiner, J., Morgan, E.H., and Remillard, R.A. (1996), *ApJ* **473**, L107.
Méndez, M. and van der Klis, M. (1997), *ApJ* **479**, 926.
Mirabel, I.F. and Rodríguez, L.F. (1994), *Nature* **371**, 46.
Miyamoto, S. et al. (1991), *ApJ* **383**, 784.
Miyamoto, S. et al. (1994), *ApJ* **435**, 398.
Oosterbroek, T. et al. (1998), *A&A* **340**, 431.
Taam, R.E., Chen, X., and Swank, J.H. (1997), *ApJ* **485**, L83.
Tanaka, Y. and Lewin, W.H.G. (1995), in W.H.G. Lewin et al. (eds.), *X-ray binaries*, Cambridge Univ. Press, p126.
Trudolyubov, S., Churazov, E. and Gilfanov, M. (1999), *Astron. Lett.* **25**, 718.
Van der Klis, M., 1995, in W.H.G. Lewin et al. (eds.), *X-ray binaries*, Cambridge Univ. Press, p252.

WEIGHING BLACK HOLES WITH THE LARGEST OPTICAL TELESCOPES

E.T. HARLAFTIS

Astronomical Institute, National Observatory of Athens
P. O. Box 20048, Athens - 118 20, Greece

AND

A.V. FILIPPENKO

Department of Astronomy, University of California,
Berkeley, CA 94720-3411, USA

Abstract. The advent of the large effective apertures of the Keck telescopes has resulted in the determination with unprecedented accuracy of the mass functions, and in most cases the mass ratios, of faint ($R \approx 21$ mag) X-ray transients (GS 2000+25, GRO J0422+32, Nova Oph 1977, Nova Vel 1993).

1. Dynamical Evidence for Black Holes in X-ray Binaries

Zel'dovich and Novikov (1966) were the first to propose the technique which is still in use for "weighing" black holes. They suggested that black holes could be detected indirectly from light emitted through the interaction with a donor star in an X-ray binary system. The motion of the donor star around the black hole would produce a radial velocity sinusoidal curve which could be detected from the Doppler shifts of the photospheric absorption lines of the donor star. The semi-amplitude (K) of the curve together with the binary period (P) determine the mass function of the black hole (a lower limit to its mass), using Kepler's third law: $f_x = PK^3/(2\pi G)$. Indeed, X-ray binaries were found in the late 1960s and the first black-hole candidate, Cyg X-1, in 1971 (Oda et al. 1971). Efforts in measuring the mass of the candidate black hole were affected by uncertainties in the evolution of the massive donor star, and with a low mass function ($f_x = 0.22 \pm 0.01 \ M_\odot$; Bolton 1975) this was not regarded as unequivocal evidence for a black hole (see Herrero et al. 1995 for the most recent work).

C. Kouveliotou et al. (eds.), The Neutron Star – Black Hole Connection, 301–305.

The observational effort was then turned to observations of X-ray novae (XRNs, a sub-group of low-mass X-ray binaries) in the 1980s as the most suitable targets since the low-mass companion star allows the mass function of the black hole to be a good approximation of the mass in a high-inclination system. The prototype target was now A0620–00, but unfortunately its mass function was close to the maximum mass of a neutron star ($f_x = 3.2 \pm 0.2$ $M_\odot$; McClintock and Remillard 1986). It was only in 1992 that a mass function of a candidate black hole in the XRN 1989 GS 2023+338 was found to be much heavier than the maximum mass of a neutron star ($f_x = 6.08 \pm 0.06$ $M_\odot$; Casares et al. 1992).

Since then, the efforts have been directed toward measuring actual masses, thus producing the first observed and theoretical mass distribution of black holes (Bailyn et al. 1998; Miller et al. 1998; Fryer 1999). The determination of the masses of stellar remnants after supernova explosions is essential for an understanding of the late stages of evolution of massive stars. Very recently, the first observational evidence for a supernova or hypernova origin with a progenitor mass $> 30 M_\odot$ that produced a black hole of 7.0 ± 0.2 $M_\odot$ in GRO J1655–40 was found with the Keck-I telescope (from high metal abundances that were presumably deposited onto the surface of the companion F5IV star by the supernova explosion; Israelian et al. 1999).

2. Keck Observations of Black Hole X-ray Binaries

In the last 8 years, X-ray satellites have found 6 XRNs with identified stars in the optical (Nova Muscae 1991, Cheng et al. 1992; Nova Persei 1992, Casares et al. 1995 and ref. therein; Nova Sco 1994, Bailyn et al. 1995; Nova Vel 1993, Filippenko et al. 1999; GRO J1719–24, e.g., Ballet et al. 1993; XTE J1550–564, e.g., Smith et al. 1998). Unlike classical novae, XRNs are accretion-driven events that show disk outbursts with a typical rise of 8–10 mag in a few days and a subsequent decline over several months. After the XRN has subsided into quiescence, the accretion disk does not dominate the observed flux, rendering the companion star visible.

Utilizing the Doppler effect produced by the shifting photospheric lines due to the orbital motion of the companion star around the black hole, but now with the 10-m Keck-I and Keck-II telescopes, Filippenko et al. (see 1999 for full list of references) have produced the four most accurate mass functions, reducing the uncertainties by almost an order of magnitude compared to previous results ($f_x = 5.0 \pm 0.1$ $M_\odot$, 1.2 ± 0.1 $M_\odot$, 4.7 ± 0.2 $M_\odot$, 3.2 ± 0.1 $M_\odot$, respectively for GS 2000+25, GRO J0422+32, Nova Oph 1977, Nova Vel 1993). Figure 1 shows the great improvement that the large aperture of the Kecks offers in relation to 4-m-class telescopes in extracting radial velocity curves of the motion of the donor star around the black hole.

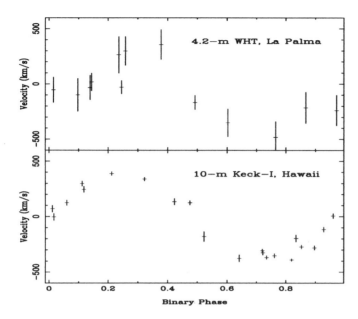

Figure 1. *Bottom:* the radial velocity curve of the companion star to the black hole GRO J0422+32 as extracted from spectra near Hα (Keck-I/LRIS) in just 1 night (Harlaftis et al. 1999). *Top:* the radial velocity curve of the companion star to the black hole GRO J0422+32 as extracted from 4.2-m WHT/ISIS near-infrared spectra (8450–8750 Å) in 3 nights (Casares et al. 1995). The reduction in the individual measurement uncertainties is a factor of four using Keck-I. The sinusoidal fit to the radial velocities gives $K = 338 \pm 39$ km s^{-1} with the WHT data and $K = 372 \pm 10$ km s^{-1} with the Keck data for the radial velocity semi-amplitude of the companion star. This yields better accuracy in the estimate of the lower limit of the black hole's mass, from $(P\,K^3)/(2\,\pi\,G) = 0.85 \pm 0.30\ M_\odot$ to $1.21 \pm 0.06\ M_\odot$ in the low-inclination system GRO J0422+32.

Moreover, determination of the mass ratio (from the rotational broadening of the photospheric lines in the companion star) and the inclination (inferred from the ellipsoidal modulations of the companion star), when combined with the mass function, can fully describe the system's parameters and the masses of the binary components. Indeed, further work on the Keck data by Harlaftis et al. (1999) has determined mass ratios for the first time for binaries as faint as 21 mag using a χ^2 optimization technique to extract the rotational broadening of the absorption lines of the donor star ($q = 0.12 \pm 0.08$, 0.042 ± 0.012, 0.014 ± 0.019, respectively for GRO J0422+32, GS 2000+25, Nova Oph 1977; see Filippenko et al. 1999 for references).

The accretion disk in its quiescent state has mainly been undetected so far by X-ray satellites but can be studied in the optical. An imaging technique, Doppler tomography, shows the accretion disks in GS 2000+25 and Nova Oph 1977 to be present. Further, mass transfer from the donor star continues vigorously to the outer disk as evidenced by the "bright spot," the impact of the gas stream onto the outer accretion disk (Figure 2; Harlaftis et al. 1996, 1997). The lithium line at 6707 Å was only detected in GS 2000+25 (see Martin et al. 1994 for lithium in X-ray binaries).

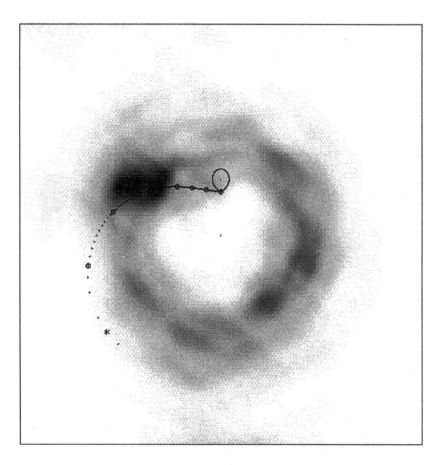

Figure 2.   The Hα Doppler image of the accretion disk surrounding the black hole GS 2000+25, as reconstructed from 13 Keck-I/LRIS spectra (1/2 hour each over the 8.3 hour binary period). By projecting the image in a particular direction, one obtains the Hα emission-line profile as a function of velocity; for example, projecting toward the top results in the profile at orbital phase 0.0, which has a blueshifted peak. The path in velocity coordinates of gas streaming from the dwarf K5 secondary star is illustrated, and the bright spot at the upper left results from collision of the gas stream with the accretion disk around the black hole. The image was reconstructed by applying Doppler tomography, a maximum entropy technique, to the phase-resolved spectra, as described by Harlaftis et al. (1996).

References

Bailyn, C. et al. (1995), *Nature* **378**, 157.
Bailyn, C. et al. (1998), *ApJ* **499**, 367.
Ballet, J. et al. (1993), IAUC No. 5874.
Bolton, C.T. (1975), *ApJ* **200**, 269.
Casares, J. et al. (1992), *Nature* **355**, 614.
Casares, J. et al. (1995), *MNRAS* **276**, 35.
Cheng, F.H. et al. (1992), *ApJ* **397**, 664.
Filippenko, A.V. et al. (1999), *PASP* **111**, 969.
Fryer, C.L. (1999), *ApJ* **522**, 413.
Harlaftis, E.T., Horne, K., and Filippenko, A.V. (1996), *PASP* **108**, 762.
Harlaftis, E.T., Steeghs, D., Horne, K., and Filippenko, A.V. (1997), *AJ* **114**, 1170.
Harlaftis, E.T., Collier, S.J., Horne, K., and Filippenko, A.V. (1999), *A&A* **341**, 491.
Herrero, A. et al. (1995), *A&A* **297**, 556.
Israelian, G. et al. (1999), *Nature* **401**, 142.
Martin, E. et al. (1994), *ApJ* **435**, 791.
McClintock, J.E. and Remillard, R.A. (1986), *ApJ* **308**, 110.
Miller, J.C., Shahbaz, T., and Nolan, L.A. (1998), *MNRAS* **294**, L25.
Oda, M. et al. (1971), *ApJ* **166**, L10.
Smith, D.A., Marshall, F.E., Smith, E.A. (1998), IAUC No. 7008.
Zel'dovich, Ya.B. and Novikov, I.D. (1966), *Sov. Physics – Uspekhi* **8**, 522.

PHYSICS OF PLASMA OUTFLOWS

S.V. BOGOVALOV

Astrophysics Institute at the
Moscow Engineering Physics Institute
Kashirskoje Shosse 31, Moscow, 115409, Russia

AND

K. TSINGANOS

University of Crete, Department of Physics
Heraklion, Crete 71003, Greece

1. Introduction

Plasma outflow from the environment of stellar or galactic objects, in the form of collimated jets, is a widespread phenomenon in astrophysics. The most dramatic illustration of such highly collimated outflows may be found in the relatively nearby regions of star formation; for example, in the Orion Nebula alone the Hubble Space Telescope (HST) has observed hundreds of aligned Herbig-Haro objects (O'Dell & Wen 1994). There is also a long catalogue of jets associated with AGN and possibly supermassive black holes (Jones and Wehrle 1994; Biretta 1996). To a lesser extent, jets are also associated with older mass losing stars and planetary nebulae, (Livio 1997), symbiotic stars (Kafatos 1996), black hole X-ray transients (Mirabel & Rodriguez 1996), supersoft X-ray sources (Kahabka & Trumper 1996), low- and high-mass X-ray binaries and cataclysmic variables (Shahbaz et al 1997). Even more exotic phenomena in the Universe such as gamma-ray bursts can be connected with jet outflows of relativistic plasma (Kulkarni et al. 1999).

The interpretation of these observations requires an understanding of basic properties of transonic flows of magnetized plasmas. In this paper we focus our attention on some of those properties of plasma outflows which are essential to the physics of radio pulsars and jets.

C. Kouveliotou et al. (eds.), The Neutron Star – Black Hole Connection, 307–312.
© 2001 *Kluwer Academic Publishers. Printed in the Netherlands.*

2. Outflows of magnetized plasma from oblique rotators

The rotation axis of radio pulsars is not aligned with their magnetic moment. The problem of the plasma flow in the magnetosphere of these objects is extremely complicated. To simplify it several models of radio pulsars have been proposed. Firstly, Goldreich & Julian (1969) considered an axisymmetrically rotating star with an initially dipolar magnetic field. However, even with such simplification the structure of the axisymmetric flow remains too complicated. There have been several attempts to solve this problem in the massless approximation (Michel 1973; Beskin, Gurevich & Istomin 1983; Lyubarskii 1990), most recently by Contopoulos, Kazanas & Fendt (1999). It is a matter of fact, however, that winds cannot be studied in the massless approximation. Michel (1969) was the first to use the MHD approximation for the investigation of the relativistic plasma flow from an axisymmetric rotator with a prescribed split-monopole poloidal magnetic field. There are no closed field lines in this split-monopole model where all lines go to infinity. This model has two important advantages. On the one hand the analysis of the flow is remarkably simplified. On the other hand the wind from any object becomes monopole-like at large distances. Therefore this model allows us to describe the wide class of real axisymmetric flows at large distances from the central objects. It is interesting that this model describes also a class of outflows from oblique rotators.

In the ideal MHD approximation (dissipationless plasma) the plasma dynamics is insensitive to the direction of the magnetic field. Therefore if we have some solution for the MHD plasma outflow and reverse the direction of the magnetic field lines in an arbitrary flux tube, this solution will also satisfy the full system of equations and thus it may describe a realistic configuration. This property of relativistic ideal plasma flows (Bogovalov 1999), holds also for nonrelativistic flows (Tsinganos & Bogovalov 2000). It allows one to easily reduce the problem of the plasma outflow from an oblique rotator to the same problem for the axisymmetric rotator in a model with an initially monopole-like magnetic field and to investigate the plasma flow in conditions more typical of real radio pulsars than occur in the axisymmetric model.

The reduction of the problem of the oblique rotator to the problem of the axisymmetric rotator can be done in the following way. Let us assume that we have a self-consistent solution for the axisymmetric problem with the initial split-monopole magnetic field which is rooted in the surface of the star, outwards in the upper hemisphere and inwards in the lower hemisphere. Let us next draw the magnetic equator inclined to the rotational equatorial plane under some angle. We can modify the direction of the field lines so that all magnetic field lines which are rooted above the magnetic

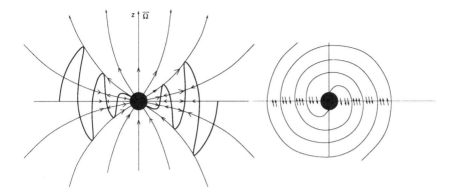

Figure 1. The left panel shows the structure of field lines and the current sheet (thick wave-like line) in the poloidal plane. The right panel shows the view from the top in the equatorial plane. Arrows show the direction of the magnetic field lines. The direction of the field lines changes on the current sheet.

equator remain outwards, while the magnetic field lines rooted below the magnetic equator are reversed such that they point towards the stellar surface. According to the invariance principle this operation does not affect the dynamics of the plasma and we obtain the solution for the oblique rotator with the initial split-monopole magnetic field.

A sketch demonstrating the structure of the cold wind from an oblique rotator is shown in Figure 1. The structure of the plasma flow is symmetrical in relation to the equator. The form of the plasma flow lines is the same as for the axisymmetric rotator. In general, there is some collimation of the plasma flow towards the axis of rotation, although the effect of the collimation depends on the parameters of the problem (Bogovalov & Tsinganos 1999). In the axisymmetric flow the current sheet dividing the magnetic fluxes of opposite direction is located at the equator. In the wind from the oblique rotator the current sheet takes the form of a wave. In the poloidal plane the poloidal magnetic field lines change direction on this current sheet. At a first glance it seems that this behavior contradicts the known property of plasma freezing on the magnetic field. The right panel of Figure1 shows the structure of the field lines in the equatorial plane. It may be seen that there is no contradiction with the magnetic flux freezing since the total magnetic field depends on the azimuthal angle φ. The velocity and the density of the plasma do not depend on the inclination and azimuthal angles and are the same as those of the axisymmetric rotator. Only the magnetic and electric fields are modulated with the period of rotation. The sign of these fields is modulated in the sector limited by the field lines which have roots on the magnetic equator. The squares of these fields are not modulated. The wave is simply the wave of the tangential discontinuity (Landau & Lifshitz 1963) which is convected with the velocity of the plasma.

This example shows that there is no magnetodipole emission from the oblique rotator ejecting dense plasma. It is actually very easy to understand why. It is well known that the magnetodipole emission of radio pulsars cannot propagate in the interstellar medium since the frequency of the wave of the magnetodipole emission is much less the plasma frequency of the medium (Lipunov 1987). But the same is valid for the ejected plasma. The magnetodipole emission does not propagate in the wind since the plasma frequency of the wind exceeds the frequency of the magnetodipole emission.

It follows from simple analysis that the rotational losses of this object do not depend on the inclination angle and there is no temporal evolution of the inclination angle (Bogovalov 1999). It is clear that the rotational losses of the object with the initially dipole magnetic field and with split-monopole magnetic field are equal to each other provided that they both produce an identical wind of relativistic plasma at large distances. An estimate of the rotational losses $E_{\rm rot}$ of radio pulsars following from this statement gives $\dot{E}_{\rm rot} \approx B_0^2 R_*^6 \Omega^4 / (6c^3)$. This expression differs from the ordinary formula for the magnetodipole losses by unimportant numerical coefficient. The deceleration of the pulsar is due primarily to the tension of the toroidal magnetic field of the wind.

3. The self-collimation property of magnetized outflows

One important property of plasmas ejected by a rotating magnetized object is their ability to be self-collimated. The asymptotical properties of magnetized outflows have been investigated by Heyvaerts & Norman (1989), Li, Chiueh & Begelman (1992) and Bogovalov (1995) while classes of self-consistent global MHD solutions have been presented in Vlahakis & Tsinganos (1998). The main results of these investigations can be formulated as the following theorem. The axisymmetric flow of plasma at large distance is always partially collimated along the axis of rotation of the central object with the formation of a cylindrically collimated jet provided that:

1. The central object is rotating (rotation can be differential).
2. The flow of the plasma is supersonic on all streamlines.
3. There is no surrounding medium which would be able to terminate the plasma flow.
4. The plasma is dissipationless.
5. The polytropic index of the equation of state δ is larger than 1.

The collimation of the plasma is provided by the Lorentz force which is directed perpendicular to the direction of motion. To demonstrate the physics of the collimation let us assume that the wind is not collimated.

The poloidal magnetic field B_p drops faster than the toroidal magnetic field. This may be seen from the frozen-in condition which at large distances where the toroidal component of the plasma velocity is negligibly small takes the form $B_\varphi = -r\Omega B_p/v_p$. With $\delta > 1$ the gas pressure also drops with distance faster than the pressure of the toroidal magnetic field. Therefore the plasma with density ρ moves with a constant speed at large distances and experiences a Lorentz force produced mainly by the toroidal magnetic field: $\rho v_p^2/R_c = j_p B_\varphi/c$ where R_c is the radius of curvature of the streamline and j_p the electric current along this poloidal streamline. It follows from Ampere's law and the frozen-in condition that $1/R_c = (B_p/4\pi\rho v_p)(\sin 2\theta\Omega^2/v_p^3)(R^2 B_p/R)$. Taking into account that the magnetic flux $r^2 B_p = const$ and the ratio of the poloidal magnetic to mass flux $B_p/4\pi\rho v_p = F = const$ along a streamline we obtain an estimate of the turning angle of the flow line $\Delta\theta = (B_p/4\pi\rho v_p)(\sin 2\theta\Omega^2/v_p^3)R^2 B_p \ln(R/R_o)$. It follows from this relationship that with the onset of motion from the central object the turning angle increases infinitely. It is clear from this simple consideration that collimation is really inevitable under the conditions formulated above. The collimation may finally be terminated by two means: if the toroidal magnetic field is balanced by the pressure of the poloidal magnetic field, or when the density of the electric currents goes to zero faster than $1/r^2$. It was argued by Bogovalov (1995) that the electric current density can not go to zero so rapidly on all field lines. Since the balance between the poloidal and the toroidal magnetic fields is possible only in a nonexpanding flow, a jet flowing exactly along the axis of rotation is formed along these field lines. Even this very simplified consideration allows us to point out the important feature of the process of self-collimation. The formation of the jet takes place at any nonzero magnetic field or angular velocity. But the distance at which the flow line turns at a finite angle increases exponentially with a decrease of the magnetic field or the angular velocity of the central object.

The characteristics of jets formed due to magnetic self-collimation become especially simple in the nonrelativistic limit with rather general assumptions about the plasma flow. Thus, by assuming that the plasma velocity v_j, the entropy per particle, and the ratio of poloidal magnetic to mass flux in the jet are constant across the jet, the dependence of the poloidal magnetic field on the distance from the axis of rotation r takes the form $B_p = B_0/(1 + (r/R_j)^2)$, where R_j can be considered as the transverse radius of the jet. This characteristic radius is $R_j = \sqrt{1 + (c_s/v_a)^2}v_j/\Omega$ where $v_a = B_0/\sqrt{4\pi\rho_0}$ is the Alfvén speed along the axis of rotation and c_s is the adiabatic sound velocity on this axis. This estimate is valid under the condition $R_j \gg R_a$, where R_a is the Alfvén radius.

4. Conclusion

We discussed in this paper two properties of plasma outflows which are essential for an understanding of the physics of these phenomena. First, the properties of outflows from oblique rotators, at large distances where the large scale poloidal magnetic field is split-monopole-like, are analogous to those of outflows from axisymmetric rotators. Therefore all properties of outflows from axisymmetric rotators are valid for a much wider class of objects. Second, the property of self-collimation is very crucial for an understanding of current observations. It follows from the dynamics of ideal plasmas that a magnetized wind ejected by a rotating object has an intrinsic mechanism for its collimation. This property makes the mechanism of magnetic collimation the most relevant explanation of observed jets.

Acknowledgments. S.V. Bogovalov is grateful to J. Ventura for his hospitality during the NATO ASI meeting. SVB was partially supported during this work by the Ministry of Education of Russia in the framework of the program "Universities of Russia - Basic Research", project N 897 and NATO Collaborative Research Grant CRGP 972857.

References

Beskin, V.S., Gurevich A.V., and Istomin Ya.N. (1983), *ZhETF* **85**, 401.
Biretta, T. (1996), in K. Tsinganos (ed.), *Solar and Astrophysical MHD Flows*, Kluwer Academic Publishers, p357.
Blandford, R.D. and Payne, D.G. (1982), *MNRAS* **199**, 883 (BP82).
Bogovalov, S.V. (1995), *Astron. Letts* **21**, 4.
Bogovalov, S.V. (1999), *A&A* **349**, 1097.
Bogovalov, S.V. and Tsinganos, K. (1999), *MNRAS* **305**, 211.
Contopoulos, I., Kazanas, D., and Fendt, C. (1999), *ApJ* **511**, 351.
Heyvaerts, J. and Norman, C.A. (1989), *ApJ* **347**, 1055.
Jones, D.L. and Wehrle, A.E. (1994), *ApJ* **427**, 221.
Kafatos, M. (1996), in K. Tsinganos (ed.), *Solar and Astrophysical MHD Flows*, Kluwer Academic Publishers, p585.
Kahabka, P. and Trumper, J. (1996), in E.P.J. Van den Heuvel and J. van Paradijs (eds.), *Compact Stars in Binaries*, Kluwer (Dordrecht), p425.
Kulkarni, S.R. et al. (1999), *Nature* **398**, 389.
Landau, L.D. and Lifshitz, E.M. (1963), *Electrodynamics of continious media*, Pergamon press (London).
Li, Z.-Y, Chiueh, T., and Begelman, M.C. (1992), *ApJ* **394**, 459.
Livio, M. (1997), in D.T. Wickramasinghe, L. Ferrario, and G.V. Bicknel (eds.), *Accretion Phenomena and Related Outflows*, ASP (San Francisco, in press).
Lipunov, V.M. (1987), *Astrophysics of Neutron stars*, Nauka (Moscow).
Lyubarskii, Yu.E. (1990), *Pis'ma v Astron. Zh.* **16**, 34.
Michel, F.C. (1969), *ApJ* **158**, 727.
Mirabel, I.F. and Rodriguez, L.F. (1996), in K. Tsinganos (ed.), *Solar and Astrophysical MHD Flows*, Kluwer Academic Publishers, p683.
O' Dell, C.R. and Wen, Z. (1994), *ApJ* **436**, 194.
Shahbaz, T., Livio, M., Southwell, K.A., and Charles, P.A. (1997), *ApJ* **484**, L59.
Tsinganos, K. and Bogovalov, S.V. (2000), *A&A* **356**, 989.
Vlahakis, N. and Tsinganos, K. (1998), *MNRAS* **298**, 777.

TIMING THE KILOHERTZ QUASI-PERIODIC OSCILLATIONS IN LOW-MASS X-RAY BINARIES

M. MÉNDEZ

Astronomical Institute "Anton Pannekoek",
University of Amsterdam,
Kruislaan 403, NL-1098 SJ Amsterdam, the Netherlands

1. Introduction

In the past 3 years the Rossi X-ray Timing Explorer has discovered kilohertz quasi-periodic oscillations (kHz QPOs) in the persistent flux of 19 low-mass X-ray binaries (LMXBs). The power density spectra of most of these sources show twin kHz peaks. Initial results seemed to indicate that the frequency separation between the twin kHz peaks, $\Delta\nu = \nu_2 - \nu_1$, remained constant even as the peaks gradually moved up and down in frequency (typically over a range of several hundred Hz) as a function of time. But thanks to precise measurements of the frequencies of the kHz QPOs, we know now that in several sources $\Delta\nu$ is not constant, but it decreases as ν_1 and ν_2 increase.

In this chapter I describe a new technique that we have been using in the past few years to get precise measurements of the frequency separation of the kHz QPOs in some LMXBs. My plan is to show how this technique (which we call "shift-and-add") works, and to present some of the results we obtained using it. It is not my purpose, however, to discuss here the details of the kHz QPO phenomenon, as this subject is extensively covered elsewhere in this book.

2. The Average Power Spectrum

The standard Fourier techniques in X-ray timing are described in detail in van der Klis (1989a); let us just recall here the definition of the power spectrum. A time series $x(t)$ is first divided into N time intervals of length τ, and for each interval the Fourier transform is calculated as:

C. Kouveliotou et al. (eds.), The Neutron Star – Black Hole Connection, 313–318.
© 2001 *Kluwer Academic Publishers. Printed in the Netherlands.*

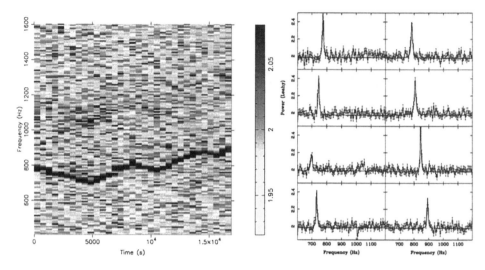

Figure 1. *Left Panel:* Dynamic power spectrum of the source 4U 1728–34. *Right Panel:* Power spectra of some selected intervals.

$$X_n(\nu) = \int_{t_0+n\tau}^{t_0+(n+1)\tau} x(t)e^{2\pi i\nu t}dt, n = 0, 1, 2, ..., N, \qquad (1)$$

where $i = \sqrt{-1}$, and t_0 is some initial time. The power spectrum of $x(t)$ is defined as the average of the square of the Fourier transforms, $P(\nu) = < X_n^2(\nu) >$. The average is computed to improve the signal-to-noise (S/N) ratio of the power spectrum (see van der Klis 1989a for details). This definition assumes that the time series $x(t)$ is *stationary*, i.e. that *all* time segments of length τ have the same statistical properties.

As I already mentioned, one property of the kHz QPOs is that their frequencies vary with time (therefore the underlying time series is non-stationary). This is shown in Figure 1, for the source 4U 1728–34. On the left panel I show a dynamical power spectrum, where the x-axis represents time (the origin of time in this case is 1997 October 1 at 06:09 UTC), the y-axis is the Fourier frequency, and the different gray levels indicate the power in each frequency bin. The dark feature that crosses the figure almost horizontally is one of the two QPOs detected in this source. The right panel shows power spectra ($P(\nu)$ vs. Fourier frequency) at some selected times: The same QPO is seen moving between ~ 700 and 900 Hz.

Obviously, the average power spectrum of the whole observation will show several peaks, at the frequencies at which the QPO stays for a while during the observation. This is shown in Figure 2 (left panel), where the average power spectrum has been calculated as described above. However, as the QPO is strong enough to allow us to measure it in short time segments,

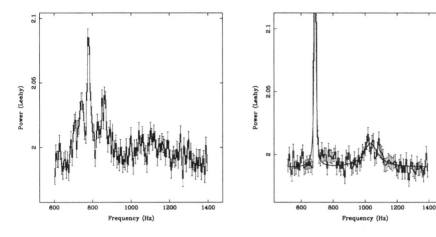

Figure 2.   Left Panel: Average (standard) power spectrum of the same data shown in Figure 1. *Right Panel:* Shifted-and-averaged power spectrum of the same data.

we can think of "shifting" the frequency axis of each individual power spectrum before we calculate the average so that the new power spectrum is $\tilde{P}(\nu) = <X_n^2(\nu - \nu_{0,n})>$. Here $\nu_{0,n}$ is the frequency shift that we need to apply to the power spectrum of each segment if we want to align the strong QPO at the same frequency before we calculate the average. Notice that this is equivalent to multiplying the function $x(t)$ by $e^{-2\pi i \nu_{0,n} t}$ before calculating the integral in eq.(1).

The right panel in Figure 2 shows the result of such a procedure (which, of course, also implies measuring $\nu_{0,n}$ for each segment; certainly this is extra work, but it comes with some rewards, as I will explain in Section 3). If it were only for the strong QPO in the average power spectrum (this QPO is stronger in the new power spectrum than in the original one) we would have not gained much by this procedure. Although there is now only one sharp peak, this shifted-and-averaged power spectrum provides no new information about the strong QPO. However, there is a second (weaker, but yet significant) QPO in the shifted power spectrum, ~ 350 Hz above the strong QPO. Although there seems to be evidence for such a weak QPO in the original power spectrum (Figure 2, left panel), there it was not significant enough.

To understand why we detect this QPO in the shifted power spectrum, while in the standard power spectrum it was not significant enough, we need to recall that the S/N-ratio of the QPO is proportional to $(T/W)^{1/2}$, where $T = N \times \tau$ is the total length of the observation, and W is the width of the QPO (see van der Klis 1989b). If the frequency separation between the 2 QPOs is roughly constant[1] (a close inspection to Figure 1 left panel shows that there might be such weak QPO in the dynamic power spectrum, "following" the strong QPO as this one moves in frequency), the shift applied to the power spectra of the individual segments to align the strong QPO also aligns the weak QPO. This makes this new QPO narrower, and therefore more significant, in $\tilde{P}(\nu)$ than in $P(\nu)$.

This is how we discovered the second QPO during the 1996 outburst of 4U 1608–52 (Méndez et al. 1998a), which we afterwards detected again (this time without requiring the sensitivity-enhancing technique described above) during another outburst in 1998 (Méndez et al. 1998b).

Not surprisingly, this same technique can be used to obtain much better measurements of the frequency separation between the kHz QPOs. This is because the error in the centroid frequency of a QPO is proportional to the width of the QPO, and inversely proportional to its S/N-ratio (see Downs & Reichley 1983 for a similar situation when measuring the arrival time of individual pulses in radio pulsars). As described above, the weak QPO is usually narrower in the shifted power spectrum, $\tilde{P}(\nu)$, than in the standard power spectrum, $P(\nu)$, and therefore its frequency can be measured more accurately in $\tilde{P}(\nu)$ than in $P(\nu)$.

One way of measuring $\Delta\nu$ using the shifted power spectrum is the following: Divide the data into N segments of length τ, and for each segment calculate the power spectrum. Measure the frequency of the strong QPO, ν_1, in each of these N power spectra, and group the data in M sets such that ν_1 is more or less constant within each set. This ensures that $\Delta\nu$ is more or less constant within each set[2] (early results in Sco X–1 showed that $\Delta\nu$ decreased monotonically as ν_1, ν_2 increased; van der Klis et al. 1997). For each set, align the power spectra of the individual time segments using ν_1 as a reference, and combine them to produce an average power spectrum. The result is M (aligned) power spectra for which both kHz QPOs are as narrow as possible. For each of these M power spectra measure ν_1 and ν_2, calculate $\Delta\nu = \nu_2 - \nu_1$, and plot $\Delta\nu$ vs. the average value of ν_1 in each set.

Figure 3 shows the results of applying such a procedure to the kHz QPOs in 4U 1608–52, 4U 1728–34, and Sco X–1. The implications of these results (especially in the case of 4U 1728–34) to the existing models for the kHz QPOs are presented elsewhere in this book (see the chapter by van der Klis), and will not be discussed here.

---

[1] Although it is not really constant, variations in $\Delta\nu$ are smaller than changes in the centroid frequencies of the QPOs.

[2] Notice that there is a trade-off here: the choice of small frequency intervals for ν_1 ensures small changes in $\Delta\nu$, preventing artificial broadening of the weak QPO when the power spectra are shifted; but it also reduces the amount of data that are averaged in each set, and therefore decreases the S/N-ratio of the weak QPO at ν_2.

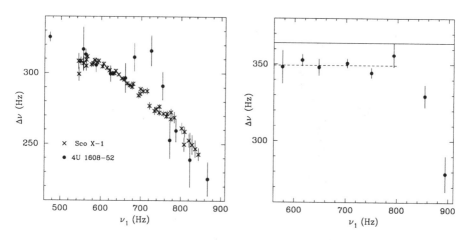

Figure 3. *Left Panel:* Frequency separation between the two simultaneous kHz QPOs in 4U 1608–52 (circles) and Sco X–1 (crosses). *Right Panel:* Frequency separation between the two simultaneous kHz QPOs in 4U 1728–34 (circles). The solid line indicates the frequency of the burst oscillations at 363.95 Hz (Strohmayer et al. 1996). The dashed line indicates the average value of $\Delta\nu$ (349.3 Hz) for the first 6 points. For the three sources $\Delta\nu$ as a function of ν_1 was calculated as explained in the text.

3. Other Results

As I mentioned in Section 2, there is a reward for the extra work of measuring ν_1 in the individual time segments: We can combine these measurements with, for instance, the source count rate or X-ray colors in those same segments. One example of this is shown in Figure 4. The left panel shows a plot of ν_1 *vs* the $2 - 16$ keV count rate for 4U 1608-52. Each point represents 128 s of data. For the same data used in the plot on the left panel, the right panel shows the relation between ν_1 and an X-ray color (see the caption of this figure for the definition of the color).

Interesting in this figure is the complex dependence of ν_1 upon X-ray count rate, which is usually assumed to be a good measure of mass accretion rate $\dot{M}$, and how this complexity is reduced to a single track in the frequency *vs* hard color diagram. As before, the interested reader can find an extensive discussion of this subject elsewhere in this book. I just want to point out here that these are "side" results of the technique described in Section 2.

4. Conclusion

I presented a new technique aimed at increasing the detection sensitivity of weak, moving QPOs in the power density spectra of LMXBs. I showed some examples of the results obtained by applying this technique to data obtained with the Rossi X-ray Timing Explorer. In particular, this technique provides

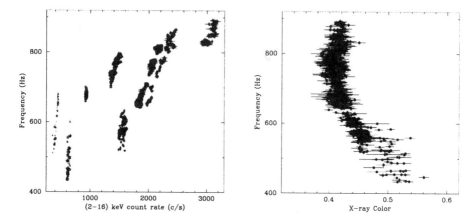

Figure 4. Left Panel: Relation between ν_1 and the $2-16$ keV count rate in 4U 1608–52. Right Panel: For the same data as on the left panel, ν_1 vs. X-ray hard color, defined as the ratio of the count rate in the $6.0-9.7$ keV band to the count rate in the $9.7-16.0$ keV band.

more precise measurements of the frequencies of the kHz QPOs, which can be used to set constraints on the models so far proposed to explain this new phenomenon.

Acknowledgements

This work was supported by the Netherlands Research School for Astronomy (NOVA), the Netherlands Organization for Scientific Research (NWO) under contract number 614-51-002, the Leids Kerkhoven-Bosscha Fonds (LKBF), and the NWO Spinoza grant 08-0 to E.P.J. van den Heuvel. MM is a fellow of the Consejo Nacional de Investigaciones Científicas y Técnicas de la República Argentina. This research has made use of data obtained through the High Energy Astrophysics Science Archive Research Center Online Service, provided by the NASA/Goddard Space Flight Center.

References

Downs, G.S. and Reichley, P.E. (1983), *ApJS* **53**, 169.
Ford, E. et al. (1997), *ApJ* **475**, L123.
Méndez, M. et al. (1998a), *ApJ* **494**, L65.
Méndez, M. et al. (1998b), *ApJ* **505**, L23.
Strohmayer, T.E. et al. (1996), *ApJ* **469**, L9.
Van der Klis, M. (1989a), in H. Ögelman and E.P.J. van den Heuvel (eds.), *Timing Neutron Stars*, NATO ASI Ser C262 (Kluwer, Dordrecht), p27.
Van der Klis, M. (1989b), *ARA&A* **27**, 517.
Van der Klis, M. et al. (1997a), *ApJ* **481**, L97.
Zhang, W. et al. (1998a), *ApJ* **495**, L9.

MODELING THE TIME VARIABILITY OF BLACK HOLE CANDIDATES

Light Curves, PSD, Lags

D. KAZANAS

Laboratory for High Energy Astrophysics
NASA/GSFC, Code 661 Greenbelt, MD 20771

Abstract. I present a model for the aperiodic variablity of accreting Black Hole Candidates (BHC) along with model light curves. According to the model this variability is the combined outcome of random (Poisson) injection of soft photons near the center of an extended *inhomogeneous* distribution of hot electrons (similar to those advocated by the ADAF or ADIOS flows) and the stochastic nature of Compton scattering which converts these soft photons into the observed high energy radiation. Thus, the timing properties (PSD, lags, coherence) of the BHC light curves reflect, to a large extent, the properties of the scattering medium (which in this approximation acts as a combination of a *linear* amplifier/filter) and they can be used to probe its structure, most notably the density profile of the scattering medium. The model accounts well for the observed PSDs and lags and also the reduction in the RMS variability and the increase in the characteristic PSD frequencies with increasing source luminosity. The electron density profiles obtained to date are consistent mainly with those of ADIOS but also with pure ADAF flows.

1. Introduction

The study of the physics of accretion-powered sources, whether on galactic (X-ray binaries) or extragalactic systems (AGN), involves length scales much too small to be resolved by current technology. As such, this study is conducted mainly through the theoretical interpretation of their spectral and temporal properties. Until recently studies of this class of objects focused, for mainly technical reasons, on their X-ray spectra. These are generally very well fit with those of Comptonization of soft photons by hot ($T_e \lesssim 10^9$ K) electrons, a process explored in great depth over the past

C. Kouveliotou et al. (eds.), The Neutron Star – Black Hole Connection, 319–324.

twenty or so years [11]. Because the electron temperatures of matter accreting onto a black hole are expected to be similar to those necessary to produce the observed spectra, it has been considered that detailed spectral fits of these sources would lead to insights on the dynamics of accretion onto the compact object.

However, the determination of accretion dynamics requires the knowledge of the density and size of the emitting region, neither of which is provided by radiative transfer and spectral fitting considerations (the equations of radiative transfer involve the optical depth as the independent variable). Indeed, as shown explicitly in [6] and [5], plasmas of very different radial extent and density profiles can yield identical Comptonization spectra. The degeneracy of this situation can be lifted with the additional information provided by timing observations.

The timing properties of BHC, however, suggest length scales inconsistent with the prevailing notion that the observed X-rays are emitted from a region of size a few Schwarzschild radii, R_S: The power spectra (PSD) of BHC exhibit most of their power at scales ~ 1 s, far removed from the characteristic time scales associated with the dynamics in the vicinity of the black hole horizon, $R_S/c \sim 10^{-3}$ s. Until recently rather little attention has been paid to this time scale discrepancy, generally attributed to the (unknown) mechanism "fueling" the black hole, presumably operating at much larger radii; rather, more attention was paid to the flicker noise–type ($\propto f^{-1}$) PSDs of this class of sources. Novel insights into the variability properties were introduced by [8], who showed that both the magnitude and the Fourier period dependence of the lags in the X-ray light curves at two different energies was inconsistent with Comptonization by a plasma of size $\sim$ a few R_S; the magnitude of time lags (~ 0.1 s) suggested an emission region of much larger size, thus precluding an explanation of the observed PSDs as being due to a modulation of the accretion rate onto the black hole.

These discrepancies between the expected and the observed variability of GBHC led [6], [3], [5] and [7] to propose that contrary to the prevailing notions, the size of the scattering region responsible for the X-ray emission is not $R \simeq 3 - 10\ R_S$ but rather $R \gtrsim 10^3 R_S$, as implied by the PSDs and lag observations. Furthermore, the scattering medium (corona) is *inhomogeneous*, with the electron density following the law,

$$n(r) = \begin{cases} n_1 & \text{for } r \leq r_1 \\ n_1(r_1/r)^p & \text{for } r_2 > r > r_1 \end{cases} \tag{1}$$

where r is the radial distance from the center of the corona (assumed to be spherical) and r_1, r_2 are its inner and outer radii respectively. The index $p > 0$ is a free parameter whose value depends on the specific dynamical

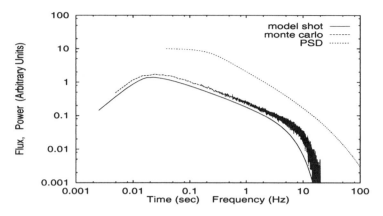

Figure 1.   The response function $g(t)$ of a corona with $r_1 = 10^{-2}$ light s, $r_2 = 10$ l. s, $p = 1$, $\tau_T = 1$. The solid line is a fit to the MC data (dashed line). The dotted line is the Fourier spectrum $|G(\omega)|^2$ of $g(t)$.

model that determines the electron density. For example, the ADAF of [10] suggest $p = 3/2$ and $T_e \lesssim 10^9$ at radii as large as $r \simeq (m_p/m_e)R_S$, while models combining inflow and outflow [2], [1] (ADIOS) allow, in addition, values $p \simeq 1$. As pointed out in [6], [5] most of the data analyzed to date suggest $p = 1$, with $p = 3/2$ also acceptable in certain cases.

It is intuitively obvious that scattering in the extended configuration of Eq. (1) produces time lags over a range of Fourier periods similar to the range of radii span by the hot corona: Scattering at a given radius R increases the X-ray energy and introduces a lag $\Delta t \lesssim R/c$ between the scattered and unscattered photons, the lag appearing at a Fourier period $P = R/c$ ([5]; N. Kylafis, these proceedings). To compare this to the lags given in [8] (which are the average over all such photon pairs scattered in a given decade in R, as a function of R) one has to multiply Δt by the probability of scattering at a given radius, $\mathcal{P}(R) \simeq \tau(R)$, with $\tau(R)$ the scattering depth over the radius R. For the density profile of Eq. (1), $\tau(R) \propto R^{-p+1}$ and since the Fourier period $P \propto R$, then $\langle \Delta t \rangle \propto R^{-p+2} \propto P^{-p+2}$. Monte Carlo simulations and analytical considerations have shown that these arguments are essentially correct [5]. In addition, these simulations showed that the configuration of Eq. (1) produces light curves of high coherence over the range $[r_2/c, r_1/c]$ of the Fourier period, provided the soft photons are injected near its center.

2.   The Model Light Curves

Based on these considerations one can easily produce model light curves, whose properties in the time or the Fourier domain can then be compared

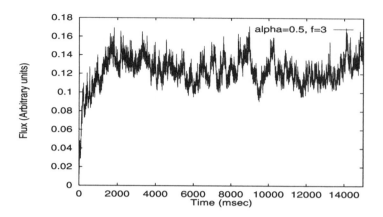

Figure 2. The light curve $F(t)$ of Eq.(2) for the $g(t)$ of Figure 2 and $f = 3$.

to observations for consistency. The high coherence of the observations [12] suggests injection of the soft photons at $r \ll r_2$. The response function, $g(t)$, of a corona with density profile given by Eq. (1) to an instantaneous release of soft photons at $r \leq r_1$ has been computed by a Monte Carlo simulation [6]; [7]. The result of a particular case is shown in Figure 1 along with an analytic fit. As pointed out in [5] these functions can be well approximated by a Gamma function distribution i.e. $g(t) \propto t^{\alpha-1} e^{-t/\beta}$ ($0 < \alpha < 1$, $\beta > 0$).

Assuming linearity, i.e. that T_e is not affected by the photon flux, model light curves can be produced by an incoherent, random (Poisson) injection of shots of the form $g(t)$. The prescription for such a light curve is

$$F(t) = \sum_{i=1}^{N} Q_i \theta \left(t - \sum_{i=1}^{N} t_i \right) g(t - t_i) . \tag{2}$$

Q_i are the shot amplitudes, assumed constant, $\theta(t - t_1)$ is the Heaviside function and t_i is a collection of Poisson distributed time intervals obtained from the expression $t_i = -f \cdot t_0 \cdot \log R_i$; R_i is a random number uniformly distributed between 0 and 1; and f is a real number, indicating the mean time between shots in terms of their rise time t_0. Figure 2 shows the light curve obtained using the above prescription with $g(t)$ as given in Figure 1 and $f = 3$, while Figure 3 shows the light curve from an identical sequence of R_i's and $f = 10$.

These light curves look similar to those of BHC sources in their low, hard state [9]. Due to the Poisson nature of t_i's and the identical shape of the shots that make up these light curves, their PSDs are those of an individual shot, also shown in Figure 1 (dotted line). The form of the PSD is also quite similar to those of BHC (see [9]), with the high and low frequency

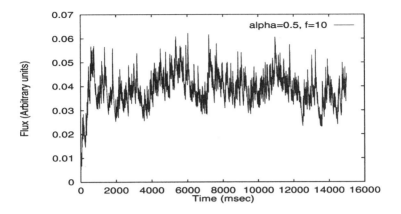

Figure 3. Same as Figure 2, with $f = 10$.

breaks associated respectively with the inner r_1 and outer r_2 edges of the scattering corona.

One can easily produce, in addition, light curves for different photon energies, noting that for photons of a higher energy, $g(t)$ is approximately of the same form with only a small increase in the parameters α and β [5]. While the resulting light curves are visually identical to those at lower energies, their difference is nonetheless easily seen in the phases of the corresponding FFT. I have generated in this way model light curves appropriate to three different energies and then computed the corresponding pairs of phase lags (a)-(b), (a)-(c) as a function of the Fourier frequency measured in Hz; the results are shown in Figure 4. The phase lags presented in this figure are very similar to those of the sources analyzed in [9] and in particular the X-ray transient GRO J0422+32 [4].

A correlation between the RMS variability and the value of f is the most prominent feature of Figures 2 and 3. Owing to the fact that, in the linear regime, smaller f is equivalent to higher luminosity (more photons per unit time), this correlation provides a natural account of the *observed* anticorrelation between the BHCs luminosity (in their hard spectral states) and their RMS variability. Furthermore, consideration of this model within its natural dynamic framework, namely that of ADAF, provides a direct correspondence between the dynamical and the timing properties of these systems: in ADAF all scales (and also r_1 and r_2) scale proportionally to $v_{\text{ff}} \tau_{\text{cool}}$ with v_{ff} the free-fall velocity and τ_{cool} the local cooling time. For τ_{cool} inversely proportional to the local density (as is the case with ADAF), all length scales (and therefore frequencies) associated with the corona should depend on $\dot{m}/\dot{m}_{\text{Edd}}$, decreasing with increasing value of $\dot{m}/\dot{m}_{\text{Edd}}$; hence the corresponding frequencies should increase with increasing luminosity, a behavior which has apparently been observed in most accreting sources

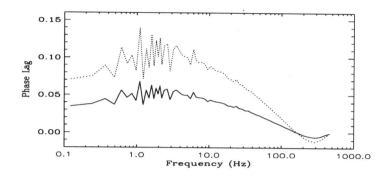

Figure 4. The phase lags of two sets of model light curves with the following parameters (a) $\alpha = 0.5$, $\beta = 16$ s; (b) $\alpha = 0.55$, $\beta = 16$ s; (c) $\alpha = 0.6$, $\beta = 16$ s. The two curves correspond to the lags between: (a) - (b) (solid line) and (a) - (c) (dotted line).

(whether neutron stars or black holes).

3. Conclusions

(a) The model of the extended inhomogeneous corona provides in a simple, well-understood fashion model light curves for BHC which have morphology, PSDs and phase (or time) lags very similar to those observed. (b) The dependence of the phase lags on Fourier period allows the determination of the density profile of the corona; the data are consistent both with $p = 3/2$ (ADAF) and $p = 1$ (ADIOS) flows. (c) The dependence of the RMS variability and the PSD break frequencies of BHC on their luminosity are also in general agreement with these dynamical considerations and provide a well defined framework within which these ideas could be tested through more detailed comparison with observations.

References

1. Blandford, R.D. and Begelman, M.C. (1999), *MNRAS* **303**, L1.
2. Contopoulos, J. and Lovelace, R.V.E. (1994), *ApJ* **429**, 139.
3. Hua, X.-M., Kazanas, D., and Titarchuk, L. (1997a), *ApJ* **482**, L57.
4. Grove, J.E. et al. (1998), *ApJ* **502**, L45.
5. Hua, X.-M., Kazanas, D., and Cui, W. (1999) *ApJ* **512**, 793.
6. Kazanas, D., Hua, X.-M., and Titarchuk, L. (1997), *ApJ* **480**, 735.
7. Kazanas, D. and Hua, X.-M. (1999), *ApJ* **519**, 750.
8. Miyamoto, S. et al. (1988), *Nature* **336**, 450.
9. Miyamoto, S. et al. (1992), *ApJL* **391**, L21.
10. Narayan, R. and Yi, I. (1994), *ApJ* **428**, L13.
11. Sunyaev, R.A. and Titarchuk, L.G. (1980), *A&A* **86**, 121.
12. Vaughan, B.A. and Nowak, M.A. (1997), *ApJ* **474**, L43.

RELATIVISTIC MODELS OF KHZ QPOS

W. KLUŹNIAK

Copernicus Astronomical Center
ul. Bartycka 18, 00-716 Warszawa, Poland
wlodek@camk.edu.pl

Abstract. After reviewing the general-relativistic "gap" model of accretion, I discuss its relation to the high frequency quasi-periodic oscillations observed in low-mass X-ray binaries. The "300" Hz frequency seen in some X-ray bursts may be a relativistic signature of keplerian rotation of the neutron star.

It is easy to see how much the field has advanced in the past decade by comparing the topics under discussion here in Elounda, with those discussed at that previous conference of this series which also took place in Crete, in Agia Pelagia, at the beginning of the decade.

Back then, magnetic fields were all the rage. Gamma-ray bursts supposedly showed in their spectra cyclotron absorption lines suggesting (to many) that the sources are Galactic neutron stars, a view completely ruled out in the decade of Compton GRO, Beppo SAX and the observations of afterglows (described carefully in Dr. Fishman's talk here in Elounda). Another view much discussed at Agia Pelagia was that low-mass X-ray binaries contain direct counterparts of millisecond pulsars, i.e., 10^9 to 10^{10} Gauss neutron stars, rotating at periods of a few milliseconds, and therefore accreting through a disk in which the orbital frequency (supposedly) differed from the stellar rotational frequency by about 50 Hz. Today, after the discovery of kHz QPOs, there is hardly any doubt that the characteristic orbital frequency in the inner disk is at least 1 kHz, very different from the value of about 300 Hz promoted at Agia Pelagia.

So let us forget about magnetic fields in LMXBs and ask what would be expected then. The answer depends on the equation of state (e.o.s.) of matter at supranuclear densities and on the mass of the neutron star, as well as its angular momentum. Here I will only discuss the general relativistic

C. Kouveliotou et al. (eds.), The Neutron Star – Black Hole Connection, 325–330.
© 2001 *Kluwer Academic Publishers. Printed in the Netherlands.*

"gap" regime of accretion, in which the accretion disk does not extend to the stellar surface—there are good reasons for that.

1. The relativistic gap regime

LMXBs are old accreting systems, so *a priori* one would expect that the central neutron stars have each gained a few tenths of a solar mass since their early days. Now, as pointed out some time ago (Kluźniak and Wagoner 1985), for all e.o.s., at sufficiently high stellar mass (which need not be very large), a slowly rotating neutron star is within the innermost stable circular orbit (ISCO) allowed by general relativity (GR), a.k.a. the marginally stable orbit. For rapidly rotating neutron stars this is not always so, but according to the tables of Cook et al. (1994), for most e.o.s. the maximally rotating models are also within the ISCO. For strange (quark) stars this is also true (Stergioulas et al. 1999). In short, it seems reasonable to assume that in LMXBs, the compact object is inside the ISCO, so let us do so.

The three-dimensional flow in accretion disks is still poorly understood (it may resemble the flow of waves crashing on the beach, particularly the rip tide dreaded by ocean swimmers—see the figure from Kita's 1995 thesis reproduced in Kluźniak 1998b). But in any case, in the relativistic gap regime, the disk should be terminated by GR effects, as in the black hole disks, whose essential properties were discussed in numerous papers, e.g., of the Warsaw school some two decades ago (by Paczyński, Abramowicz, Sikora, Muchotrzeb and others, in various combinations). Without further ado, let us accept the view that the maximum observable frequency in LMXB disks is close to the ISCO frequency and that this frequency may modulate the X-ray flux (Kluźniak et al. 1990). Then it will be easy to believe that the saturation (at 1.07 kHz) of QPO frequency in 4U 1820-30 is a signature of the ISCO (Zhang 1998), and that the e.o.s. is severely constrained by the observed maximum frequency value (Kluźniak 1998a).

What happens to the matter which leaves the disk through its inner edge (assumed to be close to the ISCO radius)? It goes into free-fall and approaches the surface at a rather shallow angle. Under these conditions, a sheared atmosphere heated by the incoming fluid is set up in the equatorial regions (or even the tropics, as in Dr. Sunyaev's talk), whose vertical structure has been found in a 1+1–d calculation with full radiative transfer (Kluźniak and Wilson 1991): the atmosphere is hot and gives off radiation with a power law spectrum extending to about 200 keV. This would agree with reports of hard radiation from several X-ray bursters. Of course this spectrum may be downgraded as the radiation interacts with the (relatively) cool disk and the accretion stream; this interaction has not yet been computed, but it is clear that on such a picture one would expect the down-scattered softer photons to lag in time the harder photons (as has been reported in SAX J 1808.4–3658).

My prejudiced view is that the balance of observations is in favour of the gap regime. It really seems that several separate facts (especially these: the presence of kHz QPOs and their frequency values, hard spectra, soft lags) suggest that the disk is terminated outside the stellar surface by effects of general relativity.

2. Model independent conclusions about QPOs?

The power spectrum of both neutron star and candidate black hole systems has been studied over the whole range of observed frequencies, and the phenomenology of QPOs in both types of systems was found to be remarkably similar (Psaltis et al. 1999). The neutron star systems show two kHz QPOs, the one with lower frequency has a clear counterpart in black hole sources (for example, in both types of systems it has identical correlations with lower frequency features in the spectrum), only the highest frequency QPO has a different phenomenology (Psaltis et al. 1999).

If the black hole candidates are indeed black holes, all black hole QPOs and their counterparts in neutron star systems must be accretion disk phenomena, reflecting fundamental properties of flow in the gravitational field of the compact object. A logical conclusion would be that in the neutron star systems it is the lower frequency kHz QPO which could be connected with orbital motion, with its characteristic cut-off frequency in the ISCO. This would make constraints on the e.o.s. even more stringent than the ones inferred from the higher frequency QPO, and discussed in the previous section (but in either case, these constraints would be relaxed if the QPO frequency were lower than the orbital frequency). The same model should then describe the power spectra of neutron stars in LMXBs as of accreting black holes, including those in AGNs (where the frequency would be scaled down in inverse proportion to the black hole mass).

On this picture, it would be the highest frequency kHz QPO alone, which would need a special explanation for neutron star systems. A special topic of attention must be the similar value of the difference in the two "kHz" QPO frequncies, to the $\sim 300\,\mathrm{Hz}$ frequency of the coherent peak in the power spectrum seen in X-ray bursts (Strohmayer et al. 1997).

3. Keplerian rotation?

While accreting mass in LMXBs, the neutron stars are also accreting angular momentum, a lot of it. Exact models (Cook et al 1994) show that maximally rotating neutron stars have angular momentum $J \approx 0.6GM^2/c$, this amount of momentum can be accreted already with $\sim 0.2M_\odot$ in mass

(Kluźniak and Wagoner 1985). Several instabilities are known which can limit the spin rate of a neutron star (mostly through emission of gravitational radiation), but it is not known whether they actually operate in practice. The most recently discovered r-mode instability could, in principle, limit neutron star periods to values even as long as a few milliseconds in LMXBs (Andersson et al. 2000).

On the observational side, there is no compelling evidence of the periodicity of persistent accretors in LMXBs (with the exception of the one or two strongly magnetized X-ray pulsars which have a low mass companion, but have nothing to do with the atoll, banana, and other LMXBs sources so colourfully described by Michiel van der Klis). This is why the 2.5 ms coherent period discovered in the transient SAX J 1808.4–3658 gave rise to much excitement. Of course, the famous "300 Hz periodicity" discovered in several X-ray bursts has been interpreted as the stellar rotational frequency (Strohmayer et al. 1997), but there seems to be no good model for its observed properties (the "hot spot model" has been criticized here in Elounda, on different grounds, by Fred Lamb and Rashid Sunyaev). The argument for rotation seems to be: what else could it be? Clearly there is a clock in the system with a very good memory of frequency, and yet one which wanders on short time scales.

The coherent peak in the power spectrum of bursters appears during the X-ray burst, and persists for several seconds, during which the frequency increases usually, and yet from burst to burst the frequency is amazingly stable. Usually these are hallmarks of an (anharmonic) oscillator. Can we find one in the system?

Marek Abramowicz and I think that we have found an anharmonic oscillator on the road to Knossos, at least in our minds. In the equatorial plane outside of (even a spinning) gravitating body the description of test-particle motion can be reduced to one dimensional motion in an effective potential, V. The characteristic shape of the effective potential in GR is shown in Figure 1. Stable motion in circular orbits, as in the Newtonian case, is possible in the minimum of the potential—for the metric and angular momentum chosen in the figure, this orbit would have radius $r_o = 8GM/c^2$; for a different angular momentum of the test particle in the same metric the potential would have a minimum at a different radius, but always outside the ISCO, which has a radius r_{ms} uniquely fixed by the metric. A characteristic feature of GR metrics is the existence of a maximum of the effective potential (in the figure at $r_u = 5GM/c^2$), at which unstable circular motion is possible. In all cases, $r_u \leq r_{ms} \leq r_o$, with the equality occuring at the minimum value of angular momentum possible in circular motion of a test particle in the (fixed) metric.

The maximum ("Keplerian") rotation rate of a star with radius R inside

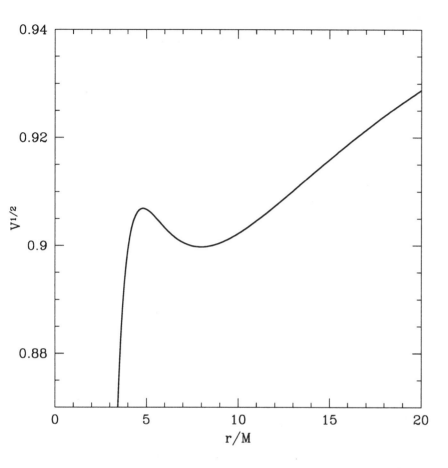

Figure 1.   The effective potential in GR (the detailed shape of the curve will vary with the metric; here a Schwarzschild potential is shown).

the ISCO (i.e., $R < r_{ms}$) occurs, when $R = r_u$. Imagine, then, that the effective potential in the figure is that of test particles in the external metric of a maximally rotating neutron star of radius $R = 5M$, or just a little bit less, for the same value of specific angular momentum as that of matter on the equator. Now imagine that an explosion (an X-ray burst) lifts some matter off the surface. The radial motion of the matter is an oscillation in the potential well. If the energy of the matter were constant, it would travel to a turning point (at about $r = 12M$) and return to the maximum of the potential. However, if a little energy is removed, the matter oscillates back and forth between the right turning point, and the left one, just below the maximum of V. This is the anharmonic oscillator suggested as the origin of the "300 Hz burst oscillation," with the frequency increasing as the amplitude of motion (range in r) decreases. At last, the matter settles in circular orbit at the bottom of the potential well.

3.1. PREDICTION

Clearly, the model presented here requires maximal rotation of the neutron star, which is expected to be higher than 600 Hz (because two radio pulsars with 1.6 ms periods have aleady been observed).

We (Marek and I), would then predict that no X-ray burst oscillation will be seen in X-ray bursts of sources with a clearly detected (phase connected solution) rotational period of more than 1.6 ms. For instance, in the transient SAX J 1808.4–3658, where a period of $P = 2.5$ ms has been measured, no such burst oscillation should be discovered.

4. Summary

It seems that the relativistic gap regime—the expected basic mode of accretion onto neutron stars with very weak magnetic fields—fits most observations of LMXBs, including the essential phenomenology of QPOs. The regime allows for neutron star rotation rates higher than orbital frequencies in the disk. For maximally rotating neutron stars, an oscillation in the relativistic "potential well" is possible, with properties similar to those of the "300 Hz" oscillation observed in some X-ray bursts.

This research was supported in part by KBN grant 2 P03D01816.

Chryssa and Jan, thank you for the conference, we missed you, and we will always miss Jan.

References

Andersson, N., Jones, D.I., Kokkotas, K.D., and Stergioulas, N. (2000), *ApJ* **534**, L75.
Cook, G.B., Shapiro, S.L., and Teukolsky, S.A. (1994), *ApJ* **424**, 823.
Kita, D. (1995), *Ph.D. Thesis*, University of Wisconsin.
Kluźniak, W. (1998a), *ApJ* **509**, L37.
Kluźniak, W. (1998b), in K. Chan, K.S. Cheng, and H.P. Singh (eds.), *Pacific Rim Conference on Stellar Astrophysics* Astron. Soc. Pac. Series 138, pp161-168.
Kluźniak, W., Michelson, P., and Wagoner, R.V. (1990), *ApJ* **358**, 538.
Kluźniak, W. and Wagoner, R.V. (1985), *ApJ* **297**, 548.
Kluźniak, W. and Wilson, J.R. (1991), *ApJ* **372**, L87.
Psaltis, D., Belloni, T., and van der Klis, M. (1999), *ApJ* **520**, 262.
Stergioulas, N., Kluźniak, W., and Bulik, T. (1999), *A&A* **352**, L116.
Strohmayer, T.E., Jahoda, K., Giles, A.B., and Lee, U. (1997), *ApJ* **486**, 355.
Zhang, W., Smale, A.P., Strohmayer, T.E., and Swank, J.H. (1998), *ApJ* **500**, L171.

PHENOMENOLOGY OF THE 35-DAY CYCLE OF HERCULES X-1

N.I. SHAKURA

Max-Planck-Institut für Astrophysik, Karl-Schwarzschild-Str. 1 85740 Garching, Germany
Sternberg Astronomical Institute, 119899 Moscow, Russia

AND

N.A. KETSARIS, K.A. POSTNOV, M.E. PROKHOROV
Sternberg Astronomical Institute, 119899 Moscow, Russia

Abstract. A strong X-ray illumination of the optical star atmosphere in Her X-1, asymmetric because of a partial shadowing by the tilted warped accretion disk around the central neutron star, leads to the formation of matter flows coming out of the orbital plane and crossing the line of sight before entering the disk. We suggest that the absorption of hard emission by these flows leads to the formation of pre-eclipse dips. Some anomalous X-ray dips (of type I) are formed by the same mechanism. Inner parts of the warped twisted disk emit extreme UV radiation so the absorption dips could be observable in the extreme UV even at phases where the central X-ray source is screened by the disk. Anomalous dips observed after the turn-on and post-eclipse recoveries are due to the tidal disk wobbling.

1. Introduction

X-ray dips observed in the light curve of Her X-1 remain the notable feature of this system. They appear in the first several orbits (up to 7) after X-ray turn-on (pre-eclipse dips (P)) (Giacconi et al. 1973) and march from the eclipse toward earlier orbital phase in successive orbits. Such dips are also clearly visible in the short-on state (Ricketts et al. 1982; Shakura et al. 1998; Scott and Leahy 1999). In the first several orbits after turn-on, anomalous dips (A) are found around orbital phase 0.5–0.6. Occasionally, post eclipse recoveries (R) are observed (Crosa and Boynton 1980).

C. Kouveliotou et al. (eds.), The Neutron Star – Black Hole Connection, 331–336.

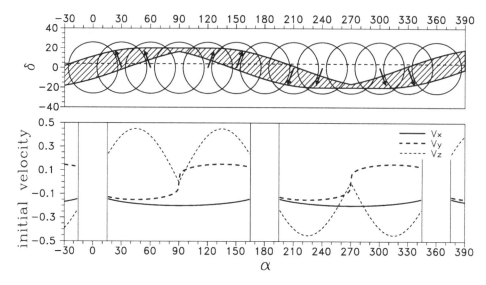

Figure 1.   Upper panel: The passage of HZ Her through the shadow formed by a tilted warped accretion disk. The coordinates α and δ count from the line of nodes of the middle part of the disk along the orbital motion and from the orbital plane, respectively. The dashed line indicates the position of an observer. The arrows show the projection of the initial accretion stream velocity on the plane YZ perpendicular to the orbit. Bottom panel: The initial stream velocity components at the L_1 point in units of the orbital velocity 250 km/s. The streams disappear within the shadow sector between the outer and inner disk node lines.

To explain these dips, we suggested a model of a tilted, warped, precessing accretion disk producing an appreciable shadow (Shakura et al. 1999). The optical star periodically enters this shadow during its orbital motion (Fig. 1). The shadowed region is such that not all the optical star surface is screened by the disk – there always should exist regions illuminated by the X-ray source with a photospheric temperature of 15,000–20,000 K, whereas the photospheric temperature of the unheated regions is $\sim 8,400$ K. Even higher temperatures (up to 10^6 K) due to soft X-ray absorption by heavy elements are attainable in the chromospheric layers over the photosphere. The X-ray heating of these layers is so strong that NV $\lambda\lambda 1238.8, 1242.8$ doublet with a FWHM of 150 km/s is observed (Boroson et al. 1996).

The shadow causes powerful pressure gradients to emerge in the chromospheric layers near the boundary separating illuminated and obscured parts of the optical star. This effect initiates large-scale motions of matter near the inner Lagrangian point L_1 with a large velocity component perpendicular to the orbital plane.

The matter streams are non-complanar with the orbital plane and supply the accretion disk with angular momentum non-parallel to the orbital one. The streams from the L_1 point intersect the observer's line of sight at

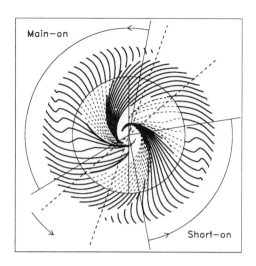

Figure 2. The projection of accretion streams onto the orbital plane. The central circle indicates the projection of the outer parts of tilted warped accretion disk intersecting the orbit along the vertical node line. Within the sector (30°) shown by the dashed lines the matter outflow is absent. The observer is above the orbital plane.

some orbital phases shortly before the X-ray eclipse and shift slowly toward earlier phases as the precession progresses. This is exactly the observed behaviour of the pre-eclipse X-ray dips. Some streams intersect the line of sight at $\phi_{\mathrm{orb}} \sim 0.45 - 0.65$ (see Fig. 2) and produce anomalous dips (type I anomalous dips).

2. Origin of pre-eclipse and type I anomalous dips

The problem of matter outflow from an asymmetrically illuminated stellar atmosphere is essentially three-dimensional and requires sophisticated numerical calculations. We exploit some trial functions for the initial outflow velocity components outside the shadowed sector (with the optimal width about ±15° from the mean disk node line, while the angle between the inner and outer disk node lines is 70°; see Fig. 1,2). In a corotating frame with the X-axis directed from the X-ray source to the optical star, the Y-axis pointing along the orbital motion, and the Z-axis perpendicular to the orbital plane, $v_{\mathrm{x}} = -v_{\mathrm{x}^\circ}| \sin \alpha|^{n_{\mathrm{x}}}$, $v_{\mathrm{y}} = \pm v_{\mathrm{y}^\circ}| \sin 2\alpha|^{n_{\mathrm{y}}}$, $v_{\mathrm{z}} = \pm v_{\mathrm{z}^\circ}| \sin 2\alpha|^{n_{\mathrm{z}}}$, with the angle α counted along the orbit from the mean disk node line (Fig. 1,2). The outflow rate is $\dot{M} \propto |\vec{v}|$. The initial velocities of particles were determined from the conditions that, on the one hand, the dynamical action of the streams on the accretion disk keeps it tilted by some angle (20° for both inner and outer parts) to the orbit, and on the other hand, the calculated pre-eclipse dips lie as close as possible in the region of the

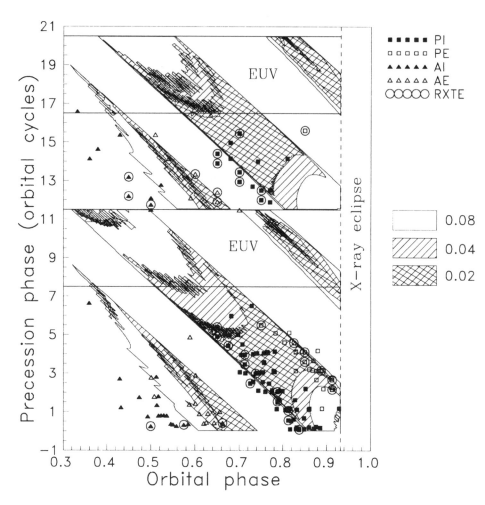

Figure 3. The observed and calculated P− and A−dips in the main-on (on the bottom) and short-on (on the top) states. Filled and open quadrangles and triangles indicate ingress to and egress from P (PI, PE) and type 1 A (AI, AE) dips, respectively. RXTE data are encircled. Contours of calculated dip location are for different minimal distances between the stream center and the line of sight. At precession phases marked with EUV the central source is screened by the disk (low states), but dips should be present in the extreme UV light curve. EUV dips are due to absorption of emission from the central parts of the disk by the streams (like type I X-ray P-dips). For clarity, the EUV dips are shown as if the disk is transparent for UV radiation. The jumps in the dip phase twice the precession period occur when the outer disk edge crosses the line of sight.

$\phi_{\rm orb} - \phi_{\rm prec}$ diagram occupied by the observed dips.

The accretion disk parameters in units of orbital separation $a \simeq 9 R_\odot$ and orbital velocity $v_{\rm orb} = 2\pi a / P_{\rm orb} \simeq 270$ km/s are: the orbit inclination to the line of sight $i_{\rm b} = 86°$, the disk tilt to the orbit $\theta \simeq 20°$, and the disk radius $R_{\rm d} \simeq 0.3$.

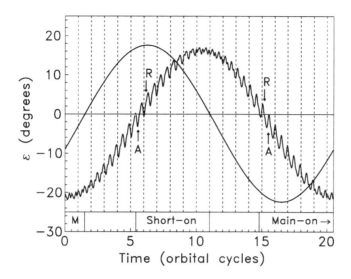

Figure 4.   The angle ϵ between the line of sight and the outer accretion disk (waiving line) and the inner accretion disk (smooth line) as a function of time. The small vertical arrows indicate the anomalous type II X-ray dips (A) and post-eclipse recovery (R). The outer disk radius was taken 0.3 a.

In a wide range of velocities v_i^o and parameters n_i the stream crosses the line of sight before the X-ray eclipse. We found that a good coincidence is attained for $v_x^o = 0.2$, $v_z^o = 0.45$, $v_y^o = 0.15$ and $n_x = n_y = 0.25$, $n_z = 1$.

In Fig. 3 we plot the calculated and observed P and type I A dip positions on the $\phi_{orb} - \phi_{pr}$ plane (data from Crosa and Boynton (1980), Ricketts et al. (1982), Shakura et al. (1998a)). The contours include the calculated P and type I A dip positions for different minimal distances between the stream centre and the line of sight (0.02, 0.04, and 0.08). The stream generated when the star again enters the X-ray illuminated sector (see Fig. 2), must also intersect the line of sight during the short-on state. This gives rise to P and type I A dips during the short-on state as well. Fig. 3 demonstrates qualitative agreement of the observed and calculated P and type I A dip positions. Note that in our model additional P X-ray dips appear at the end of on-states when the stream from another illuminated sector crosses the line of sight (see Fig. 2 and 3). In Fig. 3 we also plot the position of dips that can be observed in two low states between main-on and short-on states in extreme UV. In these low-states, the central X-ray source is screened by the disk body, however the innermost part of the tilted twisted accretion disk is a powerful source of the extreme UV and it can be seen at some phases where the central X-ray source is invisible.

3. Type II anomalous dips and post-eclipse recoveries

Type II anomalous dips and post-eclipse recoveries are formed by another mechanism. These features appear only once after the turn-on. The vector of the disk angular momentum moves along a precession cone and undergoes an oscillating (wobbling) motion at twice the orbital frequency. This causes the turn-on to occur at ~ 0.25 and ~ 0.75 orbital phases (Levine and Jernigan 1982, Katz et al. 1982). The oscillating part of the dynamical torques acts almost synchronously with the tidal wobbling. This somewhat increases the net wobbling effect. As seen from Fig. 4, X-ray turn-ons may be followed by a short time interval in which the X-rays are again occulted by the outer edge of the disk. Our modelling suggests that this may be identified with either a type II A dip ($\phi_{\mathrm{orb}} \simeq 0.5 - 0.6$) or a post-eclipse recovery ($\phi_{\mathrm{orb}} \simeq 0.1 - 0.2$). However, these features appears episodically, and moreover when type II A-dips is present, no post-eclipse recovery is observed, or vice versa (see Shakura et al 1999 for more detail).

Acknowledgements

The work was supported by the grant "Universities of Russia", No5559 and Russian Fund for Basic Research through Grant No 98-02-16801.

References

Boroson, B., Vrtilek, S.D., McCray, R., Kallman, T., and Nagase, F. (1996) *ApJ* **473**, 1079.

Crosa, L. and Boynton, P.E. (1980), *ApJ* **235**, 999.

Gerend, D. and Boynton, P.E. (1976), *ApJ* **209**, 562.

Giacconi, R., Gursky, H., Kellogg, E., Levinson, R., Schreier, E., and Tananbaum, H. (1973), *ApJ* **184**, 227.

Katz, J.I., Anderson, S.F., Margon, B., and Grandy, S.A. (1982), *ApJ* **260**, 780.

Levine, A.M. and Jernigan, J.G. (1982), *ApJ* **262**, 294.

Ricketts, M.J., Stanger, V., and Page, C.G. (1982), in W. Brinkmann, J. Trümper (eds.), *Accreting Neutron Stars*, ESO, Garching bei München, pp100-105.

Scott, D.M. and Leahy, D.A. (1999), *ApJ* **510**, 974.

Shakura, N.I., Ketsaris, N.A., Prokhorov, M.E., and Postnov, K.A. (1998), *MNRAS* **300**, 992.

Shakura, N.I., Prokhorov, M.E., Postnov, K.A., and Ketsaris, N.A. (1999), *A&A*, **348**, 917.

SPIN-ORBIT COUPLINGS IN X-RAY BINARIES

Detailed calculations of mass-transfer including tidal forces

T.M. TAURIS AND G.J. SAVONIJE

Center for High-Energy Astrophysics, University of Amsterdam

Abstract. We discuss the influence of tidal spin-orbit interactions on the orbital dynamics of close intermediate-mass X-ray binaries. In particular we consider here a process in which spin angular momentum of a contracting RLO donor star, in a synchronous orbit, is converted into orbital angular momentum and thus helps to stabilize the mass transfer by widening the orbit. Binaries which would otherwise suffer from dynamically unstable mass transfer (leading to the formation of a common envelope and spiral-in evolution) are thus shown to survive a phase of extreme mass transfer on a sub-thermal timescale. Furthermore, we discuss the orbital evolution prior to RLO in X-ray binaries with low-mass donors, caused by the competing effects of wind mass loss and tidal effects due to expansion of the (sub)giant.

1. Introduction

Tidal torques act to establish synchronization between the spin of the non-degenerate companion star and the orbital motion. Whenever the spin angular velocity of the donor is perturbed (by a magnetic stellar wind; or change in its moment of inertia due to either expansion or mass loss in response to RLO) the tidal spin-orbit coupling will result in a change in the orbital angular momentum leading to orbital shrinkage or expansion.

We have performed detailed numerical calculations of the non-conservative evolution of ~ 200 close binary systems with $1.0 - 5.0\,M_\odot$ donor stars and a $1.3\,M_\odot$ accreting neutron star. Rather than using analytical expressions for simple polytropes, we calculated the thermal response of the donor star to mass loss, using an updated version of Eggleton's numerical computer code, in order to determine the stability and follow the evolution of the mass transfer. We refer to Tauris & Savonije (1999) for a more detailed description of the computer code and the binary interactions considered.

337

C. Kouveliotou et al. (eds.), The Neutron Star – Black Hole Connection, 337–342.
© 2001 *Kluwer Academic Publishers. Printed in the Netherlands.*

2. The orbital angular momentum balance equation

Consider a circular[1] binary with an (accreting) neutron star and a companion (donor) star with mass M_{NS} and M_2, respectively. The orbital angular momentum is given by: $J_{orb} = (M_{NS} M_2 /M) \Omega a^2$, where $M = M_{NS} + M_2$ and $\Omega = \sqrt{GM/a^3}$ is the orbital angular velocity. A simple logarithmic differentiation of this equation yields the rate of change in orbital separation:

$$\frac{\dot{a}}{a} = 2\frac{\dot{J}_{orb}}{J_{orb}} - 2\frac{\dot{M}_{NS}}{M_{NS}} - 2\frac{\dot{M}_2}{M_2} + \frac{\dot{M}_{NS} + \dot{M}_2}{M} \tag{1}$$

where the total change in orbital angular momentum can be expressed as:

$$\frac{\dot{J}_{orb}}{J_{orb}} = \frac{\dot{J}_{gwr}}{J_{orb}} + \frac{\dot{J}_{mb}}{J_{orb}} + \frac{\dot{J}_{ls}}{J_{orb}} + \frac{\dot{J}_{ml}}{J_{orb}} \tag{2}$$

The first term on the right side of this equation governs the loss of J_{orb} due to gravitational wave radiation (Landau & Lifshitz 1958). The second term arises due to a combination a magnetic wind of the (low-mass) companion star and a tidal synchronization (locking) of the orbit. This mechanism of exchanging orbital into spin angular momentum is referred to as magnetic braking (see e.g. Verbunt & Zwaan 1981; Rappaport et al. 1983).

2.1. TIDAL TORQUE AND DISSIPATION RATE

The third term in eq.(2) was recently discussed by Tauris & Savonije (1999) and describes possible exchange of angular momentum between the orbit and the donor star due to its expansion or mass loss (note, we have neglected the tidal effects on the gas stream and the accretion disk). For both this term and the magnetic braking term we estimate whether or not the tidal torque is sufficiently strong to keep the donor star synchronized with the orbit. We estimate the tidal torque due to the interaction between the tidally induced flow and the convective motions in the stellar envelope by means of the simple mixing-length model for turbulent viscosity $\nu = \alpha H_p V_c$, where the mixing-length parameter α is adopted to be 2 or 3, H_p is the local pressure scaleheight, and V_c the local characteristic convective velocity. The rate of tidal energy dissipation can be expressed as (Terquem et al. 1998):

$$\frac{dE}{dt} = -\frac{192\pi}{5}\Omega^2 \int_{R_i}^{R_o} \rho r^2 \nu \left[\left(\frac{\partial \xi_r}{\partial r}\right)^2 + 6\left(\frac{\partial \xi_h}{\partial r}\right)^2 \right] dr \tag{3}$$

where the integration is over the convective envelope and Ω is the orbital angular velocity, i.e. we neglect effects of stellar rotation. The radial and

---

[1]This is a good approximation since tidal effects acting on the near RLO giant star will circularize the orbit on a short timescale of $\sim 10^4$ yr, cf. Verbunt & Phinney (1995).

horizontal tidal displacements are approximated here by the values for the adiabatic equilibrium tide:

$$\xi_r = f r^2 \rho \left(\frac{dP}{dr}\right)^{-1} \qquad \xi_h = \frac{1}{6r}\frac{d(r^2 \xi_r)}{dr} \tag{4}$$

where for the dominant quadrupole tide ($l = m = 2$) $f = -GM_2/(4a^3)$. The locally dissipated tidal energy is taken into account as an extra energy source in the standard energy balance equation of the star, while the corresponding tidal torque follows as: $\Gamma = -(1/\Omega)(dE/dt)$. The thus calculated tidal angular momentum exchange $dJ = \Gamma dt$ between the donor star and the orbit during an evolutionary timestep dt is taken into account in the angular momentum balance of the system. If the so calculated angular momentum exchange is larger than the amount required to keep the donor star synchronous with the orbital motion of the compact star we adopt a smaller tidal angular momentum exchange (and corresponding tidal dissipation rate) that keeps the donor star exactly synchronous.

2.2. SUPER-EDDINGTON ACCRETION AND ISOTROPIC RE-EMISSION

The last term in eq.(2) is the most dominant contribution and is caused by loss of mass from the system (see e.g. van den Heuvel 1994; Soberman et al. 1997). We have adopted the "isotropic re-emission" model in which all of the matter flows over, in a conservative way, from the donor star to an accretion disk in the vicinity of the neutron star, and then a fraction, β of this material is ejected isotropically from the system with the specific orbital angular momentum of the neutron star. If the mass-transfer rate exceeds the Eddington accretion limit for the neutron star, $\beta > 0$. In our calculations we assumed $\beta = max[0, 1 - \dot{M}_{\rm Edd}/\dot{M}_2]$ and $\dot{M}_{\rm Edd} = 1.5 \times 10^{-8} M_\odot$ yr^{-1}.

3. Evolution neglecting spin-orbit couplings

Assuming $\dot{J}_{\rm gwr} = \dot{J}_{\rm mb} = \dot{J}_{\rm ls} = 0$ and $\dot{J}_{\rm ml}/J_{\rm orb} = \beta q^2 \dot{M}_2/(M_2(1+q))$ one obtains easily analytical solutions to eq.(1). In Figure 1 we have plotted

$$-\frac{\partial \ln(a)}{\partial \ln(q)} = 2 + \frac{q}{q+1} + q\frac{3\beta - 5}{q(1-\beta)+1} \tag{5}$$

as a function of the mass ratio $q = M_2/M_{\rm NS}$. The sign of this quantity is important since it tells whether the orbit expands or contracts in response to mass transfer (note $\partial q < 0$). We notice that the orbit always expands when $q < 1$ and it always decreases when $q > 1.28$ [solving $\partial \ln(a)/\partial \ln(q) = 0$ for $\beta = 1$ yields $q = (1 + \sqrt{17})/4 \approx 1.28$]. If $\beta > 0$ the orbit can still expand for $1 < q \leq 1.28$. Note, $\partial \ln(a)/\partial \ln(q) = 2/5$ at $q = 3/2$ independent of β.

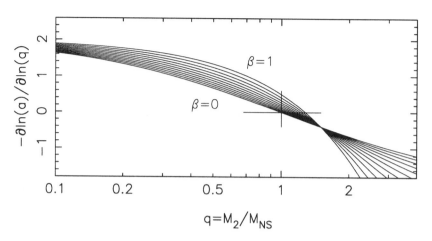

Figure 1. $-\partial \ln(a)/\partial \ln(q)$ as a function of q for X-ray binaries. The different curves correspond to different constant values of β in steps of 0.1. Tidal effects were not taken into account here. A cross is shown to highlight the case of $q = 1$ or $\partial \ln(a)/\partial \ln(q) = 0$. The evolution during the mass-transfer phase follows these curves from right to left since M_2 and q are decreasing with time (though β need not be constant).

4. Results including tidal spin-orbit couplings

In Figure 2 we have plotted the orbital evolution of an X-ray binary. The solid lines show the evolution including tidal spin-orbit interactions and the dashed lines show the calculations without these interactions. In all cases the orbit will always decrease initially as a result of the large initial mass ratio ($q = 4.0/1.3 \simeq 3.1$). But when the tidal interactions are included the effect of pumping angular momentum into the orbit (at the expense of spin angular momentum) is clearly seen. The tidal locking of the orbit acts to convert spin angular momentum into orbital angular momentum causing the orbit to widen (or shrink less) in response to mass transfer/loss. The related so-called Pratt & Strittmatter (1976) mechanism has previously been discussed in the literature (e.g. Savonije 1978). Including spin-orbit interactions many binaries will survive an evolution which may otherwise end up in an unstable common envelope and spiral-in phase. An example of this is seen in Figure 2 where the binary with initial $P_{\rm orb} = 2.5$ days (solid line) only survives as a result of the spin-orbit couplings. The dashed line terminating at $M_2 \sim 3.0\,M_\odot$ indicates the onset of a run-away mass-transfer process ($\dot{M}_2 > 10^{-3}\,M_\odot\,{\rm yr}^{-1}$) and formation of a common envelope and possible collapse of the neutron star into a black hole. In fact, many of the systems with $2.0 < M_2/M_\odot < 5.0$ recently studied by Tauris, van den Heuvel & Savonije (2000) would not have survived the extreme mass-transfer phase if the spin-orbit couplings had been neglected.

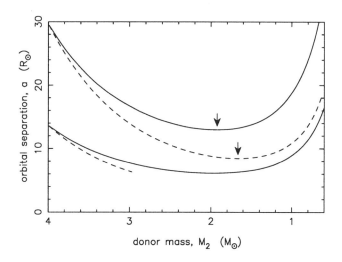

Figure 2.   Evolution of orbital separation as a function of donor star mass during the RLO phase in a binary with $M_2 = 4.0\,M_\odot$ (X=0.70, Z=0.02, α=2.0), $M_{NS} = 1.3\,M_\odot$ and $P_{orb} = 8.0$ and 2.5 days, top and bottom lines respectively. The lifetime of these X-ray binaries are only $t_X = 1.2$ and 2.1 Myr, respectively. The solid evolutionary tracks were calculated including tidal interactions and the dashed lines without. See text for details.

The location of the minimum orbital separations in Figure 2 are marked by arrows in the case of $P_{orb} = 8.0$ days. Since the mass-transfer rates in such an intermediate-mass X-ray binary are shown to be highly super-Eddington (Tauris, van den Heuvel & Savonije 2000) we have $\beta \approx 1$. Hence in the case of neglecting the tidal interactions (dashed line) we expect to find the minimum separation when $q = 1.28$ (cf. Section 3). Since the neutron star at this stage only has accreted $\sim 10^{-4}\,M_\odot$ we find that the minimum orbital separation is reached when $M_2 = 1.28 \times 1.30\,M_\odot = 1.66\,M_\odot$. Including tidal interactions (solid line) results in an earlier spiral-out in the evolution and the orbit is seen to widen when $M_2 \leq 1.92\,M_\odot$ ($q \approx 1.48$).

4.1. LOW-MASS DONORS AND PRE-RLO ORBITAL EVOLUTION

For low-mass ($\leq 1.5\,M_\odot$) donor stars there are two important consequences of the spin-orbit interactions which result in a reduction of the orbital separation: magnetic braking and expansion of the (sub)giant companion star. In the latter case the conversion of orbital angular momentum into spin angular momentum is a caused by a reduced rotation rate of the donor. However, in evolved stars there is a significant wind mass loss (Reimers 1975) which will cause the orbit to widen and hence there is a competition between this effect and the tidal spin-orbit interactions for determining the

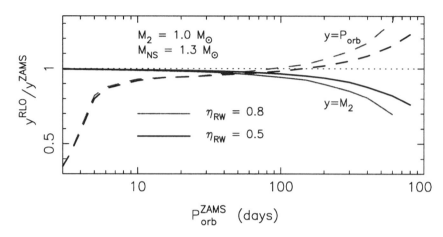

Figure 3. The changes of donor mass, M_2 (full lines) and orbital period, P_{orb} (dashed lines), due to wind mass loss and tidal spin-orbit interactions, from the ZAMS until the onset of the RLO as a function of the initial orbital period of a circular binary.

orbital evolution prior to the RLO-phase. This is demonstrated in Figure 3. We assumed $\dot{M}_{2\,wind} = -4 \times 10^{-13} \, \eta_{RW} \, L \, R_2/M_2 \, M_\odot \, yr^{-1}$ where the mass, radius and luminosity are in solar units and η_{RW} is the mass-loss parameter. It is seen that only for binaries with $P_{orb}^{ZAMS} > 100$ days will the wind mass loss be efficient enough to widen the orbit. For shorter periods the effects of the spin-orbit interactions dominate (caused by expansion of the donor) and loss of orbital angular momentum causes the orbit to shrink. This result is very important e.g. for population synthesis studies of the formation of millisecond pulsars, since P_{orb} in some cases will decrease significantly prior to RLO. As an example a system with $M_2 = 1.0 \, M_\odot$, $M_{NS} = 1.3 \, M_\odot$ and $P_{orb}^{ZAMS} = 3.0$ days will only have $P_{orb}^{RLO} = 1.0$ days at the onset of the RLO.

References

Landau, L.D. and Lifshitz, E. (1958), *The Classical Theory of Fields*, Pergamon Press (Oxford).

Pratt, J.P. and Strittmatter, P.A. (1976), *ApJ* **204**, L29.

Rappaport, S.A., Verbunt, F., and Joss, P.C. (1983), *ApJ* **275**, 713.

Reimers, D. (1975), in B. Bascheck, W.H. Kegel, and G. Traving (eds.), *Problems in Stellar Atmospheres and Envelopes*, Springer (New York), p229.

Savonije, G.J. (1978), *A&A* **62**, 317.

Soberman, G.E., Phinney, E.S. and van den Heuvel, E.P.J. (1997), *A&A* **327**, 620.

Tauris, T.M. and Savonije, G.J. (1999), *A&A* **350**, 928.

Tauris, T.M., van den Heuvel, E.P.J., and Savonije, G.J. (2000), *ApJ* **530**, L93.

Terquem, C., Papaloizou, J.C.B., Nelson, R.P., and Lin, D.N.C, (1998), *ApJ* **502**, 788.

Van den Heuvel, E.P.J. (1994), in H. Nussbaumer and A. Orr (eds.), *Interacting Binaries*, Saas-Fee 1992 lecture notes, Springer-Verlag.

Verbunt, F. and Zwaan, C. (1981), *A&A* **100**, L7.

Verbunt, F. and Phinney, E.S. (1995), *A&A* **296**, 709.

HIGH MASS BLACK HOLES IN SOFT X-RAY TRANSIENTS

Gap in Black Hole Masses?

G.E. BROWN, C.-H. LEE
Department of Physics and Astronomy,
SUNY at Stony Brook, NY 11794, USA

AND

H.A. BETHE
Floyd R. Newman Laboratory of Nuclear Studies,
Cornell University, Ithaca, New York 14853, USA

Abstract. We suggest that high-mass black holes, i.e., black holes of several solar masses, can be formed in binaries with low-mass main-sequence companions, provided that the hydrogen envelope of the massive star is removed in common envelope evolution which begins only after the massive star has finished He core burning. Our evolution scenario naturally explains the gap (low probability region) in the observed black hole masses.

1. Introduction

We suggest that high-mass black holes, i.e., black holes of several solar masses, can be formed in binaries with low-mass main-sequence companions, provided that the hydrogen envelope of the massive star is removed in common envelope evolution which begins only after the massive star has finished He core burning (Brown, Lee, & Bethe 1999). That is, the massive star is in the supergiant stage, which lasts only $\sim 10^4$ years, so effects of mass loss by He winds are small. Since the removal of the hydrogen envelope of the massive star occurs so late, it evolves essentially as a single star, rather than one in a binary. Thus, we can use evolutionary calculations of Woosley & Weaver (1995) of single stars.

We find that high-mass black holes can be formed in the collapse of stars with Zero-Age Main Sequence (ZAMS) mass $\gtrsim 20$ M$_\odot$. Mass loss by winds in stars sufficiently massive to undergo the LBV (luminous blue variable) stage may seriously affect the evolution of stars of ZAMS $> 35 - 40$ M$_\odot$, we take the upper limit for the evolution of the so-called transient sources

C. Kouveliotou et al. (eds.), The Neutron Star – Black Hole Connection, 343–348.
© 2001 *Kluwer Academic Publishers. Printed in the Netherlands.*

TABLE 1. Parameters of suspected black hole binaries in soft X-ray transients with measured mass functions (Brown, Lee, & Bethe 1999). N means nova, XN means X-ray nova. Numbers in parentheses indicate errors in the last digits.

X-ray names	other name(s)	compan. type q (M_{opt}/M_X)	P_{orb} (d) K_{opt} (km s^{-1})	$f(M_X)$ (M$_\odot$) i (degrees)	M_{opt} (M$_\odot$) M_X (M$_\odot$)
XN Mon 75	V616 Mon	K4 V	0.3230	2.83-2.99	0.53-1.22
A 0620−003	N Mon 1917	0.057-0.077	443(4)	37-44	9.4-15.9
XN Oph 77	V2107 Oph	K3 V	0.5213	4.44-4.86	0.3-0.6
H 1705−250			420(30)	60-80	5.2-8.6
XN Vul 88	QZ Vul	K5 V	0.3441	4.89-5.13	0.17-0.97
GS 2000+251		0.030-0.054	520(16)	43-74	5.8-18.0
XN Cyg 89	V404 Cyg	K0 IV	6.4714	6.02-6.12	0.57-0.92
GS 2023+338	N Cyg 1938, 1959	0.055-0.065	208.5(7)	52-60	10.3-14.2
XN Mus 91		K5 V	0.4326	2.86-3.16	0.41-1.4
GS 1124−683		0.09-0.17	406(7)	54-65	4.6-8.2
XN Per 92		M0 V	0.2127(7)	1.15-1.27	0.10-0.97
GRO J0422+32		0.029-0.069	380.6(65)	28-45	3.4-14.0
XN Sco 94		F5-G2	2.6127(8)	2.64-2.82	1.8-2.5
GRO J1655−40		0.33-0.37	227(2)	67-71	5.5-6.8
XN	MX 1543-475	A2 V	1.123(8)	0.20-0.24	1.3-2.6
4U 1543−47			124(4)	20-40	2.0-9.7
XN Vel 93		K6-M0	0.2852	3.05-3.29	0.50-0.65
		0.137± 0.015	475.4(59)	∼ 78	3.64-4.74

to be ~ 35 M$_\odot$ ZAMS mass. Both Portegies Zwart, Verbunt & Ergma (1997) and Ergma & Van den Heuvel (1998) have suggested that roughly our chosen range of ZAMS masses must be responsible for the transient sources. We believe that the high-mass black hole limit of ZAMS mass ~ 40 M$_\odot$ suggested by Van den Heuvel & Habets (1984) and later revised to ≥ 50 M$_\odot$ (Kaper et al. 1995) applies to massive stars in binaries, which undergo RLOF (Roche Lobe Overflow) early in their evolution.

The most copious high-mass black holes of masses $\sim 6 - 7$ M$_\odot$ have been found in the transient sources such as A0620−003. These have low-mass companions, predominantly of $\lesssim 1$ M$_\odot$, such as K- or M-stars. In the progenitor binaries the mass ratios must have been tiny, say $q \sim 1/25$. Following the evolutionary scenario for the black hole binary of De Kool et al. (1987), we show that the reason for this small q−value lies in the

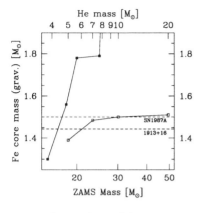

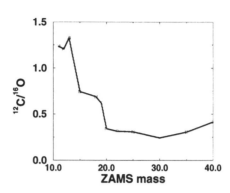

Figure 1. Comparison of the compact core masses resulting from the evolution of single stars (filled symbols, Case B of Woosley & Weaver 1995), and naked helium stars (Woosley, Langer & Weaver 1995) with masses equal to the corresponding He core mass of single stars.

Figure 2. Ratio of production rates of ^{12}C and ^{16}O resulting from the evolution of single stars (filled symbols), case of solar metallicity of Woosley & Weaver (1995).

common envelope evolution of the binary. The smaller the companion mass, the greater the radius R_g the giant must reach before its envelope meets the companion. This results because the orbit of a low-mass companion must shrink by a large factor in order to expel the envelope of the giant, hence the orbit must initially have a large radius. (Its final radius must be just inside its Roche Lobe, which sets a limit to the gravitational energy it can furnish.) A large radius R_g in turn means that the primary star must be in the supergiant stage. Thus it will have completed its He core burning while it is still "clothed" with hydrogen. This prevents excessive mass loss so that the primary retains essentially the full mass of its He core when it goes supernova. We believe this is why K– and M–star companions of high-mass black holes are favored.

2. Formation of High-Mass Black Holes

We find that the black holes in transient sources can be formed from stars with ZAMS masses in the interval $20-35$ M$_\odot$ (Brown, Lee, & Bethe 1999). The black hole mass is only slightly smaller than the He core mass, typically ~ 7 M$_\odot$ (Bethe, Brown, & Lee 2000).

Crucial to our discussion here is the fact that single stars evolve very differently from stars in binaries that lose their H-envelope either on the main sequence (Case A) or in the giant phase (Case B). However, stars that transfer mass or lose mass after core He burning (Case C) evolve, for our purposes, as single stars, because the He core is then exposed too close to its death for wind mass loss to significantly alter its fate. The core masses of single stars and binary stars are summarized in Fig. 1. Single stars above a ZAMS mass of about 20 $M_\odot$ skip convective carbon burning following core He burning, with the result, as we shall explain, that their Fe cores are substantially more massive than stars in binaries, in which H-envelope has been transferred or lifted off before He core burning. These latter "naked" He stars burn ^{12}C convectively, and end up with relatively small Fe cores. The reason that they do this has to do chiefly with the large mass loss rates of the "naked" He cores, which behave like Wolf-Rayet stars. In the ZAMS mass range $\sim 20 - 35\ M_\odot$, it is clear that many, if not most, of the single stars go into high-mass black holes, whereas stars in binaries which burn "naked" He cores go into low-mass compact objects. In this region of ZAMS masses the use of high He-star mass loss rates does not cause large effects (Wellstein & Langer 1999).

The convective carbon burning phase (when it occurs) is extremely important in pre-supernova evolution, because this is the first phase in which a large amount of entropy can be carried off in $\nu\bar{\nu}$-pair emission, especially if this phase is of long duration. The reaction in which carbon burns is $^{12}C(\alpha,\gamma)^{16}O$ (other reactions like $^{12}C +^{12} C$ would require excessive temperatures). The cross section of $^{12}C(\alpha,\gamma)^{16}O$ is still not accurately determined; the lower this cross section the higher the temperature of the ^{12}C burning, and therefore the more intense the $\nu\bar{\nu}$ emission. With the relatively low $^{12}C(\alpha,\gamma)^{16}O$ rates determined both directly from nuclear reactions and from nucleosynthesis by Weaver & Woosley (1993), the entropy carried off during ^{12}C burning in the stars of ZAMS mass $\leq 20\ M_\odot$ is substantial. The result is rather low-mass Fe cores for these stars, which can evolve into neutron stars. Note that in the literature earlier than Weaver & Woosley (1993) often large $^{12}C(\alpha,\gamma)^{16}O$ rates were used, so that the ^{12}C was converted into oxygen and the convective burning did not have time to be effective. Thus its role was not widely appreciated.

Of particular importance is the ZAMS mass at which the convective carbon burning is skipped. In Fig. 2, this occurs at ZAMS mass 19 $M_\odot$ but with a slightly lower $^{12}C(\alpha,\gamma)^{16}O$ rate it might come at 20 $M_\odot$ or higher. As the progenitor mass increases, it follows from general polytropic arguments that the entropy at a given burning stage increases. At the higher entropies of the more massive stars the density at which burning occurs is lower, because the temperature is almost fixed for a given fuel. Lower densities decrease the rate of the triple-α process which produces ^{12}C relative to the two-body $^{12}C(\alpha,\gamma)^{16}O$ which produces oxygen. Therefore, at the higher entropies in the more massive stars the ratio of ^{12}C to ^{16}O at the end of He burning is lower. The star skips the long convective carbon burning and goes on to the much shorter oxygen burning. Oxygen burning goes via

$^{16}O + ^{16}O$ giving various products, at very much higher temperature than $C(\alpha, \gamma)$ and much faster. Since neutrino cooling during the long carbon-burning phase gets rid of a lot of entropy of the core, skipping this phase leaves the core entropy higher and the final Chandrasekhar core fatter. We believe that the above discussion indicates that single stars in the region of ZAMS masses $\sim 20 - 35$ M$_\odot$ end up as high mass black holes.

Arguments have been given that SN 1987A with progenitor ZAMS mass of ~ 18 M$_\odot$ evolved into a low-mass black hole (Brown & Bethe 1994). We believe from our arguments that just above the ZAMS mass of ~ 20 M$_\odot$, single stars go into high-mass black holes without return of matter to the Galaxy. Thus, the region of masses for low-mass black hole formation in single stars is narrow, say $\sim 18 - 20$ M$_\odot$ (although we believe it to be much larger in binaries).

3. Quiet Black Hole - Main Sequence Star Binaries

We believe that there are many main sequence stars more massive than the $\lesssim 1$ M$_\odot$ we used in our schematic evolution, which end up further away from the black hole and will fill their Roche Lobe during only subgiant or giant stage. From our evolution, we see that a 2 M$_\odot$ main sequence star will end up about twice as far from the black hole as the 1 M$_\odot$, a 3 M$_\odot$ star, three times as far, etc. Two of the 9 systems in our Table 1 have subgiant donors (V404 Cyg and XN Sco). These have the longest periods, 6.5 and 2.6 days and XN Sco is suggested to have a relatively massive donor of ~ 2 M$_\odot$. It seems clear that these donors sat inside their Roche Lobes until they evolved off the main sequence, and then poured matter onto the black hole once they expanded and filled their Roche Lobe. For a 2 M$_\odot$ star, the evolutionary time is about a percent of the main-sequence time, so the fact that we see two subgiants out of nine transient sources means that many more of these massive donors are sitting quietly well within their Roche Lobes. Indeed, we could estimate from the relative time, that there are $2/9 \times 100 = 22$ times more of these latter quiet main sequence stars in binaries.

4. Discussion

We have shown that it is likely that single stars in the range of ZAMS masses $\sim 20 - 35$ M$_\odot$ evolve into high-mass black holes without return of matter to the Galaxy. This results because at mass ~ 20 M$_\odot$ the convective carbon burning is skipped and this leads to substantially more massive Fe cores. Even with more realistic reduced mass loss rates on He stars, however, it is unlikely that stars in this mass range in binaries evolve into high-mass black holes, because the progenitor of the compact object when stripped of its hydrogen envelope in either Case A (during main sequence) or Case B (RLOF) mass transfer will burn as a "naked" He star, ending up as an Fe core which is not sufficiently massive to form a high-mass black hole.

In the region of ZAMS mass $\sim 40\ M_\odot$, depending sensitively on the rate of He-star wind loss, the fate of the single star or the primary in a binary may be a low-mass black hole. In our estimates we have assumed the Brown & Bethe (1994) estimates of $1.5\ M_\odot$ for maximum neutron star mass and $1.5 - 2.5\ M_\odot$ for the range in which low-mass black holes can result.

In our evolution of the transient sources using Case C (during He shell burning) mass transfer, almost the entire He core will collapse into a high-mass black hole (Bethe, Brown, & Lee 2000), explaining the more or less common black hole mass of $\sim 7\ M_\odot$ for these objects, with the possible exception of V404 Cygni where the mass may be greater. Our evolution gives an explanation for the seemingly large gap in black-hole masses, between the $\gtrsim 1.5\ M_\odot$ for the black hole we believe was formed in 1987A and the $\sim 1.8\ M_\odot$ black hole we suggest in 1700-37 and the $\sim 7\ M_\odot$ in the transient sources.

We note that following the removal of the H envelope by Case C mass transfer, the collapse inwards of the He envelope into the developing black hole offers the Collapsar scenario for the most energetic gamma ray bursters of MacFadyen & Woosley (1999).

Acknowledgements

We would like to thank Charles Bailyn and Stan Woosley for useful discussions. We were supported by the U.S. Department of Energy under Grant No. DE–FG02–88ER40388.

References

Bethe, H.A., Brown, G.E., and Lee, C.-H. (2000), *ApJ* **541**, 918.
Brown, G.E. and Bethe, H.A. (1994), *ApJ* **423**, 659.
Brown, G.E., Lee, C.-H., and Bethe, H.A. (1999), *New Astronomy* **4**, 313.
De Kool, M., van den Heuvel, E.P.J., and Pylyser, E. (1987), *A&A* **183**, 47.
Ergma, E. and van den Heuvel, E.P.J. (1998), *A&A* **331**, L29.
Kaper, L., Lamers, H.J.G.L.M., van den Heuvel, E.P.J., and Zuiderwijk, E.J. (1995), *A&A* **300**, 446.
MacFadyen, A. and Woosley, S.E. (1999), *ApJ* **524**, 262.
Portegies Zwart, S.F., Verbunt, F., and Ergma, E. (1997), *A&A* **321**, 207.
Van den Heuvel, E.P.J. and Habets, G.M.H.J. (1984), *Nature* **309**, 598.
Weaver, T.A. and Woosley, S.E. (1993), *Phys. Rept.* **227**, 65.
Wellstein, S. and Langer, N. (1999), *A&A* **350**, 148.
Woosley, S.E. and Weaver, T.A. (1995), *ApJS* **101**, 181.

5. Magnetars

THE ANOMALOUS X-RAY PULSARS

S. MEREGHETTI

Istituto di Fisica Cosmica G.Occhialini - CNR
via Bassini 15, I-20133 Milano, Italy

1. Introduction

In the last few years it has been recognized that a few X–ray pulsars, which are not rotation powered, have peculiar properties that sets them apart from the majority of accreting pulsars in X–ray binaries. These objects, initially suggested as a homogeneous new class of pulsators in 1995 (Mereghetti & Stella 1995), have been named in different ways, reflecting our ignorance on their true nature: Very Low Mass X–ray Pulsars, Braking Pulsars, 6-sec Pulsars, Anomalous X–ray Pulsars. The latter designation (AXP) has become the most popular and will be used here.

Though we can be reasonably confident that the AXP are rotating neutron stars without massive companions, it is unclear whether they are solitary objects or are in binary systems with very low mass stars. As a consequence, different mechanisms for powering their X–ray emission have been proposed, involving either accretion or other less standard processes such as, e.g., the decay of magnetic energy.

The properties that distinguish the AXP from the more common pulsars found in High Mass X–Ray Binaries (HMXRB) are the following:

a) spin periods in a narrow range ($\sim$6–12 s), compared to the much broader one (0.069 $-$ $\sim$10^4 s) observed in HMXRB pulsars (see Figure 1)

b) no identified optical counterparts, with upper limits excluding the presence of normal massive companions, like OB (super)giants and/or Be stars

c) very soft X–ray spectra (characteristic temperature $\lesssim$ 1 keV and/or power-law photon index $\gtrsim$ 3)

d) relatively low X–ray luminosity ($\sim$ 10^{34}–10^{36} erg s^{-1}) compared to that of HMXRB pulsars (see Figure 1)

e) little or no variability (on timescales from hours to years)

C. Kouveliotou et al. (eds.), The Neutron Star – Black Hole Connection, 351–367.
© 2001 *Kluwer Academic Publishers. Printed in the Netherlands.*

TABLE 1. AXP and related objects

SOURCE	P (s)	$\dot{P}$ (s s^{-1})	SNR d (kpc)/age (kyr)	SPECTRUM kT$_{BB}$/α_{ph}
Anomalous X–ray Pulsars (AXP)				
1E 1048.1–5937	6.45 [1]	$[1.5-4]\times10^{-11}$ [2,3]	–	BB+PL [3] ~0.64 keV / ~2.5
1E 2259+586	6.98 [4]	$\sim5\times10^{-13}$ [5,6]	G109.1–0.1 [7,8,9] 4–5.6 / 3–20	BB+PL [9] ~0.44 keV / ~3.9
4U 0142+61	8.69 [10]	$\sim2\times10^{-12}$ [11]	–	BB+PL [11,12] ~0.4 keV / ~4
RXSJ170849–4009	11.00 [13]	2×10^{-11} [14]	–	BB+PL [13] ~0.41 keV/ 2.92
1E 1841–045	11.77 [15]	4.1×10^{-11} [16]	Kes 73 [17,18] 6–7.5 / $\lesssim$ 3	PL [19] – / ~3.4
AX J1845.0–0300	6.97 [20]	–	G29.6+0.1 [21] <20 / <8	BB [20] ~0.7 keV / –
Pulsed Soft Gamma-ray Repeaters (SGR)				
SGR 0526–66	8.1 [22]	–	N49 in LMC [23]	uncertain [24]
SGR 1806–20	7.48 [25]	$\sim8.3\times10^{-11}$ [25]	G10.0–0.3 [26]	PL [27] ~2.2
SGR 1900+14	5.16 [28]	$\sim[5-14]\times10^{-11}$ [29,30]	G42.8+0.6 [31]	BB+PL [32] ~0.5 keV / 1.1
(Candidate) Radio-Quiet Neutron Stars				
1E 1207–5209 [33]	–	–	G296.5+10	BB [33] ~0.25 keV
1E 1614–5055 [34]	–	–	RCW 103	BB [35] ~0.6 keV
1E 0820–4247 [36]	0.075 ? [37]	$1.5\ 10^{-13}$? [37]	Puppis A	BB [35] ~0.3 keV
RX J0720.4–3125	8.39 [38]	–	–	BB [38] ~0.08 keV
RXJ1856.5–3754 [39]	–	–	–	BB [39] ~0.06 keV

[1] Seward et al. 1986; [2] Mereghetti 1995; [3] Oosterbroek et al. 1998; [4] Fahlman & Gregory 1981; [5] Baykal & Swank 1996; [6] Kaspi et al. 1999; [7] Hughes et al. 1984; [8] Rho & Petre 1997; [9] Parmar et al. 1998; [10] Israel et al. 1994; [11] Israel et al. 1999a; [12] White et al. 1996; [13] Sugizaki et al. 1997; [14] Israel et al. 1999b; [15] Vasisht & Gotthelf 1997; [16] Gotthelf et al. 1999; [17] Sanbonmatsu & Helfand 1992; [18] Helfand et al. 1994; [19] Gotthelf & Vasisht 1997; [20] Torii et al. 1998; [21] Gaensler et al. 1999; [22] Mazets et al. 1979; [23] Cline et al. 1982; [24] Marsden et al. 1996; [25] Kouveliotou et al. 1998; [26] Kulkarni et al. 1994; [27] Sonobe et al. 1994; [28] Hurley et al. 1999; [29] Kouveliotou et al. 1999; [30] Woods et al. 1999b; [31] Vasisht et al. 1994; [32] Woods et al. 1999a; [33] Mereghetti et al. 1996; [34] Tuohy & Garmire 1980; [35] Gotthelf et al. 1997; [36] Petre et al. 1996; [37] Pavlov et al. 1999; [38] Haberl et al. 1997; [39] Walter et al. 1996;

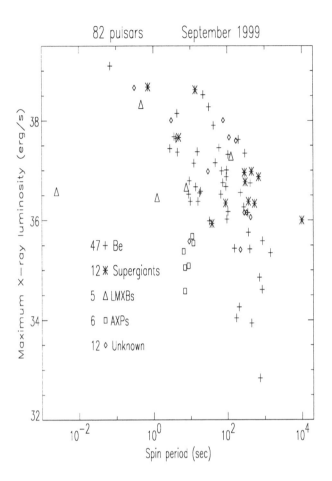

Figure 1. Spin period and maximum X–ray luminosity of different classes of X–ray pulsars (adapted from Tiengo 1999).

f) relatively stable spin period evolution, with long term spin-down trend

g) a few of them are associated with supernova remnants.

There are now six members of the AXP class (section 2). This review is mainly focussed on their observational properties (section 3), while the models are briefly discussed in section 4.

2. The AXP sample

Table 1 lists the 6 pulsars that share the above characteristics and form the current AXP sample. For comparison, also the properties of other objects that might be related to the AXP are reported in Table 1. The soft gamma-

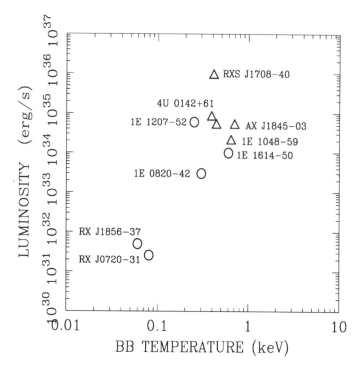

Figure 2. "Hertzsprung-Russell diagram" for AXP (triangles) and isolated neutron stars (circles). The temperatures refer to the blackbody spectral components.

ray repeaters (SGR) have P and $\dot{P}$ values very similar to those of the AXP. As discussed below, the magnetar model, originally developed to explain the SGR, has also been applied to the AXP. A few other (candidate) isolated neutron stars have some similarities with the AXP (see Figure 2), but more observations are needed to establish their nature.

On the basis of a better understanding of the AXP properties and/or of new observational results, we exclude from the AXP group a few sources that have been previously considered as part of this class of objects. 4U 1626–67 was originally included in the AXP class (Mereghetti & Stella 1995), but several authors pointed out its different nature: it has a harder spectrum, an optical identification, there is clear evidence for a binary nature, and showed an extended period of spin-up (van Paradijs et al. 1995, Ghosh et al. 1997).

The presence of pulsations at 5.45 s in the ROSAT source RXJ 1838.4–0301 (Schwentker 1994) has not been confirmed by more sensitive ASCA observations. Furthermore, optical observations of its possible counterparts revealed the presence of a main sequence K5 star with V∼ 14.5 (Mereghetti, Belloni & Nasuti 1997). This star could be responsible for the observed X–ray flux, since the implied X–ray to optical flux ratio (f_x/f_{opt}) is compatible with the level of coronal emission expected in late type stars. Thus it seems very likely that RXJ 1838.4–0301 is not a pulsar – i.e. the statistical significance of the periodicity was overestimated (Schwentker 1994).

The 8.4 s pulsar RX J0720.4–3125 (Haberl et al. 1997) has also been sometimes included in the AXP group, on the basis of its period value, high f_x/f_{opt}, and soft spectrum. Indeed its spectrum is even softer than that of AXP and it can only be detected thanks to very low interstellar absorption ($N_H \sim 10^{20}$ cm^{-2}). Since this is taken as evidence for a very small distance (d$\sim$100 pc), the implied luminosity of $\sim 3 \times 10^{31}$ erg s^{-1} is much smaller than that of the AXP. It has been suggested that RX J0720.4–3125 is an old neutron star accreting from the interstellar medium, but the possibility of a medium age neutron star, still emitting through dissipation of its internal heat, cannot be excluded.

3. Observational Properties of the AXP

3.1. SPECTRA

The AXP are characterized by soft X-ray spectra, clearly different from those of the pulsars in HMXRB. The latter have relatively hard spectra in the $2-10$ keV range (i.e. power law photon index $\alpha_{ph} \sim 1$) that steepen with an exponential cut-off above $\sim$20 keV. On the contrary, since their first observations, AXP showed very soft power law spectra, with $\alpha_{ph} \gtrsim 3-4$. Reports of possible cyclotron features at low energy ($\sim$5–10 keV) in 1E 2259+586 (Iwasawa et al. 1992) have not been confirmed.

Recent observations with ASCA and BeppoSAX, have shown that in most cases a single power law is not sufficient to describe the spectra of AXP. All the AXP for which good quality observation are available (White et al. 1996, Parmar et al. 1998, Oosterbroek et al.1998, Israel et al. 1999a) require the combination of a blackbody-like component with kT$\sim$0.5 and a steep power law ($\alpha_{ph} \sim$3–4). A single power law is adequate to describe the spectrum of 1E 1841–045 (Gotthelf & Vasisht 1998), but the analysis is complicated by the presence of the underlying emission from the SNR that might hamper the detection of the blackbody component. In AX J1845.0–0300 a blackbody with kT$\sim$0.7 keV gives a good fit without the need for an additional power law component (Gotthelf & Vasisht 1997).

The spectral parameters for all the AXP are summarized in Table 2. The emitting area inferred from the blackbody components, that account up to $\sim$40–50% of the observed luminosity, is compatible with a large fraction of a neutron star surface.

Some evidence for spectral variations as a function of the spin-period phase has been reported for several AXP: 1E 2259+586 (Iwasawa et al. 1992, Corbet et al. 1995, Parmar et al. 1988), 4U 0142+61 (Israel et al.

1999a), 1E 1048.1−5937 (Corbet & Mihara 1997, Oosterbroek et al. 1988) and 1RXS J170849−400910 (Sugizaki et al. 1997). Unfortunately, the relatively poor energy resolution, and the limited statistics, do not allow to unambiguously characterize the spectral variations in the two separate components.

It is possible that this two-component model is an oversimplified description of the true underlying spectra resulting from the current instrumental limitations. Future observations with XMM should resolve this issue, possibly leading to the discovery of narrow spectral features that so far escaped detection. Note in particular that the energy of cyclotron lines from ions lies in the 0.1−10 keV range for the high values of the magnetic field (B∼10^{14}) expected for the magnetar model (see section 4.2).

3.2. DISTANCES AND LUMINOSITIES

Due to the lack of optical identifications, the distances of AXP are quite uncertain (with the exception of the two in SNR, section 3.8). However, some constraints can be derived from their location in the Galaxy.

The low distribution on the galactic plane (< |b| >=0.35°), indicates that, as a population, they are unlikely to be nearby ($\lesssim$ 1 kpc) objects. Such a conclusion is also consistent with the relatively high column density derived from the X−ray spectral fits (Table 2).

1E 1048.1−5937 lies in the direction of the Carina Nebula, which is thought to contribute to the high absorption measured in its spectrum, giving a lower limit to the distance of 2.8 kpc (Seward et al. 1986). A similar argument can be made for 4U 0142+61 that probably lies behind a local (d $\lesssim$ 1 kpc) molecular cloud clearly visible in absorption on the Palomar Sky Survey plate (Israel, Mereghetti & Stella 1994). On the other hand, a distance much in excess of ∼5 kpc, would place this source outside the Galaxy.

The two AXP associated with SNR have better distance estimates: 6-7.5 kpc for 1E 1841−045 in Kes 73 (Sanbonmatsu & Helfand 1992) and 5.6 kpc for 1E 2259+586 in G109.1-0.1 (Hughes et al. 1984).

1RXS J170849−400910 is in the general direction of the galactic center region and has a highly absorbed X−ray spectrum, which suggests a distance of the order of 8 kpc or more.

According to Torii et al. (1998), AX J1845.0−0300 could be located in the Scutum arm, at d∼8.5 kpc. Also this source is very absorbed and its distance could be larger. More information will be obtained if its association with the new radio SNR found by Gaensler et al. (1999) is confirmed.

Based on these distances and the observed fluxes, luminosities in the ∼ 10^{34}−10^{36} erg s^{-1} range are obtained for the AXP (see Table 2).

TABLE 2. Spectral properties of AXP

SOURCE	$L_x{}^{(a)}10^{35}$ erg s^{-1} $d^{(b)}$ kpc	$kT_{BB}^{(c)}$ keV $R_{BB}^{(d)}$ km	$\alpha_{ph}^{(e)}$ $N_H 10^{22}$ cm^{-2}	L_{BB}/L_{tot} Pulsed Fraction
1E 1048.1–5937	0.2	0.64	2.5	0.55
	5	1	0.5	~70%
1E 2259+586	0.5	0.44	3.9	0.4
	5	4.1	0.9	~30%
4U 0142+61	0.8	0.4	4	0.4
	1	2.4	1.1	~10%
1RXSJ170849–4009	9	0.41	2.92	0.17
	8	3.2	1.4	~30%
1E 1841–045	3	–	3.4	–
	7		3	~35%
AX J1845.0–0300	0.5	0.7	–	–
	8	1.5	4.6	~50%

(a) corrected for interstellar absorption
(b) assumed values, see section 3.2 for the uncertainties
(c) temperature of blackbody component
(d) equivalent radius of blackbody component
(e) photon index of power law component

Another uncertainty affecting the AXP luminosity estimates is the correction for the (model dependent) X–ray absorption. In principle, this could be a relevant factor, due to the steepness of the observed spectra. Note in fact that for a power law spectrum that extends down to low energy with, e.g., α_{ph} ~4, the flux in the 0.5−2 keV range is ~15 times the 2−10 keV one. However, for the blackbody plus power law spectra discussed in section 3.1 this correction is much smaller. It seems therefore well established that AXP have X–ray luminosities smaller than those typically observed in persistent HMXRB pulsars.

3.3. VARIABILITY

In general, AXP have relatively steady X–ray fluxes, compared with the kind of variability displayed by other classes of accreting compact objects. Most AXP have been detected at similar flux levels by all the satellites that looked at them. There are, however, some interesting exceptions.

The best evidence for flux variability has been so far obtained for AX J1845.0−0300. This source was discovered at a flux level of 4.2×10^{-12} erg cm^{-2} s^{-1} (2−10 keV) in an ASCA pointing performed in December 1993, but it was not visible 3.5 years later, implying a flux decrease greater than a factor 14 (Torii et al. 1998). A further ASCA observation revealed only a weak source at a position consistent with that of the AXP (Gaensler et al. 1999). Though a search for pulsations could not be performed, due to the small number of counts, it is likely that this source is AX J1845.0−0300 in a low state, a factor ∼10 fainter than the 1993 level.

In a GINGA observation performed in 1990 (Iwasawa et al. 1992), 1E 2259+586 was a factor ∼2 brighter than in previous measurements with the same instrument. During the higher intensity state a change in the double-peaked pulse profile (a larger difference in the relative intensity of the two pulses) was also observed, as well as a variation in the spin-down rate. Most of the other observations of 1E 2259+586, obtained with different satellites, yielded flux measurements of $\sim2-3 \times 10^{-11}$ erg cm^{-2} s^{-1}, consistent with the lower intensity state (see Corbet et al. 1995, Parmar et al. 1998 and references therein).

The flux measurements available for 1E 1048.1−5937 have been summarized by Oosterbroek et al. (1998). They show long term variations within a factor ∼5 (possibly more if a rather uncertain upper limit obtained with the Einstein Observatory is also considered, Seward et al. 1986). However, the comparison of these flux measurements is affected by the uncertainties deriving from the use of different instruments.

No evidence for significant variability has been reported for the three remaining AXP: 4U 0142+61, 1E 1841−045 and 1RXS J170849−400910. However, since most of the relevant observations have been obtained with different instruments (sometimes also in different energy ranges) the limits that one can infer on the absence of variabilty are subject to considerable uncertainties. Several measurements were obtained with non-imaging instruments, and the fluxes must be corrected for the (poorly known) contribution from other components in the field of view (e.g. SNRs, diffuse galactic ridge emission, other sources, etc.), which introduce further uncertainties.

The level of variability in AXP is of interest since it is expected that some emission processes (e.g. thermal emission from the neutron star surface), produce less variability than other models (e.g. those involving mass accretion, which is in general subject to intensity fluctuations). More detailed searches for correlations between luminosity changes and spin-down variations can support accretion models, in which fluctuations in the mass accretion rate produce different torques on the rotating neutron star. Finally, the possible existence of many transient AXP with low quiescent luminosities, similar to AX J1845.0−0300, has important implications for the total number of AXP in the Galaxy and their inferred birthrate.

3.4. SPIN PERIOD DISTRIBUTION

As shown in Figure 1, X–ray pulsars in massive binaries have spin periods spanning several orders of magnitude, from 69 ms (A 0538–67) to about 3 hr (2S 0114+65). The concentration of periods in the narrow ~6−12 s interval was one of the properties that led to the identification of the AXP as a possibly distinct class of objects. It is clear, however, that a period in this range is not enough to qualify a pulsar as an AXP (in fact there are several HMXRB with periods similar to those of the AXP). If we define the AXP as "pulsars with a very soft spectrum, that are neither HMXRB nor rotationally powered neutron stars, and have luminosity $\sim 10^{34}$-10^{36} erg s^{-1}", it turns out remarkably that all the known objects satisfying this definition have periods of a few seconds and a secular spin-down (when measured).

Why no AXP are seen with much longer, or much shorter, periods? There are no obvious selection effects explaining this narrow period distribution. Though a chance result due to the statistics of small numbers cannot be ruled out, this could be a real effect related to the particular characteristics and evolution of these objects. If this period clustering reflects the fact that the AXP are (close to) equilibrium rotators, one has to invoke similar magnetic fields and accretion rates in all the AXP.

3.5. PERIOD EVOLUTION

One of the distinctive peculiarities of AXP is their long term period evolution. In general, accreting neutron stars are expected to spin-up, due to the angular momentum transferred from the accreting material, often forming an accretion disk (see, e.g. Henrichs 1983). Indeed this is observed in many HMXRB pulsars in which there is evidence for an accretion disk. Other pulsars show alternating episodes of spin-up and spin-down, the origin of which is not completely understood. On the contrary, the spin periods of AXP are increasing at a nearly constant rate (on timescales ranging from ~2,000 to ~4×10^5 yrs). This behaviour has now been observed in a few AXP for an extended period, spanning more than two decades.

It can immediately be seen that for these values of P and $\dot{P}$, and assuming the canonical value for the momentum of inertia of a neutron star I=10^{45} g cm^2, the rotational energy loss is orders of magnitude too small to power the observed luminosity of AXP.

Accurate timing measurements have shown that the spin-down of AXP is not constant, but is subject to small fluctuations (see, e.g., Iwasawa et al. 1992, Mereghetti 1995). Baykal & Swank (1996) showed that the level of $\dot{P}$ fluctuations in 1E 2259+586, the AXP with the largest number of period measurements, is similar to that typically observed in neutron stars accreting in X–ray binaries, which is several orders of magnitude greater than that of radio pulsars. More recently, Kaspi et al. (1999) have been able to obtain a phase-coherent timing solution for RXTE observations of 1E 2259+586 spanning 2.6 years. These data show a very low level of timing noise, contrary to the previous results that were based on sparse (not phase-coherent) observations spanning ∼20 years. Also 1RXS J170849−400910, monitored with RXTE for 1.4 yrs, was found to have a similar level of timing noise (Kaspi et al. 1999), while an even more stable rotator is 1E 1841−045 (Gotthelf et al. 1999). It seems therefore that, at least on timescales of a few years, some AXP can be very stable rotators, with a timing noise similar to that of radio pulsars – a finding that supports the magnetar interpretation (see section 4.2).

3.6. SEARCHES FOR ORBITAL PERIODS

No periodic intensity variations, like eclipses or dips, that might indicate the presence of a binary system, have been detected in AXP. Another clear signature of binarity, that has been of extreme importance in the study of HMXRB pulsars, is the presence of orbital Doppler shifts in the pulse frequency. The most sensitive searches for orbital Doppler shifts in AXP have been carried out with the RXTE satellite. Searches for orbital periods between a few minutes and one day gave negative results, yielding upper limits on the projected semi-major axis $a_x \sin i$ of ∼30 and ∼60 light-ms for 1E 2259+586 and 1E 1048.1−5937, respectively (Mereghetti, Israel & Stella 1998). Similar results were obtained by Wilson et al. (1998) for 4U 0142+61.

For any assumed value of the inclination angle i, these limit constrain the possible values of the companion mass M_c and orbital period (Figure 3). As discussed by Mereghetti, Israel & Stella (1998), except for the unlikely possibility that all these systems are seen nearly face-on, main sequence companion stars can be ruled out. Helium burning stars with mass $M \lesssim 0.8 \, M_\odot$ cannot be excluded, but the accretion rate produced by Roche lobe overflow would give a luminosity much greater than observed. A possibility is that of a He-burning companion, underfilling the Roche lobe and providing a low rate of accretion through a stellar wind, as suggested by Angelini et al. (1995) for 4U 1626−67. Another possibility that cannot be ruled out by the current limits on $a_x \sin i$, and on the inferred mass accretion rates, is that of a white dwarf companion. For example, a white dwarf with $M \sim 0.02 \, M_\odot$, filling its Roche lobe for an orbital period of the order of ∼30 min, would give an $\dot{M}$ of a few $\times 10^{-11} \, M_\odot$ year^{-1}, consistent with the observed luminosities. Mainly due to the lack of suitable observations, no similar searches for orbital Doppler shifts have been performed for the three remaining AXP: 1RXS J170849−400910, AX J1845.0−0300 and

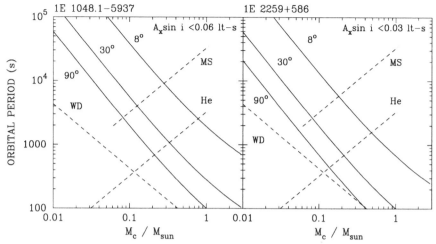

Figure 3.   Orbital constraints from the $a_x \sin i$ limit for 1E 1048.1−5937 and 1E 2259+586 (Mereghetti, Israel & Stella 1998). The limits on orbital period, P_{orb}, versus mass of the companion, M_c, are plotted assuming three different values for the unknown inclination angle. The dashed lines indicate the positions of Roche-lobe filling companions under the assumption of conservative mass transfer driven by angular momentum losses due only to gravitational radiation. They refer to the cases of a main sequence, a He burning star and a fully degenerate hydrogen white dwarf. Values of P_{orb} and M_c below the corresponding dashed line are excluded, while those above the lines require accretion through stellar wind.

1E 1841−045.

3.7. OPTICAL AND INFRARED COUNTERPARTS

The error box of 1E 1048.1−5937 has a radius of 15″ and contains several stars (Mereghetti, Caraveo & Bignami 1992). Spectroscopy of the 3 brightest objects (V $\gtrsim$ 19) did not yield a plausible counterpart showing the classical emission lines considered a signature of accreting objects. More objects were studied by Corbet & Mihara (1997), again with negative results. These studies are complicated by the presence of diffuse H_α emission from the Carina nebula, which affects the sky subtraction from the stellar spectra.

   1E 2259+586 is the AXP for which more extensive searches for counterparts have been carried out (Davies & Coe 1991, Coe & Jones 1992, Coe et al. 1994), sometime leading to possible identifications later disclaimed by better observations. The latest error box (5″ radius), reported by Coe & Pightling (1998) contains only three faint objects with K-band magnitudes of ∼18 and V>24.

   A different situation is found for 4U 0142+61, since in this case no objects are present within the small error box (∼3″ radius). The current best limits are V>24 (Steinle et al. 1987) and K>17 (Coe & Pightling 1998).

Optical observations of the field of 1RXS J170849−400910 have been reported by Israel et al. (1999b). These authors found that the possible counterparts cannot be massive early type stars, being too faint and blue (very distant and/or absorbed OB stars should appear more reddened by the interstellar dust absorption). No detailed reports on optical/IR observations of 1E 1841−045 and AX J1845.0−0300 have been published so far.

Though in general the limits on the possible optical/IR counterparts of AXP allow to rule out the presence of massive companion stars, more work is needed to explore different possibilities, especially because it is not clear which kind of properties one should expect from the AXP counterparts. Due to the crowding of these low galactic latitude fields, more precise localizations are also needed.

3.8. ASSOCIATION WITH SUPERNOVA REMNANTS (SNR)

The fact that two (possibly three) AXP are found at the center of SNR is very important, since it gives information on their origin, age and distance.

1E 2259+586 is located close to the geometrical center of G109.1−0.1 (also known as CTB 109), a partial radio/X-ray shell with an angular diameter of ~30′ (see, e.g., Rho & Petre 1997). As discussed in Parmar et al. (1998), the estimated age for this SNR is subject to a considerable uncertainty, ranging from ~3,000 yrs to 20,000 yrs. The other AXP clearly associated to a SNR is 1E 1841−045. It was discovered as an unresolved source at the center of Kes 73, a young (~ 2000 yr) SNR at a distance of ~7 kpc (Helfand et al. 1994). Gaensler et al. (1999) have recently reported the discovery of a radio SNR around AX J1845.0−0300. These three AXP are found close to the geometrical center of the respective SNR, implying relatively small transverse velocities for these objects.

One should not forget that three AXP (4U 0142+61, 1E 1048.1−5937, and 1RXS J170849−400910) lack visible SNRs. This might indicate that the lifetime of AXP is much longer than several 10^4 years.

There are also a few unresolved X-ray sources within SNRs that, apart for the lack of pulsations, share the same properties of the AXP (see Table 1). The sources in RCW 103 (Gotthelf, Petre & Hwank 1997), G296.5+10.0 (Mereghetti et al. 1996), and Puppis A (Petre et al. 1996) have high f_x/f_{opt}, soft spectra (characteristic blackbody temperatures kT $\lesssim$ 0.6 keV), and low luminosity, similar to the AXP (see Figure 2). More sensitive searches for pulsations in these sources (so far hampered by the poor statistics) might reveal in the future new AXP (this does not apply to the source in Puppis

A if the possible periodicity at 75 ms reported by Pavlov et al. (1999) is confirmed by better data).

4. Models

Though the absence of a massive companion and the presence of a neutron star are observationally well established, the AXP remain one of the more enigmatic classes of galactic high energy sources. Also the main mechanism responsible for the observed X–ray luminosity is still unclear. Having excluded models powered by the rotational energy loss of isolated neutron stars (see section 3.5), the remaining explanations advanced for the AXP fall into two main classes: models based on accretion (with or without a binary companion of very low mass) and those invoking highly magnetized neutron stars powered by the decay of the magnetic field and/or internal heat dissipation.

Binary models have the advantage of naturally providing accretion as a source of energy. However, the tight limits on the possible companion stars (sections 3.6, 3.7) have also led to interpretations based on accretion unto isolated neutron stars.

4.1. ACCRETION-BASED MODELS

In general, accretion from the interstellar medium (ISM) cannot provide the required luminosity under typical ISM parameters and neutron star velocities. In fact, the accretion luminosity is given by $L_{acc} \sim 10^{32} \ v_{50}^{-3} \ n_{100}$ erg s^{-1} where v_{50} is the relative velocity between the neutron star and the ISM in units of 50 km s^{-1} and n_{100} is the gas density in units of 100 atoms cm^{-3}. Unless all the AXP lie within nearby ($\sim$100 pc) molecular clouds, which seems very unlikely considered their distribution in the galactic plane, the accretion rate is clearly insufficient to produce the observed luminosities.

Van Paradijs et al. (1995) proposed a more efficient scenario, in which isolated neutron stars are fed from residual accretion disks, formed after the complete spiral-in of a neutron star in the envelope of a giant companion star (a Thorne-Zytkow object, TZO, Thorne & Zytkow 1977). Thus the AXP could be one possible outcome of the common envelope evolutionary phase of close HMXRB systems. The connection with massive binaries is supported by the fact that the AXP seem to be relatively young objects, being located at small distances from the galactic plane and sometimes found associated with SNR. According to van Paradijs et al. (1995), the estimated birthrate of AXP is consistent with that of TZO.

The idea that AXP are isolated neutron star accreting from a residual disk has been further developed by Ghosh et al. (1997), who put this model in the broader context of the evolution of close massive binaries. In this scenario, a HMXRB undergoing common envelope evolution can produce two kinds of objects, depending on the efficiency with which the massive star envelope is lost. Relatively wide systems have enough orbital energy to lead to the complete expulsion of the envelope before the settling of the neutron star at the center. This would result in the formation of binaries composed of a neutron star and a Helium star, like 4U 1626–67 and Cyg X–3. Closer HMXRB, on the other hand, would produce TZO due to the complete spiral in of the neutron star in the common envelope phase. These systems would subsequently evolve into AXP: isolated neutron stars undergoing accretion from two distinct flows: a disk and a spherically symmetric component, resulting from the part of the envelope with less angular momentum. According to Ghosh et al. (1997), this model would also explain the two component spectra observed in most AXP, as well as their secular spin-down: the accretion from the disk is responsible for the power-law and the long term spin-down due to the decreasing mass accretion rate, while the spherically symmetric flow gives rise to the blackbody emission from a large fraction of the neutron star surface.

Though this is certainly an interesting model, several uncertainties exist. In particular very little is known on the evolution during the common envelope phase and on the efficiency of conversion of the orbital binding energy to that of the dynamical outflow of the envelope. According to Li (1999), other problems of this model are the short lifetime of the accretion disk and the fact that in any case it would be unable to reproduce the spin-down behaviour observed in AXP.

Binary models for AXP have not been developed in detail, although we note that they cannot be completely ruled out in the case very low mass companions and/or unfavourable inclination angles (furthermore, sensitive searches for Doppler modulations have only been done for three out of six AXP). In a certain sense, this is the most conservative explanation since it does not involve new kinds of objects with relatively uncertain properties. In the context of binary systems with very low mass companions, Mereghetti & Stella (1995) proposed that the AXP are weakly magnetized neutron stars ($B\sim10^{11}$ G) rotating close to the equilibrium period. This requires accretion rates of the order of a few 10^{15} g s^{-1}, consistent with the AXP luminosities.

4.2. MAGNETARS

Models based on strongly magnetized ($B\sim10^{14}$–10^{15} G) neutron stars, or "magnetars", were originally developed to explain the peculiar properties of SGR (Duncan & Thompson 1992; Thompson & Duncan 1995,1996) and received a substantial support with the discovery of pulsations and spindown in these sources (Kouveliotou et al. 1998, 1999, Hurley et al. 1999). If one assumes that the AXP spin-down is due to magnetic dipole radiation losses, values of B $= 3.2 \times 10^{19} \, (P\dot{P})^{1/2} \gtrsim 10^{14}$ G are obtained, suggesting that also the X–ray emission from these objects could be powered by magnetic field decay (see Thompson, these proceedings).

Different authors discussed the kind of spin-down irregularities expected in the magnetar model. Heyl & Hernquist (1999) fitted the period histories of 1E 2259+586 and 1E 1048.1−5937 with glitches similar to those observed in radio pulsars. The same data were interpreted by Melatos (1999) in terms of a periodic ($\sim 5-10$ yrs) oscillation in $\dot{P}$ caused by radiative precession, an effect due to the star asphericity induced by the very strong magnetic field. Unfortunately, the sparse period measurements available for AXP do not allow for the moment to discriminate among the different possibilities.

5. Conclusions

Though the nature of the AXP is still unknown, after more than 20 years since the discovery of the prototype of this class (1E 2259+586), it is clear that these objects represent an important manifestation of neutron stars. There is growing evidence that a large fraction of neutron stars are born with properties very different from that of the Crab and Vela pulsars. This might explain why only very few energetic, rapidly spinning radio pulsars have a firm association with a SNR.

Due to their relatively low luminosity and soft spectrum (critically affected by the interstellar absorption) AXP are not easy to find. Several of the known X-ray sources, too faint for sensitive pulsation searches, could be AXP and we can expect that, thanks to the coming X–ray satellites, many more will be discovered in the near future. Furthermore, if AX J1845.0−0300 is confirmed as a "transient" AXP, the overall population of this class of objects would be even larger than assumed so far. It might well be that the "Anomalous" pulsars are indeed one of the most "normal" manifestations of young neutron stars.

References

Angelini, L. et al. (1995), *ApJ* **449**, L41.
Baykal, A. and Swank, J.H. (1996), *ApJ* **460**, 470.
Cline, T.L. et al. (1982), *ApJ* **255**, L45.
Coe, M.J. and Jones, L.R. (1992), *MNRAS* **259**, 191.
Coe, M.J. and Pightling, S.L. (1998), *MNRAS* **299**, 223.
Coe, M.J., Jones, L.R., and Letho, H. (1994), *MNRAS* **270**, 178.
Corbet, R.H.D, Smale, A.P., Ozaki, M. et al. (1995), *ApJ* **443**, 786.
Corbet, R.H.D. and Mihara, T. (1997), *ApJ* **475**, L127.
Davies, S.R. and Coe, M.J. (1991), *MNRAS* **249**, 313.
Duncan, R.C. and Thompson, C. (1992), *ApJ* **392**, L9.

Fahlman, G.G. and Gregory, P.C. (1981), *Nature* **293**, 202.
Gaensler, B.M., Gotthelf, E.V., and Vasisht, G. (1999), *ApJL* **526**, L37.
Ghosh, P., Angelini, L. and White, N.E. (1997), *ApJ* **478**, 713.
Gotthelf, E.V. and Vasisht, G. (1997), *ApJ* **486**, L133.
Gotthelf, E.V. and Vasisht, G. (1998), *New Astronomy* **3**, 293.
Gotthelf, E.V., Petre, R., and Hwang, U. (1997), *ApJ* **487**, L175.
Gotthelf, E.V., Vasisht, G., and Dotani, T. (1999), *ApJ* **522**, L49.
Haberl, F. et al. (1997), *A&A* **326**, 662.
Helfand, D. et al. (1994), *ApJ* **434**, 627.
Henrichs, H.F. (1983), in W.H.G. Lewin and E.P.J. van den Heuvel (eds.), *Accretion Driven Stellar X-ray Sources*, Cambridge University Press, p393.
Heyl, J.S. and Hernquist, L. (1999), *MNRAS* **304**, L37.
Hughes, V.A. et al. (1984), *ApJ* **283**, 147.
Hurley, K. et al. (1999), *ApJ* **510**, L111.
Israel, G.L. et al. (1999a), *A&A* **346**, 929.
Israel, G.L. et al. (1999b), *ApJ* **518**, L107.
Israel, G.L., Mereghetti, S., and Stella, L. (1994), *ApJ* **433**, L25.
Iwasawa, K., Koyama, K., and Halpern, J.P. (1992), *PASJ* **44**, 9.
Kaspi, V.M., Chakrabarty, D., and Steinberger, J. (1999), *ApJ* **525**, L33.
Kouveliotou, C. et al. (1998), *Nature* **393**, 235.
Kouveliotou, C. et al. (1999), *ApJ* **510**, L115.
Kulkarni, S.R. et al. (1994), *Nature* **368**, 129.
Li, X.-D. (1999), *ApJ* **520**, 271.
Marsden, D., Rothschild, R.E., Lingenfelter, R.E., and Puetter, R.C. (1996), *ApJ* **470**, 513.
Mazets, E.P. et al. (1979), *Nature* **282**, 587.
Melatos, A. (1999), *ApJ* **519**, L77.
Mereghetti, S. and Stella, L. (1995), *ApJ* **442**, L17.
Mereghetti, S. (1995), *ApJ* **455**, 598.
Mereghetti, S., Bignami, G.F., and Caraveo, P.A. (1996), *ApJ* **464**, 842.
Mereghetti, S., Belloni, T., and Nasuti, F. (1997), *A&A* **321**, 835.
Mereghetti, S., Caraveo, P., and Bignami, G.F. (1992), *A&A* **263**, 172.
Mereghetti, S., Israel, G.L., and Stella, L. (1998), *MNRAS* **296**, 689.
Oosterbroek, T., Parmar, A.N., Mereghetti, S., and Israel, G.L. (1998), *A&A* **334**, 925.
Parmar, A. et al., (1998), *A&A* **330**, 175.
Pavlov, G.G., Zavlin, V.E., and Truemper, J. (1999), *ApJ* **511**, L45.
Petre, R., Becker, C.M., and Winkler, P.F. (1996), *ApJ* **465**, L43.
Rho, J. and Petre, R. (1997), *ApJ* **484**, 828.
Sanbonmatsu, K.Y. and Helfand, D.J. (1992), *AJ* **104**, 2189.
Schwentker, O. (1994), *A&A* **286**, L47.
Seward, F., Charles, P.A., and Smale, A.P. (1986), *ApJ* **305**, 814.
Sonobe, T. et al. (1994), *ApJ* **436**, L23.
Steinle, H. et al. (1987), *Astroph. Space Sci.* **131**, 687.
Sugizaki, M. et al. (1997), *PASJ* **49**, L25.
Thompson, C. and Duncan, R.C. (1995), *MNRAS* **275**, 255.
Thompson, C. and Duncan, R.C. (1996), *ApJ* **473**, 322.
Thorne, K.S. and Zytkow, A.N. (1977), *ApJ* **212**, 832.
Tiengo, A. (1999), *Thesis*, University of Milano.
Torii, K. et al. (1998), *ApJ* **503**, 843.
Tuohy, I. and Garmire, G. (1980), *ApJ* **239**, L107.
Van Paradijs, J., Taam, R.E., and van den Heuvel, E.P.J. (1995), *A&A* **299**, L41.
Vasisht, G. and Gotthelf, E.V. (1997), *ApJ* **486**, L129.
Vasisht, G. et al. (1994), *ApJ* **431**, L35.
Walter, F.M. et al. (1996), *Nature* **379**, 233.
White, N.E. et al. (1996), *ApJ* **463**, L83.

Wilson, C.A. et al. (1998), *ApJ* **513**, 464.
Woods, P.M. et al. (1999a), *ApJ* **518**, L103.
Woods, P.M. et al. (1999b), *ApJ* **524**, L55.

ASTROPHYSICS OF THE SOFT GAMMA REPEATERS AND ANOMALOUS X-RAY PULSARS

C. THOMPSON
University of North Carolina,
Department of Physics and Astronomy,
Chapel Hill, NC 27599

Abstract. I summarize the recent advances in our understanding of the Soft Gamma Repeaters: in particular their spin behavior, persistent emission and hyper-Eddington outbursts. The giant flares on 5 March 1979 and 27 August 1998 provide compelling physical evidence for magnetic fields stronger than $10\,B_{\rm QED} = 4.4 \times 10^{14}$ G, consistent with the rapid spindown detected in two of these sources. The persistent X-ray emission and variable spindown of the 6-12 s Anomalous X-ray Pulsars are compared and contrasted with those of the SGRs, and the case made for a close connection between the two types of sources. Their collective properties point to the existence of *magnetars*: neutron stars in which a decaying magnetic field (rather than accretion or rotation) is the dominant source of energy for radiative and particle emissions. Observational tests of the magnetar model are outlined, along with current ideas about the trigger of SGR outbursts, new evidence for the trapped fireball model, and the influence of QED processes on X-ray spectra and lightcurves. A critical examination is made of coherent radio emission from bursting strong-field neutron stars. I conclude with an overview of the genetic connection between neutron star magnetism and the violent fluid motions in a collapsing supernova core.

1. Introduction

During the last 30 years, a comfortable picture of the Galactic neutron star population emerged: neutron stars are born with largely dipolar magnetic fields of $\sim 10^{11} - 10^{13}$ G, which do not decay significantly unless the star accretes upwards of $\sim 0.1\,M_\odot$ from a binary companion. This picture is based on observations of neutron stars whose pulsed emissions

369

C. Kouveliotou et al. (eds.), The Neutron Star – Black Hole Connection, 369–392.

are powered either by rotation, or by accretion. In the first case, there are strong selection effects against observing radio pulsations from a star whose dipole magnetic field is much stronger than the quantum electrodynamic value $B_{QED} = m_e^2 c^3/e\hbar = 4.4 \times 10^{13}$ G. At a fixed age, the spin period $P \propto B_{dipole}$ – after the magnetic dipole torque has pushed P well above its initial value – and the spindown luminosity $I\Omega\dot{\Omega} \propto B_{dipole}^{-2}$. The radio pulsations are also expected to be beamed into an increasingly narrow solid angle, a dramatic example being the 'new' 8.5 s PSR J2144-3933 (Young, Manchester, and Johnston 1999). The upper envelope of the distribution of measured pulsar dipole fields has, nonetheless, increased significantly with the recent discovery of PSRs J1119–6127 and J1814–1744, the second of which is inferred to have a polar field in excess of 10^{14} G (Camilo et al. 2000). The apparent paucity of neutron stars with $B_{dipole} > B_{QED}$ in accreting systems places tighter constraints on their birth rate *if* they have the same distribution of natal kicks as ordinary radio pulsars.

Although strong by terrestrial standards, a $\sim 10^{12}$ G magnetic field is, in a dynamical sense, quite weak. It contributes only $\sim 10^{-9}$ of the hydrostatic pressure (when the effects of proton superconductivity in the stellar core are taken into account). Much stronger magnetic fields can be generated by vigorous convective motions in a supernova core, $B \sim 10^{15}$ G (Thompson and Duncan 1993, hereafter TD93). Substantial evidence has accumulated in recent years[1] for neutron stars whose much stronger magnetic fields ($B_{dipole} \sim 10\,B_{QED} = 4.4 \times 10^{14}$ G) decay significantly on a very short timescale ($\sim 10^4$ yr). These *magnetars* were predicted to spin down much more rapidly than ordinary radio pulsars, and should be elusive (although not necessarily impossible to detect) as pulsed radio sources. They have been most cogently associated (Duncan and Thompson 1992, hereafter DT92; Paczyński 1992; Thompson and Duncan 1995, hereafter TD95) with the Soft Gamma Repeaters: a small peculiar class of neutron stars that emit extremely luminous and hard X-ray and gamma-ray bursts. The growing group of Anomalous X-ray Pulsars (Mereghetti 2000) may be closely related (Thompson and Duncan 1996, hereafter TD96).

The defining property of a magnetar is that its decaying magnetic field outstrips its rotation as a source of energy for X-ray and particle emission — by some two orders of magnitude if the core field is as strong as $\sim 10^2\,B_{QED}$

[1]Due to the very unfortunate absence of Jan van Paradijs and Chryssa Kouveliotou, this review covers the phenomenology of the Soft Gamma Repeaters and Anomalous X-ray Pulsars, as well as theoretical aspects of strong-field neutron stars. It combines two recent summaries of the bursting behavior of the SGRs (Thompson 2000b) and the persistent emission and spindown of the SGRs and AXPs (Thompson 2000c). See Norris et al. (1991) for a review of the early literature on the SGRs, and Frail (1998) and Mereghetti (2000) for a more complete review of radio-silent neutron stars and the AXPs.

(TD96). The observational signatures of this decay include persistent X-ray and particle emissions and, if $B > (4\pi\theta_{max}\mu)^{1/2} = 2 \times 10^{14} (\theta_{max}/10^{-3})^{1/2}$ G, sudden outbursts triggered by fractures of the rigid crust. (Here μ is the shear modulus and θ_{max} the yield strain in the deep crust.)

2. Soft Gamma Repeaters

The SGRs are best known for two giant flares on March 5, 1979 and August 27, 1998 (from SGR 0526–66 and SGR 1900+14, respectively). These remarkable bursts, separated by almost 20 years, are nearly carbon copies of each other. They released $\sim 4 \times 10^{44}$ and $\sim 1 \times 10^{44}$ erg in X-rays, respectively (Cline 1982; Hurley et al. 1999a; Feroci et al. 1999; Mazets et al. 1999b and references therein) and had very similar and striking morphologies. Each was initiated by a short and intense ($t \sim 0.2 - 0.5$ s) spike. The luminosity of this spike exceeded the classical Eddington luminosity – above which the outward force due to electron scattering exceeds the attractive force of gravity: $L_{edd} \simeq 2 \times 10^{38}$ erg s^{-1} for a $1.4\,M_\odot$ neutron star – by a factor $3 \times 10^6 - 10^7$ in the case of the March 5 event (Fenimore, Klebesadel, and Laros 1996). The ensuing softer emission, which lasted $200 - 400$ s and radiated somewhat more energy, had a much more stable temperature even though its luminosity exceeded $\sim 10^4 L_{edd}$. In the March 5 flare, this tail had a striking 8-second periodicity of a very large amplitude, which was inferred to be the rotation period of SGR 0526–66. The August 27 event exhibited a similar 5.16 s periodicity of an even larger amplitude.

These SGRs are more generally characterized by short (~ 0.1 s) X-ray outbursts with luminosities as high as $10^4 L_{edd}$. The statistics of the short bursts have some intriguing similarities with earthquakes and Solar flares (Cheng et al. 1996). According to the most recent analyses, the bursts of SGR 1806–20 (Göğüş et al. 1999b) and SGR 1900+14 (Göğüş et al. 1999a) have a power-law distribution of energies $dN/dE \propto E^{-1.6}$ that extends over some 4-5 decades. The distribution of waiting times is lognormal, peaking at ~ 1 day (Hurley et al. 1996; Göğüş et al. 1999a,b). In the case of SGR 1806–20, continuous monitoring in 1983 showed that the burst fluence accumulates in a piecewise-linear manner while the source is active (Palmer 1999). This indicates the existence of multiple active regions, internal to the star.

The four known SGRs are also persistent X-ray sources of luminosity $10^{35} - 10^{36}$ erg s^{-1} (Rothschild, Kulkarni, and Lingenfelter 1994; Murakami et al. 1994; Hurley et al. 1999b; Woods et al. 1999b). Although the time average of the bursting emission is uncertain due to the intermittency of SGR activity, it is in order of magnitude comparable to the persistent output. Periodicites have been convincingly detected in two sources: $P = 7.47$ s for

SGR 1806–20 (Kouveliotou et al. 1998); and $P = 5.16$ s for SGR 1900+14 (Hurley et al. 1999c). This measurement of the spin preceded the August 27 event in the case of SGR 1900+14, and agreed with the periodicity detected in the giant outburst. This narrow clustering of persistent luminosities and especially of spin periods in the SGR sources provides an important clue to the energy source that powers them.

Evidence that SGR 0526–66 is a young neutron star comes from its positional coincidence with a supernova remnant N49 (age $\sim 7 \times 10^3$ yr) in the LMC. If the star has been spun down by a magnetic dipole torque to a period of 8 s, then one infers a polar dipole field of $6 \times 10^{14} \, (t/10^4 \ \mathrm{yr})^{-1/2}$ G (Duncan and Thompson 1992). Additional evidence that the SGRs are young neutron stars comes from the association of the other three – with varying degrees of certainty – with supernova remnants (Kulkarni and Frail 1993; Vasisht et al. 1994; Hurley et al. 1999b; Woods et al. 1999b) or regions of massive star formation (Fuchs et al. 1999).

An important observational breakthrough, which supports this interpretation, came with the recent detection of rapid spindown in SGRs 1806–20 and 1900+14. Both sources have (coincidentally) nearly the same characteristic age of $P/\dot{P} = 3000$ yr (Kouveliotou et al. 1998, 1999; Marsden, Rothschild, and Lingenfelter 1999; Woods et al. 1999c; Woods et al. 2000). The inferred polar magnetic field strength exceeds 10^{15} G in both cases, in the simplest model of an orthogonal vacuum rotator. Deviations from this torque behavior are outlined in §5.2.

3. Anomalous X-ray Pulsars

The Anomalous X-ray Pulsars constitute a separate group of a half-dozen neutron stars that have been detected through their persistent X-ray pulsations but have not (yet) been observed to burst (Mereghetti and Stella 1995; Duncan and Thompson 1995; Van Paradijs, Taam, and Van den Heuvel 1995; Mereghetti 2000). The AXPs share with the SGRs remarkably similar persistent X-ray luminosities ($L_X \sim 3 \times 10^{34} - 10^{36}$ erg s^{-1}), spin periods ($P \sim 6 - 12$ s), and characteristic ages ($P/\dot{P} \sim 10^3 - 10^5$ yr). They are all consistently spinning down. At least three are associated with young supernova remnants.

This overlap between the SGRs and AXPs in a *three*-dimensional parameter space would be surprising if the two classes of sources were powered by fundamentally different energy sources. Accretion has been suggested for the AXPs – either in a low mass X-ray binary (Mereghetti and Stella 1995) or through a fossil disk (Van Paradijs et al. 1995). However, the need for an active energy source in the SGRs, almost certainly involving a strong magnetic field, combined with the variability of the X-ray output of some

AXPs, led to the suggestion that the AXPs were also powered by a decaying magnetic field (TD96; Kouveliotou et al. 1998). The possibility remains that some AXPs are strongly magnetized but otherwise passively cooling neutron stars (Heyl and Hernquist 1997a,b). The SGRs are highly intermittent as burst sources (activity in SGR 0526–66 was only detected during the years 1979 to 1983); this provides a hint that some AXPs have experienced X-ray outbursts in the past and will burst in the future. Combining both classes of sources, one roughly estimates the net birth rate of SGRs/AXPs as $\sim 1 \times 10^{-3}$ per year (Thompson et al. 2000a, hereafter T00).

The identification of AXPs with inactive magnetars resolves some of the problems associated with accretion models. There is no evidence for binarity in any of these sources (Kaspi, Chakrabarty, and Steinberger 1999; Mereghetti 2000). If their X-ray emission were powered by accretion from a low-mass binary companion, then the long orbital evolution time – $10^7 - 10^8$ yr due to gravity wave emission – combined with the presence of a few AXPs inside SNR of age $\sim 10^4$ yr would imply far more sources in the Galaxy than are in fact observed (TD96). The fossil disk model appears to be on firmer grounds, and is testable by searches for optical/UV emission reprocessed by the disk from the central X-ray continuum. (See Perna, Hernquist, and Narayan 2000 for detailed modelling.) The X-ray spectra of the AXPs are peculiarly soft by the standards of X-ray binaries, which suggests that if they are accreting then their magnetic fields are weaker than 10^{12} G. Fields that weak can barely provide the spindown torque measured in the AXPs 1E 1048.1+5937 and 1E 1841–045 (Li 1999).

Interest in the connection between the AXPs and radio pulsars has been raised by the recent discovery of PSR J1814–1744, which is positioned near the AXP 1E 2259+586 in the $P - \dot{P}$ plane (Camilo et al. 2000). It should first be noted that the spindown age of 1E 2259+586 ($P/2\dot{P} \sim 2 \times 10^5$ yr) is > 10 times the age of the supernova remnant CTB 109 near whose center it sits. Since all the other AXPs and SGRs for which this comparison can be made have *shorter* spindown ages, it seems likely that the spindown torque of 1E 2259+586 has decayed, and that over most of its history this AXP sat a factor of ~ 10 higher in the $P - \dot{P}$ plane. Several effects – alignment, field decay, or a previous phase of accelerated spindown – could explain this observation in the magnetar model (T00).

The strong-field pulsar J1814–1744 is not a conspicuous X-ray source, being at least 10–100 times weaker than the AXPs (Pivovaroff, Kaspi, and Camilo 2000). More generally, there appears to be a bifurcation in the level of internally-generated dissipation between the SGRs (and AXPs) and radio pulsars of a similarly young age. A critical parameter which could explain this bifurcation is provided by the internal (e.g. toroidal) magnetic field (TD96). Above a flux density of $\sim 100 B_{\rm QED}$ – corresponding to 5-10 times the dipole fields inferred from spindown – there is a rapid increase in the rate of ambipolar diffusion through the degenerate core (see also §7).

4. SGR Outbursts: Physical Diagnostics

The strength of the magnetic field is probably the most important parameter to be determined in the SGR and AXP sources. In this section, we review how the extreme properties of SGR outbursts directly point to flux densities higher than $10\,B_{\mathrm{QED}} = 4.4 \times 10^{14}$ G in these sources.

4.1. TRIGGER AND ENERGETICS OF SGR BURSTS

To power one giant flare like the March 5 and August 27 events, an SGR must be able to store $\sim 10^{45}$ erg of potential energy. This should be compared with the maximum elastic energy $\sim 10^{44}\,(\theta_{\max}/10^{-2})^2$ erg that can be stored by the crust of a neutron star whose yield strain is $\theta_{\max}$. It should be emphasized at this point that a magnetic field stressing the outer Coulomb lattice of the star actually contains more available energy $(\delta B^2/4\pi)$ than is stored by the lattice itself $(\sim \frac{1}{2}\theta^2\mu$ where μ is the shear modulus). The ratio is $\sim 4\pi\mu/B^2 = 10^2\,(B/10\,B_{\mathrm{QED}})^{-2}$. Nonetheless, even with this amplification it is not possible for a star composed of *strange quark matter* to retain enough potential magnetic energy to power the March 5 and August 27 giant flares. The elastic energy of its crust is smaller than that of a neutron star by at least four orders of magnitude (Alcock et al. 1986).

Why are the giant flares so rare, when short SGR bursts are relatively common? Why is there a gap of more than two orders of magnitude in fluence? One possible explanation is that the giant flares involve a large propagating fracture in the neutron star crust, whereas the short bursts require only a localized yield of the crustal lattice (TD95). A large scale motion of the crust is highly constrained compared with, e.g., a metal sheet, because the crust floats stably on the neutron star core and is very nearly incompressible. For this reason, a large fracture requires the collective and simultaneous motion of many smaller units. It is interesting to note, in this regard, that some earthquake models based on cellular automata show a bimodal distribution of events, with a secondary peak at the highest energies (Carlson, Langer, and Shaw 1994).

How precisely is energy injected into the magnetosphere of the neutron star? The very fast ($\sim$ ms) rise times both of some short SGR bursts (Kouveliotou et al. 1987) and the giant outbursts (Fenimore et al. 1996; Hurley et al. 1999a; Mazets et al. 1999b) point to a localized and direct injection of energy. Indeed, much of the energy that is eventually radiated in the burst may have been injected on a much shorter timescale than the measured duration of the X-ray pulse. Direct evidence for this behavior comes from the intense initial spikes of the March 5 and August 27 events, which released a few tens of percent of the net outburst energy over $\sim 10^{-3}$ of the duration.

A magnetic field $B > (4\pi\theta_{max}\mu)^{1/2} \sim 2 \times 10^{14} \, (\theta_{max}/10^{-3})^{1/2}$ G can fracture the crust, but is far too weak to induce anything but a horizontal motion. As a result, energy is injected in the magnetosphere, in two distinct regions. The motion will, in general, have a rotational component that creates tangential discontinuities in the magnetic field. A disturbance of the magnetosphere propagates at the speed of light, which is some 300 times the shear wave velocity c_μ in the deep crust. Thus, reconnection occurs rapidly, and induces transverse Alfvén waves of frequency $\sim c/R_{NS}$ on the connecting closed loops of magnetic flux. These waves can dissipate effectively by cascading to high wavenumber through non-linear interactions (Thompson and Blaes 1998). Because they are current-carrying, a minimal density of electrically charged particles is required to support them, which in the giant flares implies a significant optical depth to scattering across the magnetosphere. The current density (and optical depth) rises as the cascade moves to higher wavenumber, where it is finally damped by Compton drag off the electrostatically heated pairs. When the rate of transfer of wave energy exceeds a critical level $10^{42} \, (\ell/10 \text{ km}) \text{ erg s}^{-1}$ within a volume ℓ^3, no stable equilibrium between heating and radiative cooling is possible. The plasma runs away to a dense, hot fireball which cools diffusively on a much longer timescale (TD95; Thompson et al. 2000b).

The second dissipative region lies much farther out in the magnetosphere. The crustal motion (on a horizontal scale ℓ) can be expected to excite shear waves of frequency $\nu \sim c_\mu/\ell$, which in turn couple to magnetospheric Alfvén modes at a radius $R_\nu \sim c/3\nu \sim 100\,\ell$. This outer excitation may dominate if (for example) the fracture is buried deep in the crust, and has been identified with two bursts from SGR 1900+14 whose hard spectra resemble those of cosmological GRBs (Woods et al. 1999d).

4.2. FRACTURING VS. PLASTIC CREEP

Can the magnetar model accomodate two classes of sources with widely different bursting behavior, but similar levels of internally-generated dissipation? If the magnetic field in the deep crust exceeds $B_\mu \equiv (4\pi\mu)^{1/2} \sim 6 \times 10^{15}$ G, lattice stresses are not able to compensate a departure from magnetostatic equilibrium, and the crust must deform plastically (TD96). Indeed, the internal flux density above which the magnetic field is transported through the neutron star core on a timescale of $\sim 10^4$ yr lies close to this value (TD96; Heyl and Kulkarni 1998). This suggests that a magnetar is capable of two dissipative modes: one dominated by brittle fracturing and bright X-ray outbursts, and a second dominated by plastic creep. These two modes correspond naturally to the SGRs and to the Anomalous X-ray Pulsars. In principle, both modes can operate simultaneously in the same star if its magnetic field is inhomogeneous.

4.3. HARD SPIKES OF THE MARCH 5 AND AUGUST 27 EVENTS

The initial spikes of the two giant outbursts had all the appearance of an expanding $e^{\pm}$ fireball carrying $\sim 10^{44}$ erg. (In the case of the March 5 event, $L \sim 3 \times 10^6 - 10^7 \, L_{\rm edd}$ and $T \sim 500$ keV: Mazets et al. 1999b; Fenimore et al. 1996.) The peak luminosity is intermediate, on a logarithmic scale, between that of a thermonuclear X-ray flash and the bright γ-ray fireballs that are observed at cosmological distances. If the fireball contained comparable energy in radiation and in the rest energy of (baryonic) matter, then its duration could be expressed in terms of the radius $R(\tau_{\rm es} = 1) \geq (E\sigma_T/4\pi m_{\rm p}c^2)^{1/2}$ of the scattering photosphere as $\Delta t \sim R(\tau_{\rm es} = 1)/c \sim 2(E/10^{44} \, {\rm erg})^{1/2}$ s, about 10 times the observed value. We conclude that the initial fireball must have in fact expanded relativistically, and was powered by a very clean energy source.

The most obvious candidate is a magnetic field that experiences a sudden rearrangement. On energetic grounds, the (external) magnetic field must exceed $\sim 10 B_{\rm QED}$ to power $\sim 10^2$ giant outbursts over $\sim 10^4$ yr. One might consider a hybrid model in which the energy is initially released inside the neutron star, in the form of crustal shear waves or torsional Alfvén waves in the liquid core. As we now show, fields of comparable strength are required to transmit this energy to the magnetosphere. The large output of the giant outbursts requires a large scale for this energy release, and hence a low frequency for the excited mode. For example, a fracture of dimension $\ell \sim 1$ km (a conservative lower bound) will excite a shear wave of frequency $\nu \sim 10^3 \, (\ell/1 \, {\rm km})^{-1}$ Hz. The resulting harmonic displacement ξ of the crust will in turn excite oscillations of the dipolar magnetic field lines at a radius $R_\nu \sim c/3\nu \sim 10^7 \, (\nu/{\rm kHz})^{-1}$ cm. Because only a narrow bundle of the outer field is excited, the outward wave luminosity is a steep function of ξ, $dE_{\rm wave}/dt \simeq \frac{1}{2} B_{\rm dipole}^2 R_{\rm NS}^2 c \, (2\pi\xi\nu/c)^{8/3}$, or

$$\frac{dE_{\rm wave}}{dt} \simeq 2 \times 10^{44} \left(\frac{B_{\rm dipole}}{10 \, B_{\rm QED}}\right)^2 \left(\frac{\xi}{0.1 \, {\rm km}}\right)^{8/3} \left(\frac{\nu}{10^3 \, {\rm Hz}}\right)^{8/3} \quad {\rm erg \; s}^{-1},$$

$$(1)$$

(Thompson and Blaes 1998). For example, an elastic distortion of the crust of energy $\sim 10^{44}$ erg corresponds to $\xi \sim 10^{-2} \, R_{\rm NS} \sim 0.1$ km, and the luminosity approaches $10^7 \, L_{\rm edd}$ only if $B_{\rm dipole} \sim 10^{15} \, (\nu/10^3 \, {\rm Hz})^{-4/3}$ G!

Nonetheless, the short $\sim 0.2 - 0.5$ s duration of the initial hard spikes of the March 5 and August 27 outbursts provides direct evidence that internal (rather than external) magnetic stresses trigger these giant outbursts. A 10^{15} G magnetic field will move the core material at a speed $\sim B/\sqrt{4\pi\rho}$ through a distance 10 km in that period of time. By contrast, the fireball resulting from a sudden unwinding of the external field would last only $\sim R_{\rm NS}/c \sim 10^{-4}$ s (TD95).

4.4. OSCILLATORY TAILS OF THE MARCH 5 AND AUGUST 27 EVENTS

After the initial hard spike, each of the two giant outbursts released an even greater amount of energy in an extended oscillatory tail. The temperature during this phase was much more stable in spite of the hyper-Eddington flux, $L/L_{\rm edd} \sim 10^4$ (e.g. Mazets et al. 1999b). These observations suggest that a significant fraction of the initial burst of energy was trapped on closed magnetic field lines, which implies a strong lower bound to the surface dipole magnetic field, $B_{\rm dipole} > 2 \times 10^{14} (E_{\rm fireball}/10^{44} \text{ erg})^{1/2}(\Delta R/10 \text{ km})^{-3/2}$ $[(1 + \Delta R/R_{\rm NS})/2]^3$ G (TD95; Hurley et al. 1999a).

The density of this trapped energy is so high that it must form a thermal fireball, composed of $e^{\pm}$ pairs and γ-rays, at a very high temperature of ~ 1 MeV. The optical depth to scattering across this plasma bubble is huge, approaching $\sim 10^{10}$. It is clear that the plasma cannot cool by simple radiative diffusion from its center: that would take $\sim 10^3 - 10^4$ times the observed burst duration. The bubble cools instead as a sharp temperature gradient develops just inside its outer boundary, and this boundary propagates inward as a cooling wave (TD95). If the magnetic field is predominantly dipolar, then the radiative flux across the field is concentrated near the surface of the star: the opacity of the E-mode scales as $B^{-2} \propto R^6$. A cylindrical bundle of field lines containing relativistic plasma therefore has a radiative area (and luminosity) that decreases linearly with time, as is observed in a number of short SGR bursts (e.g. Mazets et al. 1999a). If higher multipoles dominate the near magnetic field, then the opacity will be much more uniform over the surface of the fireball. Parametrizing the radiative area in terms of the *remaining* fireball energy as $A \propto E_{\rm fireball}^\alpha$, the luminosity works out to

$$L_{\rm X}(t) = L_{\rm X}(0) \left(1 - \frac{t}{t_{\rm evap}}\right)^{\alpha/(1-\alpha)}. \tag{2}$$

Here $t_{\rm evap}$ is the time at which the cooling wave propagates to the center of the fireball and the fireball evaporates. This analytic law provides an excellent fit to extended lightcurve of the August 27 event for $\alpha \sim 0.7$ (Feroci et al. 2000).

The fireball temperature inferred for the short SGR bursts is lower if the confining volume is as large as in the giant flares ($T \sim 100$ keV for an energy $\sim 10^{41}$ erg). However, a recent analysis of the August 29 burst from SGR 1900+14 (which appears to have been an aftershock of the August 29 giant flare) points to an active region covering only ~ 0.1 percent of the neutron star surface (Ibrahim et al. 2000). This is consistent with a strong localization of the injected energy, leading to the formation of a fireball with a similar temperature of ~ 1 MeV.

4.5. QED PROCESSES: RADIATIVE AND SPECTRAL IMPLICATIONS

The transport of X-ray photons through a very strong (super-QED) magnetic field is determined by two coupled processes: Compton scattering and photon splitting $\gamma \to \gamma + \gamma$ (and merging $\gamma + \gamma \to \gamma$) (TD95). Even at very large scattering depth, the dielectric properties of the medium are dominated by vacuum polarization in the intense magnetic field. The normal modes of the electromagnetic field are then linearly polarized (e.g. Mészáros 1992) with $\delta \mathbf{E} \perp \mathbf{B}_0$ in one case (the extraordinary or E-mode) and $\delta \mathbf{B} \perp \mathbf{B}_0$ in the other (the ordinary mode or O-mode).

The E-mode splits because, in this situation, its energy and momentum can be conserved by dividing it into two obliquely propagating O-mode photons. (Or, at a lower rate, into a pair of E-mode and O-mode photons.) The O-mode is *not* able to split because its energy and momentum cannot be so conserved.[2] In a vacuum, neither mode is able to split for the simple reason that the two daughter photons must remain colinear to conserve energy and momentum, and there is no phase space for the process. In marked contrast with the strong B^6 scaling of the splitting rate in sub-QED magnetic fields, the splitting rate approaches a B-independent value in fields much stronger than B_{QED}, $\Gamma_{\mathrm{sp}}(\omega, B, \theta_{\mathrm{kB}}) = (\alpha_{\mathrm{em}}^3/2160\pi^2)\,(m_e c^2/\hbar)\,(\hbar\omega/m_e c^2)^5\,\sin^6\theta_{\mathrm{kB}}$ (Adler 1971; Thompson and Duncan 1992). Here, $\alpha_{\mathrm{em}} \simeq 1/137$ and θ_{kB} is the angle between the photon's wavevector and the background magnetic field. This implies immediately that a E-mode photon propagating a distance $R_{\mathrm{NS}} \sim 10$ km through a super-QED B-field will split if $\hbar\omega > 38\,(R_{\mathrm{NS}}/10\ \mathrm{km})^{-1/5}$ keV (TD95; Baring 1995).

---

[2] These selection rules depend essentially on the inequality $n_O > n_E$ between the indices of refraction of the two modes. Note that both n_O and n_E depart only very slightly from unity even in magnetic fields as strong as $\sim 10^{16}$ G. The inequality is reversed, $n_E > n_O$, when plasma dominates the dielectric properties of the medium; but in such a situation the particle density is enormous and the photons will in practice follow a Planckian distribution. The problem of calculating the emergent spectra of SGR bursts focusses on much lower temperatures and densities where departures from local thermodynamic equilibrium can occur.

Compton scattering becomes strongly anisotropic in a background magnetic field, with a strongly frequency-dependent cross-section (Mészáros 1992). In contrast with a dense plasma, both vacuum modes interact resonantly with an electron (or positron) at the Landau frequencies. Near the surface of the star, the energy of the first Landau excitation [$\simeq (2B/B_{\mathrm{QED}})^{1/2} m_e c^2$ when $B \gg B_{\mathrm{QED}}$] is much higher than the temperature of the emerging X-radiation. In this situation, there is a strong suppression of the E-mode's scattering cross-section, $\sigma_{\mathrm{E}} = (\omega m_e c/e B_0)^2 \sigma_{\mathrm{T}}$, but not of the O-mode's (e.g. Herold 1979). This suppression greatly increases the radiative flux close to the neutron star – both from its surface (Paczyński 1992, Ulmer 1994) and across the confining magnetic field lines of a trapped fireball (TD95).

However, even in the region where the E-mode is able to stream freely, the O-mode with its large cross section can still undergo many Compton scatterings and relax to a Bose-Einstein distribution. Given the strong frequency dependence of the splitting rate, there is clearly a critical temperature above which the distributions of the E- and O-modes both become thermal, which works out to $T_{\mathrm{sp}} = 11 (R_{\mathrm{NS}}/10 \text{ km})^{-1/5}$ keV (TD95). This value agrees well with the best fit black body temperature of the oscillatory tail in the August 27 giant flare (Feroci et al. 2000). (The term 'photon-splitting cascade' appearing in some of the recent literature is a misnomer. *If the Compton depth is high enough to convert the O-mode back to the E-mode and allow more than one generation of splitting, then X-rays are redistributed in frequency mainly by the Compton recoil.*)

The sharply peaked sub-pulses within the oscillatory tails of the March 5 and August 27 events have a simple interpretation in the magnetar model, and are consistent with the presence of a trapped fireball (TD95). This pattern requires a collimated, quasi-hydrodynamical outflow of the X-radiation. In an intense magnetic field, the rapid rise of the E-mode opacity with radius provides a mechanism for self-collimation: if baryonic matter is suspended in the magnetosphere by the hyper-Eddington radiative flux, then E-radiation will escape by pushing this matter to the side. In addition, a significant fraction of the E-mode flux near the E-mode photosphere is converted to the O-mode by scattering and by photon splitting (Miller 1995; TD95). The O-mode flows hydrodynamically along the magnetic field even in the presence of a tiny amount of matter, $\dot{M} c^2/L_O \sim (GM_{\mathrm{NS}}/R_{\mathrm{NS}} c^2)^{-1}(L_O/L_{\mathrm{edd}})^{-1}$. Further collimation occurs if the photosphere is aligned with extended (dipolar) magnetic field lines.

Scattering at the electron cyclotron resonance can probably be neglected during outbursts from a magnetar: the resonance sits at a large radius $8 R_{\mathrm{NS}} (B_{\mathrm{dipole}}/10 B_{\mathrm{QED}})^{1/3} (\hbar\omega/10 \text{ keV})^{-1/3}$, where the outflowing photons are sufficiently collimated to suppress the resonant scattering depth below unity. Scattering at the ion cyclotron resonance has a significant optical depth τ_{ion} if electrons and ions dominate the *electron*-scattering opacity: it is $\tau_{\mathrm{ion}} \sim (\pi/4\alpha_{\mathrm{em}})n_e\sigma_T R(B/B_{\mathrm{QED}})^{-1}$ in a dipolar magnetic field. An important effect is to convert photons between the E- and O-modes and to increase significantly the opacity of the E-mode at low frequencies. (This may be relevant to the < 7 keV suppression of the radiative flux found by Ulmer et al. 1993 in the bursts of SGR 1806–20.)

4.6. PREDICTIONS OF THE MAGNETAR MODEL

Here we mention three direct observational diagnostics of magnetars.

1) Afterglow radiation from the heated surface that is exposed to a high temperature fireball. The surface absorbs $\sim 10^{-3} - 10^{-2}$ of the fireball energy before it dissipates. This heat will be re-released on a timescale comparable to or longer than the observed duration of the SGR outburst. The resulting luminosity increases monotonically with B, and is $\sim 10^{39} \times$ [exposed area/(10 km)^2] erg s^{-1} for $T_{\mathrm{fireball}} \sim 1$ MeV and $B \sim 10\,B_{\mathrm{QED}}$ (TD95). Direct evidence for afterglow at this level is present in a ~ 4 s burst from SGR 1900+14 that followed the August 27 flare by less than 2 days (Ibrahim et al. 2000).

2) Absorption or emission in the persistent emission at the ion cyclotron resonance $\hbar\omega = 2.8\,(Z/A)(B/10\,B_{\mathrm{QED}})$ keV. A direct measurement of the surface magnetic field would be provided by the simultaneous identification of a spin-flip transition: the two frequencies are very nearly degenerate for electrons, but differ by a factor 2.8 for protons. This measurement is probably more feasible in the AXPs, whose persistent emission has a much smaller non-thermal component than the SGRs.

3) Little or no reprocessing of X-rays into the IR/optical/UV bands by an orbiting disk. The persistent X-rays of the SGRs and AXPs do not, in the magnetar model, result from accretion. Only a modest amount of material can be placed in orbit around the neutron star through hyper-Eddington winds. Detailed calculations of reprocessing in the fossil disk model for the AXPs have been performed by Perna, Hernquist, and Narayan (2000) (see also Chatterjee, Hernquist, and Narayan 2000). It is hard to avoid reprocessing $\sim 10^{-2}$ of the central X-ray source $L_X \sim 10^2\,L_\odot$ in this model.

5. Variable Spindown of the SGRs and AXPs

The four known SGRs are persistent X-ray sources, and in two cases persistent periodicites have been detected: $P = 7.47$ s for SGR 1806–20 (Kouveliotou et al. 1998); and $P = 5.16$ s for SGR 1900+14 (Hurley et al. 1999c). Together with the 8-s periodicity of the March 5 event and the 6-12 s spin periods of the AXPs, these values are clustered in a remarkably narrow range.

5.1. SPINDOWN AGES

In the magnetar model, the long spin periods of the SGRs were ascribed to large torques driven by magnetic dipole radiation (DT92), possibly amplified by a persistent flux of Alfvén waves and particles (Thompson and Blaes 1998). A key motivation for this model came from the early association between the March 5 burster and the supernova remnant N49 in the LMC (Cline 1982, and references therein): the 8-s periodicity corresponds to a magnetic dipole field of 6×10^{14} G (polar) at an age of $\sim 10^4$ yr. Of all the SGR and AXP sources, the spindown of the AXP 1841–045 is most consistent with simple magnetic dipole radiation (Gotthelf et al. 1999): the spindown age $P/2\dot{P} = 2000$ yr agrees with the estimated age of the surrounding SNR Kes 73, and the spindown is very uniform. The implied (polar) dipole field of 1.4×10^{15} G is a good candidate for the strongest yet measured in any neutron star.

However, the spindown ages of SGR 1806–20 and SGR 1900+14, measured as $P/2\dot{P} = 1400$ yr (Kouveliotou et al. 1998, 1999; Marsden, Rothschild, and Lingenfelter 1999; Woods et al. 1999c) do not appear to obey a similar correspondence. Indeed, the characteristic age of SGR 1900+14 is surprisingly short if it has been spun down by a constant external torque, and if it is physically associated with the nearby SNR G42.8+0.6: the required proper motion is $V_\perp \simeq 20,000 \, (D/7 \text{ kpc}) \, (t/1,500 \text{ yr})^{-1}$ km s^{-1}. (A spurious association leads to an equally unsatisfactory situation: a very young neutron star bereft of a progenitor supernova.) This inconsistency disappears if the spindown of SGR 1900+14 is *temporarily accelerated* with respect to the long-term trend.

5.2. DEPARTURES FROM UNIFORM SPINDOWN

One of the defining characteristics of the AXPs is that they are consistently spinning down. Nonetheless, torque variations are evident in the sources 1E 2259+586, 1E 1048.1+5937, and 4U 0142+61 (Heyl and Hernquist 1999; Mereghetti et al. 2000). Over the longest intervals measured, the spin evolution of the AXPs is unusually coherent by the standards of accreting neutron stars. Dramatically improved timing accuracy has been achieved recently through phase-connected measurements (Kaspi et al. 1999), which show that the spindown of 1E 2259+586 and RXSJ170849–4009 is, also, remarkably smooth over an interval of $\sim 10^3$ days. Thus, variations in the spindown may occur only intermittently. Recent phase-connected measurements of the spindown of SGR 1806–20 (Woods et al. 2000) point to significantly stronger torque variations than are present in the AXPs 1E 2259+586 and RXSJ170849–4009, or are characteristic of radio pulsars.

Most intriguingly, the spindown of 1E 1048.1+5937 appears to have accelerated for a $\sim$ 5 yr interval over the long term trend (Paul et al. 2000), a phenomenon that was inferred for SGR 1900+14 only indirectly from its spindown age. The main question which next arises, is whether this accelerated torque is comparable to that expected from an orthogonal vacuum rotator. If so, then in the case of SGR 1900+14 the polar dipole field is inferred to be $B_{\rm dipole} \simeq 1 \times 10^{15}$ G. A temporary acceleration *up* to the standard dipole torque is the expected consequence of a recent discharge of particles through bursting activity, if the magnetic and rotational axes of the star are almost aligned. That is because in quiescence, the outer magnetosphere of the neutron star would become charge-starved at the measured 5.16-s spin period, with a corresponding reduction in the long-term torque (T00). Independent evidence for such a torque reduction is present in the AXP 1E 2259+586, whose spindown age of $\sim 2 \times 10^5$ yr significantly *exceeds* the estimated age of its host supernova remnant CTB 109.

Also intriguingly, the spin period of SGR 1900+14 increased by $\Delta P/P = +1 \times 10^{-4}$ above the long term trend within an 80-day interval surrounding the August 27 giant outburst (Woods et al. 1999c). A transient flow of particles, photons, and Alfvén waves might provide the additional torque – by increasing the magnetic field strength at the light cylinder and by carrying off angular momentum directly – but the constraint on $B_{\rm dipole}$ is severe (T00). The net effect of such a flow (Thompson and Blaes 1998) is to increase the spindown luminosity to the geometric mean of $L_{\rm Alfven}$ and the standard magnetic dipole luminosity, $I\Omega\dot{\Omega} = \Lambda\, B_{\rm NS} R_{\rm NS} (\Omega R_{\rm NS}/c)^2\, (L_{\rm Alfven} c)^{1/2}$. Subsequent calculations have found the numerical coefficient to be $\Lambda = \sqrt{2}/3$ (Harding et al. 1999) and $\Lambda = 2/3$ (T00). Applying this formula to the August 27 outburst, and normalizing the radiated energy and duration to the observed values ($\sim 10^{44}$ erg and ~ 100 s), one finds $\Delta P/P = 1\times 10^{-5}\, (\Lambda/\tfrac{2}{3})\, (\Delta E/10^{44}\ {\rm erg})^{1/2}\, (\Delta t/100\ {\rm s})^{1/2}\, (B_{\rm dipole}/10\, B_{\rm QED})$. This falls below the measured value even for $B_{\rm dipole} \sim 10\, B_{\rm QED}$, but a more extended particle flow or an undetected soft X-ray component to the giant burst cannot be ruled out.

The long-term spindown rate of SGR 1900+14 appears not to have been perturbed by the August 27 event (Woods et al. 1999c). This observation has the important consequence that the active region of the neutron star must carry a small fraction of the external magnetic energy; hence one deduces a lower bound to the dipole field of $\sim 10\, B_{\rm QED} = 4.4 \times 10^{14}$ G (T00).

It should be kept in mind that the measured spindown luminosity $I\Omega\dot{\Omega}$

of the AXPs and SGRs is typically two orders of magnitude smaller than the persistent X-ray luminosity. As the above formula makes clear, the spindown resulting from the the release of a fixed energy increases with the duty cycle, because at a lower flux the Alfvén radius (and the lever arm) is increased. The inferred dipole fields of the two spinning-down SGRs are reduced (by a factor of ~ 4) to 4×10^{14} G if each star is a persistent source of Alfvén waves and particles with a luminosity comparable to $L_X \sim 10^{35}$ erg s^{-1} (Thompson and Blaes 1998; Kouveliotou et al. 1998, 1999; Harding, Contopoulos, and Kazanas 1999; T00). These values lie only a factor ~ 4 above the polar dipole field inferred for the new radio pulsar J1814–1744.

The evidence for a physical association between SGR 1806–20 and a conspicuous radio nebula (Kulkarni and Frail 1993; Vasisht et al. 1994) which originally motivated persistent particle winds from the SGRs has been called into question (Hurley et al. 1999d). In addition, the relatively constant long-term spindown rate of SGR 1900+14 (Woods et al. 1999c) indicates that transient surges in the persistent seismic output, if present in SGR 1900+14, must be constant over a long timescale. They do not appear to correlate directly with episodes of short outbursts.

Melatos (1999) has recently noted the intriguing possibility that the spindown torque coupled to the asymmetric inertia of the co-rotating magnetic field could be particularly effective at forcing precession in a magnetar. Free precession (which in this model is modulated by the spindown torque) has a period $\tau_{\mathrm{prec}} = P/\varepsilon_{\mathrm{B}} = 7\,(P/6\ \mathrm{s})\,(B_{\mathrm{core}}/10^2 B_{\mathrm{QED}})^{-2}$ day, where $\varepsilon_{\mathrm{B}} \simeq 1 \times 10^{-5}(B_{\mathrm{core}}/10^2 B_{\mathrm{QED}})^2$ is the dimensionless quadrupole distortion of the star by the (toroidal core) magnetic field. This period is already rather short compared with the observed spindown variations, if the magnetic field is strong enough to power the observed X-ray emission. In addition, pinning of superfluid vortex lines in the neutron star crust will force the precession period down to $\tau_{\mathrm{prec}} = P(I/I_{\mathrm{sf}})$, where I_{sf} is the fraction of the moment of inertia carried by the neutron superfluid (Shaham 1977). Detection of a spinup glitch in 1RXS J170849.0-4000910 provides direct evidence for sufficient pinning to suppress long-period precession (Kaspi, Lackey, and Chakrabarty 2000).

5.3. SUPERFLUIDITY AND GLITCHES

Superfluid-driven glitches are a potential source of spindown irregularities in isolated magnetars. SGRs 1900+14 and 1806–20 have frequency derivatives about one-tenth that of the Vela pulsar, and models of magnetic dissipation suggest internal temperatures that are comparable or higher (TD96; Heyl and Kulkarni 1998). A giant outburst like the August 27 event must involve a large fracture of the crust propagating at $\sim 10^8$ cm s^{-1}, which almost certainly unpins the $^1S_0$ neutron superfluid vortex lines from the crustal lattice. The maximum glitch that could result can be very crudely estimated (TD96) by assuming a characteristic maximum angular velocity difference $\Delta\Omega_{max}$ between the superfluid and lattice, and then scaling to the largest observed glitches (e.g. $\Delta P/P \sim -3 \times 10^{-6}$ in Vela). This gives $|\Delta\Omega/\Omega| \sim \Delta\Omega_{max}/\Omega \propto \Omega^{-1}$ and $|\Delta P/P| \simeq 3 \times 10^{-4}(P/8\text{ s})$.

In light of this, let us reconsider the observation of a transient spindown in SGR 1900+14 ($\Delta P/P = +1\times10^{-4}$) close to the August 27 giant outburst (Woods et al. 1999c). Could this be a superfluid-driven glitch in spite of the 'wrong' sign? The crust of a magnetar is deformed plastically by magnetic stresses wherever $B > (4\pi\mu)^{1/2} \sim 6 \times 10^{15}$ G (TD96). Such a deformation taking place on a timescale short compared to $P/\dot{P}$ will force the pinned vortex lines into an inhomogeneous distribution (with respect to cylindrical radius). The net effect is to *slow* the rotation of the superfluid with respect to the crust. A sudden unpinning event would then tend to *spin down* the rest of the star (T00).

Heyl and Hernquist (1999) estimated the glitch activity in a few variable AXPs, under the assumption that the spindown irregularities are entirely due to glitches of the same sign as pulsar glitches. However, recent phase-connected measurements of variable spindown in SGR 1806–20 do not support this hypothesis (Woods et al. 2000). For that reason, it is important to consider alternative mechanisms for spindown variations involving, e.g., acceleration of the torque by particle outflows from the active neutron star.

6. Persistent Emission of the SGRs and AXPs

The persistent X-ray output of the Soft Gamma Repeaters lies within a fairly narrow range of $1 - 10 \times 10^{35}$ erg/s, and has a characteristically hard spectrum with a power-law component $dN/dE \propto E^{-2}$ (Murakami et al. 1994; Hurley et al. 1999c; Woods et al. 1999b). The output of the AXPs is slightly broader but much softer spectrally, being well fit by a ~ 0.5 keV blackbody plus a (soft) non-thermal tail with photon index $\sim 3-4$ (Mereghetti 2000, and references therein). Direct evidence that the persistent X-ray output of SGR 1900+14 is not powered by accretion comes from the detection of (enhanced) emission a day after the giant August 27 flare (Murakami et al. 1999; Woods et al. 1999c). Enough radiative momentum was deposited during that outburst to excavate any accretion disk out to a very large radius, within which accretion would be established only on a much longer timescale of months (T00).

The spectral characteristics of the SGRs and AXPs suggest that i) dissipation of magnetic energy in a neutron star can produce varying persistent X-ray spectra, with hardness correlating strongly with bursting activity;

and ii) that more than one mechanism can generate persistent X-ray emission at a level of $\sim 10^{35}$ erg s^{-1}. Four such mechanisms have been proposed, all of which can be expected to operate in an SGR and at least two of which are relevant to the AXPs. We summarize them in turn:

1. *Ambipolar diffusion of a magnetic field through the neutron star core, combined with the increased transparency of the stellar envelope in a strong magnetic field* (TD96; Heyl and Kulkarni 1998). The degenerate charged electrons and protons are tied to the magnetic field lines in the neutron star core. They can be dragged across the background neutron fluid, but only very slowly (Pethick 1992; Goldreich and Reisenegger 1992). Heating of the core feeds back strongly on the rate of ambipolar diffusion (TD96). An intense magnetic field drives an imbalance between the chemical potentials of the electrons, protons and neutrons, $\Delta\mu = \mu_e + \mu_p - \mu_n \simeq B^2/8\pi n_e$, and this imbalance induces β-reactions which heat the core. Above a critical flux density, the heat produced exceeds the heat remaining in the star from its formation, and the core sits at an equilibrium temperature where heating is balanced by neutrino cooling. In practice, this balance is possible only as long as the neutrino emissivity is dominated by the modified-URCA reactions. The very strong temperature-dependence of these reactions translates into a very strong B-dependence of the diffusion rate: $t_{\rm amb} = 10^4 \ (B_{\rm core}/7 \times 10^{15} \ {\rm G})^{-14}$ yr in a normal n-p-e plasma (TD96). (This timescale depends, of course, on the *core* flux density.) Assuming a magnetized iron envelope, the resulting heat flux through the surface is $L_{\rm X}(t) = 5 \times 10^{34} \ (t/10^4 \ {\rm yr})^{-0.3}$ erg s^{-1}. Thus, there is a critical flux density above which magnetic dissipation is rapid, but below which the magnetic field is essentially frozen. *This critical flux density exceeds by a factor $\sim 4 - 10$ the dipole fields that are inferred from SGR and AXP spindown.* The absence of significant X-ray emission from the pulsar PSR J1814–1744 (Pivovaroff et al. 2000) (with a polar dipole field $\sim 10^{14}$ G) is consistent with the magnetar hypothesis as long as the internal (e.g. toroidal) field in these objects is below the critical value.

Further time-dependent calculations of ambipolar diffusion through normal n-p-e nuclear matter, including much more detailed modelling of the envelope, are reported by Heyl and Kulkarni (1998). Calculations of heat transport through the strongly magnetized envelope of a neutron star are presented in Hernquist (1985), Van Riper (1988), Potekhin and Yakovlev (1996) and Heyl and Hernquist (1998).

2. *Hall fracturing in the crust* (TD96). Protons are bound into a rigid Coulomb lattice of nuclei in the neutron star crust. In this situation, the propagation of short-wavelength magnetic irregularities through the crust is driven by the Hall electric field $\vec{E} = \vec{J} \times \vec{B}/n_e ec = (\vec{\nabla} \times \vec{B}) \times \vec{B}/4\pi n_e e$ (Goldreich and Reisenegger 1992). The polarization of a such a Hall wave rotates, which causes the crust to *yield or fracture* in a magnetic field stronger than $\sim 10^{14}$ G. A significant fraction of the wave energy is dissipated in this manner – in less than the age t_{NS} of the neutron star – if the turbulence has a short wavelength $\lambda < 0.1 \, (\delta B/B)^{-1/2} (\theta_{max}/10^{-3})^{1/2}$ $(B^2/4\pi\mu)^{-1/4} (t_{NS}/10^4 \text{ yr})^{1/2}$ km (TD96). (Recall that $(4\pi\mu)^{1/2} = 6 \times 10^{15}$ G in the deep crust.) By contract, large-scale fractures which are capable of triggering giant outbursts require more rapid transport of the magnetic field, which can occur via ambipolar diffusion through the *core*.

Each Hall fracture releases only a small energy, $\Delta E \sim 10^{36} \, (\theta_{max}/ 10^{-3})^{7/2}$ erg. The cumulative effect is to excite persistent seismic activity with a net output $L_{seismic} \sim 10^{35} \, (\delta B/B)^2 \, (B/10^{15} \text{ G})^2 \, (t_{NS}/10^4 \text{ yr})^{-1}$ erg s^{-1}. The excited seismic waves have a frequency $\nu \sim c_\mu/\lambda \sim 10^4$ Hz, where $c_\mu \sim c/300$ is the shear wave speed in the deep crust. These internal waves couple to transverse (Alfvén) excitations of the magnetosphere at a radius $R_\nu \sim c/3\nu \sim 100 \, \lambda$. Only a fraction of the wave energy need be converted to particles to support the associated electrical currents (Thompson and Blaes 1998).

3. *Twisting of the external magnetic field lines by internal motions of the star, which drives persistent electrical currents through the magnetosphere* (T00). The persistent light curve of SGR 1900+14 underwent a dramatic change following the August 27 outburst (Murakami et al. 1999): it brightened by a factor ~ 2.5 and at the same time simplified dramatically into a single large pulse. This change appeared within a day following the August 27 event, indicating that the source of the excess emission involves particle flows *external* to the star (T00). The coordinated rise and fall of the two X-ray pulses of 1E 2259+586 over a period of a few years detected by Ginga (Iwasawa, Koyama, and Halpern 1992) similarly indicates that some portion of its emission is magnetospheric (TD96).

The rate of dissipation due to a twisting of a bundle of field lines (of flux density B, radius a, twist angle θ and length L) can be estimated as follows (T00). The associated charge flow is $\dot{N} \sim \theta B a^2 c/8L$ into the magnetosphere from either end of the twisted field. The surfaces of the SGRs and AXPs are hot enough to emit thermionically for a wide range of surface compositions – even in the presence of $\sim 10^{15}$ G magnetic fields – and so the space charge very nearly cancels. An electric field $\vec{E} \cdot \vec{B} = -(Am_p/Ze)\vec{g} \cdot \vec{B}$ will compensate the gravitational force on the ions; but the same field pushes the counterstreaming electrons to bulk relativistic motion. The net luminosity in Comptonized X-ray photons is $L_{Comp} \sim 3 \times 10^{35} \theta (A/Z)(B/10 \, B_{QED}) \, (L/R_{NS})^{-1} \, (a/0.5 \, R_{NS})^2$ erg s^{-1}. This agrees with the measured value if a few percent of the crust is involved in the August 27 outburst. (Independent evidence for an active fraction this size comes from the unperturbed long-term spindown of SGR 1900+14, and from the expectation of $\sim 10^2$ giant outbursts per SGR in $\sim 10^4$ yr.) This non-thermal energy source will decay in 10–100 years, and so provides a physical motivation for *non-thermal* persistent X-ray spectra in *active* burst sources. (Note that the measured increase in the persistent L_X of SGR 1900+14 came entirely in the non-thermal component of the spectrum: Woods et al. 1999a.)

   4. Heyl and Hernquist (1997a,b) have explored the interesting possibility that the emission of some AXPs is predominantly due to *passive surface cooling, possibly enhanced by a light H or He composition*. This model is most promising for the AXP 1E 1841–045, but cannot directly accomodate the variable L_X of 1E 2259+586 or 1E 1048.1+5937. A challenge for this model comes from the very similar spin periods and persistent X-ray luminosities of the active SGRs and the quiescent AXPs. The magnetic dissipation occuring within an active SGR lengthens its *lifetime* as an bright X-ray source (TD96; Heyl and Kulkarni 1998).

6.1. RADIO EMISSION FROM MAGNETARS

The upper cutoff to the measured distribution of pulsar dipole fields lies close to B_{QED}. It is natural to ask whether this apparent cutoff results from observational selection or, instead, reflects a more fundamental physical limitation on pair cascades in very strong magnetic fields. As emphasized by Camilo et al. (2000), the discovery of three new strong-field pulsars allows for a substantial population of these objects.

   The most salient point, I believe, is that *particle flows induced by bursting activity* could short out vacuum gaps in the magnetospheres of the SGRs. The hyper-Eddington outbursts can easily blow material off the stellar surface (TD95; Miller 1995; Ibrahim et al. 2000). A modest amount of material will remain centrifugally supported outside the corotation radius (where the Keplerian orbital period equals the spin period of the star, $R_{co} = 6 \times 10^8 \, (P/6 \text{ s})^{2/3}$ cm) and still be confined by the dipole magnetic field: $\Delta M \sim (B_{NS}^2/4\pi)\Omega^{4/3}R_{NS}^6 \, (GM_{NS})^{-5/3} = 3 \times 10^{20}(B_{NS}/10\,B_{QED})^2 \times(P/6 \text{ s})^{-4/3}$ g. After an X-ray outburst, this material cools and settles into a rotationally supported disk, which spreads outward adiabatically as the centrifugal force density rises above $B^2/4\pi$ at the co-rotation radius. By spreading out to the speed-of-light cylinder, this material can maintain a relativistic flux of particles at the Goldreich-Julian density $n_{GJ} = \Omega \cdot B/2\pi ec$ for as long as $\sim 1 \times 10^5 \, (B_{NS}/10\,B_{QED}) \, (P/6 \text{ s})^{2/3}$ yr. Above this density, particles flowing downward toward the star from the light cylinder would short out the 'favorably curved' magnetic field lines that otherwise support a relativistic thermionic flow from the surface of the neutron star (e.g. Scharlemann et al. 1978). If this is happening in the SGRs, radio emission may be easier to detect from the quiescent AXPs.

Alternatively, it has been suggested that pair cascades (and thence co-herent radio emission) are *directly* suppressed in isolated strong-field neu-tron stars through QED effects: either by conversion of gamma-rays to bound positronium rather than to free pairs (Usov and Melrose 1996); or by photon splitting below the threshold for pair creation, $\gamma \to \gamma + \gamma$ (Baring and Harding 1998). Regarding the first mechanism, it should be emphasized that one polarization mode – the E-mode – converts to positronium with one particle in an excited Landau state. The energy released through the subsequent decay greatly exceeds the binding energy of the pair, which sug-gests that an unbound pair results. Notice also that the SGRs and AXPs emit a large enough flux of soft X-rays ($L_X \sim 3 \times 10^{34} - 10^{36}$ erg s^{-1}) to photo-dissociate bound positronium (cf. Usov and Melrose 1996). The other mechanism for suppressing radio emission just discussed is based on a doubtful assumption that both polarization modes can split below the threshold for single-photon pair creation (see §4.5). More generally, the splitting rate drops off so rapidly with distance from the neutron star ($\propto B^6 \sim R^{-18}$ when $B < B_{QED}$) that the region outside the 'splitting photosphere' should sustain a sufficient potential drop to induce pair cas-cades when the spin period lies well below the conventional radio death line. In other words, splitting will probably induce additional curvature in the pulsar death line at $B > B_{QED}$, but not suppress radio emission at much shorter spin periods.

7. Origins of Neutron Star Magnétism

The idea of magnetars was motivated by the realization that the violent convective motions in a collapsing supernova core can strongly amplify the entrained magnetic field (Thompson and Duncan 1993, hereafter TD93). The intense flux of neutrinos drives convection both in the central part of the core that is very thick to neutrino scattering and absorption (Pons et al. 1999, and references therein) and in a thin mass shell below the bounce shock where neutrino heating overcomes cooling (Janka and Mueller 1996, and references therein). Balancing hydrodynamic and magnetic stresses, one deduces magnetic fields of $\sim 10^{15}$ G and $\sim 10^{14}$ G respectively (TD93; Thompson 2000a). The convection inside the neutrinosphere has an over-turn time τ_{con} of a few milliseconds; the overturn time in the outer 'gain' region is somewhat longer. The inner region will support a large-scale heli-cal dynamo if the core is very rapidly rotating, with $P_{rot} < \tau_{con}$ (DT92), but not otherwise. It is also possible that rapid rotation by itself could amplify a magnetic field (Leblanc and Wilson 1970) through the magnetic shearing instability (Balbus and Hawley 1991) in the absence of convection, if the outermost parts of the collapsing core became centrifugally supported.

A newborn neutron star experiences convection with a dimensionless ratio of convective kinetic energy to gravitational binding energy ($\varepsilon_{con} \sim 10^{-4}$) that is some two orders of magnitude larger than in any previous phase driven by nuclear burning (TD93). (This is the relevant figure of merit because the gravitational binding energy and the magnetic energy are proportional under an expansion or contraction.) For this reason, neutron star magnetic fields are probably not fossils from earlier stages of stellar evolution. The intense flux of neutrinos emanating from the neutron core induces rapid heating and $n - p$ transformations, thereby allowing magnetic fields stronger than $\sim 10^{14}$ G to rise buoyantly through a thick layer of convectively stable material in less than the Kelvin time of ~ 30 s (Thompson and Murray 2000). As a result, the $10^{11} - 10^{13}$ G magnetic moments of ordinary radio pulsars, which do not appear to correlate with the axis of rotation, have a plausible origin (TD93) in a stochastic dynamo operating at slow rotation ($P_{rot} \gg \tau_{con}$). Direct amplification of a magnetic field $\langle B^2 \rangle^{1/2}$ within individual convective cells of size $\ell \sim (\frac{1}{30} - \frac{1}{10}) R_{NS}$ will generate a true dipole of magnitude $B_{dipole} \sim \langle B^2 \rangle^{1/2} (\ell^2/4\pi R_{NS}^2)^{1/2} \sim 10^{13}$ G through an incoherent superposition. A similar effect can occur during fallback as convection develops below the accretion shock (Thompson and Murray 2000).

7.1. LARGE KICKS

There is evidence that some (but not all) SGRs have proper motions approaching ~ 1000 km s^{-1}. The quiescent X-ray source associated with SGR 0526–66 (the March 5 burster) is offset from the center of N49, implying $V_\perp \sim 800 \, (t/10^4 \text{ yr})^{-1}$ km s^{-1} perpendicular to the line-of-sight (DT92). Similarly, the association of SGR 1900+14 with G42.8+0.6, if real, implies $V_\perp \sim 3000 \, (t/10^4 \text{ yr})^{-1}$. Additional indirect evidence that magnetars tend to have received large kicks comes from the paucity of accreting, strong-B neutron stars. However, it should be emphasized that the proper motions of SGR 1806–20 may be as small as ~ 100 km s^{-1}, and that the projected positions of the AXPs 1E 2259+586 and 1E 1841–045 are close to the centers of their respective remnants.

Since the SGRs already appear to have one unusual property (very strong magnetic fields), one immediately asks if a large kick could be produced by a mechanism that does not operate, or operates inefficiently, in ordinary proto-pulsars. Two mechanisms are particularly attractive if the star is initially a rapid rotator (DT92; Khokhlov et al. 1999; Thompson 2000a): anisotropy in the emission of the cooling neutrinos caused by large scale magnetic spots, which suppress convective transport within the star;

and asymmetric jets driven by late infall of centrifugally supported material. The first model is supported by observations of rotating M-dwarfs, which have deep convective zones (like proto-neutron stars) and develop large, long-lived polar magnetic spots: Vogt 1988). One estimates $M_{NS}V_{NS} \sim (E_\nu/c)(\tau_{spot}/\tau_{KH})^{1/2}(\Delta\Omega_{spot}/4\pi)$, where τ_{spot} is the coherence time of the spot(s) and τ_{KH} the Kelvin time of the star. The corresponding magnetic dipole field is $B_{dipole} \sim 5 \times 10^{14}(V_{NS}/1000 \text{ km s}^{-1})(\tau_{spot}/\tau_{KH})^{-1/2}$ G. Note that this refers to the dipole field in the *convective* neutron core, and represents an upper bound to the remnant field generated by an internal dynamo.

An asymmetric jet provides a more efficient source of linear momentum than does radiation from an off-center magnetic dipole (Harrison and Tademaru 1975), for two reasons: 1) The jet is matter-loaded and the escape speed from a proto-neutron star of radius ~ 30 km is only $\sim \frac{1}{4}$ the speed of light; and 2) a centrifugally supported disk carrying the same amount of angular momentum $(GM_{core}R_{core})^{1/2}\Delta M$ as a hydrostatically supported neutron core can provide much more energy to a directed outflow. The respective energies are $\Delta E \sim GM_{core}\Delta M/2R_{core} = 6 \times 10^{50}(\Delta M/10^{-2} M_\odot)(R_{core}/30 \text{ km})^{-1}$ erg and $\Delta E \sim \frac{5}{4}G(\Delta M)^2/R_{core} = 10^{48}(\Delta M/10^{-2} M_\odot)^2(R_{core}/30 \text{ km})^{-1}$ erg for a $1.4 M_\odot$ core. The corresponding kick velocity is $\sim 300 f(\Delta M/10^{-2} M_\odot)(R_{core}/30 \text{ km})^{-1/2}$ km s^{-1}, where f is the fractional asymmetry in the momentum. Only a very energetic jet ($\Delta E \sim 6 \times 10^{51}(f/0.3)^{-1}$ erg) can generate a kick of 1000 km s^{-1}.

Acknowledgements

I thank Chryssa Kouveliotou, Shri Kulkarni, Peter Woods, and the late Jan van Paradijs for many stimulating discussions about the peculiarities of SGRs and AXPs; Omer Blaes, Robert Duncan, and Norman Murray for theoretical collaboration; and NASA (NAG 5-3100) and the Alfred P. Sloan foundation for financial support.

References

Adler, S.L. (1971), *Ann. Phys.* **67**, 599.
Alcock, C., Farhi, E., and Olinto, A. (1986), *ApJ* **310**, 261.
Arras, P. and Lai, D. (1999), *ApJ* **519**, 745.
Balbus, S.A. and Hawley, J.F. (1991), *ApJ* **376**, 214.
Baring, M.A. (1995), *ApJ* **440**, L69.
Baring, M.A., and Harding, A.K. (1998), *ApJ* **507**, L55.
Camilo, F. et al. (2000), *ApJ* **541**, 367.
Carlson, J.M., Langer, J.S., and Shaw, B.E. (1994), *Rev. Mod. Phys.* **66**, 657.
Chatterjee, P., Hernquist, L., and Narayan, R. (2000), *ApJ* **534**, 373.
Cheng, B., Epstein, R.I., Guyer, R.A., and Young, A.C. (1996), *Nature* **382**, 518.
Cline, T.L. (1982), in R.E. Lingenfelter et al. (eds.), *Gamma Ray Transients and Related*

Astrophysical Phenomena, AIP (New York), p17.

Duncan, R.C. and Thompson, C. (1992) (DT 92), *ApJ* **392**, L9.

Duncan, R.C. and Thompson, C. (1995), in R.E. Rothschild and R.E. Lingenfelter (eds.), *High Velocity Neutron Stars and Gamma-Ray Bursts*, AIP (New York), p111.

Fenimore, E.E., Klebesadel, R.W., and Laros, J.G. (1996), *ApJ* **460**, 964.

Feroci, M. et al. (1999) *ApJ* **515**, 9.

Feroci, M., Hurley, K.H., Duncan, R.C., and Thompson C. (2001), *ApJ* **549**, 1021.

Frail, D.A. (1998), in R. Buccheri, J. van Paradijs, and M.A. Alpar (eds.), *The Many Faces of Neutron Stars*, Kluwer (Dordrecht), p179.

Fuchs, Y. et al. (1999), *A&A* **350**, 891.

Göğüş, E. et al. (1999a), *ApJ* **526**, L93.

Göğüş, E. et al. (1999b), *ApJ* **532**, L121.

Goldreich, P. and Reisenegger, A. (1992), *ApJ* **395**, 250.

Harding, A.K., Contopoulos, I., and Kazanas, D. (1999), *ApJ* **525**, L125.

Harrison, E.R. and Tademaru, E. (1975), *ApJ* **201**, 447.

Hernquist, L. (1985), *MNRAS* **213**, 313.

Herold, H. (1979), *Phys. Rev. D* **19**, 2868.

Heyl, J.S. and Hernquist, L. (1997a) , *ApJ* **489**, L67.

Heyl, J.S. and Hernquist, L. (1997b), *ApJ* **491**, L95.

Heyl, J.S. and Kulkarni, S.R. (1998), *ApJ* **506**, L61.

Heyl, J.S. and Hernquist, L. (1998), *MNRAS* **300**, 599.

Heyl, J.S. and Hernquist, L. (1999), *MNRAS* **304**, L37.

Hurley, K.J., McBreen, B., Rabbette, M., and Steel, S. (1994), *A&A* **288**, L49.

Hurley, K. et al. (1999a), *Nature* **397**, 41.

Hurley, K. et al. (1999b), *ApJ* **510**, L107.

Hurley, K. et al. (1999c), *ApJ* **510**, L111.

Hurley, K. et al. (1999d), *ApJ* **523**, L37.

Ibrahim, A. et al. (2000), *ApJ*, submitted.

Iwasawa, K., Koyama, K., and Halpern, J.P. (1992), *PASJ* **44**, 9.

Janka, H.-T. and Mueller, E. (1996), *A&A* **306**, 167.

Kaspi, V.M., Chakrabarty, D., and Steinberger, J. (1999), *ApJ* **525**, L33.

Kaspi, V.M., Lackey, J.R., Chakrabarty, D. (2000), *ApJ* **537**, L31.

Khokhlov, A.M. et al. (1999), *ApJ* **524**, L107.

Kouveliotou, C. et al. (1987), *ApJ* **322**, L21.

Kouveliotou, C. et al. (1998), *Nature* **393**, 235.

Kouveliotou, C. et al. (1999), *ApJ* **510**, L115.

Kulkarni, S.R. and Frail, D.A. (1993), *Nature* **365**, 33.

Leblanc, J.M. and Wilson J.R. (1970), *ApJ* **161**, 541.

Lewin, W.H.G., Van Paradijs, Jan, and Van den Heuvel, E.P.J. (1995), *X-ray Binaries*, Cambridge University Press.

Manchester, R.N. and Taylor, J.H. (1977), *Pulsars*, Freeman (San Francisco).

Marsden, D., Rothschild, R.E., and Lingenfelter, R.E. (1999), *ApJ* **520**, L107.

Mazets, E.P. et al. (1999a), *Astronomy Letters* **25**, 628.

Mazets, E.P. et al. (1999b), *Astronomy Letters* **25**, 635.

Melatos, A. (1999), *ApJ* **519**, L77.

Mereghetti, S. (2000), in C. Kouveliotou, J. van Paradijs, and J. Ventura (eds.), *The Neutron Star - Black Hole Connection*, in press.

Mereghetti, S. and Stella, L. (1995), *ApJ* **442**, L17.

Mészáros, P. (1992), *High-Energy Radiation from Magnetized Neutron Stars*, Chicago University Press.

Miller, M.C. (1995), *ApJ* **448**, L29.

Murakami, T. et al. (1994), *Nature* **368**, 127.

Murakami, T. et al. (1999), *ApJ* **510**, L119.

Norris, J.P., Hertz, P., Wood, K.S., and Kouveliotou, C. (1991), *ApJ* **366**, 240.

Paczyński, B. (1992), *Acta Astron* **42**, 145.

Palmer, D.M. (1999), *ApJ* **512**, L113.

Paul, B., Kawasaki, M., Dotani, T., and Nagase, F. (2000), in M. Kramer, N. Wex, and R. Wielebinski (eds.), *Pulsar Astronomy - 2000 and Beyond*, AIP (San Francisco), p695.

Perna, R., Hernquist, L., and Narayan, R. (1999), *ApJ* **541**, 344.

Pethick, C.J. (1992), in D. Pines, R. Tanagaki, and S. Tsuruta (eds.), *Structure and Evolution of Neutron Stars*, Addison-Wesley (Redwood City), p115.

Pivovaroff, M.J., Kaspi, V.M., and Camilo, F. (2000), *ApJ* **535**, 379.

Pons, J.A. et al. (1999), *ApJ* **513**, 780.

Potekhin, A.Y. and Yakovlev, D.G. (1996), *A&A* **314**, 341.

Rothschild, R.E., Kulkarni, S.R., and Lingenfelter, R.E. (1994), *Nature* **368**, 432.

Scharlemann, E.T., Arons, J., and Fawley, W.M. (1978), *ApJ* **222**, 297.

Thompson, C. (2000a), *ApJ* **534**, 915.

Thompson, C. (2000b), in P.C.H. Martens and S. Tsuruta (eds.), *Highly Energetic Physical Processes and Mechanisms for Emission from Astrophysical Plasmas*, AIP (San Francisco), p245.

Thompson, C. (2000c), in M. Kramer, N. Wex, and R. Wielebinski (eds.), *Pulsar Astronomy - 2000 and Beyond*, AIP (San Francisco), p669.

Thompson, C. and Duncan, R.C. (1992), in M. Friedlander, N. Gehrels, and R.J. Macomb (eds.), *Compton Gamma Ray Observatory*, AIP (New York), p1085.

Thompson, C. and Duncan, R.C. (1993) (TD93), *ApJ* **408**, 194.

Thompson, C. and Duncan, R.C. (1995) (TD95), *MNRAS* **275**, 255.

Thompson, C. and Duncan, R.C. (1996) (TD96), *ApJ* **473**, 322.

Thompson, C., and Blaes, O. (1998), *Phys. Rev. D* **57**, 3219.

Thompson, C. et al. (2000a) (T00), *ApJ* **543**, 340.

Thompson, C., Duncan, R.C., Feroci, M., and Hurley, K.H. (2000b), *ApJ*, submitted.

Thompson, C. and Murray, N.W. (2000), *ApJ*, submitted.

Ulmer, A. (1994), *ApJ* **437**, L111.

Ulmer, A. et al. (1993), *ApJ* **418**, 395.

Usov, V.V. and Melrose, D.A. (1986), *ApJ* **464**, 306.

Van Paradijs, J., Taam, R.E., and van den Heuvel, E.P.J. (1995), *A&A* **299**, L41.

Van Riper, K.A. (1988), *ApJ* **329**, 339.

Vasisht, G., Frail, D.A., and Kulkarni S.R. (1995), *ApJ* **440**, L65.

Vogt, S.S. (1988), in G. Cayrel de Strobel and M. Spite (eds.), *Impact of Very High S/N Spectroscopy on Stellar Physics*, Kluwer (Dordrecht), p253.

Woods, P.M. et al. (1999a), *ApJ* **518**, L103.

Woods, P.M. et al. (1999b), *ApJ* **519**, L139.

Woods, P.M. et al. (1999c), *ApJ* **524**, L55.

Woods, P.M. et al. (1999d), *ApJ* **527**, L47.

Woods, P.M. et al. (2000), *ApJ* **535**, L55.

Young, M.D., Manchester, R.N., and Johnston, S. (1999), *Nature* **400**, 848.

NEUTRON STAR ENVELOPES AND THERMAL RADIATION FROM THE MAGNETIC SURFACE

J. VENTURA

Physics Department, University of Crete, and IESL, FORTH
71003 Heraklion, Crete, Greece

AND

A.Y. POTEKHIN

Ioffe Physico-Technical Institute, 194021 St.Petersburg, Russia

Abstract. The thermal structure of neutron star envelopes is discussed with emphasis on analytic results. Recent progress on the effect of chemical constitution and high magnetic fields on the opacities and the thermal structure is further reviewed in view of the application to pulsar cooling and magnetars.

1. Introduction

Neutron stars (NS) are formed with very high internal temperatures approaching 10^{11} K in the core of a supernova explosion (see, e.g., Shapiro and Teukolski 1983 – ST83 in the following). Copious neutrino emission brings the temperature in the stellar core down to $\simeq 10^9$ K within about one day and then, more gradually, to $\simeq 10^8$ K within 10^4 years. It was realized early on that such objects were likely to have effective surface temperatures of the order of 10^6 K (Chiu and Salpeter 1964). Comparison with theoretical cooling curves can further provide information on aspects of the internal structure of NS such as superfluidity and the possible appearance of pion or kaon condensates and strange matter in their interior (e.g., Pethick 1992; Page 1997; Tsuruta 1998). It is then clear that the observation of this surface cooling X-ray emission is an objective of prime scientific importance.

The observation of such faint point sources has turned out to be difficult however, having to await the modern era of imaging X-ray telescopes. *ROSAT* observations in the 90's finally yielded improved spectral information, opening a new chapter in our ability to probe the internal structure

C. Kouveliotou et al. (eds.), The Neutron Star – Black Hole Connection, 393–414.

of superdense matter. Joachim Trümper (this meeting) already gave us an overview of the recent observations in this field (see also Caraveo et al. 1996; Trümper and Becker 1997). Increasing interest is also presently being directed on the new and important subject of magnetars, extensively discussed during this meeting.

Theoretical models of NS cooling are of necessity rather complicated, requiring detailed understanding of the stellar structure, equation of state (EOS), and thermal balance over an enormous range of densities and chemical diversity (e.g., Pethick 1992; Tsuruta 1998, and references therein). During an initial period of $10^5 - 10^6$ y the star cools principally via neutrino emission from its interior. There are many neutrino emission mechanisms, whose rates depend on the state of matter, particularly on nucleon superfluidity (for a review see Yakovlev et al. 1999). An older NS cools mainly via photon emission from its surface. Finally, the cooling may also be influenced by heating processes such as friction due to differential rotation between the superfluid and normal parts of the star (e.g., Alpar et al. 1989), β-processes arising from chemical imbalance during the spindown (Reisenegger 1997), and pulsar polar cap heating due to impinging charged particles accelerated in the magnetosphere (e.g., Halpern and Ruderman 1993).

Analysis of the NS thermal evolution is considerably simplified however by the fact that the stellar interior, from densities of $\simeq 10^{10}$ g cm^{-3} inward, is nearly isothermal because of the very high thermal conductivity in these layers. It is therefore convenient to establish a relation between this interior temperature T_i, defined as the temperature at mass density $\rho = 10^{10}$ g cm^{-3}, and the effective surface temperature T_e (Gudmundsson et al. 1983 – GPE in the following). Thus, it becomes possible to examine separately the properties and structure of the outer envelope, which turns out to be crucial in determining the ratio T_e/T_i and the nature of the emitted radiation (e.g., GPE; Hernquist and Applegate 1984 – HA84 in the following; Ventura 1989; Potekhin et al. 1997 – PCY in the following).

In the next section, we consider basic features of the mechanical and thermal structure of the outer NS envelope without magnetic field. The strong magnetic field ($B \gg 10^{10}$ G), which was found in most of the pulsars, shifts the atmosphere bottom and the region of partial ionization to higher densities. Furthermore, it strongly affects the radiative and thermal transport through the envelope. These effects will be discussed in Section 3.

2. Outer Envelope of a Neutron Star

The *outer envelope* consists mainly of electrons and ions. It extends down to a depth of a few hundred meters, where the density 4.3×10^{11} g cm^{-3}

is reached. At this density, neutrons begin to drip from the nuclei (e.g., ST83). Thus the *inner envelope*, which extends deeper down to the core, consists of electrons, atomic nuclei, and free neutrons.

The outer envelope can be divided into the atmosphere, the liquid ocean, and the solid crust. The outermost layer constitutes a thin (0.1 – 100 cm) NS atmosphere (optical depth $\tau \lesssim 1$), where the outgoing radiation is formed. The plasma density at the atmosphere bottom is about 0.001 – 0.1 g cm^{-3}, depending on temperature, surface gravity, and chemical composition. This plasma can be partially ionized and non-ideal. Bound species can be distinct until the electron Fermi energy becomes comparable with the Thomas–Fermi energy at $\rho \lesssim 10\,ZA$ g cm^{-3}, A and Z being the mass and charge numbers. At higher densities the ions are immersed in a jellium of degenerate electrons, which still strongly responds to the Coulomb fields of the ions as long as $\rho \lesssim 10\,Z^2A$ g cm^{-3}.

In the rest of the envelope (at $\rho \gg 10\,Z^2A$ g cm^{-3}) the electrons form a strongly degenerate, almost ideal gas. This gas is non-relativistic at $\rho_6 \ll 1$ and ultrarelativistic at $\rho_6 \gg 1$, where $\rho_6 \equiv \rho/10^6$ g cm^{-3}. The ions form a Coulomb gas or liquid at $\Gamma \lesssim 175$, where $\Gamma \approx 22.75\,(\rho_6/\mu_e)^{1/3}\,Z^{5/3}/T_6$ is the Coulomb coupling parameter, $T_6 = T/10^6$ K, $\mu_e = A/Z$. At $\Gamma \approx 175$, the liquid freezes into a Coulomb crystal. The pressure is almost entirely determined by degenerate electrons and thus independent of T, while the mass density is mostly determined by the ions.

While cooling of the NS during the initial neutrino dominated era is not influenced by the outer layers, it is in fact the properties of these surface layers that characterize the flux and photon spectrum emitted at the NS surface, leading in turn to estimates of T_i. At the subsequent photon cooling stage, the heat insulation by the envelope controls the cooling rate. The radiation is reprocessed in the atmosphere, which yields an emitted spectrum, in general different from that of a black body (Romani 1987). The properties of these layers are thus crucial to interpreting observations and, understandably, a lot of theoretical work has been devoted to analyzing the thermal structure of non-magnetic and magnetic NS envelopes (e.g., GPE; PCY; Heyl and Hernquist 1998) and the radiation properties of their atmospheres (e.g., Pavlov et al. 1995; Page and Sarmiento 1996; Rajagopal et al. 1997; Potekhin *et al.* 1998, and references therein).

2.1. MECHANICAL STRUCTURE

To review the cardinal properties of the NS surface layers, let us recall the enormous gravitational potential $GM/R \approx 0.148\,(M/M_\odot)\,R_6^{-1}\,c^2$ (where $R_6 \equiv R/10^6$ cm ~ 1 and $M/M_\odot \simeq 1.4$ for most typical NS), which renders effects of General Relativity appreciable. Indeed, the Schwarzschild radius

$r_g = 2GM/c^2 \approx 2.95\,(M/M_\odot)$ km is not much smaller than the stellar radius R (throughout this review, M is the gravitational mass of the star, which is 10–15% smaller than its baryon mass). The hydrostatic equilibrium is then governed by the Oppenheimer–Volkoff equation (e.g., Thorne 1977). Introducing the local proper depth $z = (R - r)(1 - r_g/R)^{-1/2}$ (where r is the radius), in the surface layers ($z \ll R$) one can rewrite this equation in the Newtonian form

$$\mathrm{d}P/\mathrm{d}z = g\rho, \tag{1}$$

where

$$g = \frac{GM}{R^2\,(1 - r_g/R)^{1/2}} \approx 1.327 \times 10^{14}\,(1 - r_g/R)^{-1/2}\,\frac{M}{M_\odot}\,R_6^{-2}\ \mathrm{cm\ s^{-2}}$$

is the local gravitational acceleration at the surface.

Since the surface gravity is huge, the atmosphere's scale height is rather small. In the non-degenerate layers, we have

$$P = (\rho/\mu m_{\mathrm{u}})\,kT, \tag{2}$$

where $m_{\mathrm{u}} = 1.6605 \times 10^{-24}$ g is the atomic mass unit and $\mu = A/(Z + 1)$. Thus, following a thin non-degenerate atmosphere of a scale height $P/(g\rho) \approx 0.626\,(T_6/\mu)\,R_6^2\,(M/M_\odot)^{-1}$ cm, electron degeneracy sets in at densities $\rho \gtrsim 6\,\mu_e T_6^{3/2}$ g cm^{-3}.

The electron kinetic energy at the Fermi surface is $kT_{\mathrm{F}} \equiv \epsilon_{\mathrm{F}} - m_e c^2$, where k is the Boltzmann constant,

$$T_{\mathrm{F}} = (\gamma_{\mathrm{F}} - 1)\,m_e c^2/k = 5.93 \times 10^9\,\chi^2/(1 + \gamma_{\mathrm{F}})\ \mathrm{K} \tag{3}$$

is the Fermi temperature, and

$$\gamma_{\mathrm{F}} = \sqrt{1 + \chi^2}, \qquad \chi = \frac{p_{\mathrm{F}}}{m_e c} = \frac{\hbar(3\pi^2 n_e)^{1/3}}{m_e c} \approx 1.009\left(\frac{\rho_6}{\mu_e}\right)^{1/3}, \tag{4}$$

are the electron Lorentz factor and *relativity parameter*, respectively.

Elementary fitting formulae to the pressure of fully ionized ion-electron plasmas as function of density at arbitrary electron degeneracy and temperature have been presented by Chabrier and Potekhin (1998). These formulae can also be extended to partially ionized atmospheric layers in the mean-ion approximation, provided the effective ion charge Z is known (cf. PCY). At sufficiently high density, however, where $T_{\mathrm{F}} \gg T$, the main contribution is that of strongly degenerate electrons with the pressure depending only on the density through the parameter χ (e.g., ST83):

$$P_e = \frac{P_0}{8\pi^2}\left[\chi\left(\frac{2}{3}\chi^2 - 1\right)\gamma_{\mathrm{F}} + \ln(\chi + \gamma_{\mathrm{F}})\right] \tag{5}$$

where $P_0 \equiv m_e c^2/\lambda^3 = 1.4218 \times 10^{25}$ dynes cm^{-2} is the relativistic unit of pressure, $\lambda = \hbar/(m_e c)$ being the Compton wavelength. This may further be approximated as $P_e \approx P_0 \chi^{3\gamma_{ad}}/(9\pi^2 \gamma_{ad})$, where γ_{ad} is the adiabatic index, equal to 5/3 at $\chi \ll 1$ and 4/3 at $\chi \gg 1$.

Note that for strongly degenerate electrons

$$dP = n_e \, d\epsilon_F = 4.93 \times 10^{17} \text{ erg g}^{-1} \, (\rho/\mu_e) \, \chi \, d\chi/\gamma_F. \tag{6}$$

From Eqs. (1) and (6), one obtains

$$1.027 \frac{\rho_6}{\mu_e} = \chi^3 = \left[\frac{z}{z_0}\left(2 + \frac{z}{z_0}\right)\right]^{3/2}, \quad z_0 = \frac{m_e c^2}{m_u g \mu_e} \approx \frac{4930 \text{ cm}}{\mu_e \, g_{14}}, \tag{7}$$

where $g_{14} = g/(10^{14} \text{ cm s}^{-2})$, and z_0 is a depth scale at which degenerate electrons become relativistic.

Let us note also that the mass ΔM contained in a layer from the surface to a given depth z is solely determined by the pressure $P(z)$ at the bottom of the given layer:

$$\Delta M(z) = 4\pi R^2 \, P(z) \, g^{-1} \, (1 - r_g/R)^{1/2} \approx 1.192 \times 10^{-9} \, g_{14}^{-2} \, M P(z)/P_0. \tag{8}$$

2.2. THERMAL STRUCTURE

From Eqs. (7) and (8), we see that the outer envelope is a very thin layer – typically within the outer 100 m, containing $\sim 10^{-7} M_\odot$, – which renders the thermal diffusion problem essentially plane-parallel and one-dimensional. Assuming a constant heat flux throughout the outer envelope, the temperature profile can be obtained by solving the heat diffusion equation:

$$F = \kappa \frac{dT}{dz} = \frac{16\sigma}{3} \frac{T^3 \, dT}{d\tau}, \quad \kappa \equiv \frac{16\sigma T^3}{3K\rho}, \tag{9}$$

where F is the heat flux, κ is thermal conductivity, σ is the Stefan–Boltzmann constant, and K is the usual Rosseland mean over the energy spectrum of the specific opacity. This leads immediately to a temperature profile $T/T_e \approx (\frac{3}{4}\tau + \frac{1}{2})^{1/4}$, where the *local* effective surface temperature T_e is defined through $F = \sigma T_e^4$ and the integration constant corresponds to the Eddington approximation ($\tau = \frac{2}{3}$ at the *radiative surface*, where $T = T_e$). A more accurate boundary condition requires solution of the radiative transfer equation in the atmosphere. A distant observer would infer from the spectrum and flux the redshifted surface temperature $T_\infty = T_e (1 - r_g/R)^{1/2}$ and apparent radius $R_\infty = R (1 - r_g/R)^{-1/2}$ (e.g., Thorne 1977).

A knowledge of the mean opacity $K(\rho, T)$ is then needed to relate the temperature to the other plasma parameters, which can then also be expressed as functions of the optical depth τ. It is also needed in order to compute the overall depth and the temperature ratio T_i/T_e.

2.2.1. Opacity

Heat is transported through the envelope mainly by radiation and by conduction electrons. In general, the two mechanisms work in parallel, hence

$$\overline{\kappa} = \kappa_r + \kappa_c, \qquad K^{-1} = K_r^{-1} + K_c^{-1}, \tag{10}$$

where κ_r, κ_c and K_r, K_c denote the radiation and conduction components of the conductivity and opacity, respectively. Typically, the radiative conduction dominates the thermal transport ($\kappa_r > \kappa_c$) in the outermost non-degenerate layers of a NS, whereas electron conduction dominates ($\kappa_c > \kappa_r$) in deeper, mostly degenerate layers. In the absence of intense magnetic fields, modern cooling calculations (e.g., PCY) make use of the Livermore library of radiative opacities OPAL (Iglesias and Rogers 1996), which also provides an EOS for the relevant thermodynamic parameters at $\rho \lesssim 10\, T_6^3$ g cm^{-3}. For the electron conduction regime, modern opacities have been worked out by Potekhin *et al.* (1999a).

Radiative opacities. In order to derive an analytic model of the NS envelope, HA84 and Ventura (1989) have written the atmospheric opacity in the form

$$K(\rho, T) = K_0 \rho^\alpha T^\beta. \tag{11}$$

In particular, this relation describes the opacity given by the Kramers formula, which corresponds to $\alpha = 1$ and $\beta = -3.5$. In a fully ionized, non-relativistic and non-degenerate plasma, the opacity provided by the free-free transitions is (e.g., Cox and Giuli 1968)

$$K_r \approx 75\, \bar{g}_{\text{eff}}\, (Z/\mu_e^2)\, \rho\, T_6^{-3.5} \text{ cm}^2/\text{g}, \tag{12}$$

where ρ is in g cm^{-3} and $\bar{g}_{\text{eff}} \sim 1$ is an effective dimensionless Gaunt factor, a slow function of the plasma parameters. For a colder plasma, where bound-free transitions dominate over free-free ones, the Kramers formula remains approximately valid, but the opacity K is about two orders of magnitude higher. An order-of-magnitude (within ≈ 0.5 in $\log K$) approximation to the realistic OPAL opacities for hydrogen in the range of parameters $T_6 \sim 10^{-1} - 10^{0.5}$ and $\rho \sim (10^{-2} - 10^1)\, T_6^3$ is given by Eq. (12) if we put formally $\bar{g}_{\text{eff}} \approx \rho^{-0.2}$. An analogous order-of-magnitude approximation to the OPAL opacities for iron at $T_6 \sim 10^0 - 10^{1.5}$ and $\rho \sim (10^{-4} - 10^{-1})\, T_6^3$ is obtained with $\bar{g}_{\text{eff}} \approx 70\, \rho^{-0.2}$. These approximations also belong to the class of functions (11), but with $\alpha = 0.8$.

Conductive opacities.   Thermal conduction of degenerate matter in deeper layers is dominated by electrons which scatter off ions. This conductivity can be written as (Yakovlev and Urpin 1980)

$$\kappa_c = \frac{\pi k^2 T m_e c^3 \chi^3}{12 Z e^4 \Lambda \gamma_F^2} \approx 2.3 \times 10^{15} \frac{T_6}{\Lambda Z} \frac{\chi^3}{1 + \chi^2} \frac{\text{erg}}{\text{cm s K}}, \tag{13}$$

where Λ is the Coulomb logarithm. Accurate analytic fitting formulas to Λ as function of ρ and T have been obtained recently by Potekhin et al. (1999a). In the solid crust, this function is reduced to small values by quantum and correlation effects in Coulomb crystals. However, we shall see shortly that in not too cold NS, the thermal profile is mainly formed in the liquid layers of the envelope. Therefore, for our purpose, it will be sufficient to note that in the NS ocean Λ is a slow function of the plasma parameters and can be approximated by a constant of the order of unity.

In the case of non-degenerate electrons, the conductivity can be found, e.g., by the method of Braginskiĭ (1957), which yields

$$\kappa_c^{\text{nd}} \approx 5 \times 10^{10} \left(F_Z / \Lambda \right) Z^{-1} T_6^{5/2} \ \text{erg}/(\text{cm s K}), \tag{14}$$

where F_Z is a slow function of Z: for example, $F_{26} = 1.34$ and $F_1 = 0.36$, whereas the Coulomb logarithm Λ is ~ 1 near the onset of degeneracy and logarithmically increases with decreasing density. Effectively, Eq. (14) may be viewed as an analog to Eq. (13) where the dimensionless Fermi momentum χ has been replaced by an appropriate thermal average ($\propto \sqrt{T}$).

2.2.2. *Temperature Profile*
The simple functional form of the opacity allows now an analytic treatment of the thermal structure (Urpin and Yakovlev 1980; HA84; Ventura 1989).

Non-degenerate regime.   Using Eqs. (1), (11), and (2) one may rewrite Eq. (9) as

$$dP = \frac{\kappa g \rho}{F} dT = \frac{16}{3} \frac{g}{K} \frac{T^3 \, dT}{T_e^4} = \frac{16}{3} \frac{g}{K_0} \left(\frac{k}{\mu m_u} \right)^\alpha \frac{T^{3+\alpha-\beta}}{P^\alpha} \frac{dT}{T_e^4}, \tag{15}$$

which is readily integrated from the surface inward to give the temperature profile. In the region far from the surface, where $T \gg T_e$, the integration constants may be dropped. Using again Eqs. (2), one can present the result as a simple power law:

$$\frac{T^{3-\beta}}{\rho^{\alpha+1}} = \frac{3}{16} \frac{4 + \alpha - \beta}{\alpha + 1} \frac{k}{\mu m_u} \frac{K_0 T_e^4}{g}. \tag{16}$$

This result depends on our implicit assumption that the effective ion charge Z remains constant, which in general is not strictly valid.

Interestingly enough, as noted by HA84, Eq. (16) establishes that the conductivity κ is constant throughout the radiative non-degenerate layer:

$$\kappa = \frac{16}{3} \frac{\sigma T^3}{K\rho} = \frac{16}{3} \frac{\sigma}{K_0} \frac{T^{3-\beta}}{\rho^{\alpha+1}} = \frac{4+\alpha-\beta}{1+\alpha} \frac{k\sigma T_e^4}{\mu m_u g}. \tag{17}$$

The constant value depends on the emitted flux, but is independent of K_0. It follows immediately from Eq. (9) that temperature grows linearly with geometrical depth. Furthermore, from Eq. (16) we obtain the T, ρ profile. In particular, substituting $\alpha = 1$ and $\beta = -3.5$, we obtain

$$K\rho = \frac{1.51}{\text{cm}} \frac{\mu g_{14}}{T_{e6}} \left(\frac{T}{T_e}\right)^3, \quad \kappa = 2 \times 10^{14} \frac{T_{e6}^4}{\mu g_{14}} \frac{\text{erg}}{\text{cm s K}}. \tag{18}$$

Substitution of K from Eq. (12) yields

$$T_6 \approx 0.284\, g_{14}\, \mu\, z \approx (50\, \bar{g}_{\text{eff}}\, q)^{2/13} (\rho/\mu_e)^{4/13}, \quad q \equiv T_{e6}^4 Z/(\mu g_{14}). \tag{19}$$

In these relations, z and ρ are measured in CGS units, and $T_{e6} = T_e/10^6$ K.

Radiative surface. It is further interesting to note that the opacity K is also slowly varying in this region. Combining Eqs. (11), and (19), we obtain $K \propto \rho/T^{3.5} \propto \rho^{-1/13}$. Invoking Eq. (2), we get $K \propto P^{-1/17}$. This justifies the assumption $K_r \approx$ const, which one often employs when determining the radiative surface from equation (e.g., GPE; PCY)

$$(K_r P)_{\text{surface}} = (2/3)\, g. \tag{20}$$

Using Eqs. (2) and (12), we obtain

$$\rho_s \approx 0.1\, \mu_e \left(\frac{\mu}{Z\, \bar{g}_{\text{eff}}} g_{14}\right)^{1/2} T_{e6}^{5/4} \text{ g cm}^{-3}. \tag{21}$$

Substituting $\bar{g}_{\text{eff}} \simeq 1$ and $\bar{g}_{\text{eff}} \simeq 200$ for hydrogen and iron, respectively, we obtain $\rho_s \sim 0.07\, \sqrt{g_{14}}\, T_{e6}^{5/4}$ g cm^{-3} and $\rho_s \sim 0.004\, \sqrt{g_{14}}\, T_{e6}^{5/4}$ g cm^{-3} for these two elements, in reasonable agreement with numerical results of PCY.

Onset of degeneracy. The solution given by Eq. (19) can be extended down to a depth where the electrons become degenerate. Let us estimate this depth from the condition $kT_d = p_F^2/2m_e$. We obtain

$$\chi_d \simeq 0.053\, (\bar{g}_{\text{eff}}\, q)^{1/7}, \quad T_d \simeq 8.5 \times 10^6\, (\bar{g}_{\text{eff}}\, q)^{2/7} \text{ K}. \tag{22}$$

Even for very high $T_e \sim 10^7$ K, we have $\chi_d \lesssim 1$, i.e. the electrons are non-relativistic at the degeneracy boundary. With decreasing T_e, the quantities χ_d and T_d decrease, i.e. the boundary shifts toward the stellar surface.

Sensitivity strip. Numerical cooling calculations (GPE) revealed early on that accurate knowledge of the opacity law is especially important within a certain "sensitivity strip" in the (ρ, T) plane. The temperature ratio T_i/T_e changes appreciably if K is modified by, say, a factor 2 within this narrow strip, while comparable changes of K outside the strip would leave this ratio unaffected within a high degree of accuracy.

The importance of the layer in the outer envelope where the opacity (10) turns from radiative to conduction dominated is now easy to demonstrate. In the non-degenerate radiative part the integral over ρ is dominated by the higher densities near the base of the layer, while in the degenerate, conductive layer it is dominated by the top, least dense part of the layer. The region where K turns from radiative to conductive is thus seen to contribute most of the resistance to heat flow. The line in the (ρ, T) plane where $K_r = K_c$ is easily determined from Eqs. (12) and (13):

$$\rho \approx 12\,\mu_e \bar{g}_{\text{eff}}^{-1/3}\, T_6^{11/6} \ \text{g\,cm}^{-3}. \tag{23}$$

On the right-hand side, we have approximated the factor $(\Lambda \gamma_F^2)^{1/3}$ by unity. Using again the solution (16), we find explicitly the temperature T_t and relativity factor χ_t at the turning point from radiative to electron conduction:

$$T_t \approx 2.3 \times 10^7\, \bar{g}_{\text{eff}}^{2/17}\, q^{6/17} \ \text{K}, \quad \chi_t \approx 0.157\, \bar{g}_{\text{eff}}^{-2/51}\, q^{11/51}. \tag{24}$$

Some caution is necessary here, however, because the approximation $K \simeq K_r$ is not justified as we approach the turning point: actually, at this point $K = K_r/2$, as seen from Eq. (10). In addition, the extrapolation of the solution (19) to the turning point is, strictly speaking, not justified, since the electron gas becomes degenerate: $\chi_t > \chi_d$ for most typical NS parameters. Nevertheless, since χ_t and χ_d are not very much different, the section of the thermal profile where our assumptions are violated is relatively small, so Eq. (24) provides a reasonable approximation. This is confirmed by a direct comparison with numerical results (PCY), which reveals an error within only a few ten percent, provided $T \gtrsim 10^{5.5}$ K.

Solution beyond the turning point. An analytic solution to the thermal profile in degenerate layers of a NS envelope has been first obtained by Urpin and Yakovlev (1980), based on the conductivity in the form (13). The hydrostatic equilibrium of the degenerate surface layers is determined by Eqs. (1) and (6), which yield $g m_u \mu_e = m_e c^2 (\chi/\gamma_F)\, d\chi/dz$. Using the heat diffusion equation (9) and Eq. (13), we obtain

$$T\frac{dT}{d\chi} = \frac{12}{\pi}\frac{FZe^4\Lambda}{m_u k^2 cg\mu_e}\frac{\gamma_F}{\chi^2} = (1.56 \times 10^7 \ \text{K})^2\,\frac{Z\Lambda T_{e6}^4}{\mu_e g_{14}}\frac{\gamma_F}{2\chi^2}. \tag{25}$$

Treating Λ, A, and Z as constants, we can integrate this equation from χ_t inwards and obtain

$$T^2(z) = T_t^2 + (1.56 \times 10^7 \text{ K})^2 \frac{Z\Lambda T_{e6}^4}{\mu_e g_{14}} \left[f(\chi) - f(\chi_t) \right], \qquad (26)$$

where $f(\chi) \equiv \ln(\chi + \gamma_F) - \gamma_F/\chi$, and the dependence of χ on z is given by Eq. (7).

Let us use the above solution to evaluate T_i. Since $\chi_t \lesssim 1$, but $\chi \gg 1$, it is easy to see that $[f(\chi) - f(\chi_t)] \approx \chi_t^{-1}$ in Eq. (26). Using Eq. (24), we obtain

$$T_i = \left(T_t^2 + T_\Delta^2 \right)^{1/2}, \quad T_\Delta \approx 4 \times 10^7 \left(\frac{Z\, T_{e6}^4}{\mu_e g_{14}} \right)^{20/51} \text{K}, \qquad (27)$$

where we have neglected some factors which are close to unity.

Figure 1 illustrates the accuracy of the above analytic solution as well as its limitations. Here, solid lines show temperature profiles of a NS with mass $M = 1.4\, M_\odot$ and radius $R = 10$ km obtained by solving the radiative transfer equation with spectral OPAL opacities in the atmosphere and integrating the thermal structure equation (9) with accurate radiative and conductive Rosseland opacities inwards in the deeper layers (Shibanov et al. 1998). Dashed curves depict the above analytic approximations. The left panel corresponds to an envelope composed of iron, and the right panel shows the thermal structure of an accreted envelope with its outermost layers composed of hydrogen, which is further burnt into heavier elements (He, C, Fe) in deeper and hotter layers. This thermo- and pycnonuclear burning is responsible for the complex shape of the upper profile. Different straight lines show the points at which thermal profiles corresponding to various heat fluxes would cross the radiative surface, the region of the onset of degeneracy, turning points $K_c = K_r$, and (on the left panel) the bottom of the ocean. The latter line is not present on the right panel, because freezing of hydrogen and helium is suppressed by large zero-point ionic vibrations (cf. Chabrier 1993).

One can see that not only the above analytic approximations correctly describe the qualitative structure of the envelope, but they also provide a reasonable quantitative estimate of the temperature at a given density. The accuracy deteriorates for lower T_e (especially when Z is high), because in this case radiative opacities are affected by bound-bound transitions and strong plasma coupling effects and thus no longer obey the simple power law (12).

Discussion.   We have seen that the internal temperature T_i is determined by two temperatures, T_t and T_Δ, related to the non-degenerate and de-

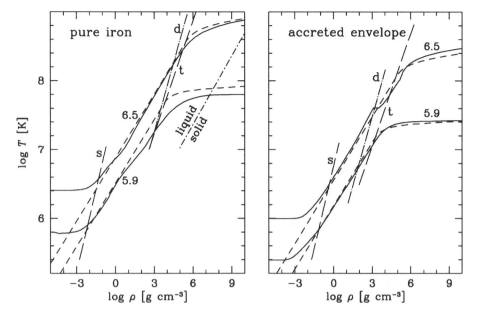

Figure 1. Thermal profiles inside non-accreted (left panel) and accreted (right panel) envelopes of NS at two effective temperatures, $\log T_e$ [K] $= 5.9$ and 6.5 (marked near the curves). Solid curves – numerical solution (Shibanov *et al.* 1998), dashed curves – analytic approximations (19) and (27). Straight lines marked "s", "d," and "t" give the values ρ and T at which the various temperature profiles cross the radiative surface [Eq. (21)], the onset of electron degeneracy [Eq. (22)], and the turning point [Eq. (24)], respectively. The Fe melting line is also shown.

generate layers, respectively, and that the temperature growth occurs in the very surface layers of the star. This justifies the separation of the NS into a blanketing envelope and internal isothermal layers used in numerical simulations (GPE; PCY). We have also shown, in agreement with GPE, that the bulk of the envelope's thermal insulation arises in the relatively low density region around the turning point defined by Eq. (24).

The dependence on Z and μ_e that enters Eqs. (27) makes the T_i/T_e ratio sensitively dependent of the chemical composition of the envelope. If we replace the iron envelope by an accreted envelope composed of hydrogen or helium, we get this ratio reduced by about a half order of magnitude, which corresponds to two orders of magnitude higher photon luminosity at a given internal temperature. Thus an envelope composed of light elements is much more transparent to heat, and this strongly affects cooling, as noted by PCY. Our analytic estimates present a tool for the fast estimation of the magnitude of such effects.

In the next section we will see how these principal properties change as a result of a strong magnetic field permeating the NS envelope.

3. Effects of Magnetic Fields

It is well known that the strong magnetic field can profoundly alter the physical properties of the NS outer layers – for reviews see Canuto and Ventura (1977), Mészáros (1992), and Yakovlev and Kaminker (1994). We will be interested here in modifications introduced by the intense field B in the EOS, the Fermi temperature T_F, and the opacity K, all of which affect the heat transfer problem. Our primary focus will be on the properties of an electron gas. Free ions give only a minor contribution to the opacities, whereas their contribution to the EOS remains unaffected by the field in the non-degenerate regime and is negligible when electrons are strongly degenerate.

Apart from the bulk properties of the electron plasma, magnetic fields also modify profoundly the properties of atoms which become very elongated and compact, having sharply increased binding energies (e.g., Canuto and Ventura 1977). This should strongly affect the emitted spectra from NS surfaces. Many works have been devoted in the past to the calculation of quantum-mechanical properties of atoms at rest in strong magnetic fields (e.g., Miller and Neuhauser 1991). Some of them have been used to construct magnetic NS atmosphere models (Rajagopal et al. 1997). However, thermal motion of the atoms at realistic NS temperatures breaks down the axial symmetry and may completely alter atomic properties. It is therefore necessary to have quantum-mechanical calculations for *moving* atoms, and their results should be included in models of EOS and opacities. Such models are available to date only for hydrogen (e.g., Potekhin et al. 1998, 1999b, and references therein). For other species, this work still remains to be done.

In the past, a lot of work has also been devoted to evaluating the conditions under which magnetic molecular chains may become stable in the surface layers; furthermore, in superstrong fields they may form a magnetically stabilized lattice (e.g., Ruderman 1971; Lai and Salpeter 1997). Such a phase transition is also expected to have observable consequences, but the field still remains largely unexplored.

3.1. ELECTRON GAS IN MAGNETIC FIELD

Motion of free electrons perpendicular to the magnetic field is quantized in Landau orbitals with a characteristic transverse scale equal to the *magnetic length* $a_m = (\hbar c/eB)^{1/2} = \lambdabar/\sqrt{b}$, where $b = \hbar\omega_c/m_ec^2 = B_{12}/44.14$ is the magnetic field strength expressed in relativistic units, $\omega_c = eB/mc$ is the electron cyclotron frequency, and $B_{12} \equiv B/10^{12}$ G. The Landau energy levels are

$$\epsilon = \epsilon_n(p_z) = c \left(m_e^2 c^2 + 2\hbar\omega_c m_e n + p_z^2 \right)^{1/2}, \tag{28}$$

with $n = 0, 1, 2, \ldots$, where the magnetic field $\boldsymbol{B}$ is assumed to be homogeneous and directed along the z axis, and p_z is the longitudinal momentum. The ground Landau level $n = 0$ is non-degenerate with respect to spin projection ($s = -1$, statistical weight $g_0 = 1$) while the levels $n > 0$ are doubly degenerate ($s = \pm 1$, $g_n = 2$). The anomalous magnetic moment of the electron, $g_e = 1.00116$, causes splitting of the energy levels $n \geq 1$ by $\delta\epsilon = (g_e - 1)\hbar\omega_c$, which, strictly speaking, removes the double spin-degeneracy. In typical NS envelopes, this splitting is negligible because $\delta\epsilon$ is smaller than either the thermal width $\sim kT$ or the collisional width of the Landau levels.

The electron's phase space is thus now a combination of an energy continuum in p_z, corresponding to the motion along the field, and a discrete spectrum (the quantum number n) corresponding to the quantized transverse motion. This property will be reflected in most of the physical processes of our interest here.

Let us denote by n^* the highest Landau excitation populated at a given energy ϵ. It equals an integer part of the combination $p_0^2(\epsilon)/(2m_e\hbar\omega_c)$, where $p_n(\epsilon) = [(\epsilon/c)^2 - (m_e c)^2 - 2m_e\hbar\omega_c n]^{1/2} = |p_z|$. Taking into account that the number of quantum states of an electron with given s and n in volume V per longitudinal momentum interval Δp_z equals $V\Delta p_z/(4\pi^2 a_m^2 \hbar)$, one can obtain the electron number density and pressure from first principles (e.g., Blandford and Hernquist 1982). For strongly degenerate electrons,

$$n_e = \frac{1}{2\pi^2 a_m^2 \hbar} \sum_{n=0}^{n^*} g_n p_n(\epsilon_F), \tag{29}$$

$$P_e = P_0 \frac{b}{4\pi^2} \sum_{n=0}^{n^*} g_n (1 + 2bn) [x_n \sqrt{1 + x_n^2} - \ln(x_n + \sqrt{1 + x_n^2})], \tag{30}$$

where $x_n = cp_n(\epsilon_F)/\epsilon_n(0)$, and P_0 is the same as in Eq. (5).

For a degenerate electron gas the thermodynamic quantities such as pressure, magnetization, and energy density exhibit quantum oscillations of the de Haas–van Alphen type whenever the dimensionless Fermi momentum reaches the characteristic values $\chi = \sqrt{2nb}$ which signify the occupation of new Landau levels. In these oscillations the various quantities typically take values around their classical $B = 0$ values, except in the limit of a strongly quantizing field ($n^* = 0$) where one often finds substantial deviations (e.g., Yakovlev and Kaminker 1994). The latter case takes place when the typical energies kT, $kT_F < \epsilon_1 - 1$ – i.e., at $T \ll T_B$ and $\rho < \rho_B$, where

$$\rho_B = m_u n_B \mu_e \approx 7045 \, B_{12}^{3/2} \mu_e \, \text{g cm}^{-3}, \tag{31}$$

$$T_B = \hbar\omega_c/k\gamma_F \;\approx\; 1.343 \times 10^8 \, (B_{12}/\gamma_F) \text{ K}, \tag{32}$$

and $n_B = 1/(\pi^2\sqrt{2}\,a_m^3)$ is the electron number density at which the Fermi energy reaches the first excited Landau level. This case is of special interest for the NS outer layers under consideration.

Strongly quantizing field.   When the electron's transverse motion is frozen in the ground state Landau level $n^* = 0$, the phase-space is effectively one-dimensional. Then $\epsilon = c\sqrt{(m_ec)^2 + p_z^2}$, and Eq. (29) simplifies to

$$p_F/\hbar = 2\pi^2 a_m^2 n_e. \tag{33}$$

We therefore see that the dimensionless Fermi momentum,

$$\chi \equiv p_F/m_ec = (2/3)\,\chi_0^3/b \approx (0.6846/b)\,\rho_6/\mu_e, \tag{34}$$

is proportional to the density ρ, in sharp contrast to the non-magnetic Eq. (4). Henceforth we denote $p_{F0} = m_ec\chi_0$ the "classical" (non-magnetic) Fermi momentum at a given density, and reserve notation $p_F = m_ec\chi$ for the same quantity modified by the magnetic field. According to Eq. (31), the strongly quantizing regime in which Eq. (34) is valid requires $\chi < \sqrt{2b}$.

The Fermi temperature T_F is again given by Eq. (3), but with the modified χ. Since $\chi = (4/3)^{1/3}(\rho/\rho_B)^{2/3}\chi_0$, T_F is strongly reduced at $\rho \ll \rho_B$. Conversely, at a given $T < T_B$, the degeneracy takes hold at much higher density than in the $B = 0$ case. An initially degenerate electron gas at $B = 0$ will thus become non-degenerate when a strong quantizing field is switched on.

Let us now consider the EOS. Since $n = 0$, Eq. (30) simplifies considerably. Given Eq. (34), this expression again takes the form of a power-law of the density, $P = P_0 b\chi^{\gamma_{\rm ad}}/(2\pi^2\gamma_{\rm ad}) \propto \rho^{\gamma_{\rm ad}}/B^{\gamma_{\rm ad}-1}$, with $\gamma_{\rm ad} = 3$ or 2 in the non-relativistic and ultrarelativistic limits, respectively.

3.2. MAGNETIC OPACITIES

In the presence of a magnetic field, the conductivity κ becomes a tensor, so that the heat fluxes along and across the field become different. Since the field varies over the NS surface, the heat transport becomes two-dimensional. Fortunately, since the crust thickness is relatively small, the one-dimensional equation (9) remains a good approximation, with $\kappa = \kappa_\parallel \cos^2\theta + \kappa_\perp \sin^2\theta$, where $\kappa_\parallel$ and $\kappa_\perp$ are the conductivities along and across $\boldsymbol{B}$ and θ is the angle between $\boldsymbol{B}$ and the normal to the surface.

Radiative opacities.   In a magnetized plasma, two propagating polarization normal modes are defined in the presence of an external field having
widely different mean free paths each in the various photon-electron interactions (e.g., Mészáros 1992; Pavlov *et al.* 1995). Silant'ev and Yakovlev
(1980) have calculated the Rosseland opacities for the cases when they are
determined by the Thomson and free-free processes. When $T_B \gg T$, κ_r
grows proportionally to $(T_B/T)^2$ – i.e., K_r decreases as $(T/T_B)^2$. In particular, at $T \ll T_B$, the free-free opacities tabulated by Silant'ev and Yakovlev
(1980) tend to

$$K_r(B) \approx (23.2\,T/T_B)^2 K_r(0) \simeq 2.2\,\bar{g}_{\text{eff}}\,(Z/\mu_e^2)\,\rho\,T_6^{-1.5}\,B_{12}^{-2}\;\mathrm{cm}^2/\mathrm{g}, \quad (35)$$

where $K_r(0)$ is given by Eq. (12). This estimate may be used if only
$T_6 \lesssim B_{12}$, $\chi \ll 1$, and the opacity is dominated by the free-free processes.

From Eqs. (2), (20), and (35) one sees immediately that the *radiative
surface* in the strong magnetic fields, $B_{12} \gtrsim T_6$, is pushed to the higher
densities, $\rho_s \propto B$.

Caution is necessary however while using the scaling law (35). The
strong magnetic field shifts the ionization equilibrium toward a lower degree
of ionization, because of the increasing binding energies. Therefore, even if
the plasma were fully ionized at some ρ and T in the absence of magnetic
field, it may be only partially ionized at the same ρ and T when B is high.
This increases the significance of the bound-bound and bound-free opacities
and may result in a total radiative opacity considerably larger than that
given by Eq. (35) (see, e.g., Potekhin *et al.* 1998).

Electron conductivities.   Unified expressions and fitting formulae for thermal and electrical electron conductivities in a fully ionized degenerate
plasma with arbitrary magnetic field have been obtained recently (Potekhin
1999). These conductivities undergo oscillations of the de Haas–van Alphen
type at $\rho \gtrsim \rho_B$. At $B \gg 10^{10}$ G, the transport across the field is suppressed
by orders of magnitude. This fact allows us to neglect $\kappa_\perp$ totally, which
will be a good approximation everywhere except at a narrow stripe near
the magnetic equator, where $\theta \approx \pi/2$.

This approximation, $\kappa \approx \kappa_\parallel \cos^2\theta$, holds despite the arguments by Heyl
and Hernquist (1998) that $\kappa_\perp$ becomes non-negligible in the strongly quantizing regime at low density because the ratio $\kappa_\perp/\kappa_\parallel$ evaluated for strongly
degenerate electrons increases without bound as $\chi \to 0$. However, the finite
thermal width of the Fermi level (neglected by these authors) removes the
divergence. At typical NS temperatures, the thermal averaging terminates
the growth of $\kappa_\perp$ and moderates the decrease of $\kappa_\parallel$, before they become
comparable. In Figure 2, we plot by solid lines an example of the thermal
conductivities calculated according to Potekhin (1999). For comparison,

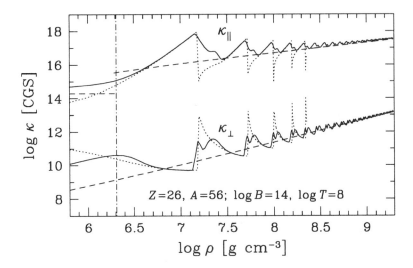

Figure 2. Longitudinal ($\|$) and transverse ($\perp$) thermal conductivities in the outer NS envelope composed of iron at $T = 10^8$ K and $B = 10^{14}$ G: comparison of accurate results (solid lines) with the classical approximation (dashed lines) and with the results without thermal averaging. The electrons are degenerate to the right of the vertical dot-dashed line, which corresponds to $T = T_F$. The dashed horizontal line in the non-degenerate region shows the conductivity given by Eq. (14) with $F_Z/\Lambda = 1$.

the dotted curves display the conductivities which would have been obtained without thermal averaging. As the electrons become non-degenerate at low densities, the solid curve for $\kappa_\|$ is seen to level off. It tends to its non-degenerate value $\kappa_\|^{nd}$ which is of the order of the non-magnetic value, Eq. (14), depending on density only logarithmically.

The dashed lines show the "classical" approximation where the quantizing nature of the field is neglected. For $\kappa_\|$, this approximation is close to the non-magnetic one. We can see that at high enough densities beyond the first oscillation, the classical approximation is good enough. At lower densities, the quantizing nature of the field must be taken into account. At $\rho < \rho_B$, the neglect of thermal averaging is justified as long as the electrons are degenerate. Then the longitudinal conductivity may be written as

$$\kappa_\| = \frac{k^2 T \, (m_e^2 c^3 b)^2}{12\pi\hbar^3 Z e^4 n_e} \frac{\chi^2}{2Q^\|} = \frac{2}{3} \frac{\kappa_0 \, (1 + \chi_0^2) \, \Lambda}{Q^\|}, \tag{36}$$

where κ_0 is the non-magnetic conductivity given by Eq. (13), in which χ must be replaced by χ_0, Λ is the non-magnetic Coulomb logarithm, and $Q^\|$ is a function of χ defined by Eq. (A9) in Potekhin (1999). In the liquid regime (far enough from the solid phase boundary), the latter function reduces to the expression (Yakovlev 1984)

$$Q^{\parallel} = \xi^{-1} - e^{\xi} E_1(\xi), \tag{37}$$

where $\xi = 2\chi^2/b + \frac{1}{2}(a_m q_s)^2$, q_s is an effective Coulomb-screening wave number, and E_1 is the standard exponential integral. A simple order-of-magnitude estimate of $\kappa_{\parallel}$ in the degenerate Coulomb liquid can be obtained if we neglect q_s and the second term in Eq. (37). In this way we obtain

$$\kappa_{\parallel} \simeq (4/3)\,\kappa_0\,\Lambda(1 + \chi_0^2)\,\chi^2/b \simeq 5 \times 10^{15}\,Z^{-1}\,T_6\,\chi^3\ \mathrm{erg/(cm\ s\ K)}. \tag{38}$$

As noted earlier, in the very strong fields considered here the onset of degeneracy is pushed into ever increasing densities as B increases. Therefore the turnover from radiative heat transfer to electron-conduction dominated transport may occur in the non-degenerate regime. In this case, Eq. (14) may be used to evaluate $\kappa_{\parallel}$.

3.3. CONSEQUENCES FOR THE HEAT TRANSPORT

We have seen that in strong magnetic fields there are several different regimes regulating the EOS and opacities. For the construction of an approximate analytic thermal profile it is sufficient to note that the non-magnetic expressions for the radiative opacity and *longitudinal* electron conductivity $\kappa_{\parallel}$ remain good approximations unless the field is strongly quantizing. Magnetic oscillations, which occur around the classical functions, will be smoothed out by integration while obtaining the thermal profile from Eq. (9); thus they are not too important.

When the field is strongly quantizing, the opacities are modified appreciably. However, in the degenerate part of NS ocean, which is of our prime interest here, the analytic expressions for $\kappa_{\parallel}$ can be again approximated as a power law, Eqs. (38) and (14). The same is true with respect to the extreme quantizing limit of radiative opacity, as follows from Eq. (35). In the non-degenerate regime, the magnetic field does not affect the EOS. In this case, we recover the solution (16) with new values of $\beta = -1.5$ and K_0, given by Eq. (35). Then

$$T_6 \approx 0.95\,(\bar{g}_{\mathrm{eff}}\,q)^{2/9}\,(\rho/\mu_e)^{4/9}\,B_{12}^{-4/9}, \tag{39}$$

with the same q as in Eq. (19). Thus the temperature is reduced (its profile becomes less steep) with increasing B as long as the field is strongly quantizing $(\rho < \rho_B)$.

It is interesting to note that the value of the constant conductivity, Eq. (17) is independent of the magnetic field, while its numerical value is only slightly lowered as a result of the changed coefficient β.

Sensitivity strip. As we have seen, the sensitivity strip is placed near the turning point from radiative transport to electron conduction, defined by $K_r = K_c$. In a strongly quantizing field, using Eqs. (35) and (38), we have

$$\rho \approx 250 \, \mu_e \, \bar{g}_{\rm eff}^{-0.2} \, T_6^{0.7} \, |\cos\theta|^{-0.4} B_{12} \quad {\rm g \, cm}^{-3} \tag{40}$$

instead of Eq. (23). With the temperature profile (39), we now obtain

$$T_t \approx \frac{3.5 \times 10^7}{|\cos\theta|^{8/31}} \, \bar{g}_{\rm eff}^{6/31} \, q^{10/31} \quad {\rm K}, \qquad \frac{\rho_t}{\mu_e} \approx \frac{3 \times 10^3 \, q^{7/31} \, B_{12}}{|\cos\theta|^{18/31} \, \bar{g}_{\rm eff}^{2/31}} \quad {\rm g \, cm}^{-3}. \tag{41}$$

If, however, the electrons are non-degenerate along the turnover line, then $\kappa_\parallel$ is represented by Eq. (14) instead of (38). In this case, we obtain the turnover at

$$\rho \approx 52 \, \sqrt{\Lambda/F_Z} \, \mu_e \, \bar{g}_{\rm eff}^{-1/2} \, T_6 \, |\cos\theta|^{-1} \, B_{12} \quad {\rm g \, cm}^{-3}. \tag{42}$$

Applying Eq. (39), we get

$$T_t \approx \frac{2.2 \times 10^7}{|\cos\theta|^{0.8}} \left(\frac{\Lambda q}{F_Z} \right)^{0.4} {\rm K}, \qquad \frac{\rho_t}{\mu_e} \approx \frac{1.1 \times 10^3 \, q^{0.4} \, B_{12}}{|\cos\theta|^{1.8} \, \bar{g}_{\rm eff}^{0.5}} \left(\frac{\Lambda}{F_Z} \right)^{0.9} \quad {\rm g \, cm}^{-3}.$$

Thus, in both cases (for degenerate and non-degenerate electrons) we have obtained similar dependences of the position of the turning point on the NS parameters. These dependences should be compared with Eq. (24). We see that T_t in both equations have similar values at $\theta = 0$, but in the magnetic field T_t increases with increasing θ. It is noteworthy that T_t is independent of B, while ρ_t grows linearly with B in the strongly quantizing field. From Eq. (31) we can evaluate the condition at which ρ_t lies in the region of strong magnetic quantization. Assuming that θ is not close to $\pi/2$ and neglecting factors about unity, we see that $\rho_t < \rho_B$ for $B_{12} \gtrsim (Z T_{e6}^4/g_{14})^{14/31}$, i.e., for the strong-field pulsars and magnetars.

Onset of degeneracy. Let us estimate the point at which the electrons become degenerate. For simplicity, let us assume that the electrons are non-relativistic. Taking into account Eqs. (3) and (34), we see that the condition $T = T_F$ in the strongly quantizing magnetic field is equivalent to $\rho \approx 608 \, \mu_e \, \sqrt{T_6} \, B_{12}$. Then from Eq. (39) we obtain $\rho_d \simeq 3700 \, (\bar{g}_{\rm eff} \, q)^{1/7} \mu_e B_{12}$. Thus, similar to the non-magnetic case, the switch between the regimes of photon and electron heat conduction occurs not far from the onset of degeneracy: $\rho_d \sim \rho_t$. Depending on θ, it can occur either in the non-degenerate domain (at $\theta \approx 0$) or in the degenerate domain (at $\theta \gtrsim 60°$).

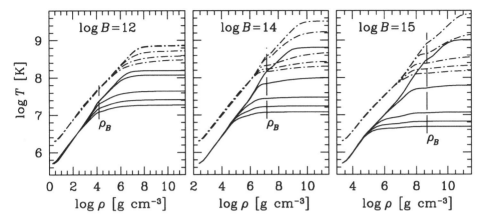

Figure 3. Temperature profiles through an iron envelope of a NS with $M = 1.4\,M_\odot$, $R = 10$ km, and with magnetic field $B = 10^{12}$ (left), 10^{14} (middle), or 10^{15} G (right panel). In every case, effective surface temperature was fixed to $T_e = 5 \times 10^5$ K (solid lines) or 2×10^6 K (dot-dashed lines). Different lines of each bunch correspond to different values of $\cos\theta$: 1 (the lowest line of a bunch), 0.7, 0.4, 0.1, and 0 (the highest line).

Beyond the turning point. The integration beyond the turning point in order to obtain T_i can be done in the same way as in the non-magnetic case. However, since the turnover occurs in the strongly quantizing field, the integration path should be divided in two parts: below ρ_B, where Eq. (38) should be used for the conductivity, and above ρ_B, where one can use Eq. (25) with the right-hand side divided by $\cos^2\theta$. The result is similar to Eq. (27), but contains a profound dependence on the inclination angle: the thermal gradient grows rapidly as θ increases toward $\pi/2$.

This dependence is illustrated in Figure 3, where we have shown several temperature profiles calculated numerically (Potekhin 2000). The curves start at the radiative surface, where $T = T_e$, and end near the neutron drip point. The onset of the profound θ-dependence signals the turning point. According to our estimates above, this point is shifted to ever higher densities with increasing B. For the radiative boundary, the above-mentioned linear dependence $\rho_s \propto B$ is confirmed. Another interesting effect is that T may continue to grow up to the bottom of the outer envelope, well beyond the "canonical" limiting density $\rho = 10^{10}$ g cm^{-3}.

4. Conclusions

In the preceding sections we have given a simplified analytic discussion of the neutron star outer envelope. It was possible in this simple picture to recover main features of more careful numerical calculations, such as the sensitivity of the temperature ratio T_i/T_e to chemical constitution and the general behavior of the (T, ρ) profile.

Magnetic fields have a strong influence on the radiative and conductive opacities in NS envelopes, as we have seen in the last section. It is remarkable though that in the region of the magnetic polar cap, in spite of the sharply reduced radiative opacities in the outer layers, the temperature ratios T_t/T_e, and T_i/T_e are only mildly affected. The effect of the reduced opacity is partly counterbalanced by the increased mass of the radiative layer. Furthermore, when the magnetic field is inclined to the surface, the temperature gradient is increased because the electron conduction is efficient mainly along the field lines. Clearly more work is still necessary to fully analyze all these details.

The possibility of magnetars, or extremely highly magnetized NS, in our Galaxy has been highlighted by recent observations of soft gamma repeaters (Kouveliotou 1998, 1999; Thompson 2000) and the special class of anomalous X-ray pulsars (e.g., Mereghetti 2000). Understanding the role of high magnetic fields in the observable properties of such objects has thus become an important theoretical task. Recent observations of compact galactic objects further suggest rather small emission areas and high temperatures, which may be attributed to magnetars (e.g., Pavlov *et al.* 2000; Dar and de Rújula 2000). The presence of an accreted hydrogen layer on the magnetar's polar cap could allow for a rather sharp temperature contrast between the hot spot and the rest of the NS surface. This configuration has been invoked to interpret the observations in the case of the central compact object in the supernova remnant Cas A (Pavlov *et al.* 2000).

Acknowledgements

We thank Dima Yakovlev for useful remarks. J.V. is deeply indebted to Vladimir Usov and Motty Milgrom for their generous hospitality and partial support from the Einstein Center for Theoretical Physics at the Weizmann Institute of Science; he also acknowledges partial support from EC grant ERB-FMRX-CT98-0195. A.Y.P. gratefully acknowledges the generous hospitality of Gilles Chabrier at CRAL, Ecole Normale Supérieure de Lyon, and partial support from grants RFBR 99-02-18099 and INTAS 96-542.

References

Alpar, M.A., Cheng, K.S., and Pines, D. (1989), *Astrophys. J.* **346**, 823.
Braginskiĭ, S.I. (1957), *Zh. Eksp. Teor. Fiz.* **33**, 645.
Blandford, R.D. and Hernquist, L. (1982), *J. Phys. C* **15**, 6233.
Canuto, V. and Ventura, J. (1977), *Fundam. Cosm. Phys.* **2**, 203.
Caraveo, P., Bignami, G., and Trümper, J. (1996), *Astron. & Astrophys. Rev.* **7**, 209.
Chabrier, G. (1993), *Astrophys. J.* **414**, 695.
Chabrier, G. and Potekhin, A.Y. (1998), *Phys. Rev. E* **58**, 4941.

Chiu, H.Y. and Salpeter, E.E. (1964), *Phys. Rev. Lett.* **12**, 413.

Cox, J.P. and Giuli, R.T. (1968), *Principles of Stellar Structure*, Gordon and Breach (New York).

Dar, A. and de Rújula, A. (2000), astro-ph/0002014.

Gudmundsson, E.H., Pethick, C.J., and Epstein, R.I. (1983) (GPE), *Astrophys. J.* **272**, 286.

Halpern, J.P. and Ruderman, M. (1993), *Astrophys. J.* **415**, 286.

Hernquist, L. and Applegate, J.H. (1984) (HA84), *Astrophys. J.* **287**, 244.

Heyl, J.S. and Hernquist, L. (1998), *Mon. Not. R. Astron. Soc.* **300**, 599.

Iglesias, C.A. and Rogers, F.J. (1996), *Astrophys. J.* **464**, 943; http://www-phys.llnl.gov/V_Div/OPAL/opal.html.

Kouveliotou, C. et al. (1998), *Nature* **393**, 235.

Kouveliotou, C. et al. (1999), *Astrophys. J.* **510**, L115.

Lai, D. and Salpeter, E.E. (1997), *Astrophys. J.* **491**, 270.

Mereghetti S. (2001), in these proceedings.

Mészáros, P. (1992), *High Energy Radiation from Magnetized Neutron Stars*, U. of Chicago Press (Chicago).

Miller, M.C. and Neuhauser, D. (1991), *Mon. Not. R. Astron. Soc.* **253**, 107.

Page, D. (1997), *Astrophys. J.* **479**, L43.

Page, D. and Sarmiento, A. (1996), *Astrophys. J.* **473**, 1067.

Pavlov, G.G., Shibanov, Yu.A., Zavlin, V.E., and Meyer, R.D. (1995), in M.A. Alpar, Ü. Kiziloğlu, and J. van Paradijs (eds.), *The Lives of the Neutron Stars*, NATO ASI Ser C450, Kluwer (Dordrecht), p71.

Pavlov, G.G., Zavlin, V.E., Aschenbach, B., Trümper, J., and Sanwal, D. (2000), *Astrophys. J.* **531**, L53.

Pethick, C.J. (1992), *Rev. Mod. Phys.* **84**, 1133.

Potekhin, A.Y. (1999), *Astron. & Astrophys.* **351**, 787.

Potekhin, A.Y. (2000), in M. Kramer, N. Wex, and R. Wielebinski (eds.), *Pulsar Astronomy – 2000 and beyond*, ASP Conf. Ser. 202, p621.

Potekhin, A.Y., Chabrier, G., and Yakovlev, D.G. (1997) (PCY), *Astron. & Astrophys.* **323**, 415.

Potekhin, A.Y., Shibanov, Yu.A., and Ventura, J. (1998), in N. Shibazaki, N. Kawai, S. Shibata, and T. Kifune (eds.), *Neutron Stars and Pulsars*, Universal Academy Press (Tokyo), p161.

Potekhin, A.Y, Baiko, D.A., Haensel, P., and Yakovlev, D.G. (1999a), *Astron. & Astrophys.* **346**, 345.

Potekhin, A.Y., Chabrier, G., and Shibanov, Y.A. (1999b), *Phys. Rev. E* **60**, 2193.

Rajagopal, M., Romani, R., and Miller, M.C. (1997), *Astrophys. J.* **479**, 347.

Reisenegger, A. (1997), *Astrophys. J.* **485**, 313.

Romani, R.W. (1987), *Astrophys. J.* **313**, 718.

Ruderman, M.A. (1971), *Phys. Rev. Lett.* **27**, 1306.

Shapiro, S.L., and Teukolski, S.A. (1983) (ST83), *Black Holes, White Dwarfs, and Neutron Stars: The Physics of Compact Objects*, Wiley (New York).

Shibanov, Yu.A., Potekhin, A.Y., Yakovlev, D.G., and Zavlin, V.E. (1998), in R. Buccheri, J. van Paradijs, and M.A. Alpar (eds.), *The Many Faces of Neutron Stars*, Kluwer (Dordrecht), p553.

Silant'ev, N.A., and Yakovlev, D.G. (1980), *Astrophys. Space Sci.* **71**, 45.

Thompson, C. (2000), in these proceedings.

Thorne, K.S. (1977), *Astrophys. J.* **212**, 825.

Trümper, J., and Becker, W. (1997), *Adv. Space Res.* **21**, 203.

Tsuruta, S. (1998), *Phys. Rep.* **292**, 1.

Urpin, V.A., and Yakovlev, D.G. (1980), *Astrophysics* **15**, 429.

Ventura, J. (1989), in H. Ögelman and E.P.J. van den Heuvel (eds.), *Timing Neutron Stars*, Kluwer (Dordrecht), p491.

Yakovlev, D.G. (1984), *Astrophys. Space Sci.* **98**, 37.
Yakovlev, D.G., and Kaminker, A.D. (1994), in G. Chabrier and E. Schatzman (eds.), *The Equation of State in Astrophysics*, Proc. IAU Coll. 147, Cambridge U. Press (Cambridge, UK), p214.
Yakovlev, D.G., and Urpin, V.A. (1980), *Sov. Astron.* **24**, 303.
Yakovlev, D.G., Levenfish, K.P., and Shibanov, Yu.A. (1999), *Physics–Uspekhi* **42**, 737.

6. Gamma-ray Bursts

GAMMA-RAY BURSTS: HISTORY AND OBSERVATIONS

G.J. FISHMAN
Space Science Department, Code SD 50
NASA-Marshall Space Flight Center
Huntsville, AL 35812 USA

Abstract. Gamma-ray bursts (GRBs) are the most luminous objects known in the Universe. Their brief, random appearance in the gamma-ray region had made their study difficult since their discovery over thirty years ago. The discovery of counterparts to gamma-ray bursts and afterglow radiation in other wavelengths has provided the long-sought breakthrough in the direct determination of their distance and luminosity scales. The observed time profiles, spectral extent and durations of gamma-ray bursts are now well-sampled, with a catalog of over 2700 GRBs. There is a rich diversity in the duration and morphology of GRB time profiles. The spectra are characterized by a smooth continuum, usually peaking in the range from 0.1 MeV to 1 MeV. This applies to the integrated burst spectra as well as to spectra over limited temporal segments. Delayed gamma-ray burst photons extending to GeV energies have been detected. Since GRBs are produced at cosmological distances, they are expected to become unique tools for studying the conditions of the early Universe.

1. Introduction

This paper gives a brief history of gamma-ray bursts and provides a description of the observed gamma-ray burst characteristics. While gamma-ray bursts have been observed for over thirty years by a wide variety of instruments on many spacecraft, their detailed study was elusive because of their random and transient nature. Furthermore, for the first twenty years or so, GRB observations were often a secondary objective of the instruments which observed them. The detectors and their associated instrumentation did not have the necessary sensitivity, processing capability or telemetry rates needed to significantly advance the field. The Burst

C. Kouveliotou et al. (eds.), The Neutron Star – Black Hole Connection, 417–429.
© 2001 *Kluwer Academic Publishers. Printed in the Netherlands.*

and Transient Source Experiment (BATSE) (Fishman et al. 1994) on the
Compton Gamma-Ray Observatory, launched in April 1991, provided the
necessary large area and versatile data system to detect a large number of
GRBs. It also had sufficient temporal and spectral resolution so that statistical properties of large numbers of bursts, correlations, intensity and sky
distributions could be finally obtained from a large sample of bursts derived
over many years of operation from a single, well-calibrated instrument. Although the location accuracy of GRBs derived by BATSE was not sufficient
for follow-up observations by narrow-field instruments, it was good enough
to derive unique measurements of the GRB sky distribution and to be of
use to wide-field optical telescopes for counterpart observations. In particular, it allowed the simultaneous optical observation of a GRB by the
wide-field robotic telescope ROTSE to be observed on January 23, 1999
(Akerlof et al. 1999). It was not until the BeppoSAX spacecraft (Boella
et al. 1997) came into operation in 1996, that precise (arc-minute) GRB
locations could be provided within hours. This soon led to rapid follow-up
observations of GRBs with narrow-field instruments and a quick succession of breakthrough observations in 1997 (see Piro, these proceedings, for
a review of BeppoSAX GRB observations). The resulting X-ray, optical,
and radio discoveries of afterglow emission, the identification of host galaxies, and the measurements of the redshift of GRBs, have opened the field
to detailed study of GRB afterglows and the GRB host galaxies at other
wavelengths. Quite understandably, this has re-energized interest in GRBs
within the theoretical community, most of whom did not previously venture
into an astrophysical realm where not even the distance to the objects under study was known within many orders of magnitude. Gamma-ray bursts
are recognized as the most luminous objects observed in the Universe. Now
that there are several dozen afterglows observed in other wavelength regions, and many of them have well-identified host galaxies associated with
them, it is becoming apparent that GRBs will be able to provide a powerful
tool for the study of the most distant and oldest regions of the Universe.

2. Discovery and Early Observations

It is quite likely that the problem of gamma-ray bursts will be considered as
one of the greatest mysteries in astronomy of the twentieth century, leading
to the recognition of a new class of objects. The gamma-ray bursts were
discovered by accident in the late 1960's by the Vela series of spacecraft.
These spacecraft were developed by the US Department of Defense and were
designed to detect clandestine nuclear test detonations in space that could
be detonated by other countries. These widely separated spacecraft carried
a large number of sensors which could provide signatures of thermonuclear

explosions. While the program itself was unclassified, the precise capabilities of the spacecraft and its sensors were classified. From the spacecraft data, it was noticed that spacecraft were simultaneously detecting bursts of gamma rays which were not coming from the direction of the earth or any other object in the solar system. These observations required considerable analysis before it could be determined for certain that they were truly of cosmic origin, and not instrumental or local in origin. After several years the first observations were published in The Astrophysical Journal in 1973 (Klebesadel et al. 1973). Although gamma-ray bursts are now being studied by hundreds of scientists in many countries of the world, comprehensive observations have been limited to relatively few satellite experiments and research groups. In the history of GRBs, three distinct distance scales, or luminosities, the GRBs have been postulated: 1) Nearby Galactic (100pc), 2) Distant Galactic, or Galactic halo (100kpc), and 3) Cosmological (3Gpc). In each of these, theorists have responded with vigor in developing detailed models of the energy source and emission mechanisms of the GRBs. Soon after the discovery of GRBs, many theories were postulated, but most of them could not be tested due to the scarcity of data and observed properties. At the Texas Symposium in 1974, Mal Ruderman gave a summary of some of the theories of GRBs which had been developed at that time (Ruderman 1975). Never before in modern astronomy had there been so great an uncertainty in the distance to a class of objects, as well as a similar level of uncertainty about the basic nature of the source and the emission mechanism. By the end of the 1970s, interest in GRBs began to wane, as new observational data were not forthcoming. In the 1980s, with the purported detection of line features in several GRBs, the high-energy astrophysical community generally believed that the GRBs were due to violent events at, or near, the surface of relatively nearby galactic neutron stars. Several conferences were held in which detailed emission models were described that were capable of producing high-energy photons with high time variability (see, for example, Harding 1991). In retrospect, it appears that the initial claims of line features were not entirely justified and that model-dependent deconvolution of the data perhaps misled some into overestimating the significance of the data. The early BATSE era (1991-1995) split the astrophysics community into two main, but quite separate camps. One group held on to the galactic neutron star hypothesis which had been modeled in great detail in numerous papers. This group had several strong and vocal supporters. Another group, initially relatively small in number, shifted to a view that the GRBs were from cosmological distances. Most were not convinced of either interpretation, and waited until more data were available.

In order to fit within the BATSE observations of an isotropic GRB dis-

2704 BATSE Gamma-Ray Bursts

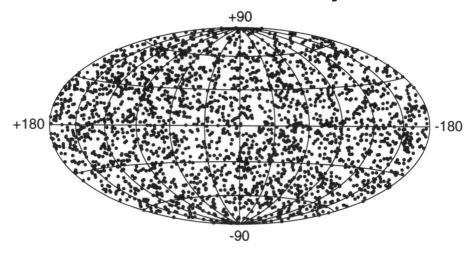

Figure 1. The sky distribution of the 2704 GRBs seen with BATSE

tribution (Meegan et al. 1992), even for those weak GRBs that showed an inhomogeneity, or confinement, the observed GRB distribution was reconciled by placing the GRB sources in a very large, uniform (homogeneous) halo. Such a distribution was not observed to exist for any other galactic component, but it was given some credibility by the observation of high-velocity neutron stars, presumably ejected from their birthplace in the galactic plane with a high velocity at their birth. By 1996, with over 1400 GRBs observed with BATSE, the isotropy held up to very strict limits. It was noted that to have such an isotropy with neither a detectable dipole or quadrupole moment was an extremely unlikely coincidence, at greater than the 3σ level. The required halo radius to meet the lack of observed dipole was nearly 200 kpc, with severe limitations on the allowable radial density gradient. No known galactic components were at that distance, nor was there observed to be any clustering of GRBs around members of the local group of galaxies. Furthermore, the energetics of a GRB at such large galactic distances required emission processes greater than those capable of being produced by most previous neutron star models. At about this time, most workers in the field were beginning to accept the new paradigm that objects at cosmological distances produced GRBs. The complete sky distribution of 2704 BATSE GRBs, plotted in Galactic Coordinates, is shown in Figure 1. This figure has not been corrected for sky exposure, which reduces the density of GRBs near the celestial equator. In 1997, BeppoSAX was able to quickly and accurately locate, for the first time, the position of a GRB so that rapid follow-up observations could be made (Costa et al. 1997). Soon afterward, a fading optical counterpart was discovered (van Paradijs et al. 1997), a radio counterpart was discovered (Frail et al. 1997) and a redshift was measured (Metzger 1997). A new and exciting era had begun in GRB research.

3. Temporal Properties

Perhaps the most striking features of the time profiles of gamma-ray bursts are their morphological diversity and the large range of burst durations. Coupled with this diversity is the inability to place many gamma-ray bursts into well-defined classes based on their time profiles, since mixes of classes are often observed. The durations of gamma-ray bursts range from about 10ms to over 1000s in the energy range in which most bursts are observed. However, EGRET observations show high energy (> 100 MeV) emission over 90 minutes after a burst trigger (Hurley, et al. 1994). Sub-millisecond structure has been detected in several bursts (e.g. Bhat et al. 1992). A sample of GRB profiles can be found in the BATSE catalog, available on the Internet (http://gammaray.msfc.nasa.gov) and in the GRB review article by Fishman and Meegan (1995). A cursory examination of burst profiles indicates that some are chaotic and spiky with large fluctuations on all time-scales, while others show rather simple structures with few peaks. However, some bursts are seen with both characteristics present within the same burst. No periodic structures have been seen from gamma-ray bursts. The duration of a gamma-ray burst is difficult to quantify since it is dependent on the intensity and background and is also somewhat dependent upon the time resolution of the experiment. The BATSE group has settled on a T-90 measure, the time over which 90% of the burst fluence is detected. Another approach is taken by Mitrofanov et al. (1999) who define an emission-time measure of GRBs that includes only time intervals when the emission is above a pre-defined level. The distribution of durations is shown in Figure 2, based on the T-90 measure for the 1637 bursts in the BATSE 4B Catalog (Paciesas et al. 1999). The shorter bursts are also seen to have harder spectra, as measured by a hardness ratio (Kouveliotou et al. 1993). The clustering in the hardness-duration diagram is seen in Figure 3, with two evident distinct classes of GRBs. In this logarithmic distribution there are two broad, un-resolved peaks at about 0.3s and 20s and a minimum at around 2s. The fall-off at short durations is due to an instrumental bias, since the minimum time-scale over which the BATSE experiment can trigger on bursts is 64ms. Recently, renewed evidence for a hard X-ray afterglow, decaying as a power-law, has been seen which overlaps

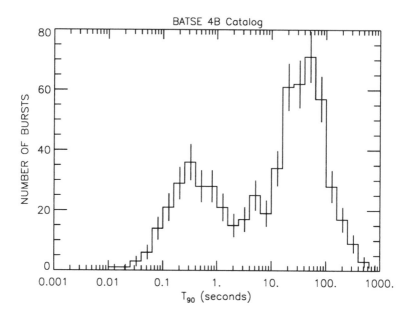

Figure 2.   The duration distribution of GRBs from the BATSE 4B Catlog

and merges into the main part of the GRB emission (Giblin et al.1999; Connaughton 2000). Figure 4 shows the lightcurve for the most prominent example of this phenomenon, GRB 980923.

No periodic structures have been seen from gamma-ray bursts. There is another general characteristic of the time profiles. At higher energies, the overall burst durations are shorter and sub-pulses within a burst tend to have shorter rise-times and fall-times (sharper spikes). Most bursts also show an asymmetry, with shorter leading edges than trailing edges. This has been quantified by Link, Epstein and Priedhorsky (1993) and by Nemiroff et al. (1994). These authors use this observation as an argument for an explosive event rather than a sweeping, beamed event, since the latter would, in general, not show such an asymmetry. However, some bursts show no asymmetry, even when high counting rates would permit such measurements to be made. There are many bursts that have similar shaped sub-pulses within the burst.

4. Spectral Properties

The unique feature of gamma-ray bursts is their high-energy emission: almost all of the power is emitted above 50 keV. Some bursts show emission as low as 1 keV, but the power is less than 1 or 2 % of the total power. Most bursts show rather simple continua spectra which appear similar in

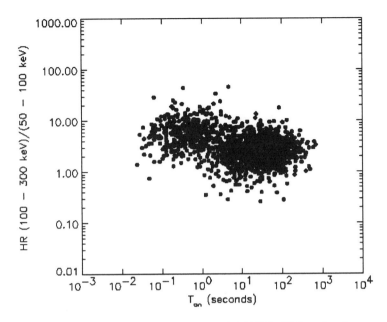

Figure 3. The hardness-duration distribution of BATSE GRBs, showing two distinct classes of GRBs; the shorter bursts tend to have a harder spectrum. The Hardness Ratio (HR) is the ratio of the two middle, broad BATSE energy bands (as indicated), integrated over the duration of the GRB. The T-90 measure of the GRB duration is the standard measure used in the BATSE catalogs.

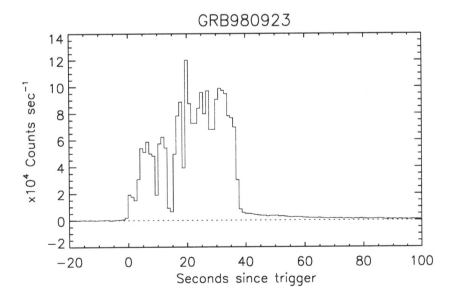

Figure 4. Extended long, soft emission seen in GRB 980923 following a period of intense, harder pulses. (Giblin et al. 1999)

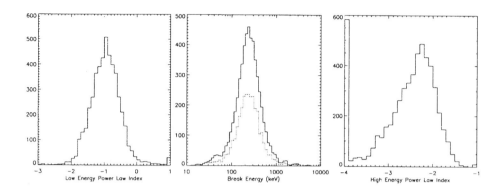

Figure 5. Distribution of Band function spectral parameters for an ensemble of bright BATSE bursts. At left is the low-energy spectral index, the higher-energy index is shown on the right, with the distribution of break energies E_{peak} displayed in the center plot (Preece et al. 2000).

shape when integrated over the entire burst and when sampled on various times-cales within a burst. Most spectra are well fit by a relatively simple analytical function, which has become known as the Band expression (Band et al. 1993). Many GRB spectra are seen to have cut-offs in which there is no detectable emission above 300 keV (Pendleton et al. 1997). This phenomenon is seen both for the duration of 25% of all GRBs and also within some sub-pulses of individual GRBs.

It was known from early observations with the gamma-ray spectrometer on the Solar Maximum Mission that the high energy emission from some bursts follows the same power law to over 80 MeV (Share et al. 1988). The EGRET instrument on the Compton Gamma Ray Observatory has seen significant flux and power into the GeV energy range from at least six bursts. Many of these high energy photons are delayed with respect to the bulk of the lower energy emission (Dingus et al. 1994). An observation with EGRET has shown that GeV photons emitted from a gamma-ray burst region are observed over one hour following the burst GRB 940217 (Hurley et al. 1994). There was no observed emission at lower energies from the burst region at that time, as observed with the BATSE experiment. A single, delayed 18 GeV photon was recorded from the burst direction, which is the highest energy photon ever recorded from a gamma-ray burst. Recent reports of TeV photons from a GRB (Atkins et al. 2000) have yet to be confirmed. Other generalizations can be made with regard to burst continua spectra. Within most (but not all) bursts, there is a hard-to-soft spectral evolution, resulting in the lower energies peaking earlier (Pendleton et al. 1994; Ford et al. 1995). Significant spectral changes on time-scales as short as tens of milliseconds have been observed in some GRBs (Ford et al. 1995). The low energy characteristics (below 20 keV) of GRBs have been examined in detail in several recent papers (Preece et al. 1998a; Strohmayer et al 1998). This is a crucial region to test emission models, such as the synchrotron shock model and the thick Comptonization model. Preece et al. (1998a) find the low-energy spectral index of many gamma-ray bursts is steeper than that allowed by GRB synchrotron shock models. From BATSE data, the high-energy spectra of GRBs which are seen to extend above 1 MeV are best fit by a power-law with a spectral index of 2, but the number of GRBs for which these data are available is limited. The particularly intense GRB of 23 January 1999, has been well fit with the Band function using data from all four CGRO experiments, over a very large energy range (Briggs, et al. 1999a). Spectra within the BATSE range from 20 keV to 2 MeV have been fit to numerous GRBs (Preece et al. 1998b). In these spectra, both the low and high-energy ends of the broad spectral distribution are fit by a power-law within the BATSE bandpass. The peaks of these νF_ν spectra are referred to as the E_{peak} energy. The COMPTEL experiment on CGRO has observed the peak power of some GRBs to be in the MeV energy range (Hanlon et al. 1994). The Band spectral parameters distribution for a large number of GRBs observed with BATSE are shown in Figure 4 (from Preece et al. 2000). A correlation has been found to exist between the E_{peak} of a GRB and its total energy fluence. This is thought to be both an intrinsic property and a possible cosmological redshift signature (Lloyd, Petrosian and Mallozzi 2000). Spectral softening is usually (but not always) seen throughout a GRB and also in sub-pulses within a burst. However, in many cases the peaks are offset, with the softer peak delayed by up to a few seconds. Liang and Kargatis (1996) have noted a correlation between the spectral evolution of sub-peaks within a given GRB and the intensity of these sub-peaks. Spectral evolution within GRBs is another GRB property which must be considered in burst emission models but is usually not, in contrast to the well-modeled spectral evolution of the GRB afterglow emission. Crider et al. (1999) find that the spectral luminosity peak energy decays linearly with energy fluence in most GRBs. They also show that the low energy spectral form is in disagreement with synchrotron shock models of GRB emission. Other relatively simple correlations between spectral properties, pulse decay time and burst intensity have been found by recent analyses (e.g. Ryde and Svensson 2000). The reported observation of spectral line features in gamma-ray bursts, and their interpretation as cyclotron lines produced in the intense magnetic fields of neutron stars had been a primary reason for associating gamma-ray bursts with neutron stars. A search for line features (either absorption or emission features) with the detectors of BATSE-Compton Observatory has thus far been unable to confirm the earlier reports of spectral line features from gamma-ray bursts (Band et al. 1998; Briggs 1999).

5. A New Era in Gamma-Ray Burst Research

Ever since the initial discovery of gamma-ray bursts, there has been a quest to discover a counterpart at any other wavelength region before, during, or after the gamma-ray event. These searches have taken many forms, including searches for statistical associations of known objects with bursts with poorly known locations as well as searches of archival plates and other data bases for transient or unusual objects within the error boxes of well-determined burst locations. The arc-minute locations available from BeppoSAX within hours of a GRB proved to be the decisive factor leading to counterpart identifications, afterglow observations, host galaxy identifications, and redshift measurements. These discoveries occurred in rapid succession in 1997 and 1998. Precise locations and the identification of afterglows have also led to an identification of the host galaxies of GRBs in almost all cases. These host galaxies are observed to be very faint; most are in the magnitude range (R band) 21 – 26. The Inter-Planetary Network (IPN) of spacecraft has now returned to a widely-spaced, three-element configuration with the Ulysses spacecraft (Hurley 1992) as the most distant node of the network. Rapid data retrieval has, for the first time, permitted a counterpart identification solely by means of the IPN. BATSE GRB locations were relatively coarse, being 1.8° for a strong burst. The BATSE locations, as measured against IPN locations (Briggs et al. 1999b; Hurley et al. 1999) have been steadily improving as various systematic errors have improved (Pendleton et al. 1999). In its final 2 years of operation, BATSE had a quick alert capability which was developed to provide burst locations. The near-realtime burst location system which was developed utilizing BATSE data, BACODINE (BAtse COordinates DIstribution NEtwork) (Barthelmy et al. 1994), has been in operation for over five years, and is now known as the Global Coordinates Network (GCN) owing to its incorporation of numerous spacecraft other than CGRO. The GCN system can provide gamma-ray burst locations to external sites within seconds of their detection by spacecraft such as HETE–2. When GCN is linked to a rapid-slewing optical telescope, there is the possibility of obtaining optical images of burst regions while the burst is in progress. This capability was used successfully for the first time in 1999, when the ROTSE system identified the optical counterpart of a BATSE-detected GRB, while it was in progress (Akerlof et al. 1999). This identification (within the very wide-field images) was made possible through the fortuitous localization of the burst by BeppoSAX, after the images were obtained. A manual burst response system was also developed to provide more accurate burst locations than the fully-automated GCN within about fifteen minutes. The Rossi X-ray Timing Explorer (RXTE) used this capability to scan a region in search of X-ray afterglow emission. The first set of counterparts derived from a GRB location provided by BATSE-RXTE-PCA was also made in 1999 (Kippen et al. 1999; Greiner 1999, and references therein).

6. Collapsars, Merging Compact Objects, and Fireballs

Most of the recent emission models for gamma-ray bursts postulate a highly relativistic fireball or jet expanding from the central object with a Lorentz factor of several hundred. Shocks within this fireball result in the prompt hard radiation, while the observed afterglows result from interactions of the relativistic material with external material which was present in the surrounding medium prior to the explosion (Rees and Meszaros 1994; Sari and Piran 1997). The details of the central engine are highly speculative, since the observed emission from the relativistic material is largely independent of the energy source. The highly relativistic, external shock will exhibit characteristic beaming and temporal structures which can uncover properties of the medium and its interaction with the blast wave (Dermer, Bottcher and Chiang 1999; Rhoads 1999; Dermer and Mitman 1999). The two most common classes of models for the central engine, or energy source, involve either the disruption of a single massive star, initiated by a central collapse (Woosley 1995; Paczyński 1998), or the coalescence (sometimes referred to as a collision) of two compact objects, such as neutron stars (Ruffert and Janka 1999, and references therein). The former object has been referred to as a collapsar or hypernova. A newly-formed black hole with a nuclear density torus, accreting at a high rate may be formed in both scenarios with sufficient energy to power the burst (Popham, Woosley and Fryer 1999). Another possible energy source for GRBs has been postulated to come from the conversion of an extremely intense magnetic field into kinetic energy. The field is thought to be produced soon after the formation of a neutron star with high rotational energy (Usov 1998; Kluzniak and Ruderman 1998).

The total energy of a GRB, calculated by various means, is estimated to be in the range $10^{52} - 3 \times 10^{53}$ erg (Schmidt 1999; Freedman and Waxman 2001) It is now recognized that GRBs, as the most luminous known objects in the Universe and observable to high redshifts, can become an important probe of star-forming regions in the early Universe (Lamb and Reichart 2000). Future GRB observations will be of interest not only in their own right, but they will also become important tools for the study of the early

Universe, just as supernovae have become in recent years.

7. Current Data and Future Observations

BATSE GRB data, now publicly available, represent the largest burst dataset obtained with a single instrument. As such, it will continue to be analyzed for many years. At least for the next decade, it will be the largest source of GRB data in the energy range from 20 keV to 2 MeV. The BeppoSAX and HETE–2 missions will continue to provide additional GRB counterpart and afterglow observations. The SWIFT mission, with a planned launch data of 2003, will usher a new era in GRB observations whereby hundreds of GRBs will be precisely and rapidly located. SWIFT has a primary detector that is much more sensitive than BATSE, although with a limited field-of-view of 3 steradians. This unique spacecraft will be capable of rapid slewing in order to rapidly observe afterglow emission with X-ray, UV and optical telescopes.

Acknowledgements

The author is grateful to V. Connaughton for her help with the preparation of this paper.

References

Akerlof, C. et al. (1999), *Nature* **398**, 400.
Atkins, R. et al. (2000), *ApJ* **533**, L119.
Band, D., Matteson, J., Ford, L. et al. (1993), *ApJ* **413**, 281.
Band, D. et al. (1998), C.A. Meegan, R.D. Preece, T. Koshut (eds.) *Proc. 4th Huntsville GRB Symposium*, AIP CP428, p329.
Barthelmy, S., Cline, T., Gehrels, N., Bialas, T. et al. (1994), in G. Fishman, J.J. Brainerd, and K. Hurley (eds.), *Proc. 2nd Huntsville GRB Symposium*, AIP CP307, p643.
Bhat, P.N. et al. (1992), *Nature* **359**, 219.
Boella, G. et al. (1997), *A& A Supp.* **122**, 299.
Briggs, M.S., Band, D.L. et al. (1999a), *ApJ* **524**, 82.
Briggs, M., Pendleton, G., Kippen, M. et al. (1999b) *ApJ Supp.* **122**, 503.
Briggs, M.S. (1999), in J. Poutanen and R. Svensson (eds.), *Gamma-Ray Bursts: The First Three Minutes*, ASP Conf.Ser. 190, p123.
Connaughton, V. (2000), *ApJ*, accepted for publication.
Costa, E. et al. (1997), *Nature* **387**, 783.
Crider, T. et al. (1999), *ApJ* **519**, 206.
Dermer, C. and Mitman, K.E. (1999), *ApJ* **513**, L5.
Dermer, C.D., Bottcher, M., and Chiang, J. (1999), *ApJ* **515**, L49.
Dingus, B.L. et al. (1994), in G. Fishman, J.J. Brainerd, and K. Hurley (eds.), *Proc. 2nd Huntsville GRB Symposium*, AIP CP307, p22.
Fishman, G.J. et al. (1994), *ApJ Supp.* **92**, 229.
Fishman, G. and Meegan, C. (1995), *Ann. Rev. Astron. Astrophys.* **33**, 415.
Ford, L.A., Band, D.L. et al. (1995), *ApJ* **439**, 307.
Frail, D.A. et al. (1997), *Nature* **389**, 261.
Freedman, D.L. and Waxman, E. (2001), *ApJ* **547**, 922.

Giblin, T. et al., (1999), *ApJ* **524**, L47.
Hanlon, L.O., Bennett, K. et al. (1994), *A&A* **285**, 161.
Harding, A.K. (1991), *Phys. Reports* **206**, 327.
Hurley, K. et al. (1992), *A&A Supp.* **92**, 401.
Hurley, K. (1994), *Nature* **372**, 652.
Hurley, K. et al. (1999), *ApJ Supp.* **122**, 497.
Kippen, R.M. et al. (1999), *GCN Circular* **463**.
Klebesadel, R.W. et al. (1973), *ApJ* **182**, L85.
Kluzniak, W. and Ruderman, M. (1998), *ApJ* **505**, L113.
Kouveliotou, C., Meegan, C., Fishman, G. et al. (1993), *ApJ* **413**, L101.
Lamb, D.Q. and Reichart, D.E. (2000), *ApJ* **536**, L1.
Liang, E. and Kargatis, V. (1996), *Nature* **381**, 49.
Link, B., Epstein, R.I., and Priedhorsky, W.C. (1993), *ApJ* **408**, L81.
Lloyd, N.M., Petrosian, V., and Mallozzi, R.S. (2000), *ApJ* **534**, 227.
Meegan, C., Fishman, G., Wilson, R., Paciesas, W., Pendleton, G. et al. (1992), *Nature* **355**, 143.
Metzger, M.R. et al. (1997), *Nature* **387**, 878.
Mitrofanov, I.G. et al. (1999), *ApJ* **522**, 1069.
Nemiroff, R.J. et al. (1994), *ApJ* **423**, 432.
Paciesas W. et al. (1999), *ApJ Supp.* **122**, L465.
Paczyński, B. (1998), *ApJ* **494**, L45.
Pendleton, G.N. et al. (1994), *ApJ* **431**, 416.
Pendleton, G.N. et al. (1997), *ApJ* **489**, 175.
Pendleton, G.N. et al. (1999), *ApJ* **512**, 362.
Piro, L. (2001), these proceedings.
Popham, R., Woosley, S.E., and Fryer, C. (1999), *ApJ* **518**, 356.
Preece, R.D., Briggs, M.S., Mallozzi, R. et al. (1998a), *ApJ* **506**, L23.
Preece, R.D., Pendleton G.N., Briggs, M.S. et al. (1998b), *ApJ* **496**, 849.
Preece, R.D., Briggs, M.S., Mallozzi, R.S. et al. (2000), *ApJ Supp.* **126**, 19.
Rees, M.J. and Meszaros, P. (1994), *ApJ* **430**, L93.
Rhoads, J. (1999), *ApJ* **525**, 737.
Ruderman, M. (1975), *Ann. N.Y. Acad. Sci.* **262**, 164.
Ruffert, M. and Janka, H.-Th. (1999), *A&A* **344**, 573.
Ryde, F. and Svensson, R. (2000), *ApJ* **529**, L13.
Sari, R. and Piran, T. (1997), *MNRAS* **287**, 110.
Schmidt, M. (1999), *ApJ* **523**, L117.
Share, G. et al. (1988), *Adv. Sp. Res.* **6**, 15.
Strohmayer, T., Fenimore, E., Murakami, T., and Yoshida, A. (1998), *ApJ* **500**, 873.
Usov, V.V. (1998), in J. Poutanen and R. Svensson (eds.), *Gamma-Ray Bursts: The First Three Minutes*, ASP Conf.Ser. 190, p153.
Van Paradijs, J. et al. (1997), *Nature* **386**, 686.
Woosley, S. (1995), *Adv.Sp.R.* **15**, 143.

THE AFTERGLOW OF GAMMA-RAY BURSTS: LIGHT ON THE MYSTERY

L. PIRO

Istituto di Astrofisica Spaziale - C.N.R.
Via Fosso del Cavaliere 100, I-00133, Roma, Italy
piro@ias.rm.cnr.it

Abstract. Gamma-ray bursts (GRB) have been one of the greatest mysteries of high energy astrophysics since their discovery in 1967. The fast and precise localization of bursts provided by BeppoSAX led to the discovery that the GRB continues to glow in X-rays and longer wavelenghts for days after the initial burst. Observations of afterglows of gamma-ray bursts are now providing a wealth of new information on these mysterious objects that, during the few seconds of their explosion, outshine any other object in the Universe. Along with their distance-scale, we are gathering information that should shed light on the origin of these explosions and their relationship with the environment of their host galaxy in the early phase of the Universe.

1. Introduction

The gamma-ray bursts are intense, short flashes of gamma-rays arriving from any direction in the sky, at unpredictable times. They usually emit most of their energy in the hard-X/gamma-ray range, but they have been detected from few keV up to tens of GeV. Gamma-ray bursts never come from the same direction of the sky, suggesting that they originate from a catastrophic phenomenon. A very few events actually violate this rule, and have been reclassified as *Soft Gamma-Ray Repeaters*. Their spectrum is markedly different from that of GRB's and they are produced in highly magnetized neutron stars (Kouveliotou et al. 1999).

GRB were discovered in 1967 by gamma-ray detectors onboard the military satellites *Vela* employed by USA to monitor potential Soviet and Chinese nuclear tests in the atmosphere, in violation of the 1963 Nuclear Test

C. Kouveliotou et al. (eds.), The Neutron Star – Black Hole Connection, 431–449.
© 2001 *Kluwer Academic Publishers. Printed in the Netherlands.*

Ban Treaty. Being a result of military investigations they were classified and announced to the scientific community only in 1973 (Klebesadel et al. 1973).

After their discovery several gamma-ray experiments were included in the scientific payload of subsequent satellites, and continued to detect GRBs. Today they are detected at a rate of one per day. Despite the many experiments and satellites devoted to the study of these elusive phenomena, a direct measurement of their distance (and therefore their luminosity) was obtained only in 1997 thanks to BeppoSAX [1] Before it, a key contribution to the study of GRBs has been gained by the launch in 1991 of the *Burst And Transient Source Experiment* (BATSE) (Fishman et al. 1994) onboard the *Compton Gamma-Ray Observatory*.

2. Global properties of gamma-ray bursts

2.1. TEMPORAL MORPHOLOGY

It is almost impossible to define a general morphology of the GRB time profiles. It changes from one to the other, passing from single pulse events to multiple pulses, through a wide variety of shapes and time-scales. The single pulse events can last from few ms to hundreds of seconds. The multiple events are usually longer than a few seconds. In Figure 1 an example of the zoo of the GRBs is shown, as detected by the Gamma-Ray Burst Monitor (Frontera et al. 1997; Feroci et al. 1997) onboard BeppoSAX.

Thanks to its high detection efficiency, the BATSE experiment has shown that there are (at least) two groups of GRB, defined by their duration, short (below 1-2 seconds) and long. This distribution has a bi-modal shape, with the two broad peaks at about 0.3 and 30 s (Fishman and Meegan 1995)

2.2. DISTRIBUTION AND POSSIBLE ORIGIN OF GRB

The distribution of gamma-ray bursts in the sky is highly isotropic. Figure 2 shows the distribution in galactic coordinates of the arrival directions of 2574 GRBs observed with BATSE (BATSE Web page). As can be seen by eye, and tested through statistical methods, there is no evidence of anisotropy in the angular distribution.

If GRB's are galactic, then they should be associated with a still unobserved class of objects. In fact, no known class of objects in our Galaxy has a similar distribution in the sky. For instance, young star populations are mainly distributed in the galactic plane. Moreover, given that our Solar

[1] BeppoSAX is a major program of the Italian Space Agency (ASI) with participation of the Netherlands Agency for Aerospace Programs (NIVR)

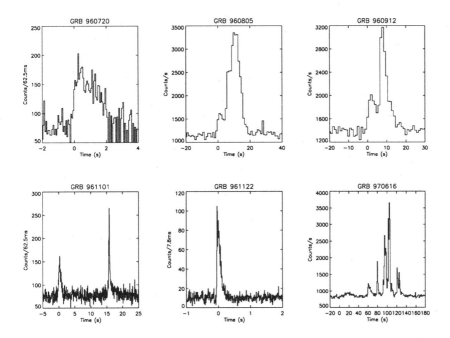

Figure 1. Sample of GRBs detected by the BeppoSAX/GRBM. It shows just an example of the possible morphology, duration and intensity variety of the GRBs

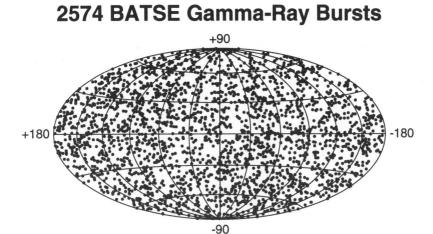

Figure 2. Sky distribution of the GRB arrival directions, as derived by BATSE

System is located about 9 kpc away from the center of the Galaxy (that has a diameter of about 23 kpc), then the high degree of isotropy implies that any galactic population should be in a shell with a minimum radius of 100 kpc from the center of the Galaxy. On the other hand, this halo cannot be too large, otherways we would observe a concentration of GRB towards the nearby Andromeda galaxy. This sets an upper limit at about 400 kpc (Hakkila et al. 1994).

Another piece of information is derived from their $\log N - \log S$ distribution. In case of homogeneous distribution of events throughout an Euclidean space (that is, a constant number density and luminosity distribution over the space), one would expect a power law distribution with index -3/2. However for low peak intensities the distribution bends from the -3/2 trend, due to a paucity of the weaker events (Hakkila et al. 1994). This is an indication that we may be observing the end of the homogeneous spatial distribution.

These evidences are in favour of the extragalactic origin of GRBs, which should reside in external galaxies at a redshift $z \simeq 1$. In fact, the galaxy distribution on large scale distances is isotropical. Furthermore, the paucity of faint events in the $\log N - \log S$ distribution is easily explained because their distance scale becomes comparable to the size of the observable Universe. In this case the intrinsic luminosity would be of the order of 10^{51} erg s^{-1}. This can only be obtained through catastrophic events, like the merging of two compact components of a binary system such as a neutron star-neutron star or a neutron star-black hole binary (Paczyński 1986). The rate needed for taking into account the observed distribution of GRBs is about 10^{-6} per galaxy per year (Fishman and Meegan 1995).

The problem of extragalactic models was to explain how such a huge luminosity in gamma-rays could be produced in a region whose size, inferred from the variability time scale, should have been at most few hundred light seconds. In this condition the region is optically thick to gamma-rays for pair production. The solution to this problem is the *fireball* model (Cavallo and Rees 1978; Rees and Mészáros 1992). The electromagnetic energy created by the initial explosion is converted, through pair-production, into kinetic energy of a small mass of baryons, which are accelerated to ultra-relativistic speeds. This expanding shell moves in an low density external medium and continues to do so until the mass of the swept up material reaches a critical value. At this point the kinetic energy is converted back into radiation. This should happen by shock acceleration of electrons that radiate X-rays (in the rest frame of the expanding shell) by synchrotron radiation. The high Lorentz factor Γ of the shell boosts these photons in the gamma-ray band. This happens at a distance r_{dec} much greater than the duration of the burst times the speed of light, such that the fireball is optically thin. The duration of this process is consistent with the typical duration of a GRB, because the observed timescale is $\sim r_{dec}/c\Gamma^2$.

In the case of a galactic origin, GRB's should be distributed in an extended halo around our Galaxy. In this case the observed intensities would imply an intrinsic luminosity of the order of 10^{42} erg s^{-1}. This amount of energy can be easily obtained from compact galactic objects (i.e. neutron stars) undergoing non-destructive phenomena, like internal reassessment or surface explosions. The most common scenario in this case would be a thermal cooling of the star surface after the GRB explosion.

3. Gamma-ray bursts in the Afterglow Era

3.1. THE BEPPOSAX MISSION

The BeppoSAX satellite (Piro, Scarsi, and Butler 1995; Boella et al. 1997) was launched on 30 April 1996 from Cape Canaveral by an Atlas-Centaur in an equatorial orbit (3.9° inclination) at an altitude over the Earth's surface of about 600 km. This orbit is particularly well suited for exploiting the shield of the geomagnetic field against the cosmic rays that are one of the primary sources of background in X-ray astronomy.

The spacecraft performs a complete orbit around the Earth in about 97 minutes. Once per orbit it passes over the ground station in Malindi (Kenya) and downloads all the data recorded onboard during the orbit, and uplinks all the telecommands needed for the operations in the next orbit. The entire radio-contact of the satellite with the ground station lasts about 10 minutes each orbit. Telecommands and data are relayed through Intelsat satellite to the Science Operation Center in Rome where they are analysed by Duty Scientists 24 hours a day.

A schematics of the scientific payload of the satellite is shown in Figure 3, were the single instruments are visible. They are basically divided in Narrow Field Instruments (NFI, with a field of view of about 1°, looking at the same direction in the sky):

- one Low Energy Concentrator Spectrometer (LECS, 0.1-10 keV) (Parmar et al. 1997);
- three Medium Energy Concentrator Spectrometers (MECS, 2-10 keV) (Boella et al. 1997b);
- one High Pressure Gas Scintillation Proportional Counter (HPGSPC, 4-100 keV) (Manzo et al. 1997);
- one Phoswich Detection System (PDS, 15-300 keV) (Frontera et al. 1997);

and Wide Field Instruments, looking at directions 90° from the NFI field of view:

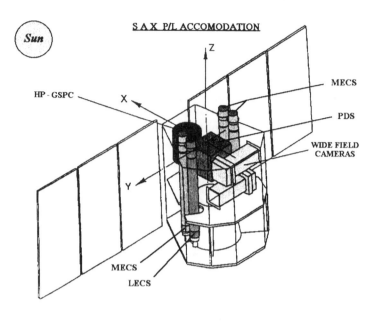

Figure 3. Schematics of the BeppoSAX satellite with the scientific instruments indicated. The GRBM is a part of the Phoswich Detection System, located in the core of the satellite.

- two Wide Field Cameras (WFC, 2-30 keV, field of view at zero response of 40°x40°) pointed at two opposite directions (Jager et al. 1997);
- one Gamma-Ray Burst Monitor (GRBM, 40-700 keV, open field of view, including the WFC field of view) (Frontera et al. 1997; Feroci et al. 1997).

The BeppoSAX instrumentation is therefore well matched both for wide band (more than three decades of energy, from 0.1 to 300 keV) pointed observations of celestial sources with good sensitivity, spatial and energy resolution, and for wide field monitoring of transient X/gamma events with the WFC and GRBM. The ground operation system has been designed to react promptly to X-ray transient sources discovered in the wide field instruments with fast follow-up observations with the more sensitive NFI (sensitivity $\sim 5 \times 10^{-14}$erg cm^{-2} s^{-1}).

For what regards specifically gamma-ray bursts, the limited positional capabilities of gamma-ray instrumentation has been overcome by coupling the Gamma-Ray Burst Monitor (GRBM) - which provides the temporal signature of the event (the so called *trigger*), with a Wide Field X-ray Camera (WFC), that provides a localization within few arcim in the X-ray band. It was indeed known that GRB's emit a small fraction ($\sim 10\%$) of their luminosity in the X-ray band (Yoshida et al. 1989). Therefore,

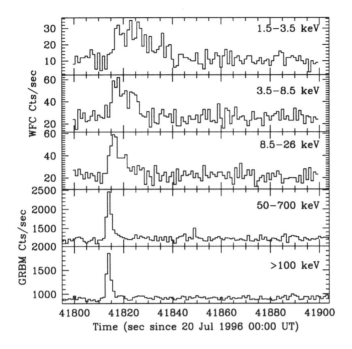

Figure 4. BeppoSAX GRBM and WFC light curves of GRB960720 (from Piro et al.1998)

whenever a GRB occurs within the field of view of one of the two WFCs, covering about 3% of the sky each, it can be promptly recognized by the simultaneous appearence of a burst in the light curves of both instruments (Figure 4) and localized by the WFC to a level of accuracy (about 3', see Figure 5) good enough to be an appropriate target for the NFI. All of this can happen in about 1 to 2 hours after the detection of the GRB onboard, because the event can happen in any part of the orbit. Soon after, the operations to point the NFI can start and the new pointing can be acquired within 3 to 10 hours. While the operations for pointing the BeppoSAX/NFI to the region of the sky in which the GRB was observed are in progress, an alert message with the coordinates of the GRB is sent to ground- and space-based telescopes working at other wavelength.

3.2. THE WIDE BAND SPECTRUM OF GAMMA-RAY BURSTS

An important feature of BeppoSAX, that has been shadowed by the break-trough results on afterglows, is its capability of providing wide band spectral information on the prompt emission of GRB, from X-rays to gamma-rays, by combining together data of the GRBM and of the WFC. In Figure 6 we

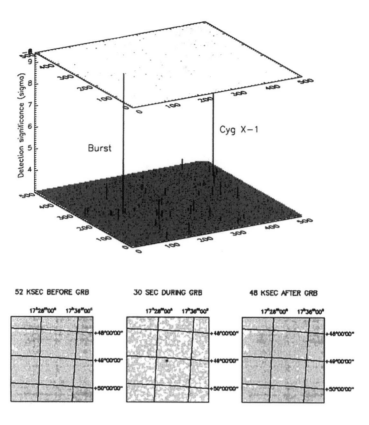

Figure 5. **a)** The 40° × 40° image in the 2 − 26 keV range of the WFC integrated over a period of 15 s on the burst. The y axis gives the significance of the detection (σ). The source close to the edge is Cyg X-1. **b)** Images of the field centered on GB960720 in time sequence. The first and last images were obtained integrating over ∼ 50,000s before and after the burst. No source was detected. The second image is a 30 s long shot which shows the sudden presence of GB960720 (from Piro et al.1998)

show the spectral evolution observed in GRB960720(Piro et al. 1998), the first GRB observed by BeppoSAX simultaneously in the WFC and GRBM and discovered in the off-line analysis of the data, during the verification phase. The vertical axis is νF_ν, that represents the energy per decade. In a few seconds the energy output moves from gamma-rays to X-rays. No known class of sources has ever show such an extreme spectral behaviour. The radiative mechanisms is likely synchrotron emission, but the paucity of X-ray photons in the early stage of the GRB emission deviates from simple optically-thin synchrotron models. This feature is present in a substantial fraction of the GRB population (> 50%) observed by BeppoSAX (Fron-

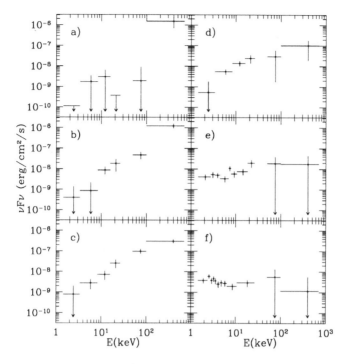

Figure 6.   GB960720: time evolution of the spectrum in νF_ν: a) to d): 1 s step; e): from 4 to 8 s; f): from 8 to 17 s (from Piro et al.1997)

tera et al. 2000) and BATSE (Preece et al. 1998) and it could be produced either by a thick medium (photoelectric absorption) or by synchrotron self-absorption. The origin of the large magnetic field that would be required to explain a self-synchrotron absorption in the X-ray band is unclear.

3.3.   THE FIRST AFTERGLOW : GB970228

Before the advent of BeppoSAX the GRB astronomy has proceeded on a statistical approach and no one had the idea of what could happen to a GRB emitter soon after the event. The operations for a prompt reaction to GRBs started on December 1996. The first opportunity was on 11 January 1997: GRB970111. The field was observed with the NFI 16 hours after the GRB. We were still considering the possible association of one of the faint sources found in the error box with the GRB (Feroci et al. 1998), when on February 28, 1997, another event, GRB970228, was detected by BeppoSAX GRBM and WFC. The NFI were pointed to the GRB location 8 hours after the burst. A previously unknown X-ray source was detected in the field of view of the LECS and MECS instruments with a flux in the 2-10 keV energy range of 3×10^{-12} erg cm^{-2} s^{-1}. The new source appeared to be fading away

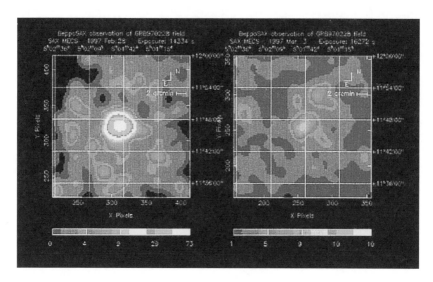

Figure 7. BeppoSAX MECS images of the GRB970228 afterglow, 8 hours after the GRB (left) and 3 days after the GRB (right) (from Costa et al.1998)

during the observation. On March 3 we performed another observation that confirmed that the source was quickly decaying : at that time its flux was a factor of about 20 lower than the first observation (Figure 7). This was the first detection of an "afterglow" of a GRB(Costa et al. 1997).

The flux of the source appeared to decrease following a power law dependence on time $(\sim t^{-\alpha})$ with index $\alpha = (1.3\pm0.1)$. Further X-ray observation with the X-ray satellites ASCA and ROSAT detected the source about one week later with a flux consistent with the same law (Yoshida et al. 1997; Frontera et al. 1997b). This kind of temporal behaviour agrees with the general predictions of the fireball models for GRBs (e.g. (Wijers et al. 1997; Mészáros and Rees 1997)). A backward extrapolation of this power law decay (Figure8) is consistent with the X-ray flux measured during the burst, suggesting that the afterglow started soon after the GRB. Another important result of the spectral analysis of the GRB970228 afterglow is that it seems to exclude a thermal origin of the emission, therefore suggesting that a model in which the radiation comes from the cooling of the surface of a neutron star cannot work in this case(?).

While the X-ray monitoring of GRB970228 was going on, an observational campaign of the same object was simultaneously started with the most important optical telescopes. This campaign led to the discovery (van Paradijs et al. 1997) of an optical transient associated with the X-ray afterglow. As in the X-ray domain, the optical flux of the source showed a decrease well described by a power law with index -1.12 (Garcia et al. 1998),

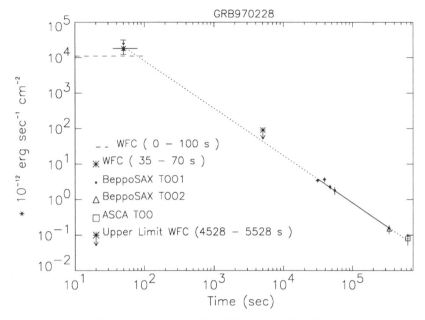

Figure 8.   BeppoSAX decay curve of the GRB970228 afterglow in the 2-10 keV energy range, obtained with the WFC and the NFI. The result of the ASCA observation is also shown at the bottom right(from Costa et al.1998).

again in agreement with the general predictions of the fireball model. The images taken with the Hubble Space Telescope (HST) (Sahu et al. 1997; Fruchter et al. 1997) showed the presence of a nebulosity around the optical transient. However the nebulosity was very weak, and it was not possible to disentangle whether it was associated with the host galaxy (extragalactic origin) or with a transient diffuse emission representing the residual of the explosion (galactic origin).

3.4. THE FIRST MEASUREMENT OF REDSHIFT: GRB970508

On 8 May 1997 the second breakthrough arrived with the burst GRB970508 (Piro et al. 1998b), detected just few minutes before the satellite was passing over the ground station in Malindi. This opportunity and the experience gained from previous events allowed us to point the BeppoSAX NFI 5.7 hours after the burst, while optical observations started 4 hours after the burst.

The early detection of the optical transient(Bond et al. 1997; Djorgovski et al. 1997; Sokolov et al. 1997; Pedersen et al. 1998; Castro-Tirado et al. 1997), and its relatively bright magnitude allowed a spectroscopical measurement of its optical spectrum with the Keck telescope(Metzger et al. 1997). The spectrum revealed the presence of FeII and MgII absorption lines at a redshift of $z = 0.835$, attributed to the presence of a galaxy between us and the GRB, and therefore demonstrated that GRB970508 was at a cosmological distance.

3.5. GAMMA-RAY BURSTS ARE NOT BRIEF AFTER ALL !

The BeppoSAX observation of GB970508 has also changed our view of the GRB *phenomenon*. The old concept of a brief - sudden release of luminosity concentrated in few seconds does not stand the new information provided by BeppoSAX. Indeed the name afterglow attributed to the X-ray emission observed after the event is somewhat misleading. This is clear when one considers the energy produced in the afterglow phase, which turns out to be comparable to that of the GRB. In order to compute the energy emitted in the afterglow phase it is necessary to integrate in time its luminosity and it is then crucial to know when the afterglow starts. A detailed analysis of the data of the WFC of GB970508 (Piro et al. 1998b) starting from 30 sec from the trigger, when the burst seemed to disappear in the light curve (Figure 9) shows instead that the source is still present(Figure 10) and remains visible for at least 2000 seconds, when the flux goes below the sensitivity of the WFC (Figure9). The energy emitted in the afterglow phase in X-rays turns out to be a substantial fraction ($\sim 40 - 50\%$) of the energy produced by the GRB.

Furthermore the light curve shows a rebursting event starting 1 day after the initial burst, an evidence that indicates that the source of the energy can re-ignite on long time scales (Piro et al. 1998b).

3.6. THE FIRST DIRECT MEASUREMENT OF THE RELATIVISTIC FIREBALL FROM RADIO OBSERVATION

GRB970508 was the first GRB in which a radio afterglow was discovered (Frail et al. 1997). In the first three weeks the radio source exhibited erratic variations with amplitude of a factor of 2 or more, after which the flux stabilized and then started to decrease. The large variations observed in the first weeks are attributed to *diffractive scintillation* (Taylor et al. 1997) of radio waves on interstellar electrons. This phenomenon is analogous to the optical twinkling of stars seen through the atmosphere, while planet images appear stable. In the first three weeks the angular size of the source was less than $3\mu as$, the typical value of the diffractive angle. Damping of the fluctuations later on indicates that the source has expanded to a value greater than the diffractive angle. At a redshift $z = 0.835$ (corresponding to a distance of 10^{28} cm), the angular size translates in a linear size of $\sim 10^{17}$ cm about three weeks post burst. This also indicates that the shell is moving with a relativistic speed.

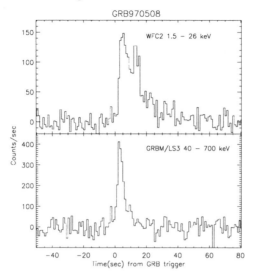

Figure 9. WFC and GRBM light curves of GRB970508 (from Piro et al.1998)

4. From the edge of the Universe the largest explosions since the Big Bang

The GRB observed by BeppoSAX on 14 December 1997, on one hand, has consolidated the extragalactic origing of GRB, on the other has underlined the problem of the energy budget and, ultimately, of the nature of the "central engine".

The chain of steps leading to the identification of the counterpart of GRB971214 (Dal Fiume et al. 1999) and its distance was the same of previous BeppoSAX observations. With a redshift z=3.42 (Kulkarni et al. 1998) this GRB and its host galaxy are at a distance that corresponds to a lookback time of about 85% of the present age of the Universe. At this distance, the luminosity would be about 3×10^{53} erg s^{-1}, were the emission isotropic. This was the highest luminosity ever observed from any celestial source. Initially the huge luminosity appeared to be not compatible with the energy available in the coalescence of neutron-star mergers (Kulkarni et al. 1998), unless beaming were invoked. Other alternative energetic models are based on the death of extremely massive stars, leading to an explosion orders of magnitude more energetic than a supernova, hence named *hypernova*

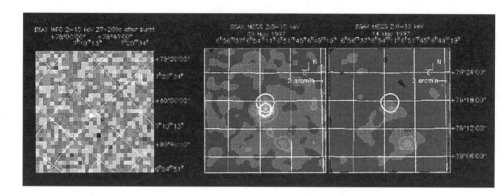

Figure 10. The X-ray afterglow of GB970508 (from Piro et al.1998): Time sequence of images of the field of GB970508 observed by the WFC2 (left image, 27–200 s after the burst), MECS(2+3) on May 9 (center image, 6 hours after the GRB), and MECS(2+3) on May 14 (after 6 days). The WFC2 show the presence of the afterglow that was then detected by the LECS and MECS (1SAXJ0653.8+7916 visible in the 99% error circle of the WFC). Note the decrease in intensity between the two MECS observations, as compared to 1RXSJ0653.8+7916, the source in the lower right corner

(Paczyński 1998). However it was shown (Mészáros and Rees 1998) that all these progenitors, whether Neutron Star - Neutron Star mergers or hypernovae, eventually go through the formation of a same Black Hole/torus system, from which the energy is extracted to form the GRB. The radiation physics and energy of all mergers and hypernovae are then, to order of magnitude the same, and still compatible with the luminosity observed in GRB971214.

The energy problem became much more severe with GRB990123, again one of the BeppoSAX GRB (Piro et al. 1999; Feroci et al. 1999; Heise et al. 1999). It was one of the brightest GRB ever observed, ranking in the 0.3% top of the BATSE flux distribution. Its distance (z=1.6 (Kulkarni et al. 1999)) would imply a total emitted energy of 1.6×10^{54} erg, assuming isotropical emission. This corresponds to $\sim 2M_\odot c^2$, at the limit of all models of mergers (Mészáros and Rees 1998). This piece of evidence is lending support to the idea that, at least in some cases, the emission is collimated. This would reduce the energy budget by $\sim \theta^2/4\pi$, where θ is the angle of the jet. A typical feature of a jet expansion (vs spherical) would be the presence of an achromatic (i.e. energy-independent) break in the light curve, that appears when the relativistic beaming angle $1/\Gamma$ becomes $\approx \theta$ (Sari, Piran, and Halpern 1999; Rhoads 1997). The presence of such a break has

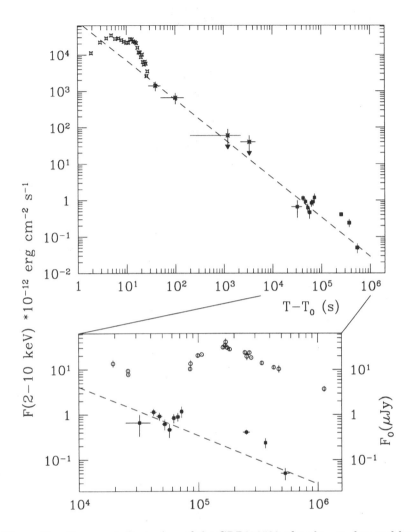

Figure 11.   Top panel: Decay law of the GRB970508 afterglow as detected by the Bep-poSAX WFC and NFI. The WFC provided the data up to 5.000 seconds after the burst. Bottom panel: enlargement of the X-ray decay law and comparison with the simultaneous time history of the optical transient (open circles) (from Piro et al.1998)

been claimed in GRB990123 (Kulkarni et al. 1999) and in another more recent case, GRB990510 (Harrison et al. 1999). With an angle $\theta \approx 10°$, the total energy would be reduced by $\approx 10^3$, therefore within the limits of the models. So far such evidence is limited to the optical range and an independent measurement confirming its presence in different regions of the spectrum is lacking or not conclusive (e.g. in X-rays (Kuulkers et al. 1999)).

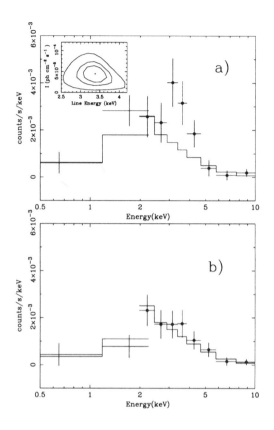

Figure 12. Spectra (in detector counts) of the X-ray afterglow of GRB970508 taken in the periods 0.2-0.6 day, (panel a) and 0.6-1 day (panel b) after the burst The continuous line represents the best fit power law. Note the feature around 3.5 keV in the first part (contour plots at 68%, 90%, 99% in the inset) that, at a redshift of 0.83, corresponds to the energy of the iron line complex (6.4-6.8 keV). The line disappears in the second part of the observation (panel b) (from Piro et al.1999)

4.1. THE NATURE OF THE PROGENITOR

Information on the nature of the progenitor can be drawn by studying the GRB environment. In the case of a hypernova progenitor the massive star should die young ($\approx 10^6$ years) and therefore GRBs should be preferentially hosted in regions near the center of star-forming galaxies. On the contrary, NS-NS coalescence happens on much longer time scales and the kick velocity given to the system by two consecutive supernova explosions should bring a substantial fraction of these systems away from the parent galaxy. So far the angular displacement of 5 optical counterparts indicates that GRBs are located within their host galaxies (Bloom et al. 1999b), favoring the association with star-forming regions. We note also that those events are not located in the very center of their galaxies, that excludes an association of GRB with AGN activity.

The other diagnostics of the progenitor is based on spectral measurements of broad and narrow features imprinted by a dusty - gas rich environment expected in the hypernova scenario (Perna and Loeb 1998; Mészáros and Rees 1998b; Böttcher et al. 1998). The absence of an optical transient in about 50% of well localized BeppoSAX GRB (18 as of May 99), in which instead an X-ray afterglow has been found in almost all the cases, may be explained by heavy absorption by dust in the optical range, which leaves almost unaffected the X-rays (Owens et al. 1998).

An exciting possibility is opened by the possible detection of X-ray iron line features in two different GRBs, one by BeppoSAX (GRB970508 (Piro et al. 1999b); Figure 12) and the other by ASCA (GRB970828 (Yoshida et al. 1999)), associated with rebursting on time scales of the order of a day. Both these temporal and spectral features betray the presence of a dense ($n \sim 10^{10}$ cm^{-3}) medium of $\approx 1 M_\odot$ near the site of the explosion ($\approx 10^{16}$ cm) (Piro et al. 1999b). Such a medium should have been pre-ejected before the GRB explosion, but the large value of the density excludes stellar winds. A possible, intriguing explanation is that the shell is the result of a SN explosion preceding the GRB (Piro et al. 1999b; Vietri et al. 1999; Vietri and Stella 1998).

Other evidence argues in favour of the association GRB-SN. In the BeppoSAX error box of GRB980425(Pian et al. 1999) two groups (Kulkarni et al. 1998b; Galama et al. 1998) found a supernova (SN1998bw) that had exploded at about the time of the GRB. The probability of a chance coincidence of the two events is $\approx 10^{-4}$. Since the majority of GRBs are not associated with SN (Graziani, Lamb, and Marion 1999), this event (if the association is true) should apparently represent an uncommon kind of GRB. However it is also possible that the two families are indeed associated: this scenario would require that the GRB are emitted by collimated jets. The majority of GRB and afterglow we see are beamed towards us, so that the contribution of the supernova to the total emission is negligible. The case of SN1998bw was then particular in that the jet producing the GRB was collimated away from our line of sight, allowing the detection of the (isotropic) SN emission at an early phase. This scenario also explains why GRB980425 was not particularly bright, notwithstanding its redshift (z=0.0085), much lower than the typical value of the other GRB ($z \approx 1$). Since the afterglow decays as a power law, it is possible that at late times the emission of the SN becomes detectable. Evidence of such emission has been claimed in at least two cases (GRB990326 (Bloom et al. 1999), GRB970228 (Reichart

1999))

5. Conclusions

The answer to the remaining problems should not wait too long. BeppoSAX is discovering and localizing GRB and X-ray afterglows at a pace of 1 per month. Other satellites have also set up successful procedures for rapid GRB localization (BATSE, XTE, ASCA and IPN). The launch of HETE–2, foreseen in early 2000 will increase substantially the number of well localized GRB. Furthermore near-future X-ray satellites, like Chandra, XMM, ASTRO-E will allow detailed spectral studies of X-ray afterglows and provide (Chandra) arcsec position of X-ray counterparts and, possibly, a direct redshift determination.

6. Acknowledgements

The BeppoSAX results presented here were obtained through the joint effort of all the members of the BeppoSAX Team.

References

See World Wide Web page: http://cossc.gsfc.nasa.gov/cossc/BATSE.html.
Bloom, J. et al.(1999), *Nature* **398**, 389.
Bloom, J.S. et al.. (1999), *Astrophys. J.* **518**, L1.
Boella, G. et al. (1997), *Astron. Astrophys. Suppl. Ser.* **122**, 299.
Boella, G. et al. (1997b), *Astron. Astrophys. Suppl. Ser.* **122**, 327.
Bond H. (1997), IAU Circular 6654.
Böttcher, M., Dermer, C.D., Crider, A.W., and Liang, E.D. (1998), *Astron. Astrophys.* **343**, 111.
Castro-Tirado, A.J. et al. (1997), *Science.*
Cavallo, G. and Rees, M.J. (1978), *MNRAS* **183**, 359.
Costa, E. et al. (1997), *Nature* **387**, 783.
Dal Fiume, D. et al.(1999), *Astron. Astrophys.* **355**, 454.
Djorgovski, G. et al. (1997), *Nature* **387**, 876.
Feroci, M. et al. (1997), *Proc. SPIE Conference* **3114**, 186;(astro-ph/9708168).
Feroci, M. et al.(1998), *Astron. Astrophys.* **332**, L29.
Fishman, G.J. et al. (1994), *Astrophys. J. Suppl. Ser.* **92**, 229.
Fishman, G.J. and Meegan, C.A. (1995), *Ann. Review Astron. Astrophys* **33**, 415.
Frail, D.A. et al. (1997), *Nature* **389**, 261.
Frontera, F. et al. (1997), *Astron. Astrophys. Suppl. Ser.* **122**, 357.
Frontera, F. et al. (1998), *Astron. Astrophys.* **334**, L69.
Frontera, F. et al.(2000), *Astrophys. J. Suppl. Ser.* **127**, 59.
Fruchter, A. et al. (1997), IAU Circular 6747.
Galama, T. et al.(1998), *Nature* **395**, 670.
Garcia, M.R. et al. (1998), *Astrophys. J.* **500**, L105.
Graziani, C., Lamb, D., and Marion, G.H. (1999), *Astron. Astrophys. Suppl. Ser.* **138**, 469.
Hakkila, J. et al. (1994), *Astrophys. J.* **422**, 659.
Harrison, F.A. et al.(1999), *Astrophys. J.* **523**, L21.

Jager, R. et al. (1997), *Astron. Astrophys. Suppl. Ser.* **125**, 557.

Klebesadel, R.W. et al. (1973), *Astrophys. J.* **182**, L85.

Kouveliotou, C. et al.(1999), *Astrophys. J.* **510**, L105.

Kulkarni, S.R. et al.(1998), *Nature* **393**, 35.

Kulkarni, S.R. et al.(1998), *Nature* **395**, 663.

Kulkarni, S.R. et al.(1999), *Nature* **398**, 389.

Kuulkers, E. et al.(1999), *Astrophys. J.* **538**, 638.

Manzo, G. et al. (1997), *Astron. Astrophys. Suppl. Ser.* **122**, 341.

Mészáros, P. and Rees, M.J. (1997), *Astrophys. J.* **476**, 319.

Mészáros, P. and Rees, M.J. (1998), *New Astronomy*;(astro-ph/9808106).

Mészáros, P. and Rees, M.J. (1998), *Mon. Not. R. Astr. Soc.* **299**, L10.

Metzger, M.R. et al. (1997), *Nature* **387**, 878.

Owens, A. et al.(1998), *Astron. Astrophys.* **339**, L37.

Paczyński, B. (1986), *Astrophys. J.* **308**, L43.

Paczyński, B. (1998), *Astrophys. J.* **494**, L45.

Parmar, A.N. et al. (1997), *Astron. Astrophys. Suppl. Ser.* **122**, 309.

Pedersen, H. et al. (1998), *Astrophys. J.* **496**, 311.

Perna, R. and Loeb, A. (1998), *Astrophys. J.* **501**, 467.

Pian, E. et al.(1999), *Astron. Astrophys. Suppl. Ser.* **138**, 463.

Piro, L., Scarsi, L., and Butler, R.C. (1995) in S. Fineschi (ed.), *X-Ray and EUV/FUV Spectroscopy and Polarimetry*, SPIE 2517, pp169-181.

Piro, L. et al. (1998), *Astron. Astrophys.* **329**, 906.

Piro, L. et al.(1998), *Astron. Astrophys.* **331**, L41.

Piro, L. et al.(1999), GCN 199; GCN 203; Feroci, M. et al.(1999), IAUC 7095; Heise, J. et al.(1999), submitted to *Nature* .

Piro, L. et al.(1999b), *Astrophys. J.* **514**, L73.

Preece, R.D. et al.(1998), in C.A. Meegan, R.D. Preece, T. Koshut (eds.) *Proc. 4th Huntsville Symposium*, AIP CP428, p319.

Reichart, D. (1999), *Astrophys. J.* **521**, L111.

Rees, M.J. and Mészáros, P. (1992), *Mon. Not. R. Astr. Soc.* **258**, 41.

Rhoads, J.E. (1997), *Astrophys. J.* **478**, L1.

Sari, R., Piran, T., and Halpern, J.P. (1999), *Astrophys. J.* **519**, L17.

Sahu, K.C. et al. (1997), *Nature* **387**, 476.

Sokolov, V.V. et al. (1998), in C.A. Meegan, R.D. Preece, T. Koshut (eds.) *Proc. 4th Huntsville Symposium*, AIP CP428, p525.

Taylor, G.B. et al. (1997), *Nature* **389**, 263.

Van Paradijs, J. et al. (1997), *Nature* **386**, 686.

Vietri, M. and Stella, L. (1998), *Astrophys. J.* **507**, L45.

Vietri, M. et al.(1999), *Mon. Not. R. Astr. Soc.* **308**, L29.

Wijers, R.A.M.J. et al. (1997), *Mon. Not. R. Astr. Soc.* **288**, L51.

Yoshida, A. et al.(1989), *Publs. Astron. Soc. Japan* **41**, 509.

Yoshida, A. et al. (1997), IAU Circular 6593.

Yoshida, A. et al.(1999), *Astron. Astrophys. Suppl. Ser.* **138**, 433.

GAMMA-RAY BURSTS AND AFTERGLOWS:
THE FIREBALL SHOCK MODEL

P. MÉSZÁROS

Dpt. of Astronomy & Astrophysics,
Pennsylvania State University,
University Park, PA 16802, USA

1. Introduction

Gamma-ray bursts (GRB) are detected about once a day, and while they are on, they outshine everything else in the gamma-ray sky, including the Sun. A major advance occurred in 1992 with the launch of the Compton Gamma-Ray Observatory, whose superb results were summarized in a review by Fishman & Meegan (1995; see also Fishman, these proceedings). The all-sky survey from the BATSE instrument showed that bursts were isotropically distributed, strongly suggesting a cosmological, or possibly an extended galactic halo distribution, with essentially zero dipole and quadrupole components. The spectra are non-thermal, typically fitted in the MeV range by broken power-laws whose energy per decade νF_ν peak is in the range 50-500 KeV (Band et al. 1993), the power law sometimes extending to GeV energies (Hurley et al. 1994). GRB appeared to leave no detectable traces at other wavelengths, except in some cases briefly in X-rays. However in early 1997 the Italian-Dutch satellite Beppo-SAX suceeded in providing accurate X-ray locations and images (Costa et al. 1997) which allowed their follow-up with large ground-based optical and radio telescopes, e.g. (van Paradijs et al. 1997). This paved the way for the measurement of redshift distances (Metzger et al. 1997; Djorgovski et al. 1998; Kulkarni et al. 1998), the identification of candidate host galaxies (Sahu et al. 1997; Bloom et al. 1998a; Odewahn et al. 1998), and the confirmation that they were at cosmological distances, comparable to those of the most distant galaxies and quasars ever measured.

The current interpretation of the gamma-ray and longer wavelength radiation is that the progenitor trigger produces an expanding relativistic fireball which can undergo both internal shocks leading to gamma-rays,

451

C. Kouveliotou et al. (eds.), The Neutron Star – Black Hole Connection, 451–466.

and (as it decelerates on the external medium) an external blast wave and a reverse shock producing a broad-band spectrum lasting much longer. A strong confirmation of the generic fireball shock model came from the correct prediction (Mészáros and Rees 1997a), in advance of the observations, of the quantitative nature of afterglows at longer wavelengths, in substantial agreement with the subsequent data (Vietri 1997a; Tavani 1997; Waxman 1997a; Reichart 1997; Wijers, Rees and Mészáros 1997). The measured γ-ray fluences imply a total energy of order $10^{54}(\Omega_\gamma/4\pi)$ erg, where $\Delta\Omega_\gamma$ is the solid angle into which the gamma-rays are beamed. Their energy output therefore needs to be stupendous. It is comparable to burning up the entire mass-energy of the Sun in a few tens of seconds, or to emit over that same period of time as much energy as our Milky Way does in a hundred years. This is somewhat reduced if collimation is indeed ocurring, evidence having been recently reported for this (Kulkarni et al. 1999; Fruchter et al. 1999; Castro-Tirado et al. 1999). In any case, such energies are possible (Mészáros and Rees 1997b) in the context of compact mergers involving neutron star-neutron star (NS–NS) or black hole-neutron star (BH–NS) binaries, or in hypernova/collapsar models involving a massive stellar progenitor (Paczyński 1998; Popham et al. 1999). In both cases, one is led to rely on MHD extraction of the spin energy of a disrupted torus and/or a central BH to power a relativistic outflow.

2. The Generic Fireball Shock Scenario

Whatever the GRB trigger is, the ultimate result must unavoidably be an $e^\pm, \gamma$ fireball, which is initially optically thick. The initial dimensions must be of order $r_{\min} \lesssim ct_{var} \sim 10^7$ cm, since variability timescales are $t_{var} \lesssim 10^{-3}$ s. Most of the spectral energy is observed above 0.5 MeV, hence the $\gamma\gamma \to e^\pm$ mean free path is very short. Many bursts show spectra extending above 1 GeV, indicating the presence of a mechanism which avoids degrading these via photon-photon interactions to energies below the threshold $m_e c^2 = 0.511$ MeV. The inference is that the flow must be expanding with a very high Lorentz factor Γ, since then the relative angle at which the photons collide is less than Γ^{-1} and the threshold for the pair production is diminished (Harding and Baring 1994). However, the observed γ-ray spectrum is generally a broken power law, i.e. highly nonthermal. In addition, the expansion would lead to a conversion of internal into kinetic energy, so even after the fireball becomes optically thin, it would be radiatively inefficient, most of the energy being kinetic, rather than in photons.

The simplest way to achieve high efficiency and a nonthermal spectrum is by reconverting the kinetic energy of the flow into random energy via shocks after the flow has become optically thin (Rees and Mészáros 1992).

Two different types of shocks may arise in this scenario. In the first case (a) the expanding fireball runs into an external medium (the ISM, or a pre-ejected stellar wind) (Rees and Mészáros 1992; Mészáros and Rees 1993; Katz 1994a; Sari and Piran 1995). The second possibility (b) is that (Rees and Mészáros 1994; Paczyński and Xu 1994), even before external shocks occur, internal shocks develop in the relativistic wind itself, faster portions of the flow catching up with the slower portions. This is a generic model, which is independent of the specific nature of the progenitor.

External shocks will occur in an impulsive outflow of total energy E_0 in an external medium of average particle density n_o at a radius

$$r_{\rm dec} \sim (3E_0/4\pi n_o m_{\rm p} c^2 \eta^2)^{1/3} \sim 10^{17} E_{53}^{1/3} n_o^{-1/3} \eta_2^{-2/3} \text{ cm} , \qquad (1)$$

and on a timescale $t_{\rm dec} \sim r_{\rm dec}/(c\Gamma^2) \sim 3 \times 10^2 E_{53}^{1/3} n_o^{-1/3} \eta_2^{-8/3}$ s, where $\eta = \Gamma = 10^2 \eta_2$ is the final bulk Lorentz factor of the ejecta. Variability on timescales shorter than $t_{\rm dec}$ may occur on the cooling timescale or on the dynamic timescale for inhomogeneities in the external medium, but generally this is not ideal for reproducing highly variable profiles (Sari and Piran 1998) (see however Dermer and Mitman 1999). It can, however, reproduce bursts with several peaks (Panaitescu and Mészáros 1998a) and may therefore be applicable to the class of long, smooth bursts.

In a wind outflow (Paczyński 1990), one assumes that a lab-frame luminosity L_0 and mass outflow $\dot{M}_0$ are injected at $r \sim r_1$ and continuously maintained over a time $t_{\rm w}$; here $\eta = L_0/\dot{M}_0 c^2$. In such wind model, internal shocks will occur at a radius (Rees and Mészáros 1994)

$$r_{\rm dis} \sim ct_{\rm var}\eta^2 \sim 3 \times 10^{14} t_{\rm var}\eta_2^2 \text{ cm}, \qquad (2)$$

on a timescale $t_{\rm w} \gg t_{\rm var} \sim r_{\rm dis}/(c\eta^2)$ s, where shells of different energies $\Delta\eta \sim \eta$ initially separated by $ct_{\rm v}$ (where $t_{\rm v} \leq t_{\rm w}$ is the timescale of typical variations in the energy at r_1) catch up with each other. In order for internal shocks to occur above the wind photosphere $r_{\rm ph} \sim \dot{M}\sigma_{\rm T}/(4\pi m_{\rm p} c\Gamma^2) = 1.2 \times 10^{14} L_{53}\eta_2^{-3}$ cm, but also at radii greater than the saturation radius (so the bulk of the energy does not appear in the photospheric quasi-thermal component) one needs to have $7.5 \times 10^1 L_{51}^{1/5} t_{\rm var}^{-1/5} \lesssim \eta 3 \times 10^2 L_{53}^{1/4} t_{\rm var}^{-1/4}$. This type of model has the advantage (Rees and Mészáros 1994) of allowing an arbitrarily complicated light curve, the shortest variation timescale $t_{\rm var} \gtrsim 10^{-3}$ s being limited only by the dynamic timescale at r_1, where the energy input may be expected to vary chaotically. Such internal shocks have been shown explicitly to reproduce (and be required by) some of the more complicated light curves (Sari and Piran 1998; Kobayashi et al. 1998; Panaitescu and Mészáros 1999d).

A potentially valuable diagnostic tool for the central engine of GRB is the power density spectrum (PDS). An analysis of BATSE light curves (Beloborodov et al. 1998) indicates that the logarithmic slope of the PDS between 10^{-2} and 2 Hz is approximately -5/3, and there is a cutoff of the average PDS above 2 Hz. Using a simple kinematical model for the ejection

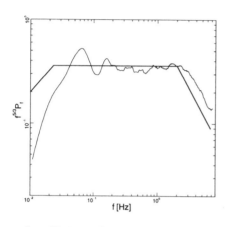

Figure 1. Average power density spectrum P_f of simulated bursts from internal shocks, compared with the observed PDS (thick line), using a square-sine modulated Lorentz factor and a cosmological distribution satisfying the observed logN-logP (Spada, Panaitescu & Mészáros 1999).

and collision of relativistic shells (Panaitescu, Spada and Mészáros 1999; Spada et al. 1999) have calculated the light curves and PDS expected for a range of total burst energies and for a total mass ejected and bulk Lorentz factor distribution compatible with the internal shock scenario (Figure 1). The redshift distribution also affects the PDS, and the observed logN-logP relation is used as a constraint. For optically thin winds, a slope approaching -5/3 requires a non-random Lorentz factor distribution, e.g. with an asymmetrical time modulation so as to produce a larger number of collisions at low frequencies (see also Beloborodov et al. 2000). A cutoff at high frequencies ($\sim$ 2 Hz) can be understood in terms of shocks which increasingly occur below the scattering photosphere of the outflow, or a deficit of energy in short pulses due to the modulation of the Lorentz factors favoring shocks arising further out.

A significant fraction of bursts appear to have low energy spectral slopes steeper than 1/3 in energy (Preece et al. 1998; Crider et al. 1997). This has motivated consideration of a thermal or nonthermal (Liang et al. 1997; Liang et al. 1999) comptonization mechanism, which can be put in the astrophysical context (Ghisellini and Celotti 1999) of internal shocks leading to self-regulated pair formation. There is also evidence that the clustering of the break energy of GRB spectra in the 50-500 keV range may not be due to observational selection (Preece et al. 1998; Brainerd et al. 1999; Dermer et al. 1999). Models using Compton attenuation (Brainerd et al. 1998) require reprocessing by an external medium whose column density adjusts itself to a few g cm^{-2}. More recently a preferred break has been attributed

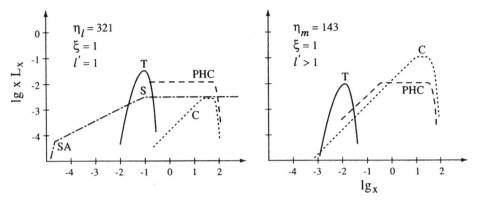

Figure 2. Luminosity per decade xL_x vs. $x = h\nu/m_ec^2$ for two values of $\eta = L/\dot{M}c^2$ and marginal (left) or large (rigt) pair compactness. T: thermal photosphere, PHC: photospheric comptonized component; S: shock synchrotron; C: shock pair dominated comptonized component (Mészáros & Rees 1999b).

to a blackbody peak at the comoving pair recombination temperature in the fireball photosphere (Eichler and Levinson 2000), the steep low energy spectral slope being due to the Rayleigh-Jeans part of the photosphere. In order for such photospheres to occur at the pair recombination temperature in the accelerating regime an extremely low baryon load is required. For very large baryon loads, a related explanation has been invoked (Thompson 1994), considering scattering of photospheric photons off MHD turbulence in the coasting portion of the outflow, which upscatters the adiabatically cooled photons up to the observed break energy. These ideas have been synthesized (Mészáros and Rees 1999b) to produce a generic scenario (see Figure 2) in which the presence of a photospheric component as well as shocks subject to pair breakdown can produce steep low energy spectra and preferred breaks.

3. Simple Standard Afterglow Model

The dynamics of GRB and their afterglows can be understood independently of any uncertainties about the progenitor systems, using a generalization of the method used to model supernova remnants. The simplest hypothesis is that the afterglow is due to a relativistic expanding blast wave, which decelerates as time goes on (Mészáros and Rees 1997a). The complex time structure of some bursts suggests that the central trigger may continue for up to 100 seconds, the γ-rays possibly being due to internal shocks. However, at much later times all memory of the initial time structure would be lost: essentially all that matters is how much energy and momentum has been injected; the injection can be regarded as instantaneous in the context of the afterglow. As indicated in Rees and Mészáros (1992), the external shock bolometric luminosity builds up as $L \propto t^2$ and decays as $L \propto t^{-(1+q)}$. Beyond the deceleration radius the bulk Lorentz factor decreases as a power law in radius, $\Gamma \propto r^{-g} \propto t^{-g/(1+2g)}$, $r \propto t^{1/(1+2g)}$, with $g = (3, 3/2)$ for the radiative or adiabatic regime (in which $\rho r^3 \Gamma \sim$ constant or $\rho r^3 \Gamma^2 \sim$ constant).

The synchrotron peak frequency in the observer frame is $\nu_m \propto \Gamma B' \gamma^2$, and both the comoving field B' and electron Lorentz factor γ are expected to be proportional to Γ (Mészáros and Rees 1993). As Γ decreases, so will ν_m, and the radiation will move to longer wavelengths. For the forward blast wave, Paczyński, B. and Rhoads, J. (1993) and Katz, J. (1994b) discussed the possibility of detecting at late times a radio or optical afterglow of the GRB. A more detailed treatment of the fireball dynamics indicates that approximately equal amounts of energy are radiated by the forward blast wave, moving with $\sim \Gamma$ into the surrounding medium, and by a reverse shock propagating with $\Gamma_r - 1 \sim 1$ back into the ejecta (Mészáros and Rees 1993). The electrons in the forward shock are hotter by a factor Γ than in the reverse shock, producing two synchrotron peaks separated by Γ^2, one peak being initially in the optical (reverse) and the other in the γ/X band (forward) (Mészáros and Rees 1993b; Mészáros and Rees 1994). Detailed calculations and predictions of the time evolution of such a forward and reverse shock afterglow model (Mészáros and Rees 1997a) preceded the observations of the first afterglow GRB970228 (Costa et al. 1997; van Paradijs et al. 1997), which was detected in γ-rays, X-rays and several optical bands, and was followed up for a number of months.

The simplest spherical afterglow model concentrates on the forward blast wave only. For this, the flux at a given frequency and the synchrotron peak frequency decay at a rate (Mészáros and Rees 1997a; Mészáros and Rees 1999a)

$$F_\nu \propto t^{[3-2g(1-2\beta)]/(1+2g)} \quad , \quad \nu_m \propto t^{-4g/(1+2g)}, \tag{3}$$

where g is the exponent of $\Gamma \propto r^{-g}$ and β is the photon spectral energy slope. The decay rate of the forward shock F_ν in eq.(3) is typically slower than that of the reverse shock (Mészáros and Rees 1997a), and the reason the "simplest" model was stripped down to its forward shock component only is that, for the first two years 1997-1998, afterglows were followed in more detail only after the several hours needed by Beppo-SAX to acquire accurate positions, by which time both reverse external shock and internal shock components are expected to have become unobservable. This simple standard model has been remarkably successful at explaining the gross features and light curves of GRB 970228, GRB 970508 (after 2 days; for early rise, see Section 4) (Wijers, Rees and Mészáros 1997; Tavani 1997; Waxman 1997a; Reichart 1997). (see Figure 3).

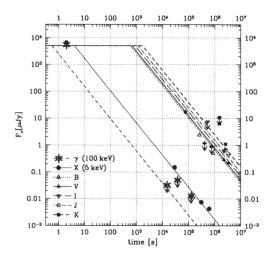

Figure 3. The light curves of the afterglow of GRB 970228 at various wavelenghts (Wijers, Rees & Mészáros 1997), compared to the simple blast wave model predictions of Mészáros & Rees (1997a)

This simplest afterglow model has a three-segment power law spectrum with two breaks. At low frequencies there is a steeply rising synchrotron self-absorbed spectrum up to a self-absorption break ν_a, followed by a $+1/3$ energy index spectrum up to the synchrotron break ν_m corresponding to the minimum energy γ_m of the power-law accelerated electrons, and then a $-(p-1)/2$ energy spectrum above this break, for electrons in the adiabatic regime (where γ^{-p} is the electron energy distribution above γ_m). A fourth segment and a third break are expected at energies where the electron cooling time becomes short compared to the expansion time, with a spectral slope $-p/2$ above that. With this third "cooling" break ν_b, first calculated in (Mészáros, Rees and Wijers 1998) and more explicitly detailed in (Sari et al. 1998), one has what has come to be called the simple "standard" model of GRB afterglows. One of the predictions of this model (Mészáros and Rees 1997a) is that the temporal decay index α, for $g = 3/2$ in $\Gamma \propto r^{-g}$, is related to the photon spectral energy index β through

$$F_\nu \propto t^\alpha \nu^\beta \quad \text{, with} \quad \alpha = (3/2)\beta \; . \tag{4}$$

This relationship appears to be valid in many (although not all) cases, especially after the first few days, and is compatible with an electron spectral index $p \sim 2.2 - 2.5$ which is typical of shock acceleration (Waxman 1997a; Sari et al. 1998; Wijers and Galama 1999). As the remnant expands the photon spectrum moves to lower frequencies, and the flux in a given band decays as a power law in time, whose index can change as breaks move through it. For the simple standard model, snapshot overall spectra have been deduced by extrapolating spectra at different wavebands and times using assumed simple time dependences (Waxman 1997b; Wijers and Galama 1999). These can be used to derive rough fits for the different physical parameters of the burst and environment, e.g. the total energy E, the magnetic and electron-proton coupling parameters ϵ_B and ϵ_e and the external density n_o. (see Figure 4).

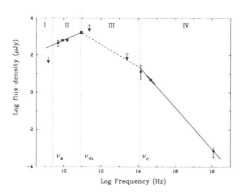

Figure 4. Snapshot spectrum of GRB 970508 at $t = 12$ days and standard afterglow model fit (Wijers & Galama 1998).

4. "Post-standard" Afterglow Models

The most obvious departure from the simplest standard model occurs if the external medium is inhomogeneous: for instance, for $n_{ext} \propto r^{-d}$, the energy conservation condition is $\Gamma^2 r^{3-d} \sim$ constant, which changes significantly the temporal decay rates (Mészáros, Rees and Wijers 1998). Such a power law dependence is expected if the external medium is a wind, say from an evolved progenitor star as implied in the hypernova scenario (such winds are generally used to fit supernova remnant models). Another departure from a simple impulsive injection approximation is obtained if the mass and energy injected during the burst duration t_w (say tens of seconds) obeys $M(> \Gamma) \propto \Gamma^{-s}$, $E(> \Gamma) \propto \Gamma^{1-s}$, i.e. more energy emitted with lower Lorentz factors at later times (but still shorter than the gamma-ray pulse duration). This would drastically change the temporal decay rate and extend the afterglow lifetime in the relativistic regime, providing a late "energy refreshment" to the blast wave on time scales comparable to the afterglow time scale (Rees and Mészáros 1998). These two cases lead to a decay rate

$$\Gamma \propto r^{-g} \propto \begin{cases} r^{-(3-d)/2} & ; \; n_{\text{ext}} \propto r^{-d}; \\ r^{-3/(1+s)} & ; \; E(>\Gamma) \propto \Gamma^{1-s}. \end{cases} \tag{5}$$

Expressions for the temporal decay index $\alpha(\beta, s, d)$ in $F_\nu \propto t^\alpha$ are obtained (Mészáros, Rees and Wijers 1998; Rees and Mészáros 1998) which now depend also on s and/or d. The result is that the decay can be flatter (or steeper, depending on s and d) than the simple standard $\alpha = (3/2)\beta$. A third non-standard effect, which is entirely natural, occurs when the energy and/or the bulk Lorentz factor injected are some function of the angle. A simple case is $E_o \propto \theta^{-j}$, $\Gamma_o \propto \theta^{-k}$ within a range of angles; this leads to the outflow at different angles shocking at different radii and its radiation arriving at the observed at different delayed times, and it has a marked effect on the time dependence of the afterglow (Mészáros, Rees and Wijers 1999), with $\alpha = \alpha(\beta, j, k)$ flatter or steeper than the standard value, depending on j, k. Thus in general, a temporal decay index which is a function of more than one parameter

$$F_\nu \propto t^\alpha \nu^\beta \quad , \text{with} \quad \alpha = \alpha(\beta, d, s, j, k, \cdots) , \tag{6}$$

is not surprising; what is more remarkable is that often, the simple relation $\alpha = (3/2)\beta$ is sufficient to describe the overall behavior at late times.

Evidence for departures from the simple standard model is provided by, for example, sharp rises or humps in the light curves followed by a renewed decay, as in GRB 970508 (Pedersen et al. 1998; Piro et al. 1998a). Detailed time-dependent model fits (Panaitescu, Mészáros and Rees 1998) to the X-ray, optical and radio light curves of GRB 970228 and GRB 970508 show that, in order to explain the humps, a *non-uniform* injection (Figure 5) or an *anisotropic* outflow is required. These fits indicate that the shock

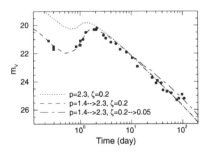

Figure 5. Optical light-curve of GRB 970508, fitted with a non-uniform injection model (a similar fit can be obtained with an off-axis jet plus a weaker isotropic component) (Panaitescu, Mészáros & Rees 1998).

physics may be a function of the shock strength (e.g. the electron index p, injection fraction ζ and/or ϵ_b, ϵ_e change in time), and also indicate that dust absorption is needed to simultaneously fit the X-ray and optical fluxes. The effects of beaming (outflow within a limited range of solid angles) can be significant (Panaitescu and Mészáros 1999a), but are coupled with other effects, and a careful analysis is needed to disentangle them.

Prompt optical, X-ray and GeV flashes from reverse and forward shocks, as well as from internal shocks, have been calculated in theoretical fireball shock models for a number of years (Mészáros and Rees 1993b; Mészáros and Rees 1994; Papathanassiou and Mészáros 1996; Mészáros and Rees 1997a; Sari and Piran 1999a), as have jets (Mészáros and Rees 1992; Mészáros, Laguna, and Rees 1993; Mészáros and Rees 1994), and in more detail (Rhoads 1997; Panaitescu, Mészáros and Rees 1998; Panaitescu and Mészáros 1999a; Rhoads 1999). However, observational evidence for these effects was largely lacking, until the detection of a prompt (within 22 s) optical flash from GRB 990123 with ROTSE (Akerlof et al. 1999), together with X-ray, optical and radio follow-ups (Kulkarni et al. 1999; Galama et al. 1999; Fruchter et al. 1999; Andersen et al. 1999; Castro-Tirado et al. 1999; Hjorth et al. 1999). GRB 990123 is so far unique not only for its prompt optical detection, but also by the fact that if it were emitting isotropically, based on its redshift $z = 1.6$ (Kulkarni et al. 1999; Andersen et al. 1999) its energy would be the largest of any GRB so far, 4×10^{54} erg. It is, however, also the first (tentative) case in which there is evidence for jet-like emission (Kulkarni et al. 1999; Fruchter et al. 1999; Castro-Tirado et al. 1999). An additional, uncommon feature is that a radio afterglow appeared after only one day, only to disappear the next (Galama et al. 1999; Kulkarni et al. 1999).

The prompt optical light curve of GRB 990123 decays initially as $\propto t^{-2.5}$ to $\propto t^{-1.6}$ (Akerlof et al. 1999), much steeper than the typical $\propto t^{-1.1}$ of previous optical afterglows detected after several hours. However, after about 10 minutes its decay rate moderates, and appears to join smoothly onto a slower decay rate $\propto t^{-1.1}$ measured with large telescopes (Galama et al. 1999; Kulkarni et al. 1999; Fruchter et al. 1999; Castro-Tirado et al. 1999) after hours and days. The prompt optical flash peaked at 9-th magnitude after 55 s (Akerlof et al. 1999), and in fact a 9-th magnitude prompt flash with a steeper decay rate had been predicted more than two years ago (Mészáros and Rees 1997a), from the synchrotron radiation of the reverse shock in GRB afterglows at cosmological redshifts (see also Sari and Piran (1999a); Mészáros, P. and Rees, M.J. (1993b); Mészáros, P., Rees, M.J., and Papathanassiou, H. (1994)). An origin of the optical prompt flash in internal shocks (Mészáros and Rees 1997a; Mészáros and Rees 1999a) cannot be ruled out, but is less likely since the optical light curve and the γ-rays appear not to correlate well (Sari and Piran 1999b; Galama et al. 1999). The subsequent slower decay agrees with predictions for the forward external shock (Mészáros and Rees 1997a; Sari and Piran

1999b; Mészáros and Rees 1999a).

The evidence for a jet is based on an apparent steepening of the light curve after about three days (Kulkarni et al. 1999; Fruchter et al. 1999; Castro-Tirado et al. 1999). If real, this steepening is probably due to the transition between early relativistic expansion, when the light-cone is narrower than the jet opening, and the late expansion, when the light-cone has become wider than the jet, leading to a drop in the effective flux (Rhoads 1997; Kulkarni et al. 1999; Mészáros and Rees 1999a; Rhoads 1999). A rough estimate leads to a jet opening angle of 3-5 degrees, which would reduce the total energy requirements to about 4×10^{52} erg. This is about two order of magnitude less than the binding energy of a few solar rest masses, which, even allowing for substantial inefficiencies, is compatible with currently favored scenarios (Popham et al. 1999; Macfadyen and Woosley 1999) based on a stellar collapse or a compact merger.

5. Location and Environmental Effects

The location of the afterglow relative to the host galaxy center can provide clues both for the nature of the progenitor and for the external density encountered by the fireball. A hypernova model would be expected to occur inside a galaxy in a high density environment $n_o > 10^3 - 10^5$ cm^{-3}. Most of the detected and well identified afterglows are inside the projected image of the host galaxy (Bloom et al 1998a), and some also show evidence for a dense medium at least in front of the afterglow (Owens et al. 1998).

In NS–NS mergers one would expect a BH plus debris torus system and roughly the same total energy as in a hypernova model, but the mean distance traveled from birth is of order several Kpc (Bloom et al. 1999), leading to a burst presumably in a less dense environment. The fits of (Wijers and Galama 1999) to the observational data on GRB 970508 and GRB 971214 in fact suggest external densities in the range of $n_o = 0.04$– 0.4 cm^{-3}, which would be more typical of a tenuous interstellar medium. These could be within the volume of the galaxy, but for NS–NS on average one would expect as many GRB inside as outside. This is based on an estimate of the mean NS–NS merger time of 10^8 years. BH–NS mergers would occur in timescales $\sim 10^7$ years, and would be expected to give bursts inside the host galaxy (Bloom et al. 1999); see however (Fryer and Woosley 1998). In at least one "snapshot" standard afterglow spectral fit for GRB 980329 (Reichart and Lamb 1998) the deduced external density is $n_o \sim 10^3$ cm^{-3}. In some of the other detected afterglows there is also evidence for a relatively dense gaseous environments, as suggested, e.g. by evidence for dust (Reichart 1998) in GRB970508; the absence of an optical afterglow and presence of strong soft X-ray absorption (Groot et al. 1997;

Murakami et al. 1997) in GRB 970828; the lack an an optical afterglow in the (radio-detected) afterglow ((Taylor et al. 1997)) of GRB980329; and spectral fits to the low energy portion of the X-ray afterglow of several bursts (Owens et al. 1998). One important caveat is that all afterglows found so far are based on positions from Beppo-SAX, which is sensitive only to long bursts $t_b \gtrsim 20$ s (Hurley 1998). This is significant, since it appears likely that NS–NS mergers lead (Macfadyen and Woosley 1999) to short bursts with $t_b \lesssim 10$ s. To make sure that a population of short GRB afterglows is not being missed will probably need to await results from HETE (HETE homepage) and from the planned Swift (SWIFT homepage) mission, which is designed to accurately locate 300 GRB/yr.

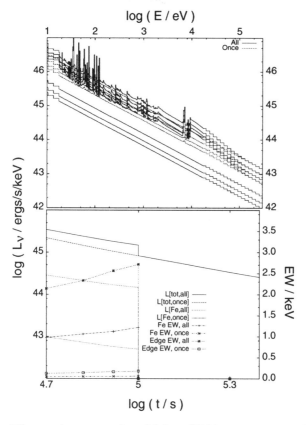

Figure 6. Spectrum (top) of a hypernova funnel model for various observer times, showing (bottom) the total and Fe light curves and equivalent widths, for $R = 1.5 \times 10^{16}$ cm, $n = 10^{10}$ cm^{-3}, and Fe abundance 10^2 times solar (Weth, Mészáros, Kallman & Rees 1999).

The environment in which a GRB occurs may also lead to specific spectral signatures from the external medium imprinted in the continuum, such as atomic edges and lines (Bisnovatyi-Kogan and Timokhin 1997; Perna and Loeb 1998; Mészáros and Rees 1998b). These may be used both to diagnose the chemical abundances and the ionization state (or local separation from the burst), as well as serving as potential alternative redshift indicators. (In addition, the outflowing ejecta itself may also contribute blue-shifted edge and line features, especially if metal-rich blobs or filaments are entrained in the flow from the disrupted progenitor debris (Mészáros and Rees 1998a), which could serve as diagnostic for the progenitor composition and outflow Lorentz factor). To distinguish between progenitors, an interesting prediction (Mészáros and Rees 1998b); see also (Ghisellini et al., 1999; Böttcher et al. 1998) is that the presence of a measurable Fe K-α X-ray *emission* line could be a diagnostic of a hypernova, since in this case one may expect a massive envelope at a radius comparable to a light-day where $\tau_T \lesssim 1$, capable of reprocessing the X-ray continuum by recombination and fluorescence. Detailed radiative transfer calculations have been performed to simulate the time-dependent X/UV line spectra of massive progenitor (hypernova) remnants (Weth et al. 2000), see Figure 6. Two groups (Piro et al. 1998b; Yoshida et al. 1998) have in fact recently reported the possible detection of Fe emission lines in GRB 970508 and GRB 970828.

An interesting case is the apparent coincidence of GRB 980425 with the unusual SN Ib/Ic 1998bw (Galama et al. 1998), which may represent a new class of SN (Iwamoto et al. 1998; Bloom et al. 1998b). If true, this could imply that some or perhaps all GRB could be associated with SN Ib/Ic (Wang and Wheeler 1998), differring only in their viewing angles relative to a very narrow jet. Alternatively, the GRB could be (Woosley et al. 1998) a new subclass of GRB with lower energy $E_\gamma \sim 10^{48}(\Omega_j/4\pi)$ erg, only rarely observable, while the great majority of the observed GRB would have the energies $E_\gamma \sim 10^{54}(\Omega_j/4\pi)$ erg as inferred from high redshift observations. The difficulties are that it would require extreme collimations by factors $10^{-3} - 10^{-4}$, and the statistical association is so far not significant (Kippen et al. 1998). However, two more GRB light curves may have been affected by an anomalous SNR (see, for example, the review of Wheeler (1999)).

6. Conclusions

The fireball shock model of gamma-ray bursts has proved quite robust in providing a consistent overall interpretation of the major features of these objects at various frequencies and over timescales ranging from the short initial burst to afterglows extending over many months. Significant progress has been made in understanding both the phenomenology and the physics of these objects, which may the most widely studied type of black hole sources. There still remain a number of mysteries, especially concerning the identity of their progenitors, the nature of the triggering mechanism, the transport of the energy, the time scales involved, and the nature and effects of beaming. However, the collective theoretical and observational understanding is vigorously advancing, and with dedicated new and planned observational missions under way, further significant progress may be expected in the near future.

Acknowledgements

I thank M.J. Rees, A. Panaitescu, R. Wijers, M. Spada and C. Weth for stimulating collaborations, NASA NAG-5 2857 for support, and the organizers of the NATO ASI Black Holes and Neutron Star workshop for hospitality.

References

Andersen et al. (1999), *Science* **283**, 2073.
Akerlof, C. et al. (1999), *Nature* **398**, 389.
Band, D. et al. (1993), *ApJ* **413**, 281.
Beloborodov, A., Stern, B., and Svensson, R. (1998) *ApJ* **508**, L25.
Beloborodov, A. et al. (2000), *ApJ* **535**, 158.
Bisnovatyi-Kogan, G. and Timokhin, A. (1997), *Astr. Rep.* **41**, 423.
Bloom, J. et al. (1998a), *A& A Supp.*,in press (Procs. Rome Conference on GRB).
Bloom J. et al. (1998a), *ApJ* **507**, L25.
Bloom, J. et al. (1998b), *ApJ* **506**, L105.
Bloom, J., Sigurdsson, S., and Pols, O. (1999), *MNRAS* **305**, 763.
Böttcher, M. et al. (1998), *A&A* **343**, 111.
Brainerd, J. et al. (1999), in *Abstr 19th Texas Symp*, Paris (astro-ph/9904039).
Brainerd, J. et al. (1998), *ApJ* **501**, 325.
Castro-Tirado, A.J. et al. (1999), *Science* **283**, 2069.
Costa, E. et al. (1997), *Nature* **387**, 783.
Crider, A. et al. (1997), *ApJ* **479**, L39.
Dermer, C. and Mitman, K.E. (1999), *ApJ* **513**, L5.
Dermer, C.D. et al. (1999), *ApJ* **515**, L49.
Djorgovski, S.G. et al. (1998), *ApJ* **508**, L17.
Eichler, D. and Levinson, A. (2000), *ApJ* **529**, 146.
Fishman, G. and Meegan, C. (1995), *Ann.Rev Astr.Ap.* **33**, 415.
Fruchter, A. et al. (1999), *ApJ* **519**, L13.
Fryer, C. and Woosley, S. (1998), *ApJ* **502**, L9.
Galama, T. et al. (1998), *Nature* **395**, 670.
Galama, T. et al. (1999), *Nature* **398**, 394.
Ghisellini, G. et al. (1999), *ApJ* **517**, 168.
Ghisellini, G. and Celotti, A. (1999), *ApJ* **511**, L93.
Groot, P. et al. (1997) in C. Meegan, R. Preece, and T. Koshut (eds.), *3rd Huntsville GRB Symposium*, AIP CP428 (New York), p557.
Harding, A.K. and Baring, M.G. (1994), in G. Fishman, J.J. Brainerd, and K. Hurley (eds.), *Proc. 2nd Huntsville GRB Symposium*, AIP CP307 (New York), p520.
HETE, http://space.mit.edu/HETE/
Hjorth, J. et al. (1999), *Science* **283**, 2073.
Hurley, K. et al. (1994), *Nature* **372**, 652.
Hurley, K. (1998), *A& A Supp.*, in press (Procs. Rome Conference on GRB).
Iwamoto, K. et al. (1998), *Nature* **395**, 672.
Katz, J. (1994a), *ApJ* **422**, 248.
Katz, J. (1994b), *ApJ* **432**, L107.
Kippen, R.M. et al. (1998), *ApJ* **506**, L27.
Kobayashi, S., Piran, T., and Sari, R. (1998), *ApJ* **490**, 92.

Kulkarni, S. et al. (1998), *Nature* **393**, 35.
Kulkarni, S. et al. (1999), *Nature* **398**, 389.
Liang, E. et al. (1997), *ApJ* **491**, L15.
Liang, E. et al. (1999), *ApJ* **519**, L21.
Macfadyen, A. and Woosley, S. (1999), *ApJ* **524**, 262.
Mészáros, P. and Rees, M.J. (1992), *ApJ* **397**, 570.
Mészáros, P. and Rees, M.J. (1993a), *ApJ* **405**, 278.
Mészáros, P. and Rees, M.J. (1993b), *ApJ* **418**, L59.
Mészáros, P., Laguna, P., and Rees, M.J. (1993), *ApJ* **415**, 181.
Mészáros, P., Rees, M.J., and Papathanassiou, H. (1994), *ApJ* **432**, 181.
Mészáros, P. and Rees, M.J. (1997a), *ApJ* **476**, 232.
Mészáros, P. and Rees, M.J. (1997b), *ApJ* **482**, L29.
Mészáros, P. and Rees, M.J. (1998a), *ApJ* **502**, L105.
Mészáros, P. and Rees, M.J. (1998b), *MNRAS* **299**, L10.
Mészáros, P., Rees, M.J., and Wijers, R. (1998), *ApJ* **499**, 301.
Mészáros, P., Rees, M.J., and Wijers, R. (1999), *New Astronomy* **4**, 313.
Mészáros, P. and Rees, M.J. (1999a), *MNRAS* **306**, L39.
Mészáros, P. and Rees, M.J. *ApJ* **530**, 292.
Metzger, M. et al. (1997), *Nature* **387**, 878.
Murakami, T. et al. (1997), in C. Meegan, R. Preece, and T. Koshut (eds.), *3rd Huntsville GRB Symposium*, AIP CP428 (New York), p435.
Odewahn, S. et al. (1998), *ApJ* **509**, L50.
Owens, A. et al. (1998), *A&A* **339**, L37.
Paczyński, B. and Rhoads, J. (1993), *ApJ* **418**, L5.
Paczyński, B. and Xu, G. (1994), *ApJ* **427** 708.
Paczyński, B. (1990), *ApJ* **363**, 218.
Paczyński, B. (1998), *ApJ* **494**, L45.
Papathanassiou, H. and Mészáros, P. (1996), *ApJ* **471**, L91.
Panaitescu, A. and Mészáros, P. (1998a), *ApJ* **492**, 683.
Panaitescu, A. and Mészáros, P. (1998b), *ApJ* **501**, 772.
Panaitescu, A. and Mészáros, P. (1999a), *ApJ* **526**, 707.
Panaitescu, A. and Mészáros, P. (1999d), *ApJ*, submitted (astro-ph/9810258).
Panaitescu, A, Mészáros, P., and Rees, M.J. (1998), *ApJ* **503**, 314.
Panaitescu, A., Spada, M., and Mészáros, P. (1999), *ApJ* **522**, L105.
Pedersen, H. et al. (1998), *ApJ* **496**, 311.
Perna, R. and Loeb, A. (1998), *ApJ* **503**, L135.
Piro, L. et al. (1998a), *A&A* **331**, L41.
Piro, L. et al. (1998b), *A&A Supp.*, in press (Procs. Rome Conference on GRB).
Popham, R., Woosley, S. and Fryer, C. (1999), *ApJ* **518**, 356.
Preece, R. et al. (1998), *ApJ* **496**, 849.
Rees, M.J. and Mészáros, P. (1992), *MNRAS* **258**, 41.
Rees, M.J. and Mészáros, P. (1994), *ApJ* **430**, L93.
Rees, M.J. and Mészáros, P. (1998), *ApJ* **496**, L1.
Reichart D. (1997), *ApJ* **485**, L57.
Reichart, D. (1998), *ApJ* **495**, L99.
Reichart, D. and Lamb, D.Q. (1998), *A&A Supp.*, in press (Procs. Rome Conf).
Rhoads, J. (1997), *ApJ* **487**, L1.
Rhoads, J. (1999), *ApJ* **525**, 737.
Sahu, K. et al. (1997), *Nature* **387**, 476.
Sari, R. and Piran, T. (1995), *ApJ* **455**, L143.
Sari, R. and Piran, T. (1998), *ApJ* **485**, 270.
Sari, R., Piran, T., and Narayan, R. (1998), *ApJ* **497**, L17.
Sari, R. and Piran, T. (1999a), *A&A Supp.* **138**, 537.
Sari, R. and Piran, T. (1999b), *ApJ* **517**, L109.
Spada, M., Panaitescu, A., and Mészáros, P. (1999) *ApJ* **537**, 824.

Swift, http://swift.gsfc.nasa.gov/
Tavani, M. (1997), *ApJ* **483**, L87.
Taylor, G.B. et al. (1997), *Nature* **389**, 263.
Thompson, C. (1994), *MNRAS* **270**, 480.
Van Paradijs, J. et al. (1997), *Nature* **386**, 686.
Vietri, M. (1997a), *ApJ* **478**, L9.
Wang, L. and Wheeler, J.C. (1998), *ApJ* **504**, L87.
Waxman, E. (1997), *ApJ* **485**, L5.
Waxman, E. (1997b), *ApJ* **489**, L33.
Weth, C., Mészáros, P., Kallman, T., and Rees, M.J. (2000), *ApJ* **534**, 581.
Wheeler, J.C. (1999), in M. Livio et al. (eds.), *Supernovae & Gamma Ray Bursts*, Cambridge U.P. (astro-ph/9909096).
Wijers, R.A.M.J. and Galama, T. (1999), *ApJ* **523**, 177.
Wijers, R.A.M.J., Rees, M.J., and Mészáros, P. (1997), *MNRAS* **288**, L51.
Woosley, S., Eastman, R. and Schmidt, B. (1998), *ApJ* **516**, 788.
Yoshida, A. et al. (1998), *A&A Supp.*, in press (Procs. Rome Conf on GRB).

GAMMA-RAY BURSTS FROM NEUTRON STARS SPUN UP IN X-RAY BINARIES

H.C. SPRUIT
Max-Planck-Institut für Astrophysik
Postfach 1523, D-85740 Garching, Germany

1. Introduction

Models for the central engine of a Gamma-Ray Burst (GRB) must satisfy a few elementary conditions. The right amount energy must be produced, of the order $10^{51} - 10^{54}$ erg. This energy must be put into an amount of mass of the order $10^{-4}M_\odot$ or less, in order for the Lorentz factor of the resulting relativistic outflow to be large enough to explain the short durations of the bursts (the 'baryon loading' constraint). Energy release in this range takes place in a supernova collapse, and several models involving supernovae or 'hypernovae' have in fact been proposed (Paczyński 1998; MacFadyen and Woosley 1999). Such explosions take place in extended envelopes containing several solar masses of matter, so that a difficulty in these models is to prevent (MacFadyen and Woosley 1999) all this mass from getting in the way of the GRB.

A neutron star in an X-ray binary (XRB) is spun up by mass transfer from its companion to rotation periods of the order of a millisecond. Evidence for this are the existence of ms pulsars (e.g. Chakrabarti 1999), believed to descend from X-ray binaries. Direct evidence of such rotation rates in XRB is the 2.5 ms rotation in SAX J1808.4-3658 (Wijnands and van der Klis 1998). The rotation energy in a neutron star spinning at 1ms is 2×10^{52} erg, in the range of that required for a GRB. The environment also satisfies the baryon loading constraint, since the accretion disk contains only of the order $10^{-9}M_\odot$, and the companion star blocks only of the order 1% of the sky, as seen from the neutron star.

Extraction of the rotation energy from a spinning neutron star to power a GRB looks very implausible at first sight. The energy must be extracted in 0.3–30 s (e.g. Fishman and Meegan 1995). This requires an extremely efficient 'brake' which does not exist in an ordinary XRB. In the following

C. Kouveliotou et al. (eds.), The Neutron Star – Black Hole Connection, 467–473.
© 2001 *Kluwer Academic Publishers. Printed in the Netherlands.*

I show, however, that such a brake can very elegantly be produced by the neutron star itself, through a sequence of straightforward events involving only known physics developed independently for other purposes (Spruit 1999a). In summary, this sequence of events is:

- 1. The neutron star's spin slowly increases by accretion of angular momentum from the companion.
- 2. At a period of the order of a ms, the star becomes unstable to an r-mode oscillation, coupled to the emission of a gravitational wave.
- 3. The negative temperature dependence of the viscosity of neutron star matter makes this instability run away in a finite time of the order of 100 years. The last few e-foldings of the oscillation amplitude take place on a time scale of hours.
- 4. The star loses angular momentum from its outer parts by the emitted gravitational wave, and starts rotating differentially by ~10-30%.
- 5. A weak initial magnetic field in the star (10^6–10^8G) gets wound up by the differential rotation into a field of $\sim 10^{17}$ G on a time scale of days to months.
- 6. This field becomes buoyantly unstable at a threshold of 10^{16} – -10^{17} G, rises through the surface on a time scale of ms to seconds, creating a nonaxisymmetric field of the order 10^{13}G at the light cylinder. (From this point on, the model is like that of Kluźniak and Ruderman 1998).
- 7. The star spins down on the pulsar spindown time scale, which for these conditions is of the order of seconds. The angular momentum is carried by a relativistic MHD wind. This powers a GRB, and extracts essentially all the rotation energy of the star (as in Usov's 1992 model).

2. Runaway gravitational wave instability

Rotating neutron stars are naturally unstable to excitation of their oscillation modes by the emission of gravitational waves (e.g. Shapiro and Teukolsky 1983). A mode is excited like a squeaking brake. The deformation of the oscillating star provides surface roughness, and the gravitational wave provides the torque coupling the star to the vacuum outside, analogous to the frictional coupling of a brake lining.

Rotational modes (Rossby waves) turn out to be much more effectively excited than the pulsation modes of the star (Andersson 1997; Andersson et al. 1999). Modes with angular quantum numbers $l = m = 2$ grow fastest. The unstable displacements in these modes are mainly along horizontal surfaces. Their frequency in an inertial frame is $2/3\,\Omega$, and in the corotating frame $-4/3\,\Omega$ (retrograde). In the absence of viscous damping, the growth rate of the most unstable mode is (Andersson et al. 1999) $\sigma_{\mathrm{GW}} = 5 \times$

$10^{-2}P_{-3}^{-6}$ s^{-1}. Including viscous damping, the growth rate is $\sigma = \sigma_{GW} - 1/\tau_v$, where τ_v is the viscous damping time $\tau_v = (\Delta R)^2/\nu$, and $\Delta R \approx 0.5R$ is the width of the unstable eigenfunction. For the kinematic viscosity ν the standard value quoted (van Riper 1991) for a neutron superfluid is $\nu \approx 2 \times 10^7 T_7^{-2}$ cm^{2}s^{-1} at an average density of 3×10^{14} g cm^{-3}, where $T = 10^7 T_7$ is the temperature. This expression holds at low temperatures; at temperatures above $\sim 3 \times 10^9$ K the viscosity starts increasing strongly with T. During the spinup by accretion, the temperature of the star is determined by the accretion rate, $T_7 \approx \dot{M}_{-9}^{1/4}$, where $\dot{M} = 10^{-9}\dot{M}_{-9}$ M$_\odot$/yr is the mass accretion rate [taking into account heating by nuclear reactions in the accreted atmosphere (Miralda-Escudé et al. 1990; Brown 1999), yields somehat higher temperatures]. With the typical value $\dot{M}_{-9} = 1$, the critical rotation period for instability to set in is 3 msec.

A star slowly spun up by accretion (on a time scale of 10^9yr) would become weakly unstable to the gravitational wave mechanism as its period decreases below the critical value. The unstable mode dissipates energy by viscous friction. This causes heating of the star, and a decrease of the viscosity. This upsets the delicate balance between the instrinsic growth and the viscous damping that exists at the time the marginal stability line is crossed, and the *growth rate* increases with the wave amplitude. This results in a runaway of the mode amplitude in a finite time (Figure 1). A few hundred years after the rotation rate has crossed the instability threshold, the mode amplitude diverges. During the final runaway phase the temperature is so high that viscous damping is weak and the mode grows on a time scale of the order of hours to minutes (depending on the initial rotation rate).

3. Differential rotation and winding-up of the magnetic field

The angular momentum loss by gravitational wave emission is largest in the outer parts of the star, where the unstable eigenfunctions typically have their largest amplitudes, and where the coupling to the gravitational wave emitted is strongest. Depending on the saturation amplitude assumed for the r-mode oscillation, this creates differential rotation on a time scale of hours to days. The differential rotation, absent before the final runaway of the instability, winds the initial magnetic field into a toroidal (azimuthal) field. Because the rotation period is so short, the nonaxisymmetric components of the initial field are quickly eliminated by magnetic diffusion (Rädler 1980; 1986; for a recent discussion see Spruit 1999b). The remaining axisymmetric and nearly azimuthal field is as sketched in Kluźniak and Ruderman (1998). To represent these processes in a simple model, I divide the star into two zones with boundary at $r \approx 0.5\,R$, such that the outer shell

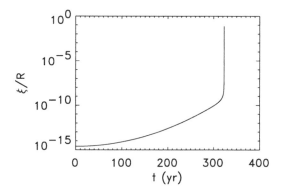

Figure 1.   Growth of the r-mode oscillation amplitude of a neutron star due to gravitational wave emission. Time in years since onset of instability, amplitude is the fractional displacement amplitude. The final runaway is due to the low viscosity of the interior, as it is heated by viscous dissipation of the mode.

has approximately the same moment of inertia as the core, and rotating at rates Ω_s and Ω_c respectively. The angular momentum loss by gravitational waves is taken to be restricted to the outer shell for simplicity. The azimuthal field increases at a rate given by the difference in rotation between the two zones. The magnetic torque between the two zones is taken into account as it modifies the differential rotation during the winding process.

The growth of the unstable mode is limited at large amplitude by nonlinear mode couplings. Since these have not been computed yet, they are parametrized by adding a damping term which becomes effective at an assumed saturation level $\xi/R = 0.03$.

4. Buoyant instability of the toroidal field

The increasing azimuthal magnetic field inside the star does not manifest itself at the surface until it becomes unstable to buoyancy instability. For this to happen the field has to become strong enough (Kluźniak and Ruderman 1998), of the order $B_c = 10^{17}$G, to overcome the stable stratification associated with the increasing neutron/proton ratio with depth. The growth time τ_B of the instability is of the order of the Alfvén travel time: $\tau_B \sim R/(V_A - V_{Ac})$, where V_{Ac} is the Alfvén speed at the critical field strength B_c. Buoyancy instability is generally nonaxisymmetric, forming loops of field lines which erupt through the surface (e.g. Matsumoto and Shibata 1992). The action of the radial velocities in the eruption process on the azimuthal field produces a strong small scale radial field component, which couples to the differential rotation and links the different parts of the star together on an Alfvén crossing time scale. The whole process is likely to yield a complex time dependent field configuration, but important aspects can be illustrated with a simple model (Spruit 1999a). This model takes into account the gradual crossing of the buoyancy stability limit, the time scale (of the order of a millisecond) of this instability, the magnetic torques coupling the star, and the external torque exerted by the spinning magnetic field.

5. Spindown and powering of the GRB

If the field emerging at the surface were a vacuum field, the energy output would be in the form of the Poynting flux of an electromagnetic wave, and would simply be given by the usual pulsar spindown formula. At a surface field strength of 10^{16}G and a spin period of 1 ms this yields a spin down time of seconds. In reality, the presence of some matter lifted into the atmosphere by the eruption of the magnetic field, and a pair plasma produced by the dissipation of some of this magnetic energy will enforce nearly ideal MHD conditions. The energy is then emitted in the form of a relativistic MHD wind instead of a pure EM wave. The energy output is still given approximately by the pulsar spindown formula (see, however, Usov 1992).

6. Discussion

The model presented is related to that of Kluźniak and Ruderman (1998), but provides a plausible scenario for the development of the strong differential rotation assumed by these authors. It has a number of properties that make it attractive as a possible GRB engine:

- The baryon loading is intrinsically low.
- The natural time scale for variations in the energy output is the Alvén travel time through the star, of the order of ms as observed.
- Burst durations as short as a second or as long as a minute can be accomodated by variation of the initial rotation rate.
- The process of emergence of magnetic loops allows for the complex time dependences observed.
- It invokes only properties of neutron stars and X-ray binaries believed to be known from current theory and observations.

The model also produces a reasonable energy output. For the largest observed energies, however, (such as the 23 January event), it is necessary to assume a significant beaming factor (as in other GRB engine models). Whether such a beaming is natural in the model depends on properties of the relativistic wind produced by a rotating neutron star which are not fully understood at present.

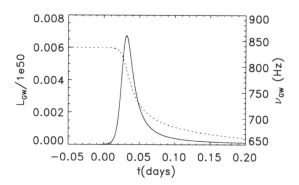

Figure 2. Typical variation of gravitational wave luminosity (solid) and frequency (dashed) during the peak of the r-mode instability (which precedes the electromagnetic GRB emission). Assumed parameters are initial spin period of 1.5 msec, and saturation amplitude of the r-modes $\xi_{max}/R = 0.03$.

The model makes some predictions that differ from engine models based on supernovae (Pazyński 1998; MacFadyen and Woosley 1998), collapsing neutron stars (Dai and Lu 1998) or neutron star-black hole mergers (Ruffert and Janka 1999). In these models, the bulk of the gravitational wave energy is emitted at the time of the GRB itself. In the present model, the gravitational waves are emitted at the peak of the gravitational wave instability, which precedes the GRB by days to months. This is because the differential rotation is generated only at this peak, and the winding up of the magnetic field that finally powers the GBR takes some time (depending on the initial field strength on the star). This difference should be obvious in gravitational wave observations whenever these become sensitive enough to detect GRB. The estimated amplitude and frequency of a gravitational wave produced near the peak of the r-mode instability are shown in Figure 2.

References

Andersson, N. (1997), *ApJ* **502**, 708.
Andersson, N., Kokkotas, K., and Schutz, B. (1999), *ApJ* **510**, 846.
Brown, E.F. (1999), *ApJ* **531**, 988.
Chakrabarti, D. (1999), HEAD meeting, American Astronomical Society **31**, 380.
Dai, Z.G. and Lu, T. (1998), *A&A* **333**, L87.
Fishman, G.J. and Meegan, C.A. (1995) *Ann. Rev. Astron. Astrophys.* **33**, 415.
Kluźniak, W. and Ruderman, M. (1998), *ApJ* **505**, L113.
MacFadyen, A. and Woosley, S.E. (1999), *ApJ* **524**, 262.
Matsumoto, R. and Shibata, K. (1992), *PASJ* **44**, 167.
Miralda-Escudé, J., Paczyński, B., and Haensel, P. (1990), *ApJ* **362**, 572.
Paczyński, B. (1998), *ApJ* **494**, L45.
Rädler, K.-H. (1980), *Astron. Nachr.* **301**, 101.

Rädler, K.-H. (1986), in *Plasma Astrophysics* (Proceedings Joint Varenna-Abastumani Workshop), ESA SP-251, p569.

Ruffert, M. and Janka, H.-T. (1998), *A&A* **338**, 535.

Shapiro, S.L. and Teukolsky, S.A. (1983), *Black Holes, White Dwarfs and Neutron Stars*, Wiley (NY).

Spruit, H.C. (1999a), *A&A* **341**, L1.

Spruit, H.C. (1999b), *A&A* **349**, 189.

Usov, V.V. (1992) *Nature* **357**, 472.

Van Riper, K.A. (1991) *ApJS* **75**, 449.

Wijnands, R. and van der Klis, M. (1998), *Nature* **394**, 344.

7. New Experiments

THE CHANDRA X-RAY OBSERVATORY (CXO)

An Overview

M.C. WEISSKOPF
NASA/MSFC SD-50
Marshall Space Flight Center, AL 35801, USA

1. Introduction

Significant advances in science inevitably occur when the state of the art in instrumentation improves. NASA's newest Great Observatory, the Chandra X-Ray Observatory (CXO) – formally known as the Advanced X-Ray Astrophysics Facility (AXAF) – launched on July 23, 1999 and represents such an advance. The CXO is designed to study the x-ray emission from all categories of astronomical objects from normal stars to quasars. Observations with CXO will therefore obviously enhance our understanding of neutron stars and black holes.

CXO has broad scientific objectives and an outstanding capability to provide high-resolution (≤ 0.5 ") imaging, spectrometric imaging and high resolution dispersive spectroscopy over the energy band from 0.1 to 10 keV. CXO, together with ESA's XMM, the Japanese-American Astro-E and ultimately the international Spectrum-X mission lead by Russia, will usher in a new age in x-ray astronomy and high-energy astrophysics.

NASA's Marshall Space Flight Center (MSFC) manages the Chandra Project, with scientific and technical support from the Smithsonian Astrophysical Observatory (SAO). TRW's Space and Electronics Group was the prime contractor and provided overall systems engineering and integration. Hughes Danbury Optical Systems (HDOS), now Raytheon Optical Systems Incorporated, figured and polished the x-ray optics; Optical Coating Laboratory Incorporated (OCLI) coated the polished optics with iridium; and Eastman Kodak Company (EKC) mounted and aligned the optics and provided the optical bench. Ball Aerospace & Technologies was responsible for the Science Instrument Module (SIM) and the CCD-based aspect camera for target acquisition and aspect determination. The scientific instruments, discussed in some detail below, comprise two sets of objective transmission gratings that can be inserted just behind the 10-m-focal-length x-ray op-

C. Kouveliotou et al. (eds.), The Neutron Star – Black Hole Connection, 477–491.
© 2001 *Kluwer Academic Publishers. Printed in the Netherlands.*

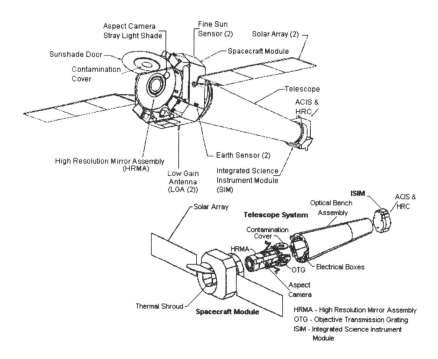

Figure 1. Schematic of the CXO fully deployed.

tics, and two sets of focal-plane imaging detectors that can be positioned by the SIM's translation table.

The fully deployed CXO, shown schematically in Figure 1, is 13.8 m long, with a 19.5 m-long solar-array wingspan. The on-orbit mass is about 4500 kg. CXO was placed in a highly elliptical orbit with a 140,000 km apogee and 10,000 km perigee by the Space Shuttle Columbia, Boeing's Inertial Upper Stage, and Chandra's own integral propulsion system. Figure 2 shows photos of the payload and the sequence of events through the deployment from the Shuttle. This particular launch gained some additional notoriety due to the Commander's (Colonel Eileen Collins) gender.

2. The X-ray Optics

The heart of the observatory is, of course, the x-ray telescope. Grazing-incidence optics function because x rays reflect efficiently if the angle between the incident ray and the reflecting surface is less than the critical angle. This critical grazing angle is approximately $10^{-2}(2\rho)^{1/2}/E$, where ρ is the density in g cm^{-3} and E is the photon energy in keV. Thus, higher energy telescopes must have dense optical coatings (iridium, platinum, gold, etc.) and smaller grazing angles. The x-ray optical elements for Chandra

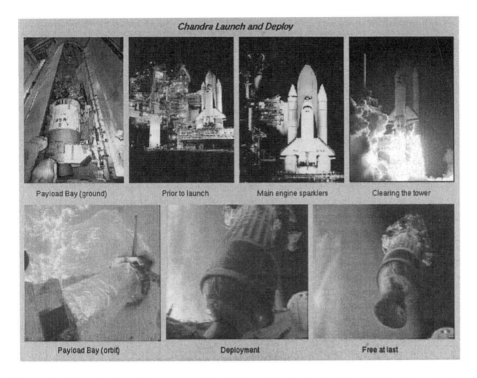

Figure 2. Chandra launch sequence.The IUS is still attached to the CXO in these photos.

and similar telescopes resemble shallow angle cones, and two reflections are required to provide good imaging over a useful field of view; the first CXO surface is a paraboloid and the second a hyperboloid – the classic Wolter-1 design. The collecting area is increased by nesting concentric mirror pairs, all having the same focal length. The wall thickness of the inner elements limit the number of pairs, and designs have tended to fall into two classes: Those with relatively thick walls achieve stability, hence angular resolution, at the expense of collecting area; those with very thin walls maximize collecting area but sacrifice angular resolution. NASA's Einstein Observatory (1978), the German ROSAT (1990), and the CXO optics are examples of the high-resolution designs, while the Japanese-American ASCA (1993) and European XMM mirrors are examples of emphasis upon large collecting area.

The mirror design for CXO includes eight optical elements comprising four paraboloid/hyperboloid pairs which have a common ten meter focal length, element lengths of 0.83 m, diameters of 0.63, 0.85, 0.97, and 1.2 m, and wall thickness between 16 mm and 24 mm. Zerodur, a glassy ceramic, from Schott was selected for the optical element material because of its low coefficient of thermal expansion and previously demonstrated capability

Figure 3.　The largest optical element being ground.

(ROSAT) of permitting very smooth polished surfaces.

Figure 3 shows the largest optical element being ground at HDOS. Final polishing was performed with a large lap designed to reduce surface roughness without introducing unacceptable lower frequency figure errors. The resulting rms surface roughness over the central 90% of the elements varied between 1.85 and 3.44 Å in the 1 to 1000-mm^{-1} band; this excellent surface smoothness enhances the encircled energy performance at higher energies by minimizing scattering.

The mirror elements were coated at OCLI by sputtering with iridium over a binding layer of chromium. OCLI performed verification runs with surrogates before each coating of flight glass; these surrogates included witness samples. The x-ray reflectivities of the witness flats were measured at SAO to confirm that the expected densities were achieved. The last cleaning of the mirrors occurred at OCLI prior to coating, and stringent contamination controls were begun at that time because both molecular

Figure 4.   The smallest parabaloid after coating.

and particulate contamination have adverse impacts on the calibration and
the x-ray performance. Figure 4 shows the smallest paraboloid in the OCLI
handling fixture after being coated.

The final alignment and assembly of the mirror elements into the High
Resolution Mirror Assembly (HRMA) was done at, and by, EKC. The completed mirror element support structure is shown in Figure 5. Each mirror
element was bonded near its mid-station to flexures previously attached
to the carbon fiber composite mirror support sleeves. The four support
sleeves and associated flexures for the paraboloids can be seen near the top
of the figure, and those for the outer hyperboloid appear at the bottom.
The mount holds more than 1000 kg of optics to sub-arcsecond precision.

The mirror alignment was performed with the optical axis vertical in
a clean and environmentally controlled tower. The mirror elements were
supported to approximate a gravity-free and strain-free state, positioned,
and then bonded to the flexures. A photograph taking during the assembly
and alignment process is shown in Figure 6. Despite the huge mass of the
system and the stringent environmental controls, the heat produced by a 50
Watt light bulb at the top of the facility caused some alignment anomalies
until detected and resolved.

The HRMA was taken to MSFC for pre-launch x-ray calibration (see
O'Dell and Weisskopf (1998) and references therein) in the fall of 1996,
and then to TRW for integration into the spacecraft. Testing at MSFC
took place in the X-Ray Calibration Facility (XRCF), shown in Figure 7.

Figure 5.   The fixture to which the eight optical elements were mounted.

The calibration facility has a number of x-ray source and detector systems and continues to be used for x-ray tests of developmental optics for such programs as Constellation-X. Details concerning the XRCF may be found in Weisskopf and O'Dell (1997) and references therein.

X-ray testing demonstrated that the CXO mirrors are indeed the largest high-resolution X-ray optics ever made; the nominal effective area (based on the ground calibrations) is shown as a function of energy in the left panel of Figure 8, along with those of their Einstein and ROSAT predecessors. The

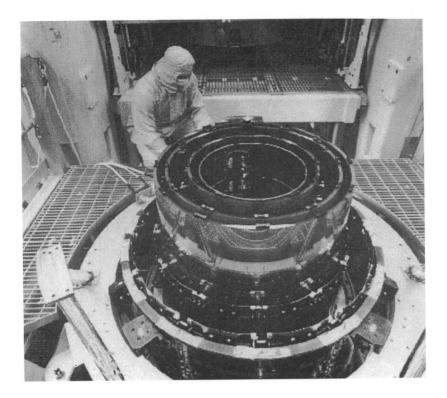

Figure 6.   A photograph of the HRMA during assembly and alignment at EKC.

CXO areas are about a factor of four greater than the Einstein mirrors. The effective areas of CXO and ROSAT are comparable at low energies because the somewhat smaller ROSAT mirrors have larger grazing angles; the smaller grazing angles of CXO yield more throughput at higher energies. The fraction of the incident energy included in the core of the expected CXO response to 1.49 keV x rays is shown as a function of image radius in the right panel of Figure 8 including early in-flight data. The responses of the Einstein and ROSAT mirrors also are shown. The improvement within 0.5" is dramatic, although it is important to note that the ROSAT mirrors bettered their specification and were well matched to the principal detector for that mission. The excellent surface smoothness achieved for the CXO (and ROSAT) mirrors result in a very modest variation of the performance as a function of energy; this reduces the uncertainties which accrue from using calibration data to infer properties of sources with different spectra, and improves the precision of the many experiments to be performed.

Figure 7. An aerial view of the X-ray Calibration Facility at MSFC.

3. The Instruments

CXO has two focal plane instruments – the High-Resolution Camera (HRC) and the Advanced CCD Imaging Spectrometer (ACIS). Each of these instruments, in turn, has two detectors, one optimized for direct imaging of x rays that pass through the optics and the other optimized for imaging x rays that are dispersed by the objective transmission gratings when the latter are commanded into place directly behind the HRMA. Each focal-plane detector operates in essentially photon counting mode and has low internal background. A slide mechanism is utilized to place the appropriate instrument at the focus of the telescope. Provision for focus adjustment is also present.

3.1. THE FOCAL PLANE INSTRUMENTS

The HRC was produced at SAO; Dr. S. Murray is the Principal Investigator. The HRC-I is a large-format, 100 mm^2 microchannel plate, coated with a cesium iodide photocathode to improve x-ray response. A conventional cross-grid charge detector reads out the photo-induced charge cloud and the electronics determine an arrival time to $16\mu s$, and the position with

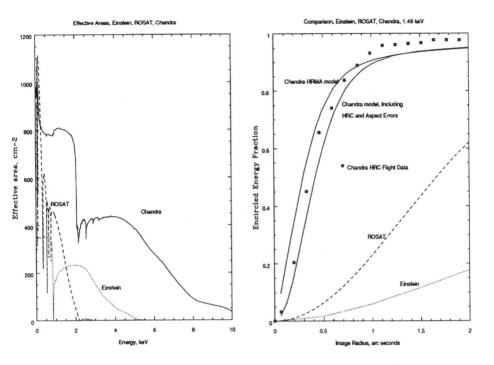

Figure 8.   Effective area and encircled energy comparisons.

a resolution of about 18 μm or 0.37". The spectroscopy readout detector (HRC-S) is a 300 mm x 30 mm, 3-section microchannel plate. Sectioning allowed the 2 outside sections to be tilted in order to conform more closely to the Rowland circle that includes the low-energy gratings.

The ACIS has 2 charge coupled-device (CCD) detector arrays: ACIS-I is optimized for high-resolution spectrometric imaging; ACIS-S is optimized for readout of the high-energy transmission gratings, although these functions are not mutually exclusive. Prof. G. Garmire of the Pennsylvania State University is the Principal Investigator. The Massachusetts Institute of Technology's Center for Space Research, in collaboration with Lincoln Laboratories, developed the detector system and manufactured the CCDs; Lockheed-Martin integrated the instrument. Stray visible light is shielded by means of baffles and an optical blocking filter (about 1500 $\overset{\circ}{A}$ aluminum on 2000 $\overset{\circ}{A}$ polyimide). The ACIS-I is a 2x2 array of CCDs. The 4 CCDs tilt slightly toward the optics to conform more closely to the focal surface.

Figure 9.   The LETG and HETG being attached to the spacecraft mounting structure.

Each CCD has 1024 x 1024 pixels of 24-μm (0.5") size. The ACIS-S is a 1x6 array with each chip tilted slightly to conform to the Rowland circle and includes two back-illuminated CCDs, one of which is at the best focus position. The back-illuminated devices cover a broader bandwidth than the front-illuminated chips and, under certain circumstances, may be the best choice for high-resolution, spectrometric imaging.

3.2. THE TRANSMISSION GRATINGS

Both sets of objective transmission gratings consist of hundreds of co-aligned facets mounted to supporting structures on 4 annuli (one for each of the four co-aligned mirror pairs) to intercept the x rays exiting the HRMA. In order to optimize the energy resolution, the grating support structure holds the facets close to the Rowland toroid that intercepts the focal plane. The two sets of transmission gratings, attached to the mounting structure are shown in Figure 9.

The Low-Energy Transmission Grating (LETG) provides high-resolution spectroscopy at the lower end of the CXO energy range. Dr. A Brinkman, of the Space Research Organization of the Netherlands, is the Principal Investigator. The LETG was developed in collaboration with the Max Planck

Figure 10.   The first observation of Capella with the HETG/ACIS-S combination.

Institut für Extraterrestische Physik, Garching. The LETG has 540 1.6 cm diameter grating facets, 3 per grating module. Ultraviolet contact lithography was used to produce an integrated all-gold facet bonded to a stainless-steel facet ring. An individual facet has 0.43 μm-thick gold grating bars with 50% filling factor and 9920 Å period, resulting in 1.15 Å/mm dispersion. The HRC-S is the primary LETG readout.

The High-Energy Transmission Grating (HETG) provides high-resolution spectroscopy at the higher end of the CXO energy range. Prof. C. Canizares of the Massachusetts Institute of Technology Center for Space Research is the Principal Investigator. This group developed the instrument in collaboration with MIT's Nanostructures Laboratory. The HETG has 336 2.5 cm^2 grating facets. Microlithographic fabrication using laser interference patterns was used to produce the facets, which consist of gold grating bars with 50% filling factor on a polyimide substrate. The HETG uses gratings with 2 different periods which are oriented to slightly different dispersion directions, forming a shallow "X" image on the readout detector as shown in Figure 10.

The Medium-Energy Gratings (MEG) have 0.40 μm-thick gold bars on 0.50 μm-thick polyimide with 4000 Å period, producing 2.85 Å/mm dispersion, and are placed behind the outer two CXO mirrors. The High-Energy gratings (HEG), placed behind the inner two CXO mirror pairs are 0.70 μm -thick-gold bars on 1.0 μm-thick polyimide with 2000 Å period, resulting in 5.7 Å/mm dispersion. The ACIS-S is the primary readout for the HETG.

4.   Science with CXO

Given the superb capabilities of the optics and associated instrumentation, the scientific possibilities are almost incredible. The most exciting investigations will no doubt result from the unexpected discoveries that the improved sensitivity, angular resolution, and energy resolution produces. The potential of Chandra is illustrated in Figure 11 which shows the official "first-light" image. The target was the supernova remnant Cas A. This image, based on only a few thousand seconds observing time, was taken with the back-illuminated chip at the best focus position on the ACIS-S. We see, for the first time, that there is a compact, x-ray emitting object at

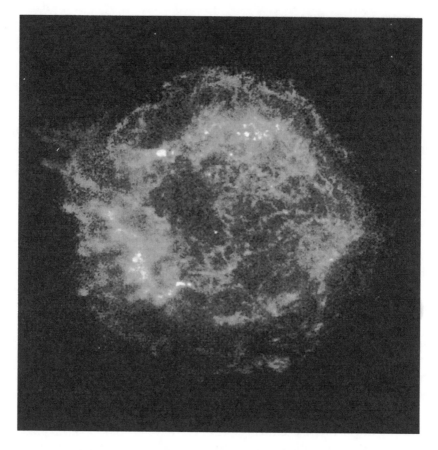

Figure 11. The official CXO first light image – the supernova remnant CasA.

the center of the 300-year-old remnant. Studies are underway to establish that the positional coincidence is no accident and to determine the nature of the compact object, possibly the long-sought after neutron star or black hole.

Perhaps the neutron star - black hole connection and the utility of CXO are best illustrated by an early observation of the Crab Nebula and pulsar taken as part of the HETG calibration. During grating observations, one also obtains an undispersed (zero order) image. The image quality is essentially that of the HRMA/detector combination and not broadened by the insertion of the grating. Figure 12 shows the zeroth-order image of the Crab Nebula. There are numerous new features, especially the inner ellipse with its bright knots. The pulsar itself is so bright that the central region is "piled up" to the point that there are no data – hence the "hole" in the image. Pile-up also is present in the data from the nebula and the study of the spectral dependence of these features ought to be the subject of a future

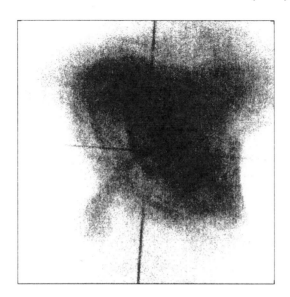

Figure 12.   The Chandra zeroth order LETG image of the Crab Nebula taken after the HETG image.

ASI. The ubiquity of the jet phenomenon clearly points to the importance of angular momentum as a physical key and critical parameter towards unlocking secrets to all or part of the emission mechanisms.

The capability to perform meaningful, high-spectral-resolution observations with the gratings is illustrated in Figure 13, which shows a portion of the x-ray spectrum from Capella around the Fe-L complex. The red is a HEG spectrum and the green spectrum was produced by the MEG. Observations such as these – with CXO, XMM, and Astro-E – will be at the center of new developments in astrophysics in the next century.

5. Conclusion

The Chandra X-Ray Observatory will have a profound influence on astronomy and astrophysics. The Observatory is open to use by scientists throughout the world who successfully propose specific investigations. Data will be available through the Chandra X-ray Center (CXC), directed by Dr. H. Tananbaum, and located at the SAO.

Acknowledgements

I would like to thank the many members of the Chandra team, especially Steve O'Dell, Leon Van Speybroeck, Harvey Tananbaum, Steve Murray, Gordon Garmire, Claude Canizares, and Bert Brinkman.

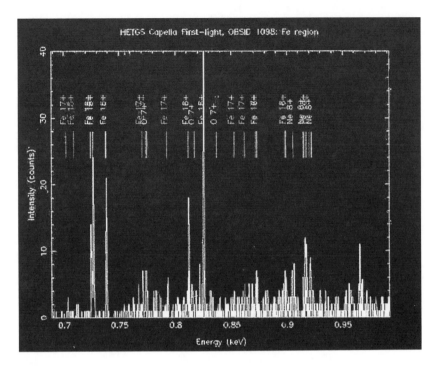

Figure 13. HEG and MEG spectra of Capella.

A. AXAF web sites

The following lists several Chandra-related sites on the World-Wide Web (WWW). Most sites are cross-linked to one another. Often you will find that these contain the best and most recent sources of detailed information; hence, the minimal number of entries in the bibliography.

http://asc.harvard.edu/ Chandra X-Ray Center (CXC), operated for NASA by the Smithsonian Astrophysical Observatory (SAO).

http://wwwastro.msfc.nasa.gov/ Chandra Project Science, at the NASA Marshall Space Flight Center (MSFC).

http://hea-www.harvard.edu/MST/ AXAF Mission Support Team (MST), at the Smithsonian Astrophysical Observatory (SAO).

http://hea-www.harvard.edu/HRC/ AXAF High-Resolution Camera (HRC) team, at the Smithsonian Astrophysical Observatory (SAO).

http://www.astro.psu.edu/xray/axaf/axaf.html Advanced CCD Imaging Spectrometer (ACIS) team at the Pennsylvania State University (PSU).

http://acis.mit.edu/ Advanced CCD Imaging Spectrometer (ACIS) team at the Massachusetts Institute of Technology (MIT).

http://www.sron.nl/ Chandra Low-Energy Transmission Grating (LETG) team at Space Research Organisation Netherlands (SRON).

http://www.rosat.mpe-garching.mpg.de/axaf/ Chandra Low-Energy Transmission Grating (LETG) team at the Max-Planck Institut für extraterrestrische Physik (MPE).

http://space.mit.edu/HETG/ Chandra High-Energy Transmission Grating (HETG) team, at the Massachusetts Institute of Technology (MIT).

References

O'Dell, S.L. and Weisskopf, M.C. (1998) *Proc SPIE* **3444**, 2.
Weisskopf, M.C. and O'Dell, S.L. (1997) *Proc SPIE* **3113**, 2.

AUTHOR INDEX

SUBJECT INDEX

510

OBJECT INDEX

518